Lecture Notes in Computer Scien

Commenced Publication in 1973
Founding and Former Series Editors:
Gerhard Goos, Juris Hartmanis, and Jan van Leeuwen

Editorial Board

Juraj Hromkovič   Manfred Nagl
Bernhard Westfechtel (Eds.)

Graph-Theoretic Concepts in Computer Science

30th International Workshop, WG 2004
Bad Honnef, Germany, June 21-23, 2004
Revised Papers

 Springer

Volume Editors

Juraj Hromkovič
Swiss Federal Institute of Technology, Department of Computer Science
ETH Zentrum, 8092 Zurich, Switzerland
E-mail: juraj.hromkovic@inf.ethz.ch

Manfred Nagl
Bernhard Westfechtel
RWTH Aachen University, Department of Computer Science III
Ahornstr. 55, 52074 Aachen, Germany
E-mail: nagl@informatik.rwth-aachen.de, westfechtel@cs.rwth-aachen.de

Library of Congress Control Number: 2004117074

CR Subject Classification (1998): F.2, G.2, G.1.6, G.1.2, E.1, I.3.5

ISSN 0302-9743
ISBN 3-540-24132-9 Springer Berlin Heidelberg New York

Springer is a part of Springer Science+Business Media

springeronline.com

© Springer-Verlag Berlin Heidelberg 2004
Printed in Germany

Typesetting: Camera-ready by author, data conversion by Olgun Computergrafik
Printed on acid-free paper   SPIN: 11369264   06/3142   5 4 3 2 1 0

Preface

During its 30-year existence, the International Workshop on Graph-Theoretic Concepts in Computer Science has become a distinguished and high-quality computer science event. The workshop aims at uniting theory and practice by demonstrating how graph-theoretic concepts can successfully be applied to various areas of computer science and by exposing new theories emerging from applications. In this way, WG provides a common ground for the exchange of information among people dealing with several graph problems and working in various disciplines. Thereby, the workshop contributes to forming an interdisciplinary research community.

The original idea of the Workshop on Graph-Theoretic Concepts in Computer Science was ingenuity in all theoretical aspects and applications of graph concepts, wherever applied. Within the last ten years, the development has strengthened in particular the topic of structural graph properties in relation to computational complexity. This workshop has become pivotal for the community interested in these areas. An aim specific to the 30th WG was to support the central role of WG in both of the prementioned areas on the one hand and on the other hand to promote its originally broader scope.

The 30th WG was held at the Physikzentrum Bad Honnef, which serves as the main meeting point of the German Physical Society. It offers a secluded setting for research conferences, seminars, and workshops, and has proved to be especially stimulating for fruitful discussions. Talks were given in the new lecture hall with a modern double rear projection, interactive electronic board, and full video conferencing equipment.

The Call for Papers received a lively response, resulting in 66 submissions. The program committee selected 31 papers for publication among a large set of high-quality contributions. In addition, two invited speakers – Derek Corneil and Roger Wattenhofer – enriched the technical program with surveys of selected fields of graph theory and applications.

We would like to thank all those who made the 30th anniversary of the WG workshop series a success: the authors who submitted their work to the workshop, the speakers, the program committee members, and all referees. We are also indebted to Dirk Bongartz, Joachim Kupke, and Walter Unger for the review organization, and to Boris Böhlen, Christian Fuß, and Ulrike Ranger for the local organization.

Special thanks go to the sponsoring institutions: the Deutsche Forschungsgemeinschaft, Bonn, proRWTH – Freunde und Förderer der RWTH Aachen e.V., the rectorate of RWTH Aachen University, and the Swiss Federal Institute of Technology, Zürich.

September 2004

Juraj Hromkovič
Manfred Nagl
Bernhard Westfechtel

The 30-Year Tradition of WG

1975	U. Pape – Berlin
1976	H. Noltemeier – Göttingen
1977	J. Mühlbacher – Linz
1978	M. Nagl, H.J. Schneider – Schloß Feuerstein, near Erlangen
1979	U. Pape – Berlin
1980	H. Noltemeier – Bad Honnef
1981	J. Mühlbacher – Linz
1982	H.J. Schneider, H. Göttler – Neunkirchen, near Erlangen
1983	M. Nagl, J. Perl – Haus Ohrbeck, near Osnabrück
1984	U. Pape – Berlin
1985	H. Noltemeier – Schloß Schwanenberg, near Würzburg
1986	G. Tinhofer, G. Schmidt – Stift Bernreid, near München
1987	H. Göttler, H.J. Schneider – Schloß Banz, near Bamberg
1988	J. van Leeuwen – Amsterdam
1989	M. Nagl – Schloß Rolduc, near Aachen
1990	R.H. Möhring – Johannesstift Berlin
1991	G. Schmidt, R. Berghammer – Richterheim Fischbackau, München
1992	E.W. Mayr – Wilhelm-Kempf-Haus, Wiesbaden-Naurod
1993	J. van Leeuwen – Sports Center Papendal, near Utrecht
1994	G. Tinhofer, E.W. Mayr, G. Schmidt – Herrsching, near München
1995	M. Nagl – Haus Eich, Aachen
1996	G. Ausiello, A. Marchetti-Spaccamela – Cadenabbia
1997	R.H. Möhring – Bildungszentrum am Müggelsee, Berlin
1998	J. Hromkovič, O. Sýkora – Castle Smolenice, near Bratislava
1999	P. Widmayer – Centro Stefano Franscini, Monte Verità, Ascona
2000	D. Wagner – Waldhaus Jacob, Konstanz
2001	A. Brandstädt – Boltenhagen, near Rostock
2002	L. Kučera – Český Krumlov
2003	H.L. Bodlaender – Mennorode, Elspeet, near Utrecht
2004	J. Hromkovič, M. Nagl – Bad Honnef

Program Committee

Hans Bodlaender	Utrecht University, The Netherlands
Andreas Brandstädt	Universität Rostock, Germany
Michael Fellows	University of Newcastle, Australia
Fedor Fomin	University of Bergen, Norway
Michel Habib	LIRMM Montpellier, France
Juraj Hromkovič	ETH Zürich, Switzerland, Co-chair
Michael Kaufmann	Universität Tübingen, Germany
Luděk Kučera	Charles University, Prague, Czech Republic
Alberto Marchetti-Spaccamela	Università di Roma, Italy
Ernst Mayr	TU München, Germany
Rolf Möhring	TU Berlin, Germany
Manfred Nagl	RWTH Aachen, Germany, Co-chair
Hartmut Noltemeier	Universität Würzburg, Germany
David Peleg	Weizmann Institute, Rehovot, Israel
Ondrej Sýkora	Loughborough University, UK
Gottfried Tinhofer	TU München, Germany
Dorothea Wagner	Universität Karlsruhe, Germany
Bernhard Westfechtel	RWTH Aachen, Germany
Peter Widmayer	ETH Zürich, Switzerland

Additional Reviewers

Michael Baur
Marc Benkert
Hajo J. Broersma
Olivier Cogis
Sabine Cornelsen
Feodor F. Dragan
Jens Ernst
Daniel Fleischer
Jan Griebsch
Pinar Heggernes
Christian Knauer
Dieter Kratsch
Nina Lehmann
Edita Máčajová
Martin Nehéz
Marc Nunkesser
Stefan Pfingstl
Thomas Schank
Frank Schulz
Jeremy Spinrad
Sebastian Stiller
Peter Ullrich
Silke Wagner
Sebastian Wernicke
Alexander Wolff

Thomas Bayer
Dirk Bongartz
Hans-Joachim Böckenhauer
Amin Coja-Oghlan
Xiaotie Deng
Stefan Eckhardt
Hazel Everett
Michael Gatto
Alexander Hall
Klaus Holzapfel
Sven Kosub
Ekkehard Köhler
Christian Liebchen
Moritz Maaß
Rolf Niedermeier
Christophe Paul
André Raspaud
Heiko Schilling
Martin Škoviera
Ladislav Stacho
Ioan Todinca
Walter Unger
Mirjam Wattenhofer
Thomas Willhalm

Luca Becchetti
Ulrik Brandes
Jana Chlebikova
Derek Corneil
Stefan Dobrev
Thomas Erlebach
Guillaume Fertin
Rodolphe Giroudeau
Toru Hasunuma
Han Hoogeveen
Jan Kratochvil
Van Bang Le
Marco Lübbecke
Steffen Mecke
Naomi Nishimura
Leon Peeters
Ulrich Rührmair
Etienne Schramm
Ines Spenke
Daniel Stefankovic
Hanjo Täubig
Imrich Vrťo
Birgitta Weber
Mark S. Withall

Table of Contents

Invited Papers

Graph Algorithms: Trees

Graph Algorithms: Recognition and Decomposition

Graph Algorithms: Various Problems

Optimization and Approximation Algorithms

Parameterized Complexity and Exponential Algorithms

Counting, Combinatorics, and Optimization

Applications (Biology, Graph Drawing)

Graph Classes and NP Hardness

Lexicographic Breadth First Search – A Survey

Derek G. Corneil

Department of Computer Science, University of Toronto,
Toronto M5S3G4, Ontario, Canada
dgc@cs.utoronto.ca

Abstract. Lexicographic Breadth First Search, introduced by Rose, Tarjan and Lueker for the recognition of chordal graphs is currently the most popular graph algorithmic search paradigm, with applications in recognition of restricted graph families, diameter approximation for restricted families and determining a dominating pair in an AT-free graph. This paper surveys this area and provides new directions for further research in the area of graph searching.

1 Introduction

Graph searching is a fundamental paradigm that pervades graph algorithms. A search of a graph visits all vertices and edges of the graph and will visit a new vertex only if it is adjacent to some previously visited vertex. Such a generic search does not, however, indicate the rules to be followed in choosing the next vertex to be visited. The two fundamental search strategies are *Breadth First Search* (BFS) and *Depth First Search* (DFS). As the names indicate, BFS visits all previously unvisited neighbours of the currently visited vertex before visiting the previously unvisited non-neighbours, whereas DFS follows unvisited edges (if possible) from the most recently visited vertex. Both searches seem to have been "discovered" in the 19th century (and probably earlier) as algorithms for maze traversal. DFS, as popularized by Tarjan [41], has been used for such diverse applications as connectivity, planarity, topological ordering and strongly connected components of digraphs. BFS has been applied to shortest path problems, network flows and the recognition of various graph classes.

In the mid 1970s, Rose, Tarjan and Lueker [42] introduced a variant of BFS called *Lexicographic Breadth First Search* (LBFS). Their application of LBFS was to the recognition of chordal graphs. This algorithm is one of the classic graph algorithms and, given the current interest in LBFS, it is somewhat surprising that little work was done on LBFS until the mid 1990s.

In this paper, we survey many of the applications of LBFS (in Section 4). Before doing so, we provide the graph theoretical background for the paper as well as a description of LBFS and its two most common variants (Section 2) and, in Section 3, present some LBFS structural results. Concluding remarks are made in the final section.

J. Hromkovič, M. Nagl, and B. Westfechtel (Eds.): WG 2004, LNCS 3353, pp. 1–19, 2004.

2   Background

Before presenting LBFS and its various variants, we give some relevant definitions. We start with standard graph theoretical definitions and then define various graph families and indicate some characterizations that will be used in the relevant LBFS algorithms. Further information regarding the definitions and families can be found in [6].

2.1   Definitions and Notation

All graphs will be assumed to be undirected and finite. For a graph $G(V, E)$, we use n to denote $|V|$ and m to denote $|E|$. K_n, C_n and P_n denote the Clique, Cycle and Path respectively on n vertices. A *House, Hole* and *Domino* are respectively: a C_4 sharing an edge with a K_3; an induced $C_k, k > 4$; a pair of C_4s sharing an edge. A subset of vertices M is a *module* if for all vertices $x, y \in M$ and $z \in V \setminus M, xz \in E$ if and only if $yz \in E$. Module M is *trivial* if $M = V, M = \emptyset$ or $|M| = 1$. A *maximal clique module* is a module that is a clique and is maximal with respect to both properties. Subset S of V is a *separator* if the graph induced on $V \setminus S$ is disconnected. A *moplex* is a maximal clique module whose neighbourhood is a minimal separator.

The *distance* between two vertices u and v is the length of a shortest path between u and v and is denoted $d(u, v)$. For vertex $v, ecc(v)$, the *eccentricity* of v is the length of a longest shortest path with v as an endpoint. The *diameter* $(diam(G))$ is the maximum eccentricity of all vertices in G. A vertex is *simplicial* if its neighbourhood is a clique. An ordering $v_1, v_2, \cdots, v_n$ of V is a *perfect elimination ordering* (PEO) if for all $i, 1 < i \leq n, v_i$ is simplicial in the graph induced on $v_1, \cdots, v_i$. A vertex v is *semisimplicial* if v is not the midpoint of any induced P_4. An ordering $v_1, v_2, \cdots, v_n$ of V is a *semiperfect elimination ordering* if for all $i, 1 < i \leq n, v_i$ is semisimplicial in the graph induced on $v_1, \cdots, v_i$. A vertex v is *2-simplicial* if there is no induced P_4 in the graph induced on $\{u : d(u, v) \leq 2\}$. An ordering $v_1, v_2, \cdots, v_n$ of V is a *2-simplicial elimination ordering* if for all $i, 1 < i \leq n, v_i$ is 2-simplicial in the graph induced on $v_1, \cdots, v_i$.

We say that path P *misses* vertex v if $P \cap N(v) = \emptyset$ (i.e., no vertex of P is adjacent to v). A path P is a *dominating path* if no vertex of G is missed by P. A pair of vertices x, y is a *dominating pair* if every path between x and y is a dominating path. Two vertices x, y are *unrelated with respect to vertex v* if there are paths P between x and v and Q between y and v such that P misses y and Q misses x. An independent triple of vertices x, y, z is an *Asteroidal Triple (AT)*, if between every pair of vertices, there is a path that misses the third. A vertex v is *admissible* if there are no unrelated vertices with respect to v. An ordering $v_1, v_2, \cdots, v_n$ of V is an *admissible elimination ordering* (AEO) if for all $i, 1 < i \leq n, v_i$ is admissible in the graph induced on $v_1, \cdots, v_i$.

For $t \geq 1$, an ordering $v_1, v_2, \cdots, v_n$ of V is a *strong t-cocomparability ordering* (strong t-CCPO) if for all $i, j, 1 \leq i < j < k \leq n, d(v_i, v_k) \leq t$ implies $d(v_i, v_j) \leq t$ or $d(v_j, v_k) = 1$. Note that a graph is a *cocomparability* graph (there is a transitive orientation of the edges of the complement) if and only if it has a strong 1-CCPO [28].

A graph is *chordal* if there is no induced cycle of length greater than 3. Fulkerson and Gross [23] showed that a graph is chordal if and only if it has a perfect elimination ordering. G is *weakly chordal* if G and $\overline{G}$ contain no induced cycle $C_k, k \geq 5$. A graph is *strongly chordal* if it is chordal and every cycle of even length at least 6 has an *odd chord*, namely a chord where the distance on the cycle between the endpoints is odd. An *interval graph* is the intersection graph of intervals of a line. If all intervals are of the same length, then G is a *unit interval graph* (equivalently known as *proper interval graphs*, where no interval is allowed to properly contain another interval). A graph G is a *distance hereditary graph* if for every connected subgraph H, $x, y \in H$ implies that $d_H(x, y) = d_G(x, y)$. Nicolai [35] has shown that a graph is distance hereditary if and only if it has a 2-simplicial elimination ordering. A graph is *HHD-free* if it contains no House, Hole or Domino, as defined above. Bipartite graphs with no induced cycles of size greater than 4 are called *chordal bipartite*.

Cographs are the graphs formed by the closure of the disjoint union and complementation operations on individual vertices. There are many equivalent characterizations of cographs including being the graphs that contain no induced P_4, and having a cotree representation. A *cotree* is a rooted tree with the leaves being the vertices of the cograph and the internal vertices alternating between being "0" and "1" nodes. Two vertices x, y of a cograph are adjacent if and only if the lowest common ancestor of x and y in the cograph is a "1" node. See Figure 1 for an example of a cograph and its related cotree. Cographs can be extended in the following ways: a graph where each vertex belongs to at most one P_4 is called a P_4-*Reducible graph*; a graph where every set of five vertices induces at most one P_4 is called a P_4-*Sparse graph*. Both P_4-Reducible and P_4-Sparse graphs (as well as distance hereditary graphs) have a tree structure representation that is an extension of cotrees. G is *AT-free* if it contains no AT. A graph is a *permutation graph* if it is the intersection graph of lines whose endpoints are on two parallel lines. Permutation graphs strictly contain cographs and are cocomparability graphs and thus are also AT-free.

A bipartite graph with bipartition (X, Y) is an *interval bigraph* if each vertex v is assigned an interval I_v and $x \in X, y \in Y$ are adjacent if and only if $I_x \cap I_y \neq \emptyset$; an interval bigraph is *proper* if no interval contains another. These graphs are also known as bipartite AT-free graphs and bipartite permutation graphs.

2.2   LBFS

As mentioned in the Introduction, BFS is one of the fundamental graph searching paradigms and can be found in any standard graph theory text. BFS uses a queue to ensure that whenever a vertex x is visited, its previously unvisited neighbours must be visited before its previously unvisited non-neighbours. A *layer* of a BFS is a set of vertices all of the same distance from the initial vertex of the BFS. LBFS is a restriction of BFS; in the following, we present the details of the generic LBFS algorithm and its implementation. We then describe two popular variants of this generic algorithm.

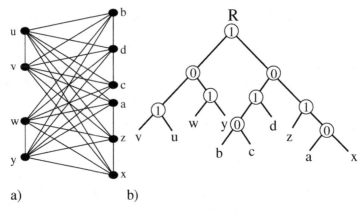

Fig. 1. a) A cograph. b) Its cotree.

Generic LBFS: Note that in the following algorithm, we require the sweep to start at given vertex x; if the algorithm is to start from an arbitrary vertex, then Step 1 is omitted and Step 2 is replaced by assigning Λ to label(y) for all $y \in V$. Note that Step 4 allows the choice of any vertex that has the lexicographically largest label. Later we will present various modifications that explicitly choose the next vertex. We warn the reader that our LBFS ordering of the vertices of the graph may seem "backwards" compared to the ordering produced by other LBFS descriptions.

Procedure LBFS(x)
{Input: Graph $G(V, E)$ and a distinguished vertex x of G;
Output: An ordering σ of the vertices of G.}

1.   label(x) $\leftarrow |V|$;
2.   **for** each vertex y in $V \setminus \{x\}$ **do** label(y) $\leftarrow \Lambda$;
3.   **for** $i \leftarrow |V|$ **downto** 1 **do**
4.   pick an unnumbered vertex y with lexicographically the largest label;
5. $\sigma(y) \leftarrow |V| + 1 - i$; {assign to y number $|V| + 1 - i$};
6.   **for** each unnumbered vertex z in $N(y)$ **do** append i to label(z).

In an LBFS σ with two arbitrary vertices u and v, if vertex u is visited before v, i.e. $u <_\sigma v$ we say that u *occurs* before v in σ or that u is *visited before v* or that u is *to the left of v*. As mentioned above, this generic LBFS algorithm allows arbitrary choice of a vertex in Step 4. We call the set of tied vertices encountered in Step 4 a *slice* and denote it by S. Note that all vertices of a slice with respect to LBFS σ appear consecutively in σ. Given two vertices u and v of an LBFS σ such that $u <_\sigma v$, $\Gamma_{u,v}^\sigma$ denotes the vertex-minimal slice with respect to σ that contains both u and v. As an example of these concepts consider the graph in Figure 2 where the boxes indicate the slices, including V itself, with respect to the LBFS σ (note that the vertices are numbered as visited by σ). $\Gamma_{9,10}^\sigma$ consists of $\{5, 6, 7, 8, 9, 10\}$.

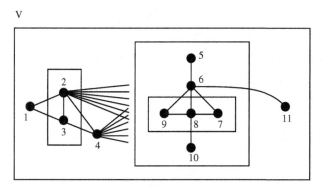

Fig. 2. A graph with its LBFS slices

To implement the generic LBFS algorithm, we use the implementation presented in [24], namely one that follows the paradigm of "partitioning". In this scheme, we start with all vertices in the same cell (i.e., slice) and choose an arbitrary vertex (for reasons that will come clear later, we will choose the first vertex in the cell). When a vertex is chosen, i.e., is chosen as the *pivot*, it is placed in its own cell and invokes a partitioning of all cells that follow it in the ordering. Under this partitioning of a cell, vertices that are adjacent to the pivot form a new cell that precedes the cell containing the vertices not adjacent to the pivot. After this partitioning is complete, a new pivot is chosen from the cell immediately following the old pivot and the process of refinement continues. We refer the reader to Figure 3 for an example of a few steps of partitioning on the graph in Figure 2.

Variants of the Generic LBFS Algorithm: We now describe two variants of the generic LBFS algorithm. In subsequent sections we will reference other variants that have appeared in the literature. In the first case, we break ties in Step 4 by referring to a previous LBFS ordering σ. This variant has been independently investigated by Simon [44] and Ma [32].

Procedure LBFS+ (σ)
{Input: Graph $G(V, E)$ and an LBFS σ of G;
Output: An ordering σ^+ of the vertices of G.}

Do an LBFS of G. When Step 4 is encountered, let S be the set of vertices with the lexicographically largest label. Now y is chosen to be the vertex in S that appears *last* in σ.

As an example, LBFS+ when given the graph in Figure 2 and that LBFS, would produce the order: 11 6 9 8 4 2 7 5 10 3 1.

As pointed out by Lanlignel [30], one of the advantages of using the partitioning implementation of generic LBFS described above, is that we immediately have an implementation of LBFS+. Once σ has been determined, we merely re-

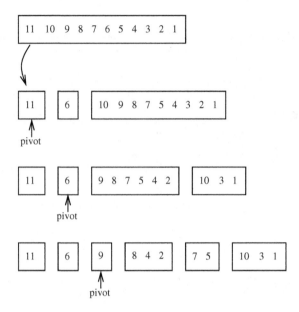

Fig. 3. The first few steps of a partitioning.

verse its ordering of V and run the generic algorithm again. Every time a slice is encountered, the last vertex from σ is automatically the vertex at the front of the list. The example in Figure 3 represents the first few steps of the LBFS+ for the sweep presented in Figure 2.

The second variant produces an LBFS of $\overline{G}$, the complement of graph G. Note that in doing so, we *do not* calculate the complement but rather, as we shall see, slightly modify the generic implementation of LBFS. Furthermore this sweep also requires a previous LBFS as input so that a specific vertex is chosen in Step 4. Note that an overbar placed on an LBFS ordering indicates that the ordering is an LBFS of $\overline{G}$.

Procedure LBFS$^-$ (σ)
{Input: Graph $G(V, E)$ and an LBFS σ of G;
Output: An ordering $\overline{\sigma}^-$ of the vertices of $\overline{G}$.}

Do an LBFS of $\overline{G}$. When Step 4 is encountered, let S be the set of vertices with the lexicographically largest label. Now y is chosen to be the vertex in S that appears *first* in σ.

An example of this algorithm will appear in Section 4, in the example of the Cograph recognition algorithm. As noted in [33, 24], an LBFS of $\overline{G}$ can be done in $O(n + m)$ time by making a slight modification of the implementation of the generic LBFS algorithm. In particular, during the partitioning, the cell containing the vertices adjacent to the pivot is placed *after* the cell containing the nonadjacent vertices. To make sure that the correct vertex in S is chosen, we merely input σ in its natural order. By choosing the first vertex of every slice as the next pivot, we automatically meet the choice requirement of the algorithm.

3 Structure Results

In this section we present some of the important structure results concerning LBFS. For most families of graphs where LBFS has been used, there are particular results that are unique to that family. The first result is the following characterization of vertex orderings that can be achieved by an LBFS. This characterization is used heavily in the various multi-sweep LBFS algorithms and it is somewhat surprising to note that, except for Maximum Cardinality Search (MCS), similar characterizations for other well known graph searches have only recently been discovered [16].

Theorem 1. *[22] An ordering $\prec$ of the vertices of an arbitrary graph $G(V, E)$ is an LBFS ordering if and only if for all vertices a, b, c of G such that $ac \in E$ and $bc \notin E$, $c \prec b \prec a$ implies the existence of a vertex d in G, adjacent to b but not to a and such that $d \prec c$.*

The following lemma establishes the existence of special paths in $\Gamma_{u,v}^{\sigma}$.

Lemma 1. *[17] (The Prior Path Lemma) Let σ be an arbitrary LBFS of a graph G. Let t be the first vertex of the connected component of $\Gamma_{u,v}^{\sigma}$ containing u. There exists a t, u-path in $\Gamma_{u,v}^{\sigma}$ all of whose vertices, with the possible exception of u, are missed by v. Moreover, all vertices on this path, other than u, occur before u in σ. (Such a path is called a* prior path*).*

As an example of this Lemma, consider the graph in Figure 2 and let $u = 7, v = 10$. Now $\Gamma_{7,10}^{\sigma} = \{5, 6, 7, 8, 9, 10\}$ and path $7 - 6 - 5$ is a prior path for this choice of u, v.

In the fundamental paper by Rose, Tarjan and Lueker [42], their chordal graph recognition algorithm was based on the following theorem.

Theorem 2. *[42] Let σ be an LBFS of a chordal graph G and let v be an arbitrary vertex of G. Let W denote the set of vertices w that occur before v in σ. Then v is simplicial in the subgraph of G induced by $W \cup \{v\}$.*

From this theorem, we immediately see that the reverse ordering of an LBFS of a chordal graph G yields a PEO of G. Berry and Bordat [2] have generalized this theorem as follows:

Theorem 3. *[2] Let σ be an LBFS of a chordal graph G and let v be an arbitrary vertex of G. Let W denote the set of vertices w that occur before v in σ. Then v belongs to a moplex in the subgraph of G induced by $W \cup \{v\}$.*

Furthermore they showed that the vertices in the moplex containing v are consecutive vertices in σ up to and including v.

Interestingly, a very similar result to Theorem 2 holds for an arbitrary LBFS in an AT-free graph. In particular,

Theorem 4. *[18] Let σ be an LBFS of an AT-free graph G and let v be an arbitrary vertex of G. Let W denote the set of vertices w that occur before v in σ. Then v is admissible in the subgraph of G induced by W ∪ {v}.*

Again, this theorem yields the result that the reverse ordering of an LBFS of an AT-free graph G yields an AEO of G. Unfortunately, the existence of an AEO for a graph G does not imply that G is AT-free. More will be said about this issue later in this section. Given any subset of vertices X in either a chordal or an AT-free graph, these two theorems show the importance of x, the last vertex of X, in any LBFS. In particular, such a vertex is guaranteed to be simplicial (respectively, admissible) in the subset of vertices that have occurred up to and including x and thus also in X itself. In many multi-sweep LBFS algorithms, we want to "break ties" by choosing a vertex with a particular property. LBFS+, the algorithm that starts a slice S with the last S vertex in the previous sweep, was developed for precisely this reason and, as we shall see in Subsection 4.1, is currently the most popular restricted version of LBFS in multi-sweep LBFS algorithms.

To formalize the notion of "last vertex" mentioned above, we define a vertex x to be an *end-vertex* of graph G if there is an LBFS of G that ends at x. Is it possible that end-vertices of a graph can be characterized? For interval graphs, the answer is affirmative as shown in the following Lemma.

Lemma 2. *[17] A vertex in an interval graph is an end-vertex if and only if it is simplicial and admissible.*

This result can be extended to arbitrary graphs in the following way.

Lemma 3. *[15] Let G be an arbitrary graph. If x is a simplicial and admissible vertex of G, then x is an end-vertex.*

Unfortunately, it seems unlikely that there is a nice characterization of end-vertices for arbitrary graphs, as shown by the following complexity result.

Theorem 5. *[15] Given a graph G and vertex x, it is NP-complete to determine whether x is an end-vertex of G.*

Furthermore, the problem remains NP-complete for weakly chordal graphs, is linearly time solvable for interval graphs (using an LBFS followed by an LBFS+) and remains unresolved for both chordal and AT-free graphs [15].

As we shall see, most LBFS based algorithms involve a number of LBFS sweeps and thus require some knowledge of the behaviour of parts of the graph in previous sweeps. Typically such arguments are based on either the behaviour of paths (where the Prior Path Lemma is fundamental) or the behaviour of slices, which we now discuss. In particular, we look at the restriction of an LBFS to a slice from some other LBFS. The strongest result of this type is for chordal graphs.

Lemma 4. *[17](The LBFS Lemma) Let G be a chordal graph and let S be a slice of an arbitrary LBFS ordering τ of G. Further let σ be another arbitrary LBFS ordering of G. Then the restriction of σ to S is an LBFS ordering of the graph induced by the vertices of S.*

As stated in [17], "to put this lemma in perspective, it is important to note that the desired property does not hold for arbitrary subsets of chordal (or even interval) graphs. For example, consider the interval graph shown in Figure 4. The numbering of the vertices indicates a legitimate LBFS ordering; however when vertex 1 is removed, the restriction of this ordering to the remaining subset is not a legitimate LBFS ordering of the subset. Also, as shown in Figure 5, the lemma does not hold for AT-free graphs. $S = \{2, 3, 4\}$ is a slice of the LBFS: 1 2 3 4 5. Now consider an arbitrary LBFS starting at 5. Vertex 3 occurs after 2 and 4, which cannot occur in an LBFS of S."

It is somewhat surprising and disappointing that the LBFS Lemma does not generalize to AT-free graphs. Nevertheless, there is something that can be said in this regard for AT-free graphs. First we note the following obvious Corollary of Theorem 2 and Lemma 4.

Corollary 1. *Let G be a chordal graph with S a slice with respect to LBFS σ. Then for every LBFS τ of G, x, the last vertex of τ_S, is an end-vertex of S.*

As shown in [15], there is a similar result for AT-free graphs.

Lemma 5. *[15] Let G be an AT-free graph with S a slice with respect to LBFS σ. Then for every LBFS τ of G, x, the last vertex of τ_S, is either an end-vertex of S or is adjacent to an end-vertex of S.*

Finally, to end this section, we mention a pair of graph families that are characterized by properties of *every* LBFS. The reader is cautioned that a statement of the form: "G is an X-graph if and only if every LBFS has property P" can be quite misleading in the sense that "every" can either be interpreted as "an arbitrary" (for example, chordal and distance hereditary graphs, as discussed in the next Section) or "all", (for example, HHD-free and AT-free graphs) as we now present.

Lemma 6. *[27] G is an HHD-free graph if and only if all LBFSs are semiperfect elimination orderings.*

Lemma 7. *[15] G is an AT-free graph if and only if all LBFSs are admissible elimination orderings.*

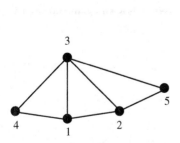

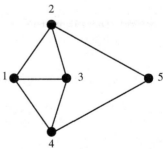

Fig. 4. The LBFS Lemma does not hold for arbitrary subsets of interval graphs.

Fig. 5. The LBFS Lemma does not hold for AT-free graphs.

4   Applications of LBFS

In this section, we survey many of the applications of LBFS. The most notable application, presented in Subsection 4.1, is the recognition of various restricted graph classes. For many graph classes, the current "best" recognition algorithm is based on LBFS, usually in the form of a multi-sweep algorithm. In Subsection 4.2, we will show other diverse applications of LBFS including diameter approximation for various graph classes and the determination of a dominating pair in an AT-free graph.

4.1   Recognition of Various Graph Classes

LBFS was discovered in the development of a simple, linear time chordal graph recognition algorithm [42]. The algorithm is based on the following fundamental result:

Theorem 6. *[42] A graph is chordal if and only if an arbitrary LBFS yields a perfect elimination ordering.*

Thus the associated chordal graph recognition algorithm is as follows:

The *Chordal graph* Recognition Algorithm[42]
{Input: Graph $G(V, E)$;
Output: A statement declaring whether or not G is a *chordal graph*.}

1.   Do an arbitrary LBFS σ.
2.   If the reverse of σ is a perfect elimination ordering, then conclude that G is a chordal graph; else, conclude that G is not a chordal graph.

Since determining whether a particular ordering is a perfect elimination ordering can be accomplished in linear time, the algorithm has a straightforward linear time implementation. In a subsequent paper, Tarjan and Yannakakis [43] extended this algorithm to be certifying by showing how to find, in linear time, an induced cycle of size greater than three if the reverse of σ is not a perfect elimination ordering.

Interestingly there is another family of graphs that has the same single LBFS recognition algorithm. In particular, Dragan and Nicolai [21] proved the following theorem:

Theorem 7. *[21] A graph is distance hereditary if and only if an arbitrary LBFS yields a 2-simplicial elimination ordering.*

As with chordal graph recognition, there is an associated distance hereditary graph recognition algorithm.

Although this is a very simple algorithm, it does not seem to have a linear time implementation since it is not clear how one can determine in linear time whether the reverse of σ is a 2-simplicial elimination ordering. Later in this subsection, we will discuss a linear time multi-sweep LBFS distance hereditary graph recognition algorithm.

We now turn our attention to LBFS recognition algorithms that require at least two LBFS sweeps. In presenting such multi-sweep recognition algorithms, we will take two "basic" algorithms for the recognition of unit interval graphs and cographs, respectively, and show how modifications of these algorithms can lead to recognition algorithms for related graph families. All of these algorithms are easily implementable in linear time. Typically, they are not the first linear time algorithm for the particular recognition problem but they are simpler than the previous non-LBFS algorithms. References to these other algorithms are contained in the appropriate reference describing the LBFS algorithm.

Unit Interval Graphs: The LBFS based unit interval graph recognition algorithm [11] is the following:

The *Unit Interval graph* Recognition Algorithm[11]
{Input: Graph $G(V, E)$;
Output: A statement declaring whether or not G is a *unit interval graph*.}

1. Do an arbitrary LBFS σ.
2. LBFS+ (σ) yielding sweep σ^+.
3. LBFS+ (σ^+) yielding sweep σ^{++}.
4. If σ^{++} satisfies a "particular condition", then conclude that G is a unit interval graph; else, conclude that G is not a unit interval graph.

In the case of unit interval graphs, the "particular condition" to be tested is the "Neighbourhood Condition", namely that $G(V, E)$ is a unit interval graph if and only if there is an ordering of V such that for all $v \in V$, $N[v]$ (the closed neighbourhood of v) is consecutive. For this and other characterizations of unit interval graphs, see [39], [40] and [31].

This algorithm is not "certifying" in the sense that if the input graph fails the Neighbourhood Condition and the algorithm concludes that the input graph is not a unit interval graph, then there is no immediate "proof" that the graph is in fact not a unit interval graph. Note, that the algorithm does certify the conclusion that the graph is a unit interval graph since it is easy to build a unit interval model if the Neighbourhood Condition is satisfied. Recently, two algorithms have been developed to provide a certificate of nonmembership. The first, by Meister [34], is similar to the above algorithm in that it uses three LBFS sweeps with the second and third sweeps using "min-LexBFS" which requires a special implementation rather than LBFS+ which, as pointed out in Subsection 2.2, has an immediate partitioning implementation. The certificate that Meister's algorithm produces is either an induced cycle of size greater than 3 (thereby showing that the graph is not chordal, and thus not interval) or an AT (showing that the graph is not AT-free, and thus not interval) or a claw (i.e., $K_{1,3}$). The second certifying algorithm, by Hell and Huang [25], augments the algorithm presented above and uses Wegner's characterization of unit interval graphs [45], namely that a graph is a unit interval graph if and only if it does not contain an induced cycle of size greater than 3, a claw, or a "3-sun" or its complement, the "net" which consists of a triangle each of whose vertices is adjacent to a unique vertex of degree 1. One of the pretty aspects of the Hell and Huang algorithm is that it incorporates the certificate steps throughout the algorithm, in the sense that it does some testing after each of the three LBFS steps, and only proceeds to the next sweep if the test has been satisfied. Furthermore, they also show that the algorithm presented above can be augmented to provide a linear time certifying recognition algorithm for proper interval bigraphs. Again the certification is distributed throughout the algorithm. Chang, Ho and Ko [9] have presented a linear time 2-sweep LBFS based algorithm for recognizing bipartite permutation graphs (equivalent to proper interval bigraphs). Their algorithm modifies the second LBFS sweep to break ties according to the value of a degree related function.

One of the early uses of LBFS appeared in the Korte - Möhring interval graph recognition algorithm [29]. By using LBFS, they were able to streamlne the first linear time interval graph recognition algorithm by Booth and Lueker [3]. We now present a second extension of the unit interval graph recognition algorithm that provides an easily implementable, linear time recognition algorithm for interval graphs. This algorithm is as follows:

The *Interval graph* Recognition Algorithm[17]
{Input: Graph $G(V, E)$;
Output: A statement declaring whether or not G is an *interval graph*.}

1.   Do an arbitrary LBFS π.
2.   LBFS+ (π) yielding sweep σ.
3.   LBFS+ (σ) yielding sweep σ^+.
4.   LBFS+ (σ^+) yielding sweep σ^{++}.
5.   LBFS* (σ^+, σ^{++}) yielding sweep σ^*.
6.   If σ^* satisfies the "Interval Graph Ordering Condition" then conclude that G is an interval graph; else, conclude that G is not an interval graph.

The "interval Graph Ordering Condition" states that a graph $G(V, E)$ is an interval graph if and only if there is a linear ordering $\prec$ on V such that for every choice of vertices u, v, w, with $u \prec v$ and $v \prec w$, $uw \in E \implies uv \in E$ [26, 36, 38, 37]. LBFS* requires *two* previous sweeps and breaks ties for a slice S by examining the last vertices of S in each of these two sweeps. Since LBFS* and the "Interval Graph Ordering Condition" can easily be implemented in linear time, the entire algorithm is easily implementable in linear time. (See [17] for further details.) Recently Choi and Farach-Colton [10] have used this algorithm to develop a new interval graph based algorithm for the sequence assembly problem that is significantly superior to existing algorithms. It is interesting to note that Simon ([44]) incorrectly claimed that terminating the algorithm after the third LBFS+ would suffice to recognize interval graphs. Ma [32], however, showed that Simon's algorithm is flawed and that for any constant c, there is an interval graph, and an initial LBFS ordering such that after c applications of LBFS+, the linear ordering of vertices is still not apparent! It is known, however, that using n applications of LBFS+ will work, but of course not in linear time [14].

Cographs: Our second basic algorithm is the one for cographs. The generic algorithm is as follows:

The *Cograph* Recognition Algorithm[8]
{Input: Graph $G(V, E)$;
Output: A cotree if G is a *cograph*, or an induced P_4 otherwise.}

1. Do an arbitrary LBFS σ.
2. LBFS$^-(\sigma)$ yielding sweep $\overline{\sigma}^-$ (of $\overline{G}$).
3. LBFS$^-(\overline{\sigma}^-)$ yielding sweep σ^- (of G).
4. If $\overline{\sigma}^-, \sigma^-$ satisfy a "particular condition", then CONSTRUCT_COTREE; else REPORT_P_4.

In the case of cograph recognition, the "particular condition" is the "Neighbourhood Subset Property", a property that can easily be checked in linear time, yielding an easily implementable linear time algorithm. See [7, 8] for further details. As an example of the algorithm, consider the cograph in Figure 6. The two LBFS$^-$ sweeps satisfy the "Neighbourhood Subset Property" and the algorithm produces the cotree.

The first extension of this cograph recognition algorithm is to P_4-Reducible graphs, namely those graphs where each vertex belongs to at most one P_4. The algorithm is as follows:

The P_4-*Reducible graph* Recognition Algorithm[7]
{Input: Graph $G(V, E)$;
Output: A P_4-R_tree, if G is a P_4-*Reducible graph*, or two P_4s containing one vertex otherwise.}

1. Do an arbitrary LBFS σ.
2. LBFS+ (σ) yielding sweep σ^+.
3. LBFS$^-(\sigma^+)$ yielding sweep $\overline{\sigma}^-$ (of $\overline{G}$).
4. LBFS$^-(\overline{\sigma}^-)$ yielding sweep σ^- (of G).
5. If $\overline{\sigma}^-, \sigma^-$ satisfy a "particular condition", then CONSTRUCT_P_4-R_TREE; else REPORT_MULTIPLE_P_4.

The appropriate "particular condition" for this algorithm is the "P_4-Reducible Neighbourhood Property" described in [7]. This condition is easily tested in linear time.

P_4-Sparse graphs, namely graphs for which no set of five vertices induces more than one P_4, generalize P_4-Reducible graphs and have a very similar recognition algorithm.

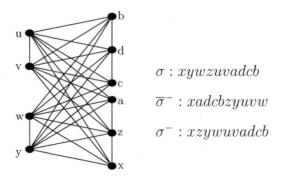

$$\sigma : xywzuvadcb$$

$$\overline{\sigma}^- : xadcbzyuvw$$

$$\sigma^- : xzywuvadcb$$

Fig. 6. The execution of the algorithm on the cograph in Figure 1.

The P_4-*Sparse graph* Recognition Algorithm[7]
{Input: Graph $G(V, E)$;
Output: A P_4-S_tree, if G is a P_4-*Sparse graph*, or a set of five vertices inducing two P_4s otherwise.}

1. Do an arbitrary LBFS σ.
2. LBFS+ (σ) yielding sweep σ^+.
3. LBFS$^-(\sigma^+)$ yielding sweep $\overline{\sigma}^-$ (of $\overline{G}$).
4. LBFS$^-(\overline{\sigma}^-)$ yielding sweep σ^- (of G).
5. If $\overline{\sigma}^-, \sigma^-$ satisfy a "particular condition", then CONSTRUCT_P_4-S_TREE; else REPORT_5_SET.

The appropriate "particular condition" for this algorithm is the "P_4-Sparse Neighbourhood Property" described in [7]. Again an easily implementable linear time algorithm is obtained.

Finally we turn to a new LBFS based recognition algorithm for distance hereditary graphs. Given Theorem 7, it is not surprising that cographs play a critical role in characterizing distance hereditary graphs. In particular, Bandelt and Mulder [1] showed that every layer of any BFS of a distance hereditary graph is a cograph and that there are specific neighbourhood intersection conditions inside and between layers. Bretscher [7] has shown that these conditions can be expressed by neighbourhood conditions on LBFS slices in BFS layers to produce an LBFS characterization of distance hereditary graphs. By using this characterization and the new LBFS cograph recognition algorithm, she has produced a new simpler linear time LBFS based distance hereditary graph recognition algorithm [7].

4.2   Other Applications

These applications vary from diameter approximation for various families of graphs to the determination of a dominating pair in an AT-free graph to common properties of powers of graphs.

Diameter Approximation: Determining the diameter of a graph is a fundamental graph property whose current best algorithm (i.e. $O(nm)$) is too slow
for very large input graphs. This naive algorithm performs a BFS from each
vertex x, thereby calculating the eccentricity, $ecc(x)$ of x. The diameter is then
determined by finding the maximum eccentricity of any vertex in G. Since any
BFS from a vertex of maximum eccentricity immediately produces the diameter
of the graph, one approach to approximating a graph's diameter is to search for
a vertex of high eccentricity. The most common way of finding such a vertex
has been to take the end-vertex of a specific search from an arbitrary vertex.
The searches that have been considered are BFS, LBFS, LL and LL+ where LL
chooses an arbitrary vertex in the last BFS layer and LL+ chooses an arbitrary
vertex in the last BFS layer that has minimum degree into the second last BFS
layer. For the restricted graph families considered in [13], none of BFS, LL or
LL+ beat LBFS. In particular, the eccentricity of an LBFS end-vertex is guaranteed: to be $diam(G)$ for interval graphs [22] and {AT,claw}– free graphs [4];
to be at least $diam(G) - 1$ for chordal [22] and AT-free [12] graphs; and to be at
least $diam(G) - 2$ for graphs that contain no induced cycles of size greater than
4 [13]. Dragan [19] presented similar LBFS results on other restricted families
of graphs. Corneil et al [12] also looked at the end-vertex of a "double-sweep"
LBFS algorithm (see the following Dominating Pair Algorithm) on chordal and
AT-free graphs. They established a forbidden subgraph structure on chordal or
AT-free graphs where $diam(G) - 1$ is the eccentricity of the end-vertex of the
second sweep. They also showed examples of chordal and AT-free graphs where
for no c, the "c-sweep" LBFS algorithm is guaranteed to find a vertex of maximum eccentricity. Furthermore, for any c there is a graph G (albeit with large
induced cycles whose size depends on c) where the eccentricity of the chosen
vertex is at least c away from the diameter of G.

In a related approach, Dragan [20] showed how particular vertex orderings
(including LBFS) can be used to appoximate the All Pairs Shortest Path problem
to within a small additive constant for various restricted families of graphs.

Dominating Pairs in AT-free Graphs: One of the first indications that
LBFS has far-reaching applications beyond families of graphs related to chordal
graphs came in the 2-sweep algorithm for finding a dominating pair in a connected AT-free graph. The algorithm is as follows:

The *Dominating Pair* Algorithm[18]
{Input: A connected AT-free graph $G(V, E)$;
Output: A pair of vertices x, y that form a dominating pair of G.}

1. Do an arbitrary LBFS σ where x is the end-vertex of σ.
2. Do an LBFS(x) τ where y is the end-vertex of τ.
3. Return x, y.

As noted in [18], this algorithm can be modified to return, in linear time, a
succinct representation of *all* dominating pairs in any connected AT-free graph of

diameter greater than 3. In particular, if $diam(G) > 3$, then there are nonempty, disjoint sets X and Y of vertices of G, such that x, y is a dominating pair if and only if $x \in X$ and $y \in Y$. The algorithm returns the sets X and Y.

Properties of Powers of Graphs: In families of graphs related to chordal graphs, considerable attention has been given to the problem of determining the graph class membership of various powers of a given graph. For example, it is known that every odd power of a chordal graph is chordal. This result is an immediate corollary of the following theorem:

Theorem 8. *[5] The reversal of every LBFS ordering of a chordal graph G is a common perfect elimination ordering of all odd powers of G.*

A similar result is captured in the following theorem:

Theorem 9. *[21] The reversal of every LBFS ordering of a distance hereditary graph G is a perfect elimination ordering of every even power $G^{2k}, k \geq 1$.*

Note that the result in Theorem 8 does not imply that the reversal of an LBFS ordering of G is also a reversal of an LBFS ordering of odd powers of G [16]. In fact, there are chordal graphs where no LBFS ordering is also an LBFS ordering of any powers of the graph. On the other hand, every LBFS ordering of a chordal bipartite graph is also an LBFS ordering of its square, a property that does not hold for bipartite graphs [16].

In a similar vein, Chang, Ho and Ko [9] with respect to LBFS orderings and strong 2-CCPOs, considered τ, the ordering produced by the second LBFS in the Dominating Pair Algorithm and proved:

Theorem 10. *[9] Given an AT-free graph G, τ is a strong 2-CCPO of G.*

5   Concluding Remarks

One of the surprising observations in the development of LBFS based algorithms is that LBFS works so well on both chordal and AT-free related families of graphs, yet these two families have very little structural similarity (other than the absence of large induced cycles). Is there some generalization of these two families that explains the success of LBFS?

Although there is now a very impressive list of graph families whose recognition is best achieved using an LBFS approach, there have been a number of graph families, especially strongly chordal, chordal bipartite and permutation graphs, that so far have resisted this approach. Currently no linear time recognition algorithm is known for the family of strongly chordal graphs. Chordal bipartite graphs are a closely related family (see [6]) that has also resisted linear time recognition. Although there are linear time recognition algorithms for permutation graphs, it is possible that there is a simpler LBFS based one.

Given the number and power of multi-sweep LBFS algorithms, it is somewhat surprising that other graph searches have seldom been used in a multi-sweep fashion. One reason for this may be that, until recently [16], most of them do not have a vertex ordering characterization similar to that for LBFS presented in Theorem 1. This also raises the possibility of multi-sweep hybrid algorithms where BFS and DFS variants are combined. Finally, given the power of the "lexicographic" extension of BFS, it is natural to wonder whether DFS would benefit from a similar extension; a candidate algorithm for Lexicographic Depth First Search is presented in [16].

Acknowledgements

Financial assistance from the Natural Sciences and Engineering Research Council of Canada was gratefully received. Many thanks to Anna Bretscher, Feodor Dragan, Ekki Koehler and Lorna Stewart for their improvements of an earlier draft of this paper.

References

1. H-K. Bandelt and H.M. Mulder: Distance-hereditary graphs, J. Combin. Theory B 41 (1986) 182–208.
2. A. Berry and J-P. Bordat: Separability generalizes Dirac's theorem, Disc. Appl. Math. 84 (1998) 43–53.
3. K.S. Booth and G.S. Lueker: Testing for the consecutive ones property, interval graphs, and graph planarity using PQ-tree algorithms, J. Comput. System Sci. 13 (1976) 335–379.
4. A. Brandstädt and F.F. Dragan: On linear and circular structure of a (claw,net)-free graph, Disc. Appl. Math. 129 (2003) 285–303.
5. A. Brandstädt, F.F. Dragan and F. Nicolai: LexBFS-orderings and powers of chordal graphs, Disc. Math. 171 (1997) 27–42.
6. A. Brandstädt, V.B. Le and J.P. Spinrad: Graph Classes: A Survey, SIAM Monographs on Disc. Math. and Applic., SIAM 1999.
7. A. Bretscher: LexBFS based recognition algorithms for cographs and related families, Ph.D. thesis in preparation, Dept. of Computer Science, University of Toronto, Toronto, Canada.
8. A. Bretscher, D.G. Corneil, M. Habib and C. Paul: A simple linear time LexBFS cograph recognition algorithm (extended abstract), LNCS 2880 (2003) 119–130.
9. J-M. Chang, C-W. Ho and M-T. Ko: LexBFS-ordering in Asteroidal Triple-free graphs, LNCS 1741 (1999) 163–172.
10. V. Choi and M. Farach-Colton: Barnacle: an assembly algorithm for clone-based sequences of whole genomes, Gene 320 (2003) 165–176.
11. D.G. Corneil, A simple 3-sweep LBFS algorithm for the recognition of unit interval graphs: Disc. Appl. Math. 138 (2004) 371–379.
12. D.G. Corneil, F.F. Dragan, M. Habib and C. Paul: Diameter determination on restricted graph families, Disc. Appl. Math. 113 (2001) 143–166.
13. D.G. Corneil, F.F. Dragan and E. Koehler: On the power of BFS to determine a graph's diameter, Networks 42 (2003) 209–222.
14. D.G. Corneil and E. Koehler: unpublished manuscript.
15. D.G. Corneil, E. Koehler and J-M. Lanlignel: On LBFS end-vertices, in preparation.

16. D.G. Corneil and R. Krueger: A unified view of graph searching, in preparation.
17. D.G. Corneil, S. Olariu, and L. Stewart: The LBFS structure and recognition of interval graphs, under revision; extended abstract appeared as The ultimate interval graph recognition algorithm? (extended abstract) in Proc. SODA 98, Ninth Annual ACM-SIAM Symposium on Discrete Algorithms (1998) 175–180.
18. D.G. Corneil, S. Olariu, and L. Stewart: Linear time algorithms for dominating pairs in asteroidal triple-free graphs, SIAM J. Comput. 28 (1999) 1284–1297.
19. F.F. Dragan: Almost diameter of a house-hole-free graph in linear time via LexBFS, Disc. Appl. Math. 95 (1999) 223–239.
20. F.F. Dragan: Estimating all pairs shortest paths in restricted graph families: a unified approach (extended abstract), LNCS 2204 (2001) 103–116.
21. F.F. Dragan and F. Nicolai: Lex-BFS-orderings of distance-hereditary graphs, Schriftenreihe des Fachbereichs Mathematik der Universität Duisburg, Duisburg, Germany, SM-DU-303 (1995).
22. F.F. Dragan, F. Nicolai, and A. Brandstädt: LexBFS-orderings and powers of graphs, LNCS 1197 (1997) 166–180.
23. D.R. Fulkerson and O.A. Gross: Incidence matrices and interval graphs, Pacific J. Math. 15 (1965) 835–855.
24. M. Habib, R. McConnell, C. Paul and L. Viennot: Lex-bfs and partition refinement, with applications to transitive orientation, interval graph recognition and consecutive ones testing, Theoret. Comput. Sci. 234 (2000) 59–84.
25. P. Hell and J. Huang: Certifying LexBFS recognition algorithms for proper interval graphs and proper interval bigraphs, to appear SIAM J. Disc. Math. (2004).
26. M.S. Jacobson, F.R. McMorris, and H.M. Mulder: Tolerance intersection graphs, in 1988 International Kalamazoo Graph Theory Conference, Y. Alavi et al, eds., Wiley 1991 705–724.
27. B. Jamison, S. Olariu: On the semi-perfect elimination, Advances in Appl. Math. 9 (1988) 364–376.
28. D. Kratsch and L. Stewart: Domination on cocomparability graphs, SIAM J. Disc. Math. 6 (1993) 400–417.
29. N. Korte and R.H. Möhring: An incremental linear-time algorithm for recognizing interval graphs, SIAM J. Comput. 18 (1989) 68–81.
30. J-M. Lanlignel: Private communications, 1999.
31. P.J. Looges and S. Olariu: Optimal greedy algorithms for indifference graphs, Computers Math. Applic. 25 (1993) 15–25.
32. T. Ma: unpublished manuscript.
33. R.M. McConnell and J. Spinrad: Linear-time modular decomposition and efficient transitive orientation of comparability graphs, Proc. SODA 94, Fifth Annual ACM-SIAM Symposium on Discrete Algorithms (1994) 536–545.
34. D. Meister: Recognizing and computing minimal triangulations efficiently, Technical Report 302, Fakultät für Mathematik und Informatik, Universität Würzburg, 2002.
35. F. Nicolai: A hypertree characterization of distance-hereditary graphs, manuscript, Gerhard-Mercator-Universität Duisburg (1996).
36. S. Olariu: An optimal greedy heuristic to color interval graphs, Inform. Process. Lett. 37 (1991) 65–80.
37. G. Ramalingam and C. Pandu Rangan: A uniform approach to domination problems on interval graphs, Inform. Process. Lett., 27 (1988) 271–274.
38. A. Raychaudhuri: On powers of interval and unit interval graphs, Congr. Numer. 59 (1987) 235–242.

39. F.S. Roberts: Indifference graphs, in: F. Harary, ed., Proof Techniques in Graph Theory, Academic Press, New York, 1969 139–146.

40. F.S. Roberts: On the compatibility between a graph and a simple order, J. Combin. Theory Ser. B 11 (1971) 28–38.

41. R.E. Tarjan: Depth first search and linear graph algorithms, SIAM J. Comput. 1 (1972) 146–160.

42. D.J. Rose, R.E. Tarjan, and G.S. Lueker: Algorithmic aspects of vertex elimination on graphs, SIAM J. Comput. 5 (1976) 266–283.

43. R.E. Tarjan and M. Yannakakis: Simple linear-time algorithms to test chordality of graphs, test acyclicity of hypergraphs, and selectively reduce acyclic hypergraphs, SIAM J. Comput. 13 (1984) 566–579.

44. K. Simon: A new simple linear algorithm to recognize interval graphs, LNCS 553 (1992) 289–308.

45. G. Wegner: Eigenschaften der Nerven homologisch-einfacher Familien in R^n, Ph.D. thesis, Universität Göttigen, Germany, 1967.

Wireless Networking: Graph Theory Unplugged

Roger Wattenhofer

ETH Zurich, 8092 Zurich, Switzerland
wattenhofer@tik.ee.ethz.ch
http://www.dcg.ethz.ch

Abstract. Wireless and mobile networks are an excellent playground for graph theoreticians. Many research challenges turn out to be variants of classic graph theory problems. In particular the rapidly growing areas of ad-hoc and sensor networks demand new solutions for timeless graph theory problems, because i) wireless devices have lower bandwidth and ii) wireless devices are mobile and therefore the topology of the network changes rather frequently. As a consequence, algorithms for wireless and mobile networks should have i) as little communication as possible and should ii) run as fast as possible. Both goals can only be achieved by developing algorithms requiring a small number of communication rounds only (so-called *local* algorithms). In this work we present a few connections between graph theory and wireless networking, such as topology control, clustering, and geo-routing. Each section is supplemented with an open problem.

1 Introduction

An ad-hoc or sensor network consists of mobile nodes featuring, among other components, a processor, some memory, a wireless radio, and a power source; physical constraints often require the power source to be feeble – a weak battery or a small solar cell.

Ad-hoc and sensor networks are emerging areas of research that have been studied intensively for a few years only. Roughly, the researchers investigating ad-hoc and sensor networks can be classified into two categories. On the one side there are the systems researchers who build real ad-hoc or sensor networks; the Berkeley Motes project [16] is a popular hardware platform marketed by Crossbow (www.xbow.com) that is used in many deployments, but alternative hardware platforms are available as well (e.g. [5], [34]). On the other hand there are the theoreticians who try to understand the fundamentals of ad-hoc and sensor networks, by abstracting away a few "technicalities" that arise in real systems.

Not surprisingly – as in other areas of computer science and engineering – there is no consensus what the technicalities are. Most theoreticians model the networks as nodes (points) in a Euclidean plane; two nodes can communicate if they are within their mutual transmission range, which in an unobstructed and homogeneous environment translates into whether their Euclidean distance

J. Hromkovič, M. Nagl, and B. Westfechtel (Eds.): WG 2004, LNCS 3353, pp. 20–32, 2004.

is at most the maximum transmission range R. This model is widely known as unit disk graph and – though not quite practical – respected as a first step by practitioners.

More surprisingly however, most theoreticians make much stronger assumptions. It seems that a majority of papers assumes that the nodes are distributed uniformly at random. At a high node density, such a postulation renders many problems trivial. Also it is not clear that a uniform node density distribution makes sense from a practical point of view. Recently deployed large-scale sensor networks report highly heterogeneous node densities – in "interesting" areas there are several nodes per square meter, whereas in other ("routing-only") areas nodes are hundreds of meters apart. For mobile ad-hoc networks (MANET's), it is often assumed that the nodes move Brownian, a behavior that is not often seen in our macroscopic world.

In this paper we advocate using more realistic *graph theoretical* models. We feel that theoretical research should drop *average-case* assumptions such as uniformly at random distributed nodes and/or Brownian motion, and instead study *worst-case* distributions and motion models. In this paper we outline a selection of the algorithms that were developed to work also in the non-uniform worst-case.

The paper is organized as follows. In Sections 2, 3, and 4, we sketch a number of algorithmic results in three key areas of ad-hoc and sensor networking. In Section 2 we discuss topology control, in Section 3 clustering, and in Section 4 geo-routing, a special but well-studied form of routing. In Section 5 we conclude the paper.

2   Topology Control

Since energy is the limiting factor for lifetime and operability of an ad-hoc network, researchers have developed a variety of mechanisms and algorithms to conserve energy. These mechanisms and algorithms are often dubbed "topology control."

For two communicating ad-hoc nodes u and v, the energy consumption of their communication grows at least quadratically with their distance. Having one or more relay nodes between u and v therefore helps to save energy. The primary target of a topology control algorithm is to abandon long-distance communication links and instead route a message over several small (energy-efficient) hops. For this purpose each node in the ad-hoc network chooses a "handful" of "close-by" neighbors "in all points of the compass" (we are going to fill in the details later). Having only near neighbors not only helps reducing energy but also interference, since fewer nodes are disturbed by high power transmissions. Clearly nodes cannot abandon links to "too many" faraway neighbors in order to prevent the ad-hoc network from being partitioned or the routing paths from becoming non-competitively long. In general there is a trade-off between network connectivity and sparseness.

Let the graph $G = (V, E)$ denote the ad-hoc network before running the topology control algorithm, with V being the set of ad-hoc nodes, and E representing the set of communication links. There is a link (u, v) in E if and only if the two nodes u and v can communicate directly. Running the topology control algorithm will yield a sparse subgraph $G_{tc} = (V, E_{tc})$ of G, where E_{tc} is the set of remaining links. The resulting topology G_{tc} should have a variety of properties:

i) Symmetry: The resulting topology G_{tc} should be symmetric, that is, node u is a neighbor of node v if and only if node v is a neighbor of node u. Asymmetric communication graphs are unpractical, because many communication primitives become unacceptably complicated [32].

ii) Connectivity/Spanner: Two nodes u and v are connected if there is a path from u to v, potentially through multiple hops. If two nodes are connected in G, then they should still be connected in G_{tc}. Although a minimum spanning tree is a sparse connected subgraph, it is often not considered a good topology, since close-by nodes in the original graph G might end up being far away in G_{tc} (G being a ring, for instance). Therefore the graph G_{tc} is generally not only being asked to be connected, but a spanner: For any two nodes u and v, if the optimal path between u and v in G has cost c, then the optimal path between u and v in G_{tc} has cost $O(c)$.

iii) Sparseness/Low Degree/Low Interference: The remaining graph G_{tc} should be sparse, that is, the number of links should be in the order of the number of nodes. More ambitiously, one might even ask that *each node* in the remaining graph G_{tc} has a low (constant) degree. Since a low degree alone does not automatically imply low interference (after all nodes might choose few but very far away neighbors!), some researchers have started studying topology control algorithms that concentrate on the interference issue.

iv) In addition to the properties i)-iii) one can often find secondary targets. For instance, it is popular to ask the remaining graph to be planar in order to run a geometric routing algorithm, such as GOAFR [28].

Since connectivity and sparseness run against each other, topology control has been a thriving research area.

The currently best algorithms feature an impressive list of properties. Wang and Li [35] present the currently most promising proposal – a distributed topology control algorithm that computes a planar constant-degree distance-spanner. (As opposed to energy-spanners as considered in earlier work [36, 17].) However, the distributed algorithm might be quite slow; in an unlikely (but possible) worst-case instance it will run for a linear number of steps. Also, like many others this algorithm makes strong assumptions: First, all the nodes need to know their exact positions, by means of a global positioning system (GPS) for example. And second, the algorithm assumes that the world is flat and without buildings (a perfect unit disk graph, so to speak). These assumptions make the algorithm unpractical.

In an almost "retro" approach [37] recently presented the XTC algorithm that works i) without GPS and ii) even in a mountainous and obstructed environment. Surprisingly the XTC algorithm features all the basic properties of topology control (symmetry, connectivity, low degree) while being faster than any previous proposals.

All known topology control algorithms including [35] and XTC [37] do not explicitly address interference, but argue that the sparseness or low degree property will take care of it[1]. In [9] it has recently been shown that the "low degree $\Rightarrow$ low interference" assumption is not correct in a worst case.

In [9] interference is formally defined as follows: Given a communication graph produced by a topology control algorithm, the *coverage* of an (undirected) edge $e = (u, v)$ is the cardinality of the set of nodes covered by the disks induced by u and v, with radius $|uv|$:

$$Cov(e) := \left|\{w \in V \,|\, w \text{ is covered by } D(u, |u, v|)\}\right.$$
$$\left.\cup \{w \in V \,|\, w \text{ is covered by } D(v, |v, u|)\}\right|.$$

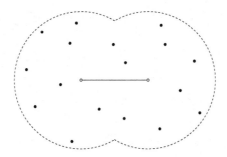

Fig. 1. Nodes covered by a communication link.

In other words the coverage $Cov(e)$ represents the number of network nodes affected by nodes u and v communicating with their transmission powers chosen such that they exactly reach each other (cf. Figure 1). Then the interference of a graph $G = (V, E)$ is

$$I(G) := \max_{e \in E} Cov(e).$$

To the best of our knowledge, all currently known topology control algorithms have in common that every node establishes a connection to at least its nearest neighbor. In other words all these topologies contain the Nearest Neighbor Forest constructed on the given network. In the following we show that by including the Nearest Neighbor Forest as a subgraph, the interference of a resulting topology can become incomparably bad with respect to a topology with optimum interference. In particular, interference of any proposed topology is $\Omega(n)$ times

---

[1] Meyer auf der Heide et al. [29] are a notable exception who study interference explicitly, however not in the context of topology control, but in relation to traffic models. They show that there are worst-case ad-hoc networks and worst-case traffic, where only one of the performance parameters congestion, energy, and dilation can be optimized at a time.

larger than the interference of the optimum connected topology, where n is the total number of network nodes.

Figure 2 depicts an example graph. In addition to a horizontal exponential node chain, each of these nodes h_i has a corresponding node v_i vertically displaced by a little more than h_i's distance to its left neighbor. Denoting this vertical distance d_i, $d_i > 2^{i-1}$ holds. These additional nodes form a second (diagonal) exponential line. Between two of these diagonal nodes v_{i-1} and v_i, an additional helper node t_i is placed such that $|h_i, t_i| > |h_i, v_i|$.

The Nearest Neighbor Forest for this given network (with the additional assumption that each node's transmission radius can be chosen sufficiently large) is shown in Figure 3. Roughly one third of all nodes being part of the horizontally connected exponential chain, interference of any topology containing the Nearest Neighbor Forest amounts to at least $\Omega(n)$. An interference-optimal topology, however, would connect the nodes as depicted in Figure 4 with constant interference.

In other words, already by having each node connect to the nearest neighbor, a topology control algorithm makes an "irrevocable" error. Moreover, it commits an asymptotically worst possible error, since the interference in any network cannot become larger than n.

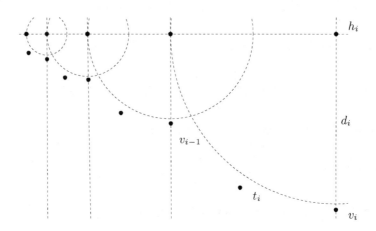

Fig. 2. Two exponential node chains.

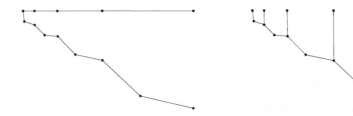

Fig. 3. The Nearest Neighbor Forest yields interference $\Omega(n)$.

Fig. 4. Optimal tree with constant interference.

Since roughly one third of all nodes are part of the horizontal exponential node chain in Figure 2, the observation also holds for an average interference measure, averaging interference over all edges[2].

In [9] three algorithm variants are presented that indeed minimize interference, and at the same time keep the symmetry and the connectivity/spanner property. These algorithms have drawbacks too: Currently only one of them is locally computable, but its running time is too slow, which makes a practical implementation impossible.

All the previously discussed algorithms work for arbitrary (worst-case) node distributions. For average-case (random) distributions there is an interesting alternative: Each node simply chooses its k best neighbors. Blough et al. [11] show that this simplest of all conceivable algorithms works surprisingly well when the nodes are distributed uniformly at random. For general distributions, clearly, [11] does not even guarantee connectivity.

Topology control has been (and still is!) a thriving research area for theoreticians. What works well in analysis and simulation has recently also been implemented on the basis of the 802.11 standard [19]. These early practical experiences proof that topology control is a technique that is paying off, and deserves more attention.

Open Problem: For the sake of concreteness, let us specify one of the many open problems. We are given n nodes in the plane. As above we must connect these nodes with a spanning tree. This time, however, we do not charge each edge by how many nodes it will disturb. Instead we charge each node by how many edges it is disturbed. The spanning tree should be chosen such that it minimizes the maximum (or average) disturbed node. Apart from a simple directed sensor-network model [13] nothing is known about the problem.

3   Clustering

Akin to topology control, clustering (a.k.a. backbone building) also aims for computing a subgraph of the original graph. In some sense however, in clustering this subgraph is not trying to optimize energy by dropping long-range neighbors, but (quite on the opposite) optimizing the number of hops by dropping short-range neighbors.

In mobile ad-hoc networks, nodes communicate without stationary server infrastructure. When sending a message from one node to another, intermediate network nodes have to serve as routers. Although a number of interesting suggestions have been made, finding efficient algorithms for the routing process remains the most important problem for ad-hoc networks. Since the topology of an ad-hoc network is constantly changing, routing protocols for ad-hoc networks differ significantly from the standard routing schemes which are used in wired networks. One effective way to improve the performance of routing algorithms

---

[2] Interestingly, the example in Figure 2 works as well for a number of other definitions of interference.

is by grouping nodes into clusters. The routing is then done between clusters. A most basic method for clustering is calculating a dominating set. Formally, in a graph G, a dominating set is a subset of nodes such that for every node v either i) v is in the dominating set or ii) a direct neighbor of v is in the dominating set. The minimum dominating set problem asks for a dominating set of minimum size. Only the nodes of the dominating set act as routers, all other nodes communicate via a neighbor in the dominating set.

Between traditional wired networks and mobile ad-hoc networks two main distinctions can be made: i) typically wireless devices have much lower bandwidth than their wired counterparts and ii) wireless devices are mobile and therefore the topology of the network changes rather frequently. As a consequence, distributed algorithms which run on such devices should have as little communication as possible and they should run as fast as possible. Both goals can only be achieved by developing algorithms requiring a small number of communication rounds only (often called local algorithms).

Most of the algorithms to compute a dominating set use the fact that a maximal independent set (MIS) is by definition already a dominating set. For unit disk graphs it can be shown that any MIS is only a constant factor larger than a minimum dominating set. Often, in a second phase of the algorithm the nodes in the MIS are then connected through two- and three-hop bridges. All these nodes (the MIS and the bridging nodes) then form the backbone. One can route from any backbone node to any other through nodes in the backbone only [2].

Unfortunately, from a worst-case standpoint, it is conjectured that computing a MIS is not as efficient as it seems at first sight. In particular in [23] it was shown that a distributed MIS construction can take as long as $\Omega(\sqrt{\log n/\log\log n})$ time in a graph with n nodes[3]. This is too slow in the setting of a mobile ad-hoc network because by the time the MIS is computed, the topology has already changed. In a paper by Gao et al. [15] it was shown that in a unit disk graph one can construct an asymptotically optimal dominating set in time $O(\log\log n)$ only. However, to do so, nodes need to know their coordinates, an assumption that is not always realistic.

Recently, algorithms to quickly compute a dominating set fast even if there the nodes do not know their coordinates have been proposed. These algorithms in fact even work if the network is not a unit disk but a general graph. In general graphs, the problem of finding a minimum dominating set has been proven to be NP-hard. The best known approximation is already achieved by the greedy algorithm: As long as there are uncovered nodes, the greedy algorithm picks a node which covers the biggest number of uncovered nodes and puts it into the dominating set. It achieves an approximation ratio of $\ln\Delta$ where Δ is the highest degree in the graph. Unless the problems of NP can be solved by deterministic $n^{O(\log\log n)}$ algorithms, this is the best possible up to lower order terms [12]. In [18] a logarithmic approximation in polylogarithmic time was proposed.

---

[3] Another lower bound is $\Omega(\log\Delta/\log\log\Delta)$, where Δ is the maximum degree (number of neighbors) in the graph.

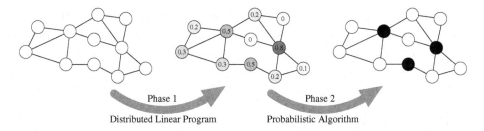

Fig. 5. Distributed dominating set approximation.

In [24] the only distributed algorithm which achieves a nontrivial approximation ratio in a constant number of rounds is given. Precisely, for an arbitrary parameter k, in $O(k)$ rounds, an expected approximation ratio of $O(\sqrt{k}\Delta^{2/\sqrt{k}} \log \Delta)$ is presented.

The algorithm consists of two phases (see Figure 5). First, an approximate solution to the fractional dominating set problem is obtained. In the fractional MDS, weights are assigned to all nodes such that the sum of weights each node sees is greater than or equal to 1. If the MDS problem is formulated as an integer program, the fractional MDS corresponds to the LP relaxation of MDS. The solution to the fractional dominating set can be summarized as follows. Initially all nodes have weight 0. As the algorithm progresses, the nodes gradually increase their weights. This is done in decreasing order of the degrees of the nodes. In order to achieve the locality, the degrees are divided into classes and the assigning of weights is done simultaneously for all nodes of the same class. We obtain a distributed algorithm for the fractional MDS which computes a $k\Delta^{2/k}$-approximation in $O(k^2)$ rounds.

In the second phase of the algorithm, based on their weights, the nodes locally decide whether they become a dominater or not. The second phase only needs two rounds of communication and it merely adds a factor $O(\log \Delta)$ to the overall approximation ratio. This is asymptotically optimal since the integrality gap of the problem is $\ln \Delta$ unless P almost equals NP. In an optional third phase (which is omitted in Figure 5) nodes can locally approximate a *connected* dominating set by building "bridges" between dominators.

Recently, with a primal-dual approach it was possible to improve the algorithm such that the first phase of the algorithm essentially constructs a local polynomial time approximation scheme (PTAS), not only for dominating sets but for more general covering and packing problems [25].

All algorithms so far assume that the scheduling of transmissions is handled by the MAC layer. In other words, they assume perfect point-to-point connections between two neighboring nodes. Since a backbone (dominating set) is often used to compute a reasonable MAC layer, many of these papers experience a severe "chicken-and-egg" problem. In [21], Kuhn et al. take a more realistic approach to clustering in ad-hoc networks. They consider a multi-hop radio network without collision detection, where nodes wake up asynchronously, and do not have access to a global clock. For this rather harsh model, they show that a

$O(1)$-approximative dominating set can be computed within $polylog(\hat{n})$ time, $\hat{n}$ being an a-priori upper bound on the number of nodes in the system.

Open Problem: Though there is some early understanding about the static version of the problem of clustering using dominating sets, the question how to efficiently maintain a clustering when the nodes are mobile, is still wide open.

4   Geo-Routing

Routing is of central importance in ad-hoc networks. With the notable exception of a link reversal [14] routing algorithm analysis by Busch et al. [10], not many worst-case results are known.

For a special case of routing known as geo-routing (a.k.a. geographic, geometric, location-, or position-based routing) however, there have been quite a few worst-case results. In geo-routing each node is informed about its own as well as its neighbors' positions. Additionally the source of a message knows the position of the destination. The first assumption becomes more and more realistic with the advent of inexpensive and miniaturized positioning systems. It is also conceivable that approximate position information could be attained by local computation and message exchange with stationary devices [4, 6] or completely autonomously [33, 30]. In order to come up to the second assumption, that is to provide the source of a message with the destination position, a (peer-to-peer) overlay network could be employed [3, 38, 1][4]. For some scenarios it can also be sufficient to reach any destination currently located in a given area ("geocasting" [31]).

The first correct geo-routing algorithm was Face Routing [20]. Face Routing routes messages along faces of planar graphs and proceeds along the line connecting the source and the destination. Besides guaranteeing to reach the destination, it does so with $O(n)$ messages, where n is the number of network nodes. Face routing was later combined with greedy routing to give better average-case performance [7].

This is unsatisfactory since already a simple flooding algorithm will reach the destination with $O(n)$ messages. Additionally it would be desirable to see the algorithm cost depend on the distance between the source and the destination. The first algorithm competitive with the shortest path between the source and the destination was AFR [27]. It basically enhances Face Routing by the concept of a bounding region restricting the searchable area. With a lower bound argument AFR was shown to be asymptotically optimal.

Despite its asymptotic optimality AFR is not practicable due to its pure face routing concept. For practical purposes there have been attempts to combine greedy approaches (always send to the message to the neighbor closest to the destination) and face routing; for example the GOAFR and GOAFR+ algorithms by Kuhn et al. [28, 26], which are variants of AFR and remain worst-case optimal. (See Figure 6.)

---

[4] Abraham et al. [1] fits well into the context of this paper, since the authors share our worst-case philosophy.

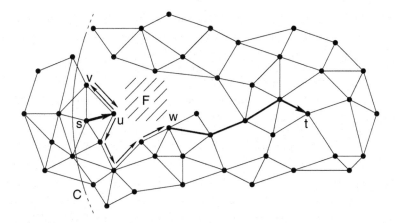

Fig. 6. The GOAFR$^+$ algorithm starts from source node s in greedy mode. At node u it reaches a local minimum, a node without any neighbors closer to destination node t. GOAFR$^+$ switches to face routing mode and begins to explore the boundary of face F (in clockwise direction). At node v the algorithm hits the bounding circle C (for details, please see [28, 26]) and turns back to continue the exploration of F's boundary in the opposite direction. At node w the algorithm decides that it made significant progress (c.f. [26]), falls back to greedy mode, and continues to finally reach destination node t.

On the other side, GOAFR+ is currently also the best geo-routing algorithm in the average-case. In this sense GOAFR+ is a success story for worst-case analysis, where an algorithm derived from a worst-case algorithm is also the best average-case algorithm.

Open Problem: Recently [33] proposed to use geo-routing algorithms in complete absense of position information. Instead, an algorithm assigns so-called "virtual coordinates" to the nodes; these virtual coordinates should model the connectivity of the nodes as well as possible. In particular, each node is assigned a coordinate in the plane, such that nodes that are neighbors in the connectivity graph have at most Euclidean distance 1 in the plane, and nodes that are *not* neighbors in the connectivity graph have at least distance 1. In other words, we would like to *embed* a given unit disk graph in the plane. Unfortunately, it was shown by Breu and Kirkpatrick [8] that this is impossible in polynomial time. Recently, there was progress in understanding the problem better by the first non-trivial lower bound [22], and also the first non-trivial approximation algorithm for the problem [30]. However, the gap between the upper and the lower bound is still glaring; we believe that this is a most challenging open problem.

5   Conclusions

In this paper we have discussed several "worst-case" algorithms for various classic problems in ad-hoc and sensor networking. Clearly, the selection of areas in this paper is highly subjective. Besides topology control, clustering, and geo-routing there are a dozen more research areas that are currently in the focus of the community (e.g. positioning, models, data gathering, multicast, ...). Moreover the selection is dreadfully skewed towards our own recent work.

However, there is not as much algorithmic work as one might think. The vast majority of ad-hoc and sensor network research follows the heuristics/simulations approach: A heuristic for solving a problem is proposed, and simulated against other heuristics. Unfortunately, this approach does rarely produce solid results, on which one can build, since the quality of the heuristics depends on the parameters of the simulation. We feel that with the field generally becoming more mature, "average-case" heuristics will make way for "worst-case" algorithms.

References

1. I. Abraham, D. Dolev, and D. Malkhi. LLS: A Locality Aware Location Service for Mobile Ad Hoc Networks. *Workshop on Discrete Algorithms and Methods for Mobile Computing and Communications (DIAL-M)*, 2004.
2. K. Alzoubi, P.-J. Wan, and O. Frieder. Message-Optimal Connected Dominating Sets in Mobile Ad Hoc Networks. In *Proc. ACM Int. Symposium on Mobile ad hoc networking & computing (MobiHoc)*, pages 157–164, EPFL Lausanne, Switzerland, 2002.
3. B. Awerbuch and D. Peleg. Concurrent online tracking of mobile users. In *SIG-COMM*, pages 221–233, 1991.
4. J. Beutel. Geolocation in a Picoradio Environment. *Master Thesis, ETH Zurich and UC Berkeley*, 1999.
5. J. Beutel, O. Kasten, and M. Ringwald. BTnodes – A Distributed Platform for Sensor Nodes. In *Prof. of the ACM Conference on Embedded Networked Sensor Systems (SenSys)*, 2003.
6. R. Bischoff and R. Wattenhofer. Analyzing Connectivity-Based Multi-Hop Ad-Hoc Positioning. In *Proc. of the Second Annual IEEE International Conference on Pervasive Computing and Communications (PerCom)*, 2004.
7. P. Bose, P. Morin, I. Stojmenovic, and J. Urrutia. Routing with Guaranteed Delivery in ad hoc Wireless Networks. In *Proc. of the 3rd International Workshop on Discrete Algorithms and Methods for Mobile Computing and Communications (DIAL-M)*, pages 48–55, 1999.
8. H. Breu and D. G. Kirkpatrick. Unit Disk Graph Recognition is NP-hard. *Comput. Geom. Theory Appl.*, 9(1-2):3–24, 1998.
9. M. Burkhart, P. von Rickenbach, R. Wattenhofer, and A. Zollinger. Does topology control reduce interference? In *Proc. ACM Int. Symposium on Mobile ad hoc networking & computing (MobiHoc)*, 2004.
10. C. Busch, S. Surapaneni, and S. Tirthapura. Analysis of Link Reversal Routing Algorithms for Mobile Ad Hoc Networks. In *15th ACM Symposium on Parallelism in Algorithms and Architectures (SPAA)*, 2003.
11. D.M.Blough, M.Leoncini, G.Resta, and P.Santi. The k-Neigh Protocol for Symmetric Topology Control in Ad Hoc Networks. In *Proc. ACM Int. Symposium on Mobile ad hoc networking & computing (MobiHoc)*, 2003.
12. U. Feige. A Threshold of ln n for Approximating Set Cover. *Journal of the ACM (JACM)*, 45(4):634–652, 1998.
13. M. Fussen, R. Wattenhofer, and A. Zollinger. On Interference Reduction in Sensor Networks. Technical Report 453, Department of Computer Science, ETH Zurich, 2004.

14. E. M. Gafni and D. P. Bertsekas. Distributed algorithms for generating loop-free routes in networks with frequently changing topology. *IEEE Transactions on Communication*, 29, 1981.

15. J. Gao, L. Guibas, J. Hershberger, L. Zhang, and A. Zhu. Discrete Mobile Centers. In *Proc. 17th Annual Symposium on Computational Geometry (SCG)*, pages 188–196. ACM Press, 2001.

16. J. Hill, R. Szewczyk, A. Woo, S. Hollar, D. E. Culler, and K.S.J. Pister. System architecture directions for networked sensors. In *Architectural Support for Programming Languages and Operating Systems (ASPLOS)*, pages 93–104, 2000.

17. L. Jia, R. Rajaraman, and C. Scheideler. On Local Algorithms for Topology Control and Routing in Ad Hoc Networks. In *Proc. of the 15th Annual ACM Symposium on Parallel Algorithms and Architectures (SPAA)*, 2003.

18. L. Jia, R. Rajaraman, and R. Suel. An Efficient Distributed Algorithm for Constructing Small Dominating Sets. In *Proc. of the 20th ACM Symposium on Principles of Distributed Computing (PODC)*, pages 33–42, 2001.

19. V. Kawadia and P. R. Kumar. Power control and clustering in ad hoc networks. In *Proc. of the Annual Joint Conference of the IEEE Computer and Communications Societies (INFOCOM)*, 2003.

20. E. Kranakis, H. Singh, and J. Urrutia. Compass Routing on Geometric Networks. In *Proc. 11th Canadian Conference on Computational Geometry (CCCG)*, pages 51–54, Vancouver, August 1999.

21. F. Kuhn, T. Moscibroda, and R. Wattenhofer. Initializing Newly Deployed Ad Hoc and Sensor Networks. In *Proceedings of the 10th Annual International Conference on Mobile Computing and Networking (MobiCom)*, 2004.

22. F. Kuhn, T. Moscibroda, and R. Wattenhofer. Unit Disk Graph Approximation. *Workshop on Discrete Algorithms and Methods for Mobile Computing and Communications (DIAL-M)*, 2004.

23. F. Kuhn, T. Moscibroda, and R. Wattenhofer. What cannot be computed locally! In *Proc. of the 23rd ACM Symposium on Principles of Distributed Computing (PODC)*, 2004.

24. F. Kuhn and R. Wattenhofer. Constant-Time Distributed Dominating Set Approximation. In *Proc. of the 22nd ACM Symposium on the Principles of Distributed Computing (PODC)*, 2003.

25. F. Kuhn and R. Wattenhofer. Distributed Combinatorial Optimization. Technical Report 426, Department of Computer Science, ETH Zurich, 2003.

26. F. Kuhn, R. Wattenhofer, Y. Zhang, and A. Zollinger. Geometric Routing: Of Theory and Practice. In *Proc. of the 22nd ACM Symposium on the Principles of Distributed Computing (PODC)*, 2003.

27. F. Kuhn, R. Wattenhofer, and A. Zollinger. Asymptotically optimal geometric mobile ad-hoc routing. In *Proc. of the International Workshop on Discrete Algorithms and Methods for Mobile Computing and Communications (DIAL-M)*, Atlanta, Georgia, USA, September 2002.

28. F. Kuhn, R. Wattenhofer, and A. Zollinger. Worst-Case Optimal and Average-Case Efficient Geometric Ad-Hoc Routing. In *Proc. ACM Int. Symposium on Mobile ad hoc networking & computing (MobiHoc)*, 2003.

29. F. Meyer auf der Heide, C. Schindelhauer, K. Volbert, and M. Grunewald. Energy, congestion and dilation in radio networks. In *Proc. of the 14th Annual ACM Symposium on Parallel Algorithms and Architectures (SPAA)*, 2002.

30. T. Moscibroda, R. O'Dell, M. Wattenhofer, and R. Wattenhofer. Virtual Coordinates for Ad hoc and Sensor Networks. *Workshop on Discrete Algorithms and Methods for Mobile Computing and Communications (DIAL-M)*, 2004.

31. J.C. Navas and T. Imielinski. GeoCast – Geographic Addressing and Routing. In *Proceedings of the Annual International Conference on Mobile Computing and Networking (MobiCom)*, pages 66–76, 1997.

32. R. Prakash. Unidirectional Links Prove Costly in Wireless Ad-Hoc Networks. In *Proc. of the 3rd International Workshop on Discrete Algorithms and Methods for Mobile Computing and Communications (DIAL-M)*, 1999.

33. A. Rao, C. Papadimitriou, S. Ratnasamy, S. Shenker, and I. Stoica. Geographic Routing without Location Information. In *Proceedings of the 10th Annual International Conference on Mobile Computing and Networking (MobiCom)*, 2003.

34. J. Schiller and et al. The scatterweb project: See http://www.scatterweb.net for more details.

35. Y. Wang and X.-Y. Li. Localized Construction of Bounded Degree Planar Spanner. In *Proc. of the DIALM-POMC Joint Workshop on Foundations of Mobile Computing*, 2003.

36. R. Wattenhofer, L. Li, P. Bahl, and Y.-M. Wang. Distributed Topology Control for Power Efficient Operation in Multihop Wireless Ad Hoc Networks. In *Proc. of the Annual Joint Conference of the IEEE Computer and Communications Societies (INFOCOM)*, 2001.

37. R. Wattenhofer and A. Zollinger. XTC: A Practical Topology Control Algorithm for Ad-Hoc Networks. In *Proceedings of 4th International Workshop on Algorithms for Wireless, Mobile, Ad Hoc and Sensor Networks (WMAN)*, 2004.

38. Y. Xue, B. Li, and K. Nahrstedt. A scalable location management scheme in mobile ad-hoc networks. In *IEEE LCN*, 2001.

Constant Time Generation of Trees
with Specified Diameter

Shin-ichi Nakano[1] and Takeaki Uno[2]

[1] Gunma University, Kiryu-Shi 376-8515, Japan
nakano@cs.gunma-u.ac.jp
[2] National Institute of Informatics, Tokyo 101-8430, Japan
uno@nii.jp

Abstract. Many algorithms to generate all trees with n vertices without repetition are already known. The best algorithm runs in time proportional to the number of trees. However, the time needed to generate each tree may not be bounded by a constant, even though it is "on average". In this paper we give a simple algorithm to generate all trees with exactly n vertices and diameter d, without repetition. Our algorithm generates each tree in constant time. It also generates all trees so that each tree can be obtained from the preceding tree by at most three operations. Each operation consists of a deletion of a vertex and an addition of a vertex. By using the algorithm for each diameter $2, 3, \cdots, n-1$, we can generate all trees with n vertices.

1   Introduction

It is useful to have the complete list of graphs for a particular class. One can use such a list to search for a counter-example to some conjecture, to find the best graph among all candidate graphs, or to experimentally measure the average performance of an algorithm over all possible input graphs.

Many algorithms to generate a particular class of graphs, without repetition, are already known [2, 6, 7, 9, 8, 10, 15]. Many excellent textbooks have been published on the subject [3, 5, 14].

Algorithms to generate all trees with n vertices without repetition are already known. The best algorithm [15] runs in time proportional to the number of trees. However, the time needed to generate each tree may not be bounded by a constant, even though it is "on average".

In this paper we give a simple algorithm to generate, without repetition, all trees with exactly n vertices and diameter d. Our algorithm generates each tree in constant time. It does not output each tree entirely, but outputs the difference from the preceding tree.

The main idea of our algorithm is to define a simple relation among the trees, that is "a family tree" of trees (see Fig. 1), and outputs trees by traversing the family tree. *The family tree*, denoted by $T_{n,d}$, is the tree such that the vertices of $T_{n,d}$ correspond to the trees with n vertices and diameter d, and each edge corresponds to some relation between trees. By traversing the family tree with

J. Hromkovič, M. Nagl, and B. Westfechtel (Eds.): WG 2004, LNCS 3353, pp. 33–45, 2004.

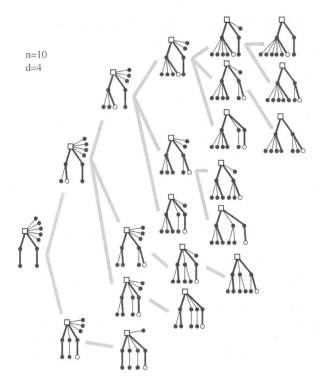

Fig. 1. The family tree $T_{10,4}$.

some ideas we can generate all trees corresponding to the vertices of the family tree, without repetition.

Furthermore, the algorithm generates all trees so that each tree can be obtained from the preceding tree by at most three operations, where each operation consists of a deletion of a vertex and an addition of a vertex. Therefore the derived sequence of trees is a kind of combinatorial Gray code [4, 12, 14] for trees with n vertices and diameter d. A Gray code [11] is a cyclic sequence of all 2^k bitstrings of length k, such that each bitstring differs from the preceding one in a small number of bit entries.

The rest of the paper is organized as follows. Section 2 gives some definitions. Section 3 introduces the family tree. Section 4 presents our first algorithm for the even diameter case. In Section 5 we sketch our algorithm for the odd diameter case. The algorithm generates each tree in $O(1)$ time on average. In Section 6 we improve the algorithm so that it generates each tree in $O(1)$ time. Finally Section 7 is a conclusion.

2   Preliminaries

In this section we give some definitions.

Let G be a connected graph with n vertices. An edge connecting vertices x and y is denoted by (x, y). The *degree* of a vertex v, denoted by $d(v)$, is

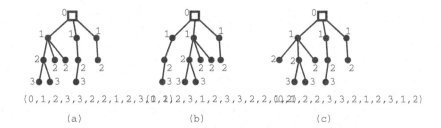

$(0,1,2,3,3,2,2,1,2,3,1,2)$ $(0,2)2,3,1,2,3,3,2,2,(0,2),2,2,3,3,2,1,2,3,1,2)$

(a)   (b)   (c)

Fig. 2. The depth sequences.

the number of neighbors of v in G. A *path* is a sequence of distinct vertices $(v_0, v_1, \cdots, v_k)$ such that (v_{i-1}, v_i) is an edge for $i = 1, 2, \cdots, k$. The *length* of a path is the number of edges in the path. The *distance* between a pair of vertices u and v is the minimum length of a path between u and v. The *diameter* of G is the maximum distance between two vertices in G.

A *tree* is a connected graph without cycles. A *rooted* tree is a tree with one vertex r chosen as its *root*. For each vertex v in a rooted tree, let $UP(v)$ be the unique path from v to the root r. If $UP(v)$ has exactly k edges then we say that the *depth* of v is k, and write $dep(v) = k$. The *parent* of $v \neq r$ is its neighbor on $UP(v)$, and the *ancestors* of $v \neq r$ are the vertices on $UP(v)$ except v. The parent of the root r and the ancestors of r are not defined. We say that if v is the parent of u then u is *a child* of v, and if v is an ancestor of u then u is a *descendant* of v. A *leaf* is a vertex that has no child.

An *ordered tree* is a rooted tree with left-to-right ordering specified for the children of each vertex. We denote by $T(v)$ the ordered subtree of an ordered tree T consisting of a vertex v and all descendants of v that preserve the left-to-right ordering for the children of each vertex.

Let T be an ordered tree with n vertices, and $(v_1, v_2, \cdots, v_n)$ be the list of the vertices of T in preorder [1]. Let $dep(v_i)$ be the depth of v_i for $i = 1, 2, \cdots, n$. Then, the sequence $L(T) = (dep(v_1), dep(v_2), \cdots, dep(v_n))$ is called the *depth sequence* of T. Some examples are shown in Fig. 2. Note that those trees in Fig. 2 are isomorphic as rooted trees, but non-isomorphic as ordered trees.

Let T_1 and T_2 be two ordered trees, and $L(T_1) = (a_1, a_2, \cdots, a_n)$ and $L(T_2) = (b_1, b_2, \cdots, b_m)$ be their depth sequences. If either (1) $a_i = b_i$ for each $i = 1, 2, \cdots, j-1$ (possibly $j = 1$) and $a_j > b_j$, or (2) $a_i = b_i$ for each $i = 1, 2, \cdots, m$ and $n > m$, then we say that $L(T_1)$ is *heavier* than $L(T_2)$, and write $L(T_1) > L(T_2)$.

3   The Family Tree

In Section 3 and 4 we only consider the case where the diameter is even.

If a tree has $n \geq 3$ vertices and diameter 2, then the number of such a tree is exactly one, which is $K_{1,n-1}$. In the rest of the section we assume that the diameter is $2k \geq 4$.

Let T be a tree with the diameter $2k$. Let $v_0, v_1, \cdots, v_{2k}$ be a path in T having length $2k$. One can observe that T may have many such paths, but the vertex v_k, called *the center* of T, is unique [13, p72]. We assign to T the rooted tree R derived from T by choosing v_k as the root. Then we assign to R a unique ordered tree as follows.

Given a rooted tree R, since we can choose many left-to-right orderings for the children of each vertex, we can observe that R corresponds to many non-isomorphic ordered trees. Let H be the ordered tree corresponding to R that has the heaviest depth sequence $L(H)$. Then we say that H is the *left-heavy embedding* of R. For example, the ordered tree in Fig. 2(a) is the left-heavy embedding of a rooted tree, however the trees in Fig. 2(b) and (c) are not, since the one in Fig. 2(a) is heavier. We assign the ordered tree H to R.

Given a tree T, we have assigned to T a unique distinct rooted tree R, and then we have assigned to R a unique distinct ordered tree H, which is the left-heavy embedding of R. Note that T, R and H have the same diameter $2k$. One can observe that the assignment is a one-to-one mapping. Let $S_{n,2k}$ be the set of all left-heavy embeddings with exactly n vertices and diameter $2k$. If we generate all ordered trees in $S_{n,2k}$, then it also means the generation of all trees with exactly n vertices and diameter $2k$. We are going to generate all ordered trees in $S_{n,2k}$.

We have the following lemma.

Lemma 1. *An ordered tree H is the left-heavy embedding of a rooted tree if and only if for every pair of consecutive child vertices v_1 and v_2, that appear in this order in the left-to-right ordering, $L(T(v_1)) \geq L(T(v_2))$ holds.*

Proof. By contradiction.   □

In the rest of the paper the condition "$L(T(v_1)) \geq L(T(v_2))$ for each consecutive child vertices v_1 and v_2", is called *the left-heavy condition*.

Let H be a left-heavy embedding in $S_{n,2k}$ with root r_k. Let $c_1, c_2, \cdots, c_{d(r_k)}$ be the children of r_k. Assume they appear in this order in the left-to-right ordering. We say that c_i, $3 \leq i \leq d(r_k)$ is a *waiting vertex* if $c_i, c_{i+1}, \cdots, c_{d(r_k)}$ are leaves. Since H has a path of lenght $2k$ with the center r_k, one can observe that c_1 and c_2 have a descendant at depth k, respectively. Thus, neither c_1 nor c_2 are leaves. We denote by $A(H)$ the ordered tree derived from H by removing all (possibly none) waiting vertices. We say that $A(H)$ is *the active tree* of H. Note that the diameter of $A(H)$ is also $2k$.

Let c_a be the rightmost child of the root r_k in $A(H)$. Let $P_{right} = (v_0 = r_k, v_1 = c_a, v_2, \cdots, v_x)$ be the path in $A(H)$ such that v_i is the rightmost child of v_{i-1} for each i, $1 \leq i \leq x$, and v_x is a leaf in $A(H)$. We call P_{right} *the right path* of H. If $v_1 = c_2$ and $H(v_1)$ is a path, then we say H is *right empty*. Note that $H(v_1)$ is the ordered subtree of H induced by v_1 and all descendants of v_1. Similarly, let $P_{left} = (u_0 = r_k, u_1 = c_1, u_2, \cdots, u_y)$ be the path in $A(H)$ such that u_1 is the leftmost child of u_0, and u_i is the rightmost child of u_{i-1} for each i, $2 \leq i \leq y$, and u_y is a leaf in $A(H)$. We call P_{left} *the left path* of H. If $H(u_1)$ is a path, then we say H is *left empty*. The right and left paths are depicted as thick lines in Fig. 1.

If H is not right empty then v_x is called *the active leaf* of H. Otherwise, if H is not left empty then u_y is called *the active leaf* of H. Otherwise, $A(H)$ is a path of lenght $2k$, and H has no active leaf.

We have the following lemma.

Lemma 2. *Let H be an ordered tree in $S_{n,2k}$ that has an active leaf. Then the ordered tree derived from H by (i) removing the active leaf of H, then (ii) adding one leaf as the rightmost child of the root, is also in $S_{n,2k}$. Moreover, H is heavier than the derived ordered tree.*

Proof. Removing the active leaf and then adding one leaf as the rightmost child of the root never destroys the left-heavy condition. And the number of vertices in the derived tree is still n. Furthermore the diameter of the derived tree is again $2k$. Thus any derived tree is also in $S_{n,2k}$.

The proof for the second half of the claim is omitted.   □

Assume that H is an ordered tree in $S_{n,2k}$ that has an active leaf. We denote by $P(H)$ the ordered tree derived from H by (i) removing the active leaf of H, then (ii) adding one leaf as the rightmost child of the root. We say that $P(H)$ is *the parent tree* of H and H is *a child tree* of $P(H)$. By the lemma above, $P(H)$ is also in $S_{n,2k}$. Given an ordered tree H in $S_{n,2k}$, we can have the unique sequence $H, P(H), P(P(H)), \cdots$ of ordered trees in $S_{n,2k}$, which eventually ends with the ordered tree that has no active leaf. That is the ordered tree consisting of a path of length $2k$ and $(n - 2k - 1)$ waiting vertices. By merging these sequences we can have *the family tree* of $S_{n,2k}$, denoted by $T_{n,2k}$, such that the vertices of $T_{n,2k}$ correspond to the trees in $S_{n,2k}$, and each edge corresponds to each relation between some H and $P(H)$. For instance, $T_{10,4}$ is shown in Fig. 1.

4   Algorithm

In this section we give an algorithm to construct $T_{n,2k}$.

If we can generate all child trees of a given ordered tree in $S_{n,2k}$, then in a recursive manner we can construct $T_{n,2k}$. This means we can generate all trees with exactly n vertices and diameter $2k$. Now we are going to generate all child trees of a given ordered tree.

Let H be an ordered tree in $S_{n,2k}$. Let $P_{right} = (v_0 = r_k, v_1, \cdots, v_x)$ be the right path of H, and $P_{left} = (u_0 = r_k, u_1, \cdots, u_y)$ be the left path of H. We construct some ordered trees by slightly modifying H as follows. Set $x' = \min\{x, k - 1\}$ and $y' = \min\{y, k - 1\}$.

If H has at least one waiting vertex and H is right empty then we define $H[i]$, $1 \le i \le y'$, as the ordered tree derived from H by (i) removing the rightmost waiting vertex, then (ii) adding a new vertex as the rightmost child of u_i. See Fig. 3 for some examples. Note that the constraint $i \le y' \le k - 1$ ensures that the diameter of $H[i]$ remains $2k$.

If H has at least one waiting vertex, then we define $H[i+]$, $1 \le i \le x'$, as the ordered tree derived from H by (i) removing the rightmost waiting vertex, then (ii) adding a new vertex as the rightmost child of v_i. See some examples in Fig. 3.

If H has at least two waiting vertices, then we define $H[+]$ as the ordered tree derived from H by (i) removing the rightmost waiting vertex, then (ii) adding a new vertex as the only child vertex of the leftmost waiting vertex. See Fig. 3.

We can observe that each child tree of H is in $\{H[1], H[2], \cdots, H[y']\} \cup \{H[1+], H[2+], \cdots, H[x'+]\} \cup \{H[+]\}$. However, not all trees in $\{H[1], H[2], \cdots, H[y']\} \cup \{H[1+], H[2+], \cdots, H[x'+]\} \cup \{H[+]\}$ are child trees of H, so we need to check whether each possible child tree is actually a child tree of H.

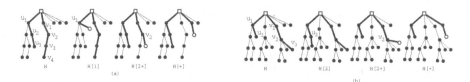

Fig. 3. The possible child trees.

We need some notations here. If vertex v_{i-1} has two or more children in the active tree $A(H)$, then we denote by v_i' the child of v_{i-1} that precedes v_i. Thus v_i' is the 2nd last child of v_{i-1} in $A(H)$. Similarly, for u_{i-1}, we denote by u_i' the 2nd last child of u_{i-1}. Note that $H(v)$ is the ordered subtree of H induced by v and all descendants of v.

We now have the following lemma.

Lemma 3. *Let H be an ordered tree in $S_{n,2k}$ with the right path $(v_0 = r_k, v_1, \cdots, \cdots, v_x)$ and the left path $(u_0 = r_k, u_1, \cdots, u_y)$.*

(1) $H[i]$, $i \leq \min\{y, k-1\}$, is a child tree of H if and only if H has at least one waiting vertex and is right empty, and for each j, $j = 1, 2, \cdots, i$, either u_{j-1} has only one child u_j in H, or $L(H(u_j')) \geq L(H(u_j))$ holds in $H[i]$.

(2) $H[i+]$, $i \leq \min\{x, k-1\}$, is a child tree of H if and only if H has at least one waiting vertex, and for each j, $j = 1, 2, \cdots, i$, either v_{j-1} has only one child v_j in H, or $L(H(v_j')) \geq L(H(v_j))$ holds in $H[i+]$.

(3) $H[+]$ is a child tree of H if and only if H has at least two waiting vertices.

Proof. (1) Since $H \in S_{n,2k}$ the left heavy condition has held in H. Then, only for vertex $u = u_0, u_1 \cdots, u_i$, $L(H(u))$ in $H[i]$ is heavier than $L(H(u))$ in H. The claim checks all of these possible changes that may destroy the left-heavy condition.

(2) (3) Omitted.   □

If we generate each tree in $\{H[1], H[2], \cdots, H[y']\} \cup \{H[1+], H[2+], \cdots, H[x'+]\} \cup \{H[+]\}$ and check whether it is actually a child tree or not based on the lemma above, then we need considerable running time. However, we can save running time as follows. We need some definitions here.

Let H be an ordered tree in $S_{n,2k}$. We define "active at depth" in the following three cases. First, assume that H is not right empty. We say that H is *active at depth i* if (i) the right path contains a vertex v_i with depth i, (ii) v_i has two or more child vertices, and (iii) $L(H(v_{i+1}))$ is a prefix of $L(H(v'_{i+1}))$. Intuitively, if H is active at depth i, then we are copying subtree $H(v_{i+1})$ from $H(v'_{i+1})$. Then, assume that H is right empty but not left empty. We say that H is *active* at depth i if (i) the left path contains a vertex u_i with depth i, (ii) u_i has two or more child vertices, and (iii) $L(H(u_{i+1}))$ is a prefix of $L(H(u'_{i+1}))$. Then assume that H is right and left empty. We say that H is *active* at depth 0. Note that $L(H(v_1))$ is a prefix of $L(H(u_1))$.

We can show that H is always active at some depth as follows. If H is not right empty, then let j be the maximum index such that vertex v_j has two or more child vertices. Since H is not right empty, H always has such a vertex. Now since H is left-heavy and $H(v_{j+1})$ is a path, $L(H(v_{j+1}))$ is a prefix of $L(H(v'_{j+1}))$. Thus H is active at depth j. Otherwise, H is right empty. Then if H is not left empty, in a similar manner as above, we can show that H is active at some depth. Otherwise, H is right and left empty. In this case H is active at depth 0. Therefore H is always active at some depth.

We say the *copy-depth* of H is c if H is active at depth c but not active at any depth in $\{0, 1, \cdots, c-1\}$.

Now we are going to generate all child trees of an ordered tree H in $S_{n,2k}$. We have the following four cases.

We assume that H has the copy-depth c, the right path $P_{right} = (v_0 = r_k, v_1, \cdots, v_x)$ and the left path $P_{left} = (u_0 = r_k, u_1, \cdots, u_y)$.

Case 1: H has no waiting vertex.

Then H corresponds to a leaf in $T_{n,2k}$. Hence H has no child tree.

Case 2: Otherwise, and if H is not right empty.

In this case, for $H[i]$, $i = 1, 2, \cdots, \min\{y, k-1\}$, the active leaf of $H[i]$ is on the right path of $H[i]$. So $H[i]$ is not a child tree of H.

If H has two waiting vertices, then $H[+]$ is defined and is a child tree of H. The copy-depth of $H[+]$ is 0. Otherwise, H has exactly one waiting vertex and $H[+]$ is not defined.

We have two subcases for $H[i+]$. Note that since Case 1 does not occur, H has a waiting vertex.

Case 2a: $L(H(v'_{c+1})) = L(H(v_{c+1}))$. (Intuitively the copy has completed.)

First we show that $H[c+]$ is a child tree of H. Since H has the copy-depth c, for $j = 1, 2, \cdots, c$, $L(H(v'_j)) > L(H(v_j))$ holds in H and $L(H(v_j))$ is not a prefix of $L(H(v'_j))$. Since, for $j = 1, 2, \cdots, c$, $L(H(v_j))$ is not a prefix of $L(H(v'_j))$, $L(H(v'_j)) > L(H(v_j))$ still holds in $H[c+]$. Thus by Lemma 4.1 $H[c+]$ is a child tree of H. The copy-depth of $H[c+]$ remains at c.

Similarly, $H[i+]$, $i = 1, 2, \cdots, c-1$, is a child tree of H, and the copy-depth of $H[i+]$ is i.

However, for each $H[i+]$, where $i = c+1, c+2, \cdots, \min\{x, k-1\}$, the left-heavy condition is destroyed because of $L(H(v'_{c+1})) < L(H(v_{c+1}))$ in $H[i+]$. Thus, they are not child trees.

Case 2b: Otherwise. (Now $L(H(v'_{c+1})) > L(H(v_{c+1}))$ holds. Intuitively the copy has not completed yet.)

Let $L(H(v'_{c+1})) = (dep(s_1), dep(s_2), \cdots, dep(s_{n'}))$, $L(H(v_{c+1})) = (dep(t_1),$ $dep(t_2), \cdots, dep(t_{n''}))$, and set $z = dep(s_{n''+1})$. (Intuitively we are copying $H(v_{c+1})$ from $H(v'_{c+1})$ and $s_{n''+1}$ is the next vertex to be copied.)

First, $H[(z-1)+]$ is a child tree of H, and the copy-depth of $H[(z-1)+]$ remains at c.

Similarly, $H[1+], H[2+], \cdots, H[(z-2)+]$ are child trees of H, and we will prove in a lemma below that the copy-depth of $H[i+]$ is i for $i = 0, 1, \cdots, z-2$.

For each of $H[i+]$, where $i = z, z+1, \cdots, \min\{x, k-1\}$, $L(H(v'_{c+1})) <$ $L(H(v_{c+1}))$ holds in $H[i]$. Therefore, they are not left-heavy.

Case 3: Otherwise, and if H is not left empty.

Now H is right empty and H has a waiting vertex. Let z' be the $(k+1)$-th depth in $L(H)$.

Then $H[i+], i = 1, 2, \cdots, z'-1$, is a child tree of H. The copy-depth of $H[i+]$ is i for $i = 1, 2, \cdots, z'-2$, and 0 for $z'-1$. On the other hand, $H[i+]$, where $i = z, z+1, \cdots, \min\{x, k-1\}$, is not a child tree of H, since $L(T(u_1)) < L(T(v_1))$ and so $H[i+]$ is not left-heavy.

If H has two waiting vertices, then $H[+]$ is a child tree of H and the copy-depth of $H[+]$ is 0. Otherwise, $H[+]$ is not defined.

We have two subcases for $H[i]$. Note that H has a waiting vertex.

Case 3a: $L(H(u'_{c+1})) = L(H(u_{c+1}))$.

$H[i]$, $i = 1, 2, \cdots, c$, is a child tree of H, and the copy-depth of $H[i]$ is i. However, $H[i]$, where $i = c+1, c+2, \cdots, y$, is not a child tree of H.

Case 3b: Otherwise.

Let $L(H(u'_{c+1})) = (dep(s_1), dep(s_2), \cdots, dep(s_{n'}))$, $L(H(u_{c+1})) = (dep(t_1),$ $dep(t_2), \cdots, dep(t_{n''}))$, and set $z = dep(s_{n''+1})$.

$H[1], H[2], \cdots, H[(z-1)]$ are child trees of H. The copy-depth of $H[i]$ is i for $i = 0, 1, \cdots, z-2$, and c for $i = z-1$.

For each of $H[i]$, where $i = z, z+1, \cdots, \min\{y, k-1\}$, $L(H(v'_{c+1})) < L(H(v_{c+1}))$ holds in $H[i]$, therefore they are not left-heavy.

Case 4: Otherwise. (Now H is right and left empty.)

$H[i+], i = 1, 2, \cdots, \min\{x, k-1\}$, is not a child tree of H.

If H has two waiting vertices, then $H[+]$ is a child tree of H and the copy-depth of $H[+]$ is 0. Otherwise, $H[+]$ is not defined.

$H[i]$, $i = 1, 2, \cdots, k-1$, is a child tree of H, and the copy-depth of $H[i+]$ is i.

Lemma 4. *In Case 2(b) the copy-depth of $H[i]$ is i for $i = 1, 2, \cdots, z-2$.*

Proof. For $i = 1, 2, \cdots, c$ the claim is obvious, so we assume otherwise. We can observe that the copy-depth of $H[i]$, $c+1 \le i \le z-2$, is never smaller than c, and $H[i]$ is active at i. So the copy-depth of $H[i]$ is somewhere between i and c.

Assume for contradiction that the copy-depth of $H[i]$ is $j < i$. Let $dep(w)$ be the last occurrence of depth j in $L(H[i])$. By the assumption above, w has two or more child vertices. Let w_1 be the rightmost child of w, and w_2 be the child vertex of w preceding w_1. See Fig. 4 for examples. Let w' be the vertex in $H(v'_{c+1})$

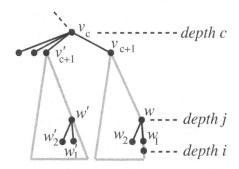

Fig. 4. Illustration for Lemma 4.2.

corresponding to w, and w_1' and w_2' be vertices in $H(v_{c+1}')$ corresponding to w_1 and w_2. (Note that we are copying $H(v_{c+1})$ from $H(v_{c+1}')$.) Now since $H \in S_{n,2k}$, $L(H(w_2')) \geq L(H(w_1'))$ holds. By the choice of i, $L(H(w_1')) > L(H(w_1))$ holds and $L(H(w_1))$ is not a prefix of $L(H(w_1'))$. Since the copy-depth of H is c, $L(H(w_2')) = L(H(w_2))$. Then $L(H(w_2)) = L(H(w_2')) \geq L(H(w_1')) > L(H(w_1))$ holds, and $L(H(w_1))$ is not a prefix of $L(H(w_1'))$. Thus $L(H(w_1))$ is not a prefix of $L(H(w_2))$, and the copy-depth of $H[i]$ is not j, a contradiction.

Thus the copy-depth of $H[i]$ is i for $i = 1, 2, \cdots, z - 2$. $\qquad\square$

Based on the case analysis above, we have the following algorithm.

Procedure find-all-children(T, c)
$\{$ T is the current tree, and c is the copy-depth of T.$\}$
begin
01 Output H $\{$ Output the difference from the preceding tree.$\}$
02 **if** H has no waiting vertices $\{$Case 1$\}$
03 **then return**
04 **else if** H is not right empty
05 **then** $\{$Case 2$\}$
06   **begin**
07   **if** H has two waiting vertices **then find-all-children**$(H[+], 0)$
08   **if** $L(H(v_{c+1}')) = L(H(v_{c+1}))$ **then** $\{$Case 2a$\}$
09   **for** $i = 1$ **to** c
10   **find-all-children**$(H[i+], i)$
11   **else** $\{$Case 2b$\}$ $\{$ $H(T(v_{c+1}')) > L(H(v_{c+1}))$ $\}$
12      $\{$ Let z be the depth of the next vertex to be copied.$\}$
13   **for** $i = 1$ **to** $z - 2$
14   **find-all-children**$(H[i+], i)$
15   **find-all-children**$(H[(z-1)+], c)$
16   **end**
17 **else if** H is not left empty
18 **then** $\{$Case 3$\}$
19   **begin**
20      $\{$ Let z' be the $(k+1)$-th depth in $L(H)$.$\}$

```
21      for  i = 1 to z' − 2
22        find-all-children(H[i+], i)
23        find-all-children(H[(z' − 1)+], 0)
24        if H has two waiting vertices then find-all-children(H[+], 0)
25        if L(H(u'_{c+1})) = L(H(u_{c+1}))        then {Case 3a}
26          for  i = 1 to c
27            find-all-children(H[i], i)
28        else  {Case 3b} {  H(T(u'_{c+1})) > L(H(u_{c+1}))  }
29          { Let z be the depth of the next vertex to be copied.}
30            for  i = 1 to z − 2
31              find-all-children(H[i], i)
32            find-all-children(H[z − 1], c)
33      end
34 else  {H is right empty and left empty.}
35    begin
36      if H has two waiting vertices then find-all-children(H[+], 0)
37        for  i = 1 to k − 1
38          find-all-children(H[i], i)
39    end
      end
```

Algorithm find-all-trees(n)
begin
    Output the tree H that consists of the path of length $2k$ and $(n − 2k − 1)$
of waiting vertices.
    find-all-children(H, 0)
end

Theorem 1. *The algorithm uses $O(n)$ space and runs in $O(f(n))$ time, where $f(n)$ is the number of nonisomorphic trees with exactly n vertices and diameter $2k$.*

Proof. Since we traverse the family tree $T_{n,2k}$ and output each ordered tree at each corresponding vertex of $T_{n,2k}$, we can generate all trees with exactly n vertex and diameter $2k$.

We maintain the last two occurrences of each depth in each subtree $T(v_1)$ and $T(u_1)$ in four arrays of length k. We record the update of the four arrays and restore the arrays if return occur. Thus we can find v_i, v'_i, u_i and u'_i in constant time for each i.

We also maintain the current copy-depth c and the vertex next to be copied. Therefore with the help of the above arrays we can check the conditions in Lines 08 and 25 in constant time. Also, we can compute the value z and z' in constant time.

Other parts of the algorithm need only constant time of computation for each edge of $T_{n,2k}$.

Thus the algorithm runs in $O(f(n))$ time. Note that the algorithm does not output each tree entirely, but the difference from the preceding tree.

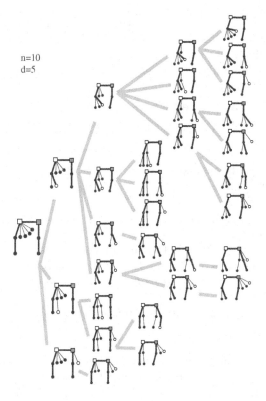

n=10
d=5

Fig. 5. The family tree $T_{10,5}$.

For each recursive call we need a constant amount of space, and the depth of the recursive call is bounded by n. Thus the algorithm uses $O(n)$ space.   □

5   The Odd Diameter Case

In this section we sketch the case where the diameter is odd.

It is known that a tree with odd diameter $2k + 1$ may have many paths of length $2k + 1$, but all of them share a unique edge, called *the center* of T [13, p72].

Intuitively, by treating the edge as the root in a similar manner to the even diameter case, we can define the family tree $T_{n,2k+1}$. The detail is omitted. We only show $T_{10,5}$ in Fig. 5 as an example of the family tree.

6   Modification

The algorithm in Section 4 generates all trees with n vertices and diameter d in $O(f(n))$ time, where $f(n)$ is the number of nonisomorphic trees with n vertices and diameter d. Thus the algorithm generates each tree in $O(1)$ time "on average". However, after generating the tree corresponding to the last vertex in a large subtree of $T_{n,d}$, we have to merely return from the deep recursive call without outputting any tree. This may take $O(n)$ time. Therefore, we cannot generate each tree in $O(1)$ time.

However, a simple modification improves the algorithm to generate each tree in $O(1)$ time. The algorithm is as follows.

Procedure find-all-children2(T, c, *depth*)
{ T is the current tree, c is the copy-depth of T, and *depth* is the depth of the recursive call.}
begin
01   **if** T has no waiting vertex
02   **then** Output T { T is a leaf.}
03   **else**
04   **begin**
05   **if** *depth* is even
06   **then** Output T { before outputting its child trees.}
07   Generate child trees $T_1, T_2, \cdots, T_x$ by the method in Section 4, and
08   recursively call **find-all-children2** for each child tree.
09   **if** *depth* is odd
10   **then** Output T { after outputting its child trees.}
11   **end**
   end

An execution of the algorithm is shown in Fig. 6.

One can observe that the algorithm generates all trees so that each tree can be obtained from the preceding tree by tracing at most three edges of $T_{n,k}$, each of which corresponds to an operation consisting of a deletion of a vertex and an addition of a vertex. Note that if T corresponds to a vertex v in $T_{n,k}$ with odd depth, then we may need to trace three edges to generate the next tree. Otherwise we need to trace at most two edges to generate the next tree. Thus, the derived sequence of the trees is a combinatorial Gray code [4, 12, 14] for rooted trees.

In Fig. 6 the added vertices are drawn as white circles, and the deleted, then added again, vertices are drawn as gray circles. (See the sixth tree in Fig. 6.) Each integer near an arrow mark is the number of edges in $T_{n,d}$ between the two vertices corresponding to the two trees. Each tree corresponding to a vertex in $T_{n,d}$ at odd depth is surrounded by a rectangle, and these trees are generated after all its child trees are generated.

Since $T_{10,4}$ has 21 vertices corresponding to the 21 trees in $S_{10,4}$, shown in Fig. 1, $T_{10,4}$ has 20 edges. In the algorithm we trace each edge twice, once for down and once for up. Therefore the sum is 40. This matches the sum of the integers near the arrow marks in Fig. 6.

7   Conclusion

In this paper we gave a simple algorithm to generate all trees with n vertices and diameter d. The algorithm generates each tree in constant time and clarifies the family tree of the trees.

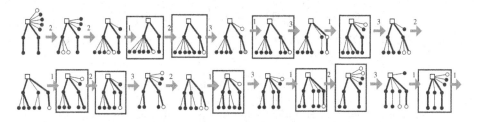

Fig. 6. An execution of the algorithm for $T_{10,4}$.

References

1. A.V. Aho and J.D. Ullman, *Foundations of Computer Science*, Computer Science Press, New York, (1995).
2. T. Beyer and S. M. Hedetniemi, *Constant Time Generation of Rooted Trees*, SIAM J. Comput., 9, (1980), pp. 706–712.
3. L.A. Goldberg, *Efficient Algorithms for Listing Combinatorial Structures*, Cambridge University Press, New York, (1993).
4. J.T. Joichi, D.E. White and S.G. Williamson, *Combinatorial Gray Codes*, SIAM J. Comput., 9, (1980), pp. 130–141.
5. D.L. Kreher and D.R. Stinson, *Combinatorial Algorithms*, CRC Press, Boca Raton, (1998).
6. Z. Li and S. Nakano, *Efficient Generation of Plane Triangulations without Repetitions*, Proc. ICALP2001, LNCS 2076, (2001), pp. 433–443.
7. G. Li and F. Ruskey, *The Advantage of Forward Thinking in Generating Rooted and Free Trees*, Proc. 10th Annual ACM-SIAM Symp. on Discrete Algorithms, (1999), pp. 939–940.
8. B.D. McKay, *Isomorph-free Exhaustive Generation*, J. of Algorithms, 26, (1998), pp. 306–324.
9. S. Nakano, *Efficient Generation of Plane Trees*, Information Processing Letters, 84, (2002), pp. 167–172.
10. R.C. Read, *How to Avoid Isomorphism Search When Cataloguing Combinatorial Configurations*, Annals of Discrete Mathematics, 2, (1978), pp. 107–120.
11. K.H. Rosen (Eds.), *Handbook of Discrete and Combinatorial Mathematics*, CRC Press, Boca Raton, (2000).
12. C. Savage, *A Survey of Combinatorial Gray Codes*, SIAM Review, 39, (1997) pp. 605–629.
13. D.B. West, *Introduction to Graph Theory, 2nd Ed*, Prentice Hall, NJ, (2001).
14. H.S. Wilf, *Combinatorial Algorithms: An Update*, SIAM, (1989).
15. R.A. Wright, B. Richmond, A. Odlyzko and B.D. McKay, *Constant Time Generation of Free Trees*, SIAM J. Comput., 15, (1986), pp. 540–548.

Treelike Comparability Graphs: Characterization, Recognition, and Applications

Sabine Cornelsen[1,*] and Gabriele Di Stefano[2]

[1] Universität Konstanz, Fachbereich Informatik & Informationswissenschaft
cornelse@inf.uni-konstanz.de
[2] Università dell'Aquila, Dipartimento di Ingegneria Elettrica
gabriele@ing.univaq.it

Abstract. An undirected graph is a treelike comparability graph if it admits a transitive orientation such that its transitive reduction is a tree. We show that treelike comparability graphs are distance hereditary. Utilizing this property, we give a linear time recognition algorithm. We then characterize permutation graphs that are treelike. Finally, we consider the PARTITIONING INTO BOUNDED CLIQUES problem on special subgraphs of treelike permutation graphs.

1 Introduction

An undirected graph is a treelike comparability graph if it admits a transitive orientation such that its transitive reduction is a tree. It is an arborescence, if its transitive reduction is a directed rooted tree. Arborescences were studied by Golumbic [9] and Wolk [15] and characterized as trivially perfect graphs or as graphs that do not contain an induced path of length four nor an induced cycle of length four, respectively. Treelike posets and their linear extension were studied by Atkinson [1].

A graph is completely separable [11] (or distance hereditary) if it can be recursively decomposed into so called splits, such that the remaining components are cliques and stars. The structure of the decomposition is represented in the so called split tree.

In this paper, we first characterize treelike comparability graphs and treelike permutation graphs and give recognition algorithms. We show that a graph is a treelike comparability graph if and only if it is distance hereditary with a special treelike orientation on its split tree. We show how to utilize the split decomposition to recognize treelike comparability graphs in linear time and show that a treelike orientation is unique. Treelike permutation graphs are characterized as paths of arborescence-like graphs and it is shown that the minimum length of such a path can be determined in linear time.

---

* Work mainly done while the author was visiting the University of L'Aquila, supported by the Human Potential Program of the EU under contract no HPRN-CT-1999-00104 (AMORE Project).

J. Hromkovič, M. Nagl, and B. Westfechtel (Eds.): WG 2004, LNCS 3353, pp. 46–57, 2004.

Motivated by train shunting problems [8], we consider the problem PARTI-TIONING INTO BOUNDED CLIQUES in a second part of this paper, i.e. the problem given $m \in \mathbb{N}$ and a graph $G = (V, E)$, is there a partition of G into cliques of size m? For general graphs, the PARTITIONING INTO BOUNDED CLIQUES-problem is $\mathcal{NP}$-complete for $m \geq 3$ [13] and polynomial time solvable for $m = 2$. It remains $\mathcal{NP}$-complete for comparability graphs and $m \geq 3$ [14], and for permutation graphs and $m \geq 6$ [12]. The complexity of the problem is open for permuta-tion graphs and $m = 3, 4$ or 5. It was shown by Lonc [14] that for fixed m the problem can be solved in linear time on interval graphs. However, it remains $\mathcal{NP}$-complete even for interval graphs if m is part of the input [2]. Bodlaender and Jansen [2] showed that the problem can be solved in $\mathcal{O}(n^{2(m-1)+1})$ time on a graph with n vertices that does not contain an induced path of length four. The problem was considered for many other graph classes. A nice overview can be found, e.g., in [12].

In this paper, we show that the PARTITIONING INTO BOUNDED CLIQUES problem is solvable in linear time for arborescences, even if m is part of the input. We then consider a special matching problem on arborescences and apply its solution to solve the PARTITIONING INTO TRIANGLES-problem in polynomial time on the arborescence-like subgraphs of treelike permutation graphs.

The paper is organized as follows. In Sect. 2, we provide some basic defini-tions. In Sect. 3, we characterize treelike comparability graphs as special distance hereditary graphs. We utilize this characterization to construct a treelike orien-tation in linear time. Sect. 4 characterizes treelike permutation graphs. Finally, we consider the PARTITIONING INTO BOUNDED CLIQUES problem on special subgraphs of treelike permutation graphs in Sect. 5.

2   Preliminaries

Let $G = (V, E)$ be an undirected graph. An orientation of E maps each el-ement $\{v, w\}$ of E on exactly one of the ordered pairs (v, w) or (w, v). We refer to the image $\boldsymbol{E}$ of E under a given orientation also as orientation. v is the tail and w is the head of an edge $(v, w) \in \boldsymbol{E}$. Let $v, w \in V$. A $(v - w)$-*path* is a sequence $v, v_1, \ldots, v_{\ell-1}, w$ with $v_1, \ldots, v_{\ell-1} \in V$ distinct vertices and $\{v, v_1\}, \{v_1, v_2\}, \ldots, \{v_{\ell-1}, w\} \in E$. Given an orientation on E, a directed $(v - w)$-path is a path $v, v_1, \ldots, v_{\ell-1}, w$ with $(v, v_1), (v_1, v_2), \ldots, (v_{\ell-1}, w) \in \boldsymbol{E}$. An *(undirected) cycle* is a sequence $v_1, \ldots, v_{\ell}$ of $\ell > 2$ distinct vertices such that $\{v_1, v_2\}, \ldots, \{v_{\ell-1}, v_{\ell}\}, \{v_{\ell}, v_1\} \in E$. A *transitive orientation* is an orientation with the property that there is a directed $(v - w)$-path between two vertices v and w if and only if $(v, w) \in \boldsymbol{E}$. The graph G is a *comparability graph* if there exists a *transitive orientation* on its edges. The *transitive reduction* of a compa-rability graph G with respect to a fixed transitive orientation $\boldsymbol{E}$ is the spanning subgraph of G that contains exactly the edges of $\boldsymbol{E}$ between two vertices v and w for which there is no directed $(v - w)$-path of length greater than one.

Suppose now that G is a connected comparability graph. A transitive orien-tation $\boldsymbol{E}$ is called *treelike* if the transitive reduction with respect to $\boldsymbol{E}$ does not

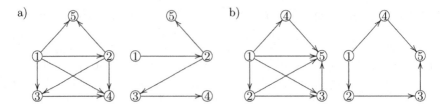

Fig. 1. Transitive reductions with respect to two significantly different transitive orientations of a permutation graph. a) With respect to a treelike orientation b) With respect to an orientation that is not treelike. These two orientations correspond to the two permutations a) $4, 3, 5, 2, 1$ and b) $5, 3, 2, 4, 1$.

contain any undirected cycle. A connected comparability graph is called treelike, if there exists a transitive orientation that is treelike. See Fig. 1 for an example of a comparability graph with two different orientations.

Let π be a permutation of $1, \ldots, n$. The *permutation graph* corresponding to π is the graph $G = (V, E)$ with $V = \{1, \ldots, n\}$ and $E = \{\{i, j\}; \ i < j \Rightarrow \pi(i) > \pi(j)\}$. It has a transitive orientation $\{(i, j); \ i < j \text{ and } \pi(i) > \pi(j)\}$. We use the representation of a permutation as the graph of the function $i \mapsto \pi(i)$ in the plane, i.e., with the points $(i, \pi(i))$. By definition, there is an edge in the corresponding permutation graph if and only if the slope of the segment between the points $(i, \pi(i))$ and $(j, \pi(j))$ is negative. A graph is a *treelike permutation graph* if it is treelike and a permutation graph.

3   Recognizing Treelike Comparability Graphs

In this section, we show how to construct a treelike orientation of an undirected graph in linear time – if it exists. The algorithm is based on the split decomposition. A *split* of a connected graph $G = (V, E)$ is a partition V_1, V_2 of V into two subsets that have at least two vertices each such that there exist subsets $W_1 \subseteq V_1, W_2 \subseteq V_2$ with the property that the set of edges of G between V_1 and V_2 corresponds to $\{\{w_1, w_2\}; \ w_1 \in W_1 \text{ and } w_2 \in W_2\}$. The *split decomposition* of G is defined recursively. Take an arbitrary split $W_1 \subseteq V_1, W_2 \subseteq V_2$. Let the graphs $G_i, i = 1, 2$ be defined as follows. First, consider the subgraph of G induced by V_i. Then add a new vertex w_i – called *special vertex* – with neighborhood W_i. Recursively decompose G_i. The split is memorized in a *special edge* $\{w_1, w_2\}$. The remaining graphs of a split decomposition are called *split components*. The *split tree* associated with a split decomposition is the graph that consists of all split components and all special edges. Let $\{w_1, w_2\}$ be a special edge of a split tree H and let G_1, G_2 be the two split components containing w_1 and w_2. The *re-composition* of G_1 and G_2 is the graph that is obtained from H by deleting w_1 and w_2 and by adding the edges $\{v_1, v_2\}$ for all adjacent vertices $v_1 \neq w_2$ of w_1 and $v_2 \neq w_1$ of w_2. A *minimal split decomposition* of G is a split decomposition of G into three types of components – cliques, stars, and graphs that do not contain a split – such that the number of components is minimized.

The minimal split decomposition of a connected graph is unique [5, 6]. G is *completely separable* [11] (or *distance hereditary*) if G can be decomposed such that all split components are cliques or stars.

Before we show that treelike comparability graphs are completely separable, we mention some properties of treelike orientations of the split tree of a completely separable graph. They follow from the facts that a) special edges are not contained in any cycle and b) that the split components of a connected graph are connected and contain at least three vertices.

Remark 1. Let G be a connected graph. Let H be the split tree of G with respect to some split decomposition and assume that H is a treelike comparability graph. The transitive reduction T with respect to a treelike orientation H has the following properties.

1. All special edges are in T.
2. A special vertex is either only the head or only the tail of its incident edges.
3. If two special vertices are adjacent in T, but incident to different special edges, then at least one of them has a degree higher than two in T.

Theorem 1. *Let G be a connected graph and let H be the split tree of G with respect to the minimal split decomposition. Then G is a treelike comparability graph if and only if G is completely separable and there is a treelike orientation with transitive reduction T on H that fulfills the following property.*

(Z) *At most one vertex of each special edge is incident to more than two edges in T.*

Proof. We will show the following properties.

1. Every treelike comparability graph has a split decomposition into components of size three such that the split tree with respect to this decomposition admits a treelike orientation with Property Z.
2. The existence of a treelike orientation on a split tree that fulfills Property Z is maintained under recomposition of two components.

Now, by Property 1, a treelike comparability graph is completely separable. The split components of a minimal split decomposition are obtained by recursively re-composing adjacent cliques or adjacent stars, respectively, in the split tree. Hence, applying recursively Property 2 to the decomposition obtained in 1 yields the only-if direction. Since we obtain the original graph by recursive re-composition, the if-direction follows immediately from Property 2. It remains to show the two properties.

1. Let $G = (V, E)$ be a treelike comparability graph. Let T be the transitive reduction of G with respect to a treelike orientation $\boldsymbol{E}$. We show Property 1 by induction on the number n of vertices of G. There is nothing to show if $n \leq 3$. So let $n > 3$.
   If there is an edge of T that is not incident to a leaf of T, let $e = (v_0, v_1)$ be such an edge. Let $v_0 \in V_0, v_1 \in V_1$ be the sets of vertices in the two

connected components of $T - e$. Let $W_i, i = 0, 1$ be the set of vertices in V_i that are adjacent to v_{1-i} in G.

If each edge of T is incident to a leave of T, i.e. if T is a star, let r be the central vertex of T. Since T has at least 4 vertices, there are two vertices $v, w \in V$ such that either $(v, r), (w, r)$ are both edges of T or $(r, v), (r, w)$ are both edges of T. Assume $(v, r), (w, r)$. Let $V_0 = \{v, w\}$, $W_0 = V_0$ and $W_1 = \{x \in V; (r, x) \text{ in } T\}$.

In either case $\{\{w_0, w_1\}; w_0 \in W_0 \text{ and } w_1 \in W_1\}$ corresponds to the set of edges of G between V_0 and V_1 and hence V_0, V_1 is a split. Let $G_i, i = 0, 1$ be the subgraphs that result from the decomposition as described above and let $w_i, i = 0, 1$ be the special vertices. Orienting the new edges $(w, w_0), w \in W_0$ and $(w_1, w), w \in W_1$, respectively results in a treelike orientation of G_i with the following new edges in the transitive reduction: (v_0, w_0) and (w_1, v_1) if (v_0, v_1) was chosen as an edge of T non-incident to a leaf and $(v, w_0), (w, w_0)$, and (w_1, r), else. Finally, we orient the special edge (w_1, w_0). Thus, a treelike orientation of the split tree with the required property is maintained in every decomposition step.

2. Let (w_1, w_2) be a special edge. Suppose that w_1 is adjacent to exactly two vertices in T. Let $v \neq w_2$ be the adjacent vertex of w_1 in its component. For each adjacent vertex $w \neq w_1$ of w_2 orient the new edges (w, v) This results in a treelike orientation on the re-composition of the two components containing w_1 and w_2, respectively. The only vertex whose degree might increase is v. But v was incident to a special vertex with degree two. So if v is a special vertex then it already had a degree higher then two. Hence, Prop. Z is maintained. □

The proof of Theorem 1 showed especially that there is the following correspondence between a treelike orientation of a treelike comparability graph and a treelike orientation of its split tree.

Remark 2. Let $\boldsymbol{E}$ be a treelike orientation of a graph G and let H be the split tree of G with respect to the minimal split decomposition. Then there is a treelike orientation $\boldsymbol{E}_H$ of H such that $(v, w) \in \boldsymbol{E}$ if and only if there is an undirected $(v - w)$-path $v = v_0, v_1, \ldots, v_\ell = w$ in H with

- $(v_i, v_{i+1}) \in \boldsymbol{E}_H$ if $\{v_i, v_{i+1}\}$ is not a special edge and
- $(v_{i+1}, v_i) \in \boldsymbol{E}_H$ if $\{v_i, v_{i+1}\}$ is a special edge.

Theorem 2. *It can be tested in linear time whether a graph is a treelike comparability graph. Moreover, let G be a connected treelike comparability graph.*

1. *The treelike orientation of G is unique up to isomorphism and reversing the whole orientation.*
2. *The treelike orientation of G as well as its transitive reduction can be found in linear time.*

Proof. Let G be a connected graph. The following algorithm applied to G outputs the transitive reduction with respect to a treelike orientation of G if G is treelike.

1. Let Q be a queue.
2. Compute the split tree with respect to the minimal split decomposition.
3. If G is not completely separable, G is not treelike. Break.
4. Choose some special edge $\{w_1, w_2\}$ and orient it arbitrarily.
5. Append w_1 and w_2 to Q.
6. While Q is not empty
   (a) Remove the first element w from Q. Suppose (w, w') is a special edge.
   (b) Let H be the split component containing w.
   (c) Orient each edge e of H that is incident to w such that w is the tail of e.
   (d) If H is a star.
      i. If both w and w' are the center of a star, G is not treelike. Break.
      ii. Orient remaining edges such that the center of H is only the head or only the tail of all its incident edges.
   (e) If H is a clique.
      i. If H contains more than two special vertices, G is not treelike. Break.
      ii. Choose an arbitrary ordering $w = v_1, \ldots, v_\ell$ of the vertices of H such that $v_2, \ldots, v_{\ell-1}$ are not special.
      iii. Orient edges (v_i, v_{i+1}) and eliminate remaining edges of H.
   (f) For all special vertices $w_1 \neq w$ of H, let $e = \{w_1, w_2\}$ be a special edge.
      i. If w_1 is the tail of an edge in H orient (w_1, w_2), else (w_2, w_1).
      ii. Append w_2 to Q.
7. Recompose G maintaining only non-transitive edges.

If the algorithm breaks then G is not completely separable (Step 3), or there cannot be a treelike orientation on the split tree that fulfills Property Z in Theorem 1 (Step 6(d)i) or Property 2 in Remark 1 (Step 6(e)i). In either case, G is not treelike.

In Steps 4-6, the algorithm constructs a treelike orientation of the split tree in a breadth first search. By Property 2+3 of Remark 1, there are only two steps in which there is a free choice for the orientation of an edge (Step 4 and Step 6(e)ii). The latter corresponds to choosing the orientation among edges between isomorphic vertices. Hence, a treelike orientation of the split graph of a completely separable graph is unique up to isomorphism and reversing the whole orientation. Thus, if the split tree has a treelike orientation that fulfills Prop. Z of Theorem 1 then the algorithm finds it. This implies the correctness of the algorithm. Uniqueness of the treelike orientation of G follows by Remark 2. □

4 Treelike Permutation Graphs and Arborescences

In this section, we will characterize treelike permutation graphs as paths of double-arborescences. On orientation E of a graph $G = (V, E)$ is an *arborescence-orientation* if the transitive reduction is a rooted tree, i.e., if E is treelike and there is a vertex $r \in V$ such that

$$V = \{r\} \cup \{v \in V; \ (v, r) \in E\} \quad \text{or} \quad V = \{r\} \cup \{v \in V; \ (r, v) \in E\}.$$

E is a *double-arborescence-orientation* if E is treelike and there is a vertex $r \in V$ such that

$$V = \{r\} \cup \{v \in V; \ (v,r) \in \boldsymbol{E}\} \cup \{v \in V; \ (r,v) \in \boldsymbol{E}\}$$

The treelike orientation in Fig. 1a is in fact an arborescence-orientation. We refer to the special vertex r as the root of an arborescence- or a double-arborescence-orientation, respectively. A connected comparability graph is called an arborescence, or a double-arborescence, if there exists an arborescence-, or a double-arborescence-orientation, respectively.

A graph G is a *path of ℓ double-arborescences* if it has a treelike transitive orientation $\boldsymbol{E}$ such that there exists a (not necessarily directed) path P of length $\ell - 1$ in the transitive reduction T that fulfills the following property. Let $V_1, \ldots, V_\ell$ be the vertex sets of the connected components of the graph that results from T by deleting the edges of P. Let $v_i \in V_i, i = 1, \ldots, \ell$ be the vertex in P. Let $G_i, i = 1, \ldots, \ell$ be the subgraphs of G that are induced by V_i. Then $\boldsymbol{E}$ induces a double-arborescence-orientation on $G_i, i = 1, \ldots, \ell$ with root v_i.

To characterize treelike permutation graphs, we apply some results about AT-free graphs. A graph is *AT-free* if it doesn't contain an *asteroidal triple*, i.e. three independent vertices with the property that for every pair of them there is a path connecting the two vertices that does not contain the neighborhood of the remaining vertex. Two vertices u and v are a *dominating pair* of a graph if each vertex of the graph is adjacent to each $(u - v)$-path.

Theorem 3. *Let G be a treelike comparability graph. Then the following are equivalent:*

1. *G is a permutation graph.*
2. *G is AT-free.*
3. *G has a dominating pair.*
4. *G is a path of double-arborescences.*

Proof. $1 \Rightarrow 2$: The complement of a permutation graph is a comparability graph. Hence, a permutation graph is AT-free [10].

$2 \Rightarrow 3$: Every AT-free graph has a dominating pair [4].

$3 \Rightarrow 4$: Let T be the transitive reduction of G with respect to a treelike transitive orientation $\boldsymbol{E}$ of G. Let v, w be a dominating pair of G. Then the unique $(v - w)$-path P in T is a dominating path of G. Hence, for each vertex u of G there has to be a directed path in T from u to a vertex of P. Hence, $\boldsymbol{E}$ induces a double-arborescence orientation on the subgraphs of G that are induced by the connected components of the graph that results from T by deleting the edges of P.

$4 \Rightarrow 1$: Let $\boldsymbol{E}$ be a treelike orientation of G and let T be the transitive reduction of G. Let $P = v_1, \ldots, v_\ell$ be a path of T such that $\boldsymbol{E}$ induces a double-arborescence orientation on the subgraphs of G that are induced by the connected components of the graph that results from T by deleting the edges of P. Let $A_V = \{v \in V; \ (v,v_1) \in \boldsymbol{E}\}$, $B_V = \{v \in V; \ (v_1,v)\}$, and $C_V = V \setminus (A_V \cup B_V \cup \{v_1\})$. It can be shown by induction on ℓ that G is a permutation graph of a permutation π and that the graph $i \mapsto \pi(i)$ has the following shape:

$$
\begin{array}{c|c}
A_V & C_V \\
& v_1 \\
\hline
& B_V
\end{array}
$$

$\square$

The previous theorem implies especially that a graph is a treelike permutation graph if and only if it is a path of double-arborescences. In the next theorem, we discuss how to find such a path of minimum length. We will use that a dominating pair of an AT-free graph can be found in linear time [3].

Theorem 4. *Let G be a treelike permutation graph. The minimum ℓ for which G is a path of ℓ double-arborescences can be determined in linear time.*

Proof. Let G be a treelike permutation graph. Let T be the transitive reduction of the unique treelike orientation of G. Let v, w be a dominating pair of G. Find the unique $(v - w)$-path P in T. Recursively remove the first vertex v_1 from P if v_1 is the head of the first edge e of P and the tail of all other edges of T that are incident to v_1 or vice versa. Under the same condition, remove recursively the last vertex of P. Let $\ell - 1$ be the length of the remaining path. Then ℓ is minimum such that G is a path of ℓ double-arborescences. $\square$

5   Partitioning into Bounded Cliques

Let $G = (V, E)$ be a graph. An m-*clique* is a subset $C \subseteq V$ of m vertices, such that $\{v, w\} \in E$ for each pair of vertices $v, w \in C$. A sequence $C_1, \ldots, C_k$ of k cliques is a partition of V into k cliques if $V = C_1 \cup \ldots \cup C_k$ and $C_i \cap C_j = \emptyset, 1 \leq i < j \leq k$. A *triangle* is a 3-clique. We consider the following problem.

PARTITIONING INTO m-CLIQUES: Given a graph $G = (V, E)$, is there a partition of G into m-cliques?

We say that a graph $G' = (V', E')$ results from a graph $G = (V, E)$ by *adding a k-clique* if there are distinct vertices $v_1, \ldots, v_k \notin V$ such that

$$
V' = V \cup \{v_1, \ldots, v_k\} \quad \text{and}
$$
$$
E' = E \cup \{\{v, v_i\}; \ v \in V, i = 1, \ldots, k\} \cup \{\{v_i, v_j\}; \ 1 \leq i < j \leq k\}.
$$

Note that for graph-classes that are closed under adding cliques the PARTITIONING INTO m-CLIQUES problem is equivalent to the following problem: Given a graph $G = (V, E)$ and a number $k \in \mathbb{N}$, is there a partition of G into k cliques of maximum size m? Examples for such graph classes are comparability graphs, permutation graphs, arborescences, and double-arborescences.

In this section, we will show that the PARTITIONING INTO m-CLIQUES problem can be solved in linear time on arborescences – even if m is part of the input. We then show how to compute the maximum number of 2-cliques in an arborescence after deleting some triangles. Finally, we demonstrate how to use these numbers to solve the PARTITIONING INTO TRIANGLES problem on double-arborescences.

5.1 Partitioning Arborescences into Bounded Cliques

Since arborescences are permutation graphs and do not contain an induced cycle
of length four, they are especially interval graphs. Recall that the PARTITIONING
INTO m-CLIQUES-problem for fixed m can be solved in linear time on interval
graphs [14], but that the problem remains $\mathcal{NP}$-complete for interval graphs if
m is part of the input [2]. In this section, we give an algorithm that solves the
problem partitioning arborescences into bounded cliques in linear time – even if
m is part of the input.

Theorem 5. *The problem* PARTITIONING INTO m-CLIQUES *can be solved in
linear time on arborescences even if m is part of the input.*

Proof. Let $G = (V, E)$ be an arborescence. Let T be the transitive reduction with
respect to an arborescence-orientation. By Theorem 2, T can be constructed in
linear time. Proceeding from the leaves to the root r of T, we assign a label
MISS to each vertex v. Let v be the next vertex of T that is considered. If v is
a leave or the only vertex of G, we set $\text{MISS}(v) = m - 1$. Else let $v_1, \ldots, v_k$ be
the children of v. If $\sum_{i=1}^{k} \text{MISS}(v_i) = 0$, set $\text{MISS}(v) = m - 1$, else set $\text{MISS}(v) = -1 + \sum_{i=1}^{k} \text{MISS}(v_i)$.

By induction on the number n of vertices of G it follows that $\text{MISS}(r) = k$ if
and only if k is the smallest non-negative integer such that adding a k-clique to
G results in a graph that has a partition into m-cliques. Hence, G has a partition
into m-cliques if and only if $\text{MISS}(r) = 0$.   □

5.2 Important Triangles for Maximum Matchings in Arborescences

Throughout this section let $G = (V, E)$ be an arborescence, let T be the transitive
reduction of G with respect to an arborescence-orientation, let r be the only sink
and let $t_{\max}$ be the maximum number of disjoint triangles of G. A *matching* of
a graph $G = (V, E)$ is a subset $M \subseteq E$ of the edge set such that each vertex
of G is adjacent to at most one edge in M. Let $\mathcal{V}$ be a set of subsets of V. We
denote by $G - \mathcal{V}$ the graph that results from G by deleting all vertices in all sets
of $\mathcal{V}$. Let $\mathcal{T}$ be a set of triangles of G. With $c_{\mathcal{T}}$ we denote the maximum size of
a matching in $G - \mathcal{T}$. Let $c^{(i)} = \max c_{\mathcal{T}}$, where $\mathcal{T}$ ranges over all disjoint sets of
i triangles of G.

Lemma 1. *A set $\mathcal{T}$ of t_{max} disjoint triangles of G with $c_{\mathcal{T}} = c^{(t_{max})}$ can be
computed in linear time.*

Proof. Let G be an arborescence. Let T be the transitive reduction with respect
to an arborescence-orientation. We proceed again from the leaves to the root r
of T. To each node v we assign a list P of 2-cliques and a list S of singletons that
is contained in the subtree rooted at v. If v is a leaf, let $P(v) = \emptyset$ and $S(v) = v$.
Else let $v_1, \ldots, v_k$ be the children of v.

1. If there is a $j = 1, \ldots, k$ such that $P(v_j)$ contains a 2-clique $\{x, y\}$, delete
   one 2-clique $\{x, y\}$ from $P(v_j)$ and add the triangle $\{x, y, v\}$ to $\mathcal{T}$.

2. If all lists $P(v_i), i = 1, \ldots, k$ are empty, but there is some $j = 1, \ldots, k$ such that $S(v_j)$ contains a vertex w, delete w from $S(v_j)$, and add $\{w, v\}$ to $P(v_1)$.
3. If all lists $P(v_i), S(v_i), i = 1, \ldots, k$ are empty, add v to $S(v_1)$.

Set $P(v) = P(v_1), \ldots, P(v_k)$ and $S(v) = S(v_1), \ldots, S(v_k)$. Based on the property that vertices in two subtrees of T rooted at distinct children of r are not adjacent in G, it can be easily verified by induction on the number n of vertices of G that $|T| = t_{\max}$ and that $c_T = c^{(t)}$. $\square$

Note that by omitting Step 1, the algorithm in the proof of Lemma 1 can be used to create a maximum matching $P(r)$. In the remainder of this section, let T be the set of triangles computed in the proof of Lemma 1.

Theorem 6.

1. *The curve $i \mapsto c^{(i)}$ has the* stair shape, *i.e. there are $1 \leq t_{flat} \leq t_{stair} \leq t_{max}$ such that $t_{stair} - t_{flat}$ is even and*
   (a) $c^{(i-1)} - c^{(i)} = 1, i = 1, \ldots, t_{flat}$
   (b) $c^{(t_{flat}+2i-2)} - c^{(t_{flat}+2i-1)} = 2, c^{(t_{flat}+2i-1)} - c^{(t_{flat}+2i)} = 1, i = 1, \ldots, \frac{t_{stair}-t_{flat}}{2}$
   (c) $c^{(i-1)} - c^{(i)} = 2, i = t_{stair} + 1, \ldots, t_{max}.$
2. *The triangles $t_1, \ldots, t_{t_{max}}$ in T can be ordered such that $c^i = c_{\{t_1,\ldots,t_i\}}, i = 1, \ldots, t_{max}.$*
3. *The sequence $c^{(i)}, i = 1, \ldots, t_{max}$ can be computed in $\mathcal{O}(n^4)$ time.*

For space reasons, we only give the algorithm for Theorem 6.3. The proof of Theorem 6 is basically an induction on the number of steps in the algorithm, but quiet complex. Let t be a triangle of G. Let v be the sink of t. The *subtree of T rooted at t* is the subtree T_v of T rooted at v. The *level* of t is the maximum number of disjoint triangles of G that are contained in T_v.

Let $i = 0$. While $i < t_{\max}$ do one of the following cases.

1. If there is a triangle $t \in T \setminus \{t_1, \ldots t_i\}$ such that $c_{\{t_1,\ldots,t_i\}} - c_{\{t_1,\ldots,t_i,t\}} = 1$, choose t_{i+1} among these triangles on a lowest level. Set $i = i + 1$.
2. Else, if there are two triangles $t, t' \in T \setminus \{t_1, \ldots t_i\}$ such that $c_{\{t_1,\ldots,t_i\}} - c_{\{t_1,\ldots,t_i,t,t'\}} = 3$, choose t_{i+1}, t_{i+2} among these pairs of triangles. Set $i = i+2$.
3. Else, choose a triangle $t_{i+1} \in T \setminus \{t_1, \ldots t_i\}$ on a lowest level. Set $i = i + 1$.

5.3   Partitioning into Triangles of Double-Arborescences

Note that double-arborescences are P_4-free (also called cographs), i.e. they do not contain an induced path of length four. Recall that the PARTITIONING INTO TRIANGLES problem can be solved in $\mathcal{O}(n^5)$ on a P_4-free graph with n vertices [2]. In this section we show how to apply the results of Section 5.2 to solve the problem in $\mathcal{O}(n^4)$ time on double-arborescences.

Theorem 7. *It can be tested in $\mathcal{O}(n^4)$ time whether a double-arborescence with n vertices has a partition into triangles.*

Proof. Let $G = (V, E)$ be a double-arborescence and let $\boldsymbol{E}$ be a a double-arborescence orientation of G with root r. Let G_1 be the subgraph of G induced by $\{r\} \cup \{w \in V;\ (r, w) \in \boldsymbol{E}\}$ and G_2 be the subgraph of G induced by $\{w \in V;\ (w, r) \in \boldsymbol{E}\}$. Then the connected components of G_1 and G_2 consist of arborescences. Hence, we can compute the maximum numbers t_1 and t_2 of triangles and the numbers $c_1^{(0)}, \ldots, c_1^{(t_1)}$ and $c_2^{(0)}, \ldots, c_2^{(t_2)}$ of remaining maximum matchings in the two subgraphs, respectively. Further let $n_{1,2}$ be the number of vertices in $G_{1,2}$. Now, note that a triangle in G is either a triangle in G_1 or G_2 or it consists of an edge in G_1 and a single vertex in G_2 or vice versa. Hence, there is a partition into triangles of G if and only if there exists a pair $(i, j), 1 \leq i \leq t_1, 1 \leq j \leq t_2$ and $\alpha_1, \alpha_2 \in \mathbb{N}$ with the property that $\alpha_1 \leq c_1^{(i)}$, $\alpha_2 \leq c_2^{(j)}$, and

$$2\alpha_1 + \alpha_2 = n_1 - 3i \quad \text{and} \quad 2\alpha_2 + \alpha_1 = n_2 - 3j \tag{1}$$

These equations describe the case that there are i triangles in G_1, j triangles in G_2, that α_1 of the maximum $c_1^{(i)}$ remaining 2-cliques of G_1 build a triangle with different singletons in G_2, and that the remaining vertices of G_1 build triangles with α_2 2-cliques of G_2. Resolving the equations for α_1 and α_2 results in the condition that n is a multiple of 3 and that

$$c_1^{(i)} \geq \frac{2}{3}n - n_2 - 2i + j \geq 0 \quad \text{and} \quad c_2^{(j)} \geq \frac{2}{3}n - n_1 - 2j + i \geq 0. \tag{2}$$

Clearly, for each pair (i, j), the conditions in (2) can be tested in constant time. Since the number of pairs (i, j) is at most quadratic in the number of vertices of G, the over all running time of the algorithm, once the values of c are computed, is quadratic. □

6   Conclusion

We characterized treelike comparability graphs as a subclass of completely separable graphs. We showed that the treelike orientation of a treelike comparability graph is unique and that it can be constructed in linear time. We characterized treelike permutation graphs as paths of double-arborescences. We showed that the minimum ℓ such that a given treelike permutation graph is a path of ℓ double-arborescences can be determined in linear time. We then considered the PARTITIONING INTO m-CLIQUES problem. We showed that the problem can be solved in linear time on arborescences even if m is part of the input. Based on an algorithm for finding the maximum size of a matching after deleting some triangles of an arborescence, we gave a polynomial time algorithm for solving the PARTITIONING INTO TRIANGLES problem on double-arborescences.

References

1. M. D. Atkinson. On computing the number of linear extensions of a tree. *Order*, 7:23–25, 1990.
2. H. L. Bodlaender and K. Jansen. On the complexity of scheduling incompatible jobs with unit-times. In A. M. Borzyszkowski and S. Sokolowski, editors, *Proceedings of the 18th International Symposium on Mathematical Foundations of Computer Science (MFCS '93)*, volume 711 of *Lecture Notes in Computer Science*, pages 291–300. Springer, 1993.
3. D. G. Corneil, S. Olariu, and L. Stewart. A linear time algorithm to compute a dominating path in an AT-free graph. *Information Processing Letters*, 54(5):253–257, 1995.
4. D. G. Corneil, S. Olariu, and L. Stewart. Astroidal triple-free graphs. *SIAM Journal on Discrete Mathematics*, 10(3):399–430, 1997.
5. W. H. Cunningham and J. Edmonds. A combinatorial decomposition theory. *Canadian Journal of Mathematics*, 32:734–765, 1980.
6. W. H. Cunningham and J. Edmonds. Decomposition of directed graphs. *SIAM Journal on Algebraic and Discrete Methods*, 3:214–228, 1982.
7. E. Dahlhaus. Efficient parallel and linear time sequential split decomposition. In P. S. Thiagarajan, editor, *Proceedings of the 14th Conference on Foundations of Software Technology and Theoretical Computer Science (FSTTCS '94)*, volume 880 of *Lecture Notes in Computer Science*, pages 171–180. Springer, 1994.
8. G. Di Stefano and M. L. Koci. A graph theoretical approach to the shunting problem. In B. Gerards, editor, *Proceedings of the Workshop on Algorithmic Methods and Models for Optimization of Railways (ATMOS 2003)*, volume 92 of *Electronic Notes in Theoretical Computer Science*, 2004.
9. M. C. Golumbic. Trivially perfect graphs. *Discrete Mathematics*, 24:105–107, 1978.
10. M. C. Golumbic, C. L. Monma, and W. T. Trotter, Jr. Tolerance graphs. *Discrete Applied Mathematics*, 9:157–170, 1984.
11. P. L. Hammer and F. Maffray. Completely seperable graphs. *Discrete Applied Mathematics*, 27:85–99, 1990.
12. K. Jansen. The mutual exclusion scheduling problem for permutation and comparability graphs. In M. Morvan, C. Meinel, and D. Krob, editors, *Proceedings of the 15th Symposium on Theoretical Aspects of Computer Science (STACS '98)*, volume 1373 of *Lecture Notes in Computer Science*, pages 287–297. Springer, 1998.
13. D. G. Kirckpatrick and P. Hell. On the completeness of a generalized matching problem. In *Proceedings of the 10th Annual ACM Symposium on the Theory of Computing (STOC '78)*, pages 240–245. ACM, The Association for Computing Machinery, 1978.
14. Z. Lonc. On complexity of some chain and antichain partition problems. In G. Schmidt and R. Berghammer, editors, *Graph Theoretic Concepts in Computer Science, 17th International Workshop, (WG '91)*, volume 570 of *Lecture Notes in Computer Science*, pages 97–104. Springer, 1992.
15. E. S. Wolk. A note on "The comparability graph of a tree". *Proceedings of the American Mathematical Society*, 16:17–20, 1965.

Elegant Distance Constrained Labelings of Trees

Jiří Fiala[1,*], Petr A. Golovach[2], and Jan Kratochvíl[1]

[1] Institute for Theoretical Computer Science[**]
and Department of Applied Mathematics,
Charles University, Prague
{fiala,honza}@kam.mff.cuni.cz
[2] Department of Applied Mathematics,
Syktyvkar State University, Syktyvkar, Russia
golovach@syktsu.ru

Abstract. In our contribution to the study of graph labelings with three distance constraints we introduce a concept of elegant labelings: labelings where labels appearing in a neighborhood of a vertex can be completed into intervals such that these intervals are disjoint for adjacent vertices. We justify introduction of this notion by showing that use of these labelings provides good estimates for the span of the label space, and also provide a polynomial time algorithm to find an optimal elegant labeling of a tree for distance constraints $(p, 1, 1)$. In addition several computational complexity issues are discussed.

1 Introduction

In the past decades graph theoretic models of telecommunication networks became natural and frequent subject both in theory and in practice. One of the possible applications considers an allocation of frequencies to transmitters, such that a possible interference is minimized. The notion of distance constrained labeling reflects the fact that interference decreases with increasing distance between transmitters, hence close frequencies should be used only on distant transmitters.

For given integral parameters $p_1 \geq \cdots \geq p_k$ called *distance constraints*, an $L(p_1, p_2, \ldots, p_k)$-*labeling* of a graph G assigns integers to vertices of G such that any pair of vertices that are at distance at most $i \leq k$ get labels that differ by at least p_i. The *span* of a labeling is the difference between the lowest and the highest labels used in the labeling. The graph invariant $\lambda_{(p_1, \ldots, p_k)}(G)$ is the minimum span among all $L(p_1, p_2, \ldots, p_k)$-labelings of G.

Clearly, $L(1)$-labelings are graph colorings, $L(1, \ldots, 1)$-labelings are colorings of the k-th distance power of the underlying graph G. A considerable attention was paid to the first "non-chromatic" collection of distance constraints, namely

---

* Participation at the conference WG 2004 has been in part generously supported by the Deutsche Forschungsgemeinschaft (DFG) and proRWTH! (Freunde und Förderer der RWTH Aachen).
** Supported by the Ministry of Education of the Czech Republic as project LN00A056.

J. Hromkovič, M. Nagl, and B. Westfechtel (Eds.): WG 2004, LNCS 3353, pp. 58–67, 2004.
© Springer-Verlag Berlin Heidelberg 2004

$(p_1, p_2) = (2, 1)$, suggested by Roberts and formally introduced by Griggs and Yeh in 1992 [8]. A variety of results appeared, among others we shall mention a nontrivial dynamic-programming algorithm for computing $\lambda_{(2,1)}(T)$ for trees by Chang and Kuo [3] and a long lasting conjecture stating that for any graph G, it holds that $\lambda_{(2,1)}(G) \leq \Delta(G)^2$, where $\Delta(G)$ stands for the maximum degree of a vertex in G.

From the computational complexity point of view it is also interesting that for an arbitrary constant c, the problem of testing whether $\lambda_{(2,1)}(G) \leq c$ is solvable in linear time when restricted to graphs of bounded treewidth, while the computational complexity of determining $\lambda_{(2,1)}(G)$ for the same class of graphs remains open.

Other collections of distance constraints were also considered by several authors. Labelings of meshes were considered in [14, 11] while $L(p_1, 1, \ldots)$-labelings of trees and interval graphs were studied in [2]. Further hardness results on $L(2, 1, \ldots, 1)$-labelings of restricted classes of graphs can be found in [7].

The computational complexity of finding $\lambda_{(p_1, p_2)}(T)$ is not fully resolved yet even for trees. For example, this problem becomes tractable when p_2 divides p_1, but the precoloring extension and the list-coloring versions of this problem are both NP-complete otherwise [6]. On the other hand, as follows from works on graph properties expressible in Monadic Second Order Logic [4, 1], if the span of a possible labeling is bounded by constant c the test whether $\lambda_{(p_1, \ldots, p_k)}(G) \leq c$ can be performed in linear time for a graph of bounded treewidth (an explicit algorithm is presented in [10]).

Distance constrained labelings can be generalized in several ways – one of the possible directions is the use of different metrics on the label space. Such labelings with constraints $(2, 1)$ were considered in [5] as special graph homomorphisms that are required to be locally injective. In our study we follow this concept and prove several of our results also for the cyclic metric on the label space.

In this paper we show that with an additional requirement on the labeling – that label space of the neighborhood of each vertex can be completed into an interval such that these intervals are disjoint for adjacent vertices – we can obtain both good estimates on the graph invariants $\lambda_{(p_1, p_2, p_3)}(T)$ for trees, but moreover an optimal so called *elegant* $L(p, 1, 1)$-labeling of a tree can be computed in a polynomial time.

Besides the results on computational complexity we provide also a necessary and a sufficient conditions for a tree to allow an elegant $C(2, 1, 1)$-labeling of the minimal possible span. The main motivation of this study is our belief that further exploration of properties of elegant and non-elegant labelings of trees might bring a new insight and new methods to finally resolve the computational complexity of the problem of determining $\lambda_{(p_1, \ldots, p_k)}$ and in particular $\lambda_{(p_1, p_2)}$ on this class of graphs.

Our results on trees are finally accompanied with an NP-hardness proof of the $L(2, 1, 1)$-labeling problem on general graphs, which is presented in the appendix.

2   Preliminaries

All graphs considered in this paper are simple, i.e. without loops and multiple edges. For a vertex $u \in V_G$ the set of all neighbors of u in G is denoted by $N(u)$, the size of $N(u)$ is the *degree* $\deg(u)$ of the vertex u.

A connected graph without a cycle as a subgraph is called a *tree*, its vertices of degree one are called *leaves*, the other are *inner* vertices. A *star* is a graph isomorphic to the complete bipartite graph $K_{1,n}$, $n \geq 1$. The symbol $\omega(G)$ denotes the size of a maximum complete subgraph of G.

The graph distance $\text{dist}(u, v)$ is the number of edges in a shortest path connecting vertices u and v. The k-th distance power G^k of a graph G is the graph on the same vertex set $V_{G^k} = V_G$ where edges of G^k connect distinct vertices that are at distance at most k in G, i.e. $E_{G^k} = \{(u, v) : 1 \leq \text{dist}_G(u, v) \leq k\}$.

For integers $0 \leq a \leq b \leq t$, we define *discrete intervals* $(\bmod\, t + 1)$ in the following way: $[a, b] = \{a, a + 1, \ldots, b\}$ and $[b, a] = \{b, b + 1, \ldots, t, 0, 1, \ldots a\}$.

The term $[t]$-*labeling of* G stands for a mapping $V_G \to [0, t]$.

For our purposes we use both linear and cyclic metric spaces in the definition of distance constrained labelings.

Definition 1. *Let* $p_1 \geq p_2 \geq \cdots \geq p_k \geq 1$ *be a* k-*tuple of integral distance constraints. A* $[t]$-*labeling* f *of* G *is said to be an* $L(p_1, p_2, \ldots, p_k)$-*labeling of span* t *if* $|f(u) - f(v)| \geq p_i$ *whenever* $1 \leq \text{dist}(u, v) \leq i \leq k$.

A $[t]$-*labeling* f *is called a* $C(p_1, p_2, \ldots, p_k)$-*labeling of span* t *if for any pair of distinct vertices* u, v *at distance at most* $i \leq k$, *it holds that* $p_i \leq |f(u) - f(v)| \leq t + 1 - p_i$.

For both kinds of labelings we introduce an additional property of elegance:

Definition 2. *A* $[t]$-*labeling* f *is called* elegant *if for every vertex* u, *there exists an interval* I_u $(\bmod\, k + 1)$, *such that* $f(N(u)) \subseteq I_u$ *and for every edge* $(u, v) \in E_G : I_u \cap I_v = \emptyset$.

Fig. 1. An example of a tree T with $c_{(2,2,1)}(T) = 7 < 10 = c^*_{(2,2,1)}(T)$.

Observe that only triangle-free graphs may admit elegant labelings. On the other hand, it is not hard to deduce that every tree allows an elegant labeling for an arbitrary collection of distance constraints. An example of a $C(2, 2, 1)$-labeling and of an elegant $C(2, 2, 1)$-labeling of a tree T is depicted in Fig. 1.

The minimum t for which a graph G allows an $L(p_1, p_2, \ldots, p_k)$-labeling, and $C(p_1, p_2, \ldots, p_k)$-labeling resp., of span t is denoted by $\lambda_{(p_1, \ldots, p_k)}(G)$ and $c_{(p_1, \ldots, p_k)}(G)$, resp. The corresponding parameters for elegant labelings are indicated by asterisks (and are left to be $+\infty$ if no elegant labeling exists). Note that $\lambda_{(1)}(G) = c_{(1)}(G) = \chi(G) - 1$, where $\chi(G)$ denotes the chromatic number of G.

Observation 1 *For any distance constraints* $(p_1, \ldots, p_k)$ *and any graph* G, *it holds that*

$$p_k(\omega(G^k) - 1) \leq \lambda_{(p_1,\ldots,p_k)}(G) \leq p_1(\chi(G^k) - 1),$$

$$\lambda_{(p_1,\ldots,p_k)}(G) \leq c_{(p_1,\ldots,p_k)}(G) \leq c^*_{(p_1,\ldots,p_k)}(G),$$

$$\lambda_{(p_1,\ldots,p_k)}(G) \leq \lambda^*_{(p_1,\ldots,p_k)}(G) \leq c^*_{(p_1,\ldots,p_k)}(G).$$

Proof. The proof follows from the fact that every labeling with respect to the cyclic metric is also a valid labeling for linear metric, and that elegant labelings are also valid labelings. Moreover vertices of every complete subgraph of G^k should get labels pairwise at least p_k apart and a coloring of G^k can be transformed to an $L_{(p_1,\ldots,p_k)}$-labeling by using labels that form an arithmetic progression of difference p_1 as colors.

3   Tree Labelings with 3 Distance Constraints

The concept of elegant labelings became useful in considering three distance constraints. The reason is, that in this case it is enough to maintain separation p_3 only between intervals associated to adjacent vertices instead of checking every pair of vertices at distance three.

Observe first that all $[t]$-colorings of a star $K_{1,n}$ (including labelings with at least one constraint) are elegant $(\bmod\, t + 1)$, since only two intervals play a role – the interval for the center $I_c = [f(c) + 1, f(c) - 1](\bmod\, t + 1)$ and all other intervals can be chosen as $[f(c), f(c)]$.

3.1   An Upper Bound for Elegant $C(p_1, p_2, p_3)$-Labelings

We present an upper bound on distance constrained labelings of a tree with circular metric. It is well known that powers of trees are chordal graphs (see [9, 12]) and that $\chi(T^k) = \omega(T^k)$. Observe that in contrary to the general upper bound of Observation 1 for the parameter $\lambda_{(p_1,p_2,p_3)}(G)$, the coefficient by the main term $\omega(T^3)$ becomes p_2 instead of p_1 and hence it provides an essential improvement when $p_2 \ll p_1$ and $\omega(T^3)$ is sufficiently large.

Theorem 2. *For any* $p_1 \geq p_2 \geq p_3 \geq 1$ *and any tree* T *different from a star, it holds that* $c^*_{(p_1,p_2,p_3)}(T) \leq p_2\omega(T^3) + p_1 + \max\{p_1 - p_2, p_3\} - 3$.

Proof. By induction on the number s of inner vertices of T we construct an elegant labeling of T such that for each vertex u, $f(N(u))$ is an arithmetic progression of length $\deg(u)$ and difference p_2.

When $s = 2$, let u and u' be the two inner vertices of T of degrees $d, d' \geq 1$. We choose

$$t = \omega(T^3)p_2 + p_1 + \max\{p_1 - p_2, p_3\} - 2p_2 - 1$$

and define a $[t]$-labeling f of the tree T explicitly as $f(N(u)) = \{0, p_2, 2p_2, \ldots, (d-1)p_2\}$ where $f(u') = (d-1)p_2$ and the other labels are distributed on leaves of $N(u)$ arbitrarily. For $r = (d-1)p_2 + p_1$ we similarly lay out labels $\{r, r+p_2, r+2p_2, \ldots r+(d'-1)p_2\}$ on $N(u')$ such that $f(u) = r$.

To show that f is a valid $C(p_1, p_2, p_3)$-labeling we denote first by v, v' the two vertices of the minimum and the maximum label, i.e. $f(v) = 0$ and $f(v') = r + (d'-1)p_2 = \omega(T^3)p_2 + p_1 - 2p_2$.

Since $\text{dist}(v, v') = 3$ we need $\omega(T^3)p_2 + p_1 - 2p_2 \leq t+1-p_3$, which is assured by the choice of t. For the adjacent vertices v and u we need $(d-1)p_2 + p_1 \leq t+1-p_1$, which holds as well, because $t \geq \omega(T^3)p_2 + p_1 - 2p_2 - 1 + p_1 - p_2 \geq \omega(T^3)p_2 + p_1 - 2p_2 - 1 + p_1 - (d'-1)p_2 = (d-1)p_2 + 2p_1 - 1$. The same inequality can be analogously derived for the labels of v' and u. Observe, that these conditions on u, u', v and v' imply, that the distance constraints are valid also for other pairs of vertices.

Now suppose that T has at least three inner vertices. Since inner vertices induce a subtree of T called the *inner tree of T*, it is possible to choose a pair (u, v) of adjacent inner vertices such that v is a leaf in the inner tree and the sum $\deg(u) + \deg(v)$ is minimized. We remove all vertices adjacent to v with exception of u and denote the resulting tree by T'. By the choice of (u, v) we have $\omega((T')^3) = \omega(T^3) \geq \deg(u) + \deg(v)$.

By the induction hypothesis the tree T' allows an elegant labeling f' of span $t = \omega(T^3)p_2 + p_1 + \max\{p_1 - p_2, p_3\} - 3$. Now assume that the arithmetic progression on $f'(N(u))$ is of form $a, a+p_2, \ldots, a+(\deg(u)-1)p_2, \pmod{t+1}$. Then the vertices of $N(v)$ should avoid interval $I_1 = [a - p_3 + 1, a + (\deg(u) - 1)p_2 + p_3 - 1]$ due to the constraint on distance three as well as the interval $I_2 = [f'(v) - p_1 + 1, f'(v) + p_1 - 1]$.

Since $f'(v)$ is at distance at least $p_3 - 1$ from the boundary of I_1, and similarly at least $p_1 - 1$ points apart from the boundary of I_2 we get that $|I_1 \cap I_2| = p_3 + \max\{(\deg(u) - 1)p_2 + p_3, p_1\} - 1 \geq p_3 + \max\{p_2 + p_3, p_1\} - 1$.

Then $I = [0, t] \setminus (I_1 \cup I_2)$ is an interval of size

$$
\begin{aligned}
|I| &= t + 1 - |I_1| - |I_2| + |I_1 \cap I_2| \\
&\geq \deg(u)p_2 + \deg(v)p_2 + p_1 + \max\{p_1 - p_2, p_3\} - 3 - \\
&\quad - \deg(u)p_2 + p_2 - p_3 - 2p_1 + 2 + \max\{p_2 + p_3, p_1\} \\
&= \deg(v)p_2 + p_2 - 1
\end{aligned}
$$

an hence can accommodate an arithmetic progression A of length $\deg(v)$ and difference p_2, which contains $f'(u)$ as one of its elements.

We extend the labeling f' into a labeling f of T by using elements of $A \setminus f'(u)$ as the labels of the leaf vertices adjacent to v in T. This concludes the proof.

For a particular choice of $(p_1, p_2, p_3) = (2, 1, 1)$, we have obtained an almost a tight bound:

Corollary 1. *Every tree T satisfies*

$$\omega(T^3) - 1 \leq \lambda_{(2,1,1)}(T) \leq \lambda^*_{(2,1,1)}(T) \leq \omega(T^3),$$

and for any tree T different from a star it holds

$$\omega(T^3) - 1 \leq c_{(2,1,1)}(T) \leq c^*_{(2,1,1)}(T) \leq \omega(T^3).$$

Proof. If T is a star then it can be easily seen that $\lambda_{2,1,1}(T) = \omega(T^3)$ and it was already mentioned that any of its labelings is elegant.

The bound $c^*_{2,1,1}(T) \leq \omega(T^3)$ when the tree T is different from a star follows from Theorem 2. All other inequalities and bounds were shown in Observation 1.

3.2   An Algorithm to Compute $c^*_{(p,1,1)}(T)$

The proof of Theorem 2 was constructive, hence it can be straightforwardly converted into a polynomial-time algorithm which finds a $C(p_1, p_2, p_3)$-labeling within the claimed upper bound.

For the special choice of distance constraints $p_2, p_3 = 1$ the computation of $\lambda^*_{(p,1,1)}(T)$ and $c^*_{(p,1,1)}(T)$ can be resolved in a polynomial time. We describe here an algorithm for deciding whether $c^*_{(p,1,1)} \leq k$. The algorithm for linear metric differs only in minor details. We use a dynamic programming approach, similarly as it was used in the algorithm for computation of $\lambda_{(2,1)}(T)$ (see [3,6]).

Let T be a tree and k be a positive integer. Our algorithm tests the existence of an elegant $C(p, 1, 1)$-labelling of T of span k. We may assume that $k \leq n + 2p - 4$, where n is the number of vertices of T, since if $k > n + 2p - 4$, such a labeling always exists due to Theorem 2.

We first choose a leaf r as the root of T, which defines the parent-child relation between every pair of adjacent vertices. For any edge (u, v) such that u is a child of v, we denote by T_{uv} the subtree of T rooted in v and containing u and all descendants of u. For every such edge and for every pair of integers $i, j \in [0, k]$ and an interval I (mod$(k + 1)$) such that $j \in I$, we introduce a boolean function $\phi(u, v, i, j, I)$, which is evaluated true if and only if T_{uv} has an elegant $C(p, 1, 1)$-labelling f where $f(u) = i$, $f(v) = j$ and $I_u = I$. This function ϕ can be calculated as follows:

1. Set an initial value $\phi(u, v, i, j, I) = $ false for all edges (u, v), integers $i, j \in \{0, 1, \ldots, k\}$ and intervals I $(j \in I)$.
2. If u is a leaf adjacent to v then we set $\phi(u, v, i, j, I) = $ true for all integers $i, j \in [0, k] : p \leq |i - j| \leq k - p$ and intervals I such that $j \in I$ and $i \notin I$.
3. Let us suppose that ϕ is already calculated for all edges of T_{uv} except (u, v). Denote by $v_1, v_2, \ldots, v_m$ children of u. For all pairs of integers $i, j \in [0, k] : p \leq |i - j| \leq k - p$ and for all intervals $I : j \in I, i \notin I$ we consider the set system $\{M_1, M_2, \ldots, M_m\}$, where

$$M_t = \{s : s \in I \setminus \{j\}, \exists \text{ interval } J : \phi(v_t, u, s, i, J) = \text{true}, i \in J, I \cap J = \emptyset\}$$

We set $\phi(u, v, i, j, I) = $ true if the set system $\{M_1, M_2, \ldots, M_m\}$ allows a system of distinct representatives, i.e. if there exists an injective function $r : [1, m] \to [0, k]$ such that $r(t) \in M_t$ for all $t \in [1, m]$.

The correctness of calculation of the function ϕ follows by an easy inductive argument. The only nontrivial point is that in the constructed entry $f(v)$ differs from $f(x)$ for every child x of v_t, because $f(v) = j \in I$, and $f(x) \in J$, where $I \cap J = \emptyset$.

Now we evaluate the complexity of computation of this function. It is calculated for $n-1$ edges. Since each interval I is defined by the pair of it's endpoints, the set of arguments has the cardinality $O(nk^4)$. Computation of ϕ for leafs (see step 2) demands $O(1)$ operation for each argument. The recursive step (see item 3) takes time $O(mk^3)$ for constructing the sets M_t and then $O((m+k)^2 mk)$ for the testing of the existence of the system of distinct representatives (we have m sets of cardinality of no more than k). Since $m \leq n$ and $k \leq n + 2p - 4$, this step demands $O(n^3 k)$ operations for a single collection of arguments. So the total time of computation of ϕ is equal to $O(n^4 k^5)$ and this function can be calculated for all sets of arguments polynomially.

To finish the description of the algorithm we have only to note that an elegant $C(p, 1, 1)$-labelling of span k exists if and only if there are integers $i, j \in [0, k]$ and a interval I ($j \in I$), for which $\phi(r, w, i, j, I) = \mathsf{true}$ where w is the only child of the root r.

It suffices to test at most $O(n)$ values of k, which provides the total $O(n^{10})$ time complexity. Observe that for linear metric the algorithm basically remains the same, with the exception that also pairs i, j such that $|i - j| > k - p$ are allowed in steps 2) and 3).

Thus we proved following theorem:

Theorem 3. *For any tree T, $\lambda^*_{(p,1,1)}(T)$ and $c^*_{(p,1,1)}(T)$ can be computed in a polynomial time.*

For the computation of $\lambda^*_{(2,1,1)}(T)$ (or $c^*_{(2,1,1)}(T)$) it is necessary to run this algorithm only once for $k = \omega(T^3) - 1$. If the algorithm returns positive answer, then $\lambda^*_{(2,1,1)}(T) = \omega(T^3) - 1$, else $\lambda^*_{(2,1,1)}(T) = \omega(T^3)$.

Finally note, that if we wanted to generalize the above algorithm to arbitrary distance constraints (p_1, p_2, p_3), it would require resolving of a system of *distant representatives* in the step 3), which is an NP-hard problem in general [6], and moreover it is exactly the same bottleneck of a possible polynomial algorithm for computing $\lambda_{(p_1,p_2)}$ on trees for a general pair of distance constraints $p_1 > p_2 > 1$ [6].

3.3   Perfect Labelings

In order to illustrate the above notions, we notice that for any tree we are able to show that either $c_{(2,1,1)}(T) = \omega(T^3) - 1$ and find such a labeling, called *perfect*, or we find an elegant labeling of span $\omega(T^3)$, leaving the possibility that T may allow a perfect labeling but no such labeling can be elegant (we leave as an open question whether a tree with this property exists). It would certainly be interesting to characterize the trees that satisfy $c_{(2,1,1)}(T) = c^*_{(2,1,1)}(T) = \omega(T^3) - 1$.

We present a necessary condition that a tree must satisfy to allow a perfect elegant labeling. We first classify edges of the tree with respect to the fact whether their neighborhood induces a maximum clique in T^3 or not. Hence, an edge $(u, v) \in E_T$ will be called saturated if $\deg(u) + \deg(v) = \omega(T^3)$, and it will be called unsaturated otherwise.

Theorem 4. *If a tree allows a prefect elegant labeling, then every inner vertex is incident with at least two unsaturated edges.*

Proof. Assume for the contrary that an inner vertex v is incident with at most one unsaturated edge. For any neighbor u incident with v along a saturated edge it holds that $\deg(u) + \deg(v) = \omega(T^3)$, hence for any perfect elegant labeling follows $I_u = [0, \omega(T^3) - 1] \setminus I_v$.

Since $I_v = [a, b]$ is an interval of length $\deg(v)$, each element of I_v is used as a label of some $u \in N(v)$. As v is incident with at most one unsaturated edge, at least one of a or b is used as a label of a neighbor w connected to v via a saturated edge. But then the label of w is one unit away from I_w, a contradiction.

If we interpret this condition in the construction of Theorem 2, we get:

Corollary 2. *A tree allows a perfect elegant labeling if it can be rooted such that each inner vertex has at least two children connected to it by unsaturated edges.*

There exist trees with at least two unsaturated edges incident with each inner vertex, but which allow no labeling of span $\omega(T^3) - 1$ (neither elegant nor not elegant). An example of such a tree is depicted in Fig. 2

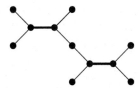

Fig. 2. A tree with $c_{(2,1,1)}(T) = \omega(T^3)$ (saturated edges indicated in bold).

4   Computational Complexity of the $L(2, 1, 1)$-Labeling Problem

To complete the picture we shortly present a full computational complexity characterization of the decision problem whether $\lambda_{(2,1,1)} \leq k$ for general graphs.

Theorem 5. *The decision problem whether $\lambda_{(2,1,1)} \leq k$ is NP-complete for every $k \geq 5$ and it is solvable in polynomial time for all $k \leq 4$.*

Proof. We start with the second part of the theorem and prove that the labeling problem is tractable for $k \leq 4$. Only finitely many connected graphs allow a $\lambda_{(2,1,1)}$-labeling of span at most 3. So, without loss of generality we may consider only the case $k = 4$.

It can be easily seen, that if G is a graph for which an $L(2, 1, 1)$-labelling of span 4 exists, then it can not contain as a subgraph any of the graphs depicted in Fig. 3. Clearly, the maximum degree of G is at most 3 and each connected component of G is formed by a path or by a cycle, where some vertices are equipped with an additional leaf, or two consecutive vertices may also be joined

Fig. 3. Some graphs of $\lambda_{(2,1,1)}(G) > 4$.

by a path of length 2. It is not difficult to observe that such graphs have treewidth bounded by 3, and hence the existence of an $L(2,1,1)$-labelling of span 4 can be tested in linear time by dynamic programming (e.g., [10]).

For $k \geq 5$, we reduce the NOT-ALL-EQUAL p-SATISFIABILITY (NAE p-SAT) problem. An instance of NAE p-SAT is a formula Φ in conjunctive normal form with p positive literals in each clause (no negations). It is well known [13] that for all $p \geq 3$, the decision problem whether such Φ allows a satisfying assignment where each clause contains also a negatively valued literal is NP-complete.

For each variable x_i we construct a gadget consisting of a chain of m_i copies of the graph depicted in Fig 4, where m_i is the number of occurrences of x_i and $p = \lceil \frac{k}{2} \rceil$, $r = \lceil \frac{p-1}{2} \rceil$. In the figure the symbol E_n stands for an independent set with n vertices, K_n for a complete graph, and M_n for a matching on n edges.

It can be explored by a case analysis that any $L(2,1,1)$ labeling of span k of the constructed variable gadget satisfies:

- All vertices u_i are labelled by the same label, either by 0 or by k.
- The vertices v_i are given labels either from the set $L = \{0, 2, 4, \ldots, k - 4 + (k \mod 2)\}$, when u_i's are labeled by k, or otherwise from the set $L' = \{k - l, l \in L\}$.

We finalize the construction of the graph G_Φ such that for each clause C of the formula Φ we insert an extra new vertex w_C and for each variable x which appears in the clause we link w_C with one of the vertices v of the vertex gadgets associated with x. (Each v-type vertex is adjacent to only one w_C).

The properties of the variable gadgets assure that G_Φ allows an $L(2,1,1)$-labeling of span k if and only if Φ has a required assignment. These labelings are related to assignments e.g. by letting $x =$ true whenever the vertices u_i of the gadget for x are all labeled by k, and $x =$ false if u_i get 0.

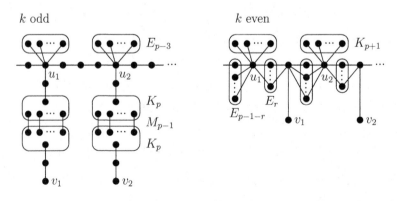

Fig. 4. Variable gadgets.

Clearly, as for any clause vertex w_C it holds $\deg(w_C) \geq |L| = |L'|$, these labelings indicate only valid assignments, i.e., at least one of the adjoining gadgets represents positively valued variable and at least one stands for a negatively valued one.

In the opposite direction, each assignment for Φ can be converted into an $L(2,1,1)$-labeling of G_Φ in a straightforward way.

References

1. ARNBORG, S., LAGERGREN, J., AND SEESE, D. Easy problems for tree-decomposable graphs. *J. Algorithms 12*, 2 (1991), 308–340.

2. BERTOSSI, A. A., PINOTTI, M. C., AND RIZZI, R. Channel assignment on strongly-simplicial graphs. In *International Parallel and Distributed Processing Symposium, 17th IPDPS '03, Nice* (2003), IEEE Computer Society, p. 222.

3. CHANG, G. J., AND KUO, D. The $L(2,1)$-labeling problem on graphs. *SIAM Journal of Discrete Mathematics 9*, 2 (May 1996), 309–316.

4. COURCELLE, B. The monadic second-order logic of graphs. I: Recognizable sets of finite graphs. *Inf. Comput. 85*, 1 (1990), 12–75.

5. FIALA, J., AND KRATOCHVÍL, J. Partial covers of graphs. *Discussiones Mathematicae Graph Theory 22* (2002), 89–99.

6. FIALA, J., KRATOCHVÍL, J., AND PROSKUROWSKI, A. Distance constrained labelings of precolored trees. In *Theoretical Computer Science, 7th ICTCS '01, Torino* (2001), no. 2202 in Lecture Notes in Computer Science, Springer Verlag, pp. 285–292.

7. GOLOVACH, P. A. Systems of pair of q-distant representatives and graph colorings. *Zap. nau. sem. POMI 293* (2002), 5–25. in Russian.

8. GRIGGS, J. R., AND YEH, R. K. Labelling graphs with a condition at distance 2. *SIAM Journal of Discrete Mathematics 5*, 4 (Nov 1992), 586–595.

9. KEARNEY, P. E., AND CORNEIL, D. G. Tree powers. *Journal of Algorithms 29*, 1 (Oct. 1998), 111–131.

10. KOHL, A., SCHREYER, J., TUZA, Z., AND VOIGT, M. Choosability problems for (d, s)-colorings. in preparation, 2003.

11. LEESE, R. A. Radio spectrum: a raw material for the telecommunications industry. 10th Conference of the European Consortium for Mathematics in Industry, Goteborg, 1998.

12. LIN, Y. L., AND SKIENA, S. S. Algorithms for square roots of graphs. *SIAM Journal on Discrete Mathematics 8*, 1 (Feb. 1995), 99–118.

13. SCHAEFER, T. J. The complexity of the satisfability problem. In *Proceedings of the 10th Annual ACM Symposium on Theory of Computing* (1978), ACM, pp. 216–226.

14. VAN DEN HEUVEL, J., LEESE, R. A., AND SHEPHERD, M. A. Graph labeling and radio channel assignment. *Journal of Graph Theory 29*, 4 (1998), 263–283.

Collective Tree Spanners and Routing in AT-free Related Graphs

Feodor F. Dragan[1], Chenyu Yan[1], and Derek G. Corneil[2]

[1] Department of Computer Science, Kent State University, Kent, Ohio, USA
{dragan,cyan}@cs.kent.edu
[2] Department of Computer Science, University of Toronto, Toronto, Ontario, Canada
dgc@cs.toronto.edu

Abstract. In this paper we study collective additive tree spanners for families of graphs that either contain or are contained in AT-free graphs. We say that a graph $G = (V, E)$ *admits a system of μ collective additive tree r-spanners* if there is a system $\mathcal{T}(G)$ of at most μ spanning trees of G such that for any two vertices x, y of G a spanning tree $T \in \mathcal{T}(G)$ exists such that $d_T(x, y) \leq d_G(x, y) + r$. Among other results, we show that AT-free graphs have a system of two collective additive tree 2-spanners (whereas there are trapezoid graphs that do not admit any additive tree 2-spanner). Furthermore, based on this collection of trees, we derive a compact and efficient routing scheme for those graphs. Also, any DSP-graph (there exists a dominating shortest path) admits one additive tree 4-spanner, a system of two collective additive tree 3-spanners and a system of five collective additive tree 2-spanners.

1 Introduction

Given a graph $G = (V, E)$, a spanning subgraph H is called a *spanner* if H provides a "good" approximation of the distances in G. More formally, for $t \geq 1$, H is called a *multiplicative t–spanner* of G [1, 14, 13] if $d_H(u, v) \leq t \cdot d_G(u, v)$ for all $u, v \in V$. If $r \geq 0$ and $d_H(u, v) \leq d_G(u, v) + r$ for all $u, v \in V$, then H is called an *additive r–spanner* of G [8]. The parameters t and r are called, respectively, the *multiplicative* and the *additive stretch factors*. Clearly, every additive r-spanner of G is a multiplicative $(r + 1)$-spanner of G (but not vice versa). In this paper, we continue the approach taken in [4] of studying *collective tree spanners*. We say that a graph $G = (V, E)$ *admits a system of μ collective additive tree r-spanners* if there is a system $\mathcal{T}(G)$ of at most μ spanning trees of G such that for any two vertices x, y of G a spanning tree $T \in \mathcal{T}(G)$ exists such that $d_T(x, y) \leq d_G(x, y) + r$ (a multiplicative variant of this notion can be defined analogously). Clearly, if G admits a system of μ collective additive tree r-spanners, then G admits an additive r-spanner with at most $\mu \times (n - 1)$ edges (take the union of all those trees), and if $\mu = 1$ then G admits an additive tree r-spanner. Note also that any graph on n vertices admits a system of at most $n-1$ collective additive tree 0-spanners (take $n - 1$ Breadth-First-Search–trees rooted at different vertices of G). In particular, we examine the problem of finding small

J. Hromkovič, M. Nagl, and B. Westfechtel (Eds.): WG 2004, LNCS 3353, pp. 68–80, 2004.
© Springer-Verlag Berlin Heidelberg 2004

systems of collective additive tree r-spanners for small values of r on classes of graphs that are related to the well known *asteroidal triple-free (AT-free) graphs*, notably the restricted families: permutation graph and trapezoid graphs and the generalizations: DSP-graphs and graphs with bounded asteroidal number.

Once one has determined a system of collective additive tree spanners, it is interesting to see how such a system can be used to design compact and efficient routing schemes for the given graph. Following [12], one can give the following formal definition. A family $\Re$ of graphs is said to have an $l(n)$-bit *routing labeling scheme* if there is a function L labeling the vertices of each n-vertex graph in $\Re$ with distinct labels of up to $l(n)$ bits, and there exists an efficient algorithm, called the *routing decision*, that given the label of a source vertex v and the label of the destination vertex (the header of the packet), decides in time polynomial in the length of the given labels and using only those two labels, whether this packet has already reached its destination, and if not, to which neighbor of v to forward the packet. The quality of a routing scheme is measured in terms of its *additive stretch*, called *deviation*, (or *multiplicative stretch*, called *delay*), namely, the maximum surplus (or ratio) between the length of a route, produced by the scheme for some pair of vertices, and their distance.

1.1   Our Results

After introducing the notation and definitions used throughout the paper, we examine various families of graphs related to AT-free graphs from the perspective of determining whether they have a small constant number of collective additive tree r-spanners for small constant r. In Section 2 we show that AT-free graphs have a system of two collective additive tree 2-spanners, permutation graphs have a single additive tree 2-spanner but there are trapezoid graphs that do not admit any additive tree 2-spanner (thereby disproving a conjecture of [15]). All of these tree spanners can be easily constructed in linear time. For families that strictly contain AT-free graphs, we prove that any DSP-graph admits one additive tree 4-spanner, a system of two collective additive tree 3-spanners and a system of five collectible additive tree 2-spanners. Furthermore, any graph G with asteroidal number $\mathsf{an}(G)$ admits a system of $\mathsf{an}(G)(\mathsf{an}(G) - 1)/2$ collective additive tree 4-spanners and a system of $\mathsf{an}(G)(\mathsf{an}(G) - 1)$ collective additive tree 3-spanners. In Section 3, we show how the system of two collective additive tree 2-spanners for AT-free graphs can be used to derive a compact and efficient routing scheme. In particular we will show that any AT-free graph with diameter $\mathcal{D}$ and maximum vertex degree Δ admits a $(3\log_2\mathcal{D} + 6\log_2\Delta + O(1))$-bit routing labeling scheme of deviation at most 2. Moreover, the scheme is computable in linear time, and the routing decision is made in constant time per vertex.

1.2   Basic Notions and Notation

All graphs occurring in this paper are connected, finite, undirected, loopless and without multiple edges. In a graph $G = (V, E)$ the *length* of a path from a vertex v to a vertex u is the number of edges in the path. The *distance* $d_G(u, v)$ between the vertices u and v is the length of a shortest path connecting u and

v. The *eccentricity $ecc(v)$* of a vertex v of G is $\max_{u \in V} d_G(u, v)$. The *diameter $diam(G)$* of G is $\max_{v \in V} ecc(v)$. The *ith neighborhood* of a vertex v of G is the set $N_i(v) := \{u \in V : d_G(v, u) = i\}$. For a vertex v of G, the sets $N(v) := N_1(v)$ and $N[v] := N(v) \cup \{v\}$ are called the *open neighborhood* and the *closed neighborhood* of v, respectively. For a set $S \subseteq V$, by $N[S] := \bigcup_{v \in S} N[v]$ we denote the *closed neighborhood* of S and by $N(S) := N[S] \setminus S$ the *open neighborhood* of S. A set $D \subseteq V$ is called a *dominating set* of a graph $G = (V, E)$ if $N[D] = V$.

An independent set of three vertices such that each pair is joined by a path that avoids the neighborhood of the third is called an *asteroidal triple*. A graph G is an *AT-free graph* if it does not contain any asteroidal triples [2]. In [7], the notion of asteroidal triple was generalized. An independent set $A \subseteq V$ of a graph $G = (V, E)$ is called an *asteroidal set* of G if for each $a \in A$ the vertices of $A \setminus \{a\}$ are contained in one connected component of $G - N[a]$, the graph obtained from G by removing vertices of $N[a]$. The maximum cardinality of an asteroidal set of G is denoted by $an(G)$, and called the *asteroidal number* of G. The class of *graphs of bounded asteroidal number* extends naturally the class of AT-free graphs; AT-free graphs are exactly the graphs with asteroidal number at most two.

Let P be a shortest path of G. If every vertex z of G belongs to the neighborhood $N[P]$ of P, then we say that P is a *dominating shortest path* of G. A graph G is called a *Dominating-Shortest-Path–graph* (or *DSP–graph*, for short), if it has a dominating shortest path. By the Dominating Pair Theorem given in [2], any AT-free graph is a DSP-graph.

The class of AT-free graphs contains many intersection families of graphs, among them the permutation graphs, the trapezoid graphs and the cocomparability graphs. These three families of graphs can be defined as follows. Consider two parallel lines (upper and lower) in the plane. Assume that each line contains n points, labeled 1 to n, and each two points with the same label define a segment with that label. The intersection graph of such a set of segments between two parallel lines is called a *permutation graph*. Assume now that each line contains n intervals, labeled 1 to n, and each two intervals with the same label define a trapezoid with that label (a trapezoid can degenerate to a triangle or to a segment). The intersection graph of such a set of trapezoids between two parallel lines is called a *trapezoid graph*. Clearly, every permutation graph is a trapezoid graph, but not vice versa. The class of cocomparability graphs (which contains all trapezoid graphs as a subclass) can be defined as the intersection graphs of continuous function diagrams, but for this paper it would be more convenient to them via the existence of a special vertex ordering. A graph G is a *cocomparability graph* if it admits a vertex ordering $\sigma = [v_1, v_2, \ldots, v_n]$, called a *cocomparability ordering*, such that for any $i < j < k$, if v_i is adjacent to v_k then v_j must be adjacent to v_i or to v_k. According to [11], such an ordering of a cocomparability graph can be constructed in linear time. Note also that, given a permutation graph G, a *permutation model* (i.e., a set of segments between two parallel lines, defining G) can be found in linear time [11]. A *trapezoid model* for a trapezoid graph can be found in $O(n^2)$ time [9].

2   Collective Additive Tree Spanners

2.1   AT-free Graphs

It is known [15] that any AT-free graph admits one additive tree 3-spanner. In this subsection we show that any AT-free graph admits a system of two collective additive tree 2-spanners.

As a consequence of the Dominating Pair Theorem given in [2], any AT-free graph has a dominating shortest path which can be found in linear time by $2 \times LexBFS$ [3]. The $2 \times LexBFS$ method first starts a *lexicographic breadth-first search* (*LexBFS*) from an arbitrary vertex x of G and then starts a second LexBFS from the vertex x_0 last visited by the first LexBFS. Let x_l be the vertex of G last visited by the second LexBFS. As shown in [3], every shortest path $(x_0, x_1, \ldots, x_l)$, connecting x_0 and x_l, is a dominating shortest path of G. Next we demonstrate how to use such a dominating shortest path in an AT-free graph to show that every AT-free graph admits a system of two collective additive tree 2-spanners. We will need the following result from [6].

Lemma 1. *[6] Let $P := (x_0, x_1, \ldots, x_l)$ be a dominating shortest path of an AT-free graph $G = (V, E)$ constructed by $2 \times LexBFS$. Then, for every $i = 1, 2, \ldots, l$, every vertex $z \in N_i(x_0)$ is adjacent to x_i or x_{i-1}.*

Using this lemma, we construct a first spanning tree $T_1 = (V, E_1)$ for an AT-free graph $G = (V, E)$ as follows: put into initially empty E_1 all edges of the path $P := (x_0, x_1, \ldots, x_l)$, and then for each vertex $z \in N_i(x_0)$, put edge zx_{i-1} into E_1, if z is adjacent to x_{i-1} in G, and put edge zx_i into E_1, otherwise. We call this spanning tree the *caterpillar-tree* of G (with *spine P*). According to [15], this caterpillar-tree gives already an additive tree 3-spanner for the AT-free graph G. To get a collective additive stretch factor 2 for G, we construct a second spanning tree $T_2 = (V, E_2)$ for G as follows. Set $L_i := N_i(x_0)$ for each $i = 1, 2, \ldots, l$.

set $E_2 := \{$all edges of the path $P := (x_0, x_1, \ldots, x_l)\}$;
set $dev(x_i) := 0$ for each vertex x_i of the path P;
for $i = 1$ to l do
   for each vertex $z \in L_i \setminus \{x_i\}$ do
      among all neighbors of z in L_{i-1} choose a neighbor w with minimum
         deviation $dev(w)$;
      add edge zw to E_2 and set $dev(z) := dev(w) + 1$;
   enddo
enddo.

We call spanning tree T_2 the *cactus-tree* of G (with *stem P*). It is evident, by construction, that the cactus-tree T_2 is a special kind of *breadth-first-search–tree* of G. The value $dev(z)$ (called the *deviation of z from stem P*) gives the distance in T_2 between vertex z and path P. In Figure 1 we show an AT-free graph G along with its caterpillar-tree T_1 and cactus-tree T_2.

Lemma 2. *Spanning trees $\{T_1, T_2\}$ are collective additive tree 2-spanners of AT-free graph G.*

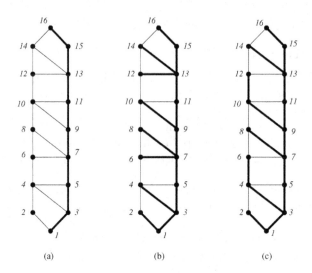

Fig. 1. (a) An AT-free graph G with a dominating path P, (b) the caterpillar-tree T_1 of G and (c) the cactus-tree T_2 of G

Proof. Consider two arbitrary vertices $x \in L_i$ and $y \in L_j$ ($y \neq x$) of G, where $j \leq i$. If $i = j$, i.e., both x and y lie in the same layer $L_i = L_j$, then the distance in T_1 between x and y is at most 3, since in the worst case one of them is adjacent to x_i in T_1 and the second to x_{i-1}. Thus, $d_{T_1}(x, y) \leq 3 \leq d_G(x, y) + 2$ holds when $i = j$, and therefore, we may assume that $i > j$.

We know that $d_G(x, y) \geq i - j$. By the construction of the caterpillar-tree T_1, we have $d_{T_1}(y, x_j) \leq 2$ and $d_{T_1}(x, x_{i-1}) \leq 2$. Hence, $d_{T_1}(x, y) \leq d_{T_1}(x, x_{i-1}) + d_{T_1}(x_{i-1}, x_j) + d_{T_1}(y, x_j) \leq 2 + i - 1 - j + 2 \leq d_G(x, y) + 3$, and equality $d_{T_1}(x, y) = d_G(x, y) + 3$ holds if and only if $d_G(x, y) = i - j$, vertex x is adjacent to x_i in T_1 (and thus in G, vertex x is not adjacent to x_{i-1}) and vertex y is adjacent to x_{j-1} in T_1 but does not coincide with x_j. We will show that in this case in the cactus-tree T_2, $d_{T_2}(x, y) \leq d_G(x, y) + 2$.

Consider in G a shortest path ($y = y_0, y_1, \ldots, y_{i-j} = x$) connecting vertices y and x. Clearly, $y_k \in L_{j+k}$ for each $k = 0, 1, \ldots, i - j - 1$, and since y_k is a neighbor of y_{k+1} in layer L_{j+k}, by construction of T_2, we have $dev(y_0) = 1$ and $dev(y_{k+1}) \leq dev(y_k) + 1 \leq k + 2$. Hence, the deviation of vertex x is at most $i - j + 1$. That is, there is a path in T_2 between x and a stem vertex x_s ($j - 1 \leq s \leq i - 2$) of length $i - s$. The latter implies the existence in T_2 of a path of length $i - j + 1$ between vertices x and x_{j-1}. Therefore, $d_{T_2}(x, y) \leq d_{T_2}(x, x_{j-1}) + 1 = i - j + 1 + 1 = d_G(x, y) + 2$.   □

From this lemma we immediately conclude.

Theorem 1. *Any AT-free graph admits a system of two collective additive tree 2-spanners, constructable in linear time.*

In the next subsection, we will show that to get a collective additive stretch factor 2 for some AT-free graphs, one needs at least two spanning trees. Therefore, the result given in Theorem 1 is best possible. Furthermore, to achieve a collective additive stretch factor 1 or 0 for some AT-free graphs, one needs $\Omega(n)$ spanning trees.

2.2   Permutation Graphs and Trapezoid Graphs

It is known [10] that any permutation graph admits a multiplicative tree 3-spanner. In this subsection, we show that any permutation graph admits an additive tree 2-spanner and any system of collective additive tree 1–spanners must have $\Omega(n)$ spanning trees for some permutation graphs. Here also we disprove a conjecture given in [15], that any cocomparability graph admits an additive tree 2-spanner. We show that there exists even a trapezoid graph which does not admit any additive tree 2-spanner.

Let $G = (V, E)$ be a permutation graph given together with a permutation model. In what follows, "u.p." and "l.p." refer to a vertex's point on the upper and lower, respectively, line of the permutation model. Construct BFS-layers $(\{L_0, L_1, \cdots\})$ and the spine $\{x_1, x_2, \cdots\}$ of G as follows (the process continues until $L_i = \emptyset$).

set $x_0 :=$ the vertex whose u.p. is as far left as possible;
set $L_0 := \{x_0\}$;
set $L_1 := \{$vertices whose l.p.s are to the left of the l.p. of $x_0\}$;
set $x_1 :=$ the vertex in L_1 with the u.p. as far right as possible;
set $L_2 := \{$vertices whose u.p.s are between the u.p.s of x_0 and $x_1\} \setminus L_1$;
set $x_2 :=$ the vertex in L_2 with the l.p. as far right as possible;
for $i = 3$ to n do
  if i is odd then
      set $L_i := \{$vertices with l.p. between the l.p.s of x_{i-3} and $x_{i-1}\} \setminus L_{i-1}$;
      set $x_i :=$ the vertex in L_i with the u.p. as far right as possible;
  else
      set $L_i := \{$vertices whose u.p.s are between the u.p.s of x_{i-3} and $x_{i-1}\} \setminus L_{i-1}$;
      set $x_i :=$ the vertex in L_i with the l.p. as far right as possible;
enddo.

It is straightforward to show that for all $i \geq 0$, if $y \in L_{i+1}$ then $x_i y \in E$. Now by forming T to include all edges from x_i to L_{i+1} for appropriate i, we can conclude: (The details will be in the journal version of the paper.)

Theorem 2. *Every permutation graph admits an additive tree 2-spanner, constructable in linear time.*

In the journal version we show that there exists a trapezoid graph which does not admit any additive tree 2-spanner, thereby disproving a conjecture from [15] that any cocomparability graph admits an additive tree 2-spanner. We show also that there are bipartite permutation graphs on $2n$ vertices for which any system of collective additive tree 1–spanners will need to have at least $\Omega(n)$ spanning trees.

2.3   DSP-graphs

It follows from a result in [15] that any DSP-graph admits one additive tree
4-spanner. In this subsection we show that any DSP-graph admits a system of
two collective additive tree 3-spanners and a system of five collective additive
tree 2-spanners.

Let $G = (V, E)$ be a DSP-graph and $P := (v = x_0, x_1, \ldots, x_l = u)$ be a dominating shortest path of G. We will build five spanning trees $\{T_1, T_2, T_3, T_4, T_5\}$ for
G, all containing the edges of P, in such a way that for any two vertices $x, y \in V$,
there will be a tree $T' \in \{T_1, T_2, T_3, T_4, T_5\}$ with $d_{T'}(x, y) \leq d_G(x, y) + 2$.

Our first three trees T_1, T_2, T_3 are very similar to the trees constructed for
AT-free graphs. The tree $T_1 = (V, E_1)$ is constructed as follows. Add to initially
empty set E_1 all edges of path P. Then, for each vertex $z \in V \setminus P$ choose an
arbitrary neighbor w_z in P and add edge zw_z to E_1. The tree T_1 is an analog
of the caterpillar-tree constructed for an AT-free graph. The second and third
trees are analogs of the cactus-tree considered for an AT-free graph. The tree
$T_2 = (V, E_2)$ is a special breadth-first-search-tree T_v with vertex v as the root,
the tree $T_3 = (V, E_3)$ is a special breadth-first-search-tree T_u with vertex u as
the root. For construction of T_2 we can use the algorithm given in Subsection
2.1 with one additional line at the end: for each $z \in N_{l+1}(v)$, add edge zu to
E_2 and set $dev(z) := 1$. T_3 is constructed similarly; we simply reverse the order
of vertices of P and consider u instead of v and E_3 instead of E_2. The detailed
algorithm will appear in the journal version.

Our tree $T_4 = (V, E_4)$ is a generalization of the tree T_2 and is constructed as
follows.

set $E_4 := \{$all edges of the path $P := (v = x_0, x_1, \ldots, x_l = u)\}$;
set $dev(x_i) := 0$ for each vertex x_i of the path P;
for $i = 1$ to l do
    for each vertex $z \in N_i(v) \setminus \{x_i\}$ do case
        case (z is adjacent to x_{i-1} in G)
            add edge zx_{i-1} to E_4 and set $dev(z) := 1$;
        case (z is adjacent to x_i in G)
            add edge zx_i to E_4 and set $dev(z) := 1$;
        case (z is adjacent to a vertex $w \in N_i(v)$ which is adjacent to x_{i-1})
            choose such a w and add edge zw to E_4 and set $dev(z) := 2$;
        otherwise /* none of above */
            among all neighbors of z in $N_{i-1}(v)$ choose a neighbor w with
                minimum deviation $dev(w)$ (break ties arbitrarily);
            add edge zw to E_4 and set $dev(z) := dev(w) + 1$;
    endcase
enddo
for each $z \in N_{l+1}(v)$, add edge zu to E_4 and set $dev(z) := 1$.

It is an easy exercise to show by induction that for any vertex $z \in N_i(v)$,
the vertex of P closest to z in T_4 is either x_s or x_{s-1} with $s = i - dev(z) + 1$.
Moreover, the length of the path of T_4 between z and P is $dev(z)$. Our last tree
$T_5 = (V, E_5)$ is a version of the tree T_4, constructed downwards.

set $E_5 := \{$all edges of the path $P := (v = x_0, x_1, \ldots, x_l = u)\}$;
set $dev(x_i) := 0$ for each vertex x_i of the path P and $dev(z) := \infty$ for any $z \in V \setminus P$;
for $i = l - 1$ down to 1 do
    for each vertex $z \in N_i(v) \setminus \{x_i\}$ do case
        case (z is adjacent to x_{i+1} in G)
            add edge zx_{i+1} to E_5 and set $dev(z) := 1$;
        case (z is adjacent to x_i in G)
            add edge zx_i to E_5 and set $dev(z) := 1$;
        case (z is adjacent to a vertex $w \in N_i(v)$ which is adjacent to x_{i+1})
            choose such a w and add edge zw to E_5 and set $dev(z) := 2$;
        otherwise
            if z has neighbors in $N_{i+1}(v)$ then
                among all neighbors of z in $N_{i+1}(v)$ choose a neighbor w with
                    minimum deviation $dev(w)$ (break ties arbitrarily);
                if $dev(w) < \infty$ then add edge zw to E_5 and set $dev(z) := dev(w) + 1$;
    endcase
enddo
for each vertex z with $dev(z)$ still ∞ do
    let $z \in N_i(v)$;
    if $i = l$ and z is adjacent to x_l then add edge zx_l to E_5 and set $dev(z) := 1$;
    else add edge zx_{i-1} to E_5;
    /* this edge exists in G since P is a dominating path
        and z is adjacent in G neither to x_{i+1} nor x_i */
enddo.

Again, it is easy to see that for any vertex $z \in N_i(v)$ with finite deviation $dev(z)$, the vertex of P closest to z in T_5 is either x_s or x_{s+1} with $s = i + dev(z) - 1$. The length of the path of T_5 between z and P is $dev(z)$.

We are ready to present the main result of this subsection.

Lemma 3. *Let G be a DSP-graph with a dominating shortest path $P := (v = x_0, x_1, \ldots, x_l = u)$ and spanning trees T_1, T_2, T_3, T_4, T_5 constructed starting from P as described above. Then, for any two vertices $x, y \in V$:*

1. *$d_{T_1}(x, y) \le d_G(x, y) + 4$;*
2. *there is a tree $T' \in \{T_1, T_2\}$ such that $d_{T'}(x, y) \le d_G(x, y) + 3$;*
3. *there is a tree $T'' \in \{T_1, T_2, T_3, T_4, T_5\}$ such that $d_{T''}(x, y) \le d_G(x, y) + 2$.*

Proof. The proof of this lemma is quite technical and will appear in the journal version of the paper. □

Theorem 3. *Any DSP-graph admits one additive tree 4-spanner, a system of two collective additive tree 3-spanners and a system of five collective additive tree 2-spanners. Moreover, given a dominating shortest path of G, all trees are constructable in linear time.*

Note that there is a DSP-graph for which two trees are necessary to get a collective additive stretch factor 3. However, it is an open question whether to achieve a collective additive stretch factor 2, one really needs five spanning trees.

2.4   Graphs with Bounded Asteroidal Number

It is known [7] that any graph G with asteroidal number $\mathsf{an}(G)$ admits an additive tree $(3\mathsf{an}(G) - 1)$-spanner. In this subsection we show that any graph with asteroidal number $\mathsf{an}(G)$ admits a system of $\mathsf{an}(G)(\mathsf{an}(G) - 1)/2$ collective additive tree 4-spanners and a system of $\mathsf{an}(G)(\mathsf{an}(G) - 1)$ collective additive tree 3-spanners.

In what follows we will use the following definitions and results from [7]. An asteroidal set A of a graph G is *repulsive* if for every vertex $v \in V \setminus N[A]$, not all vertices of A are contained in one connected component of $G - N[v]$. According to [7], any graph has a repulsive asteroidal set. A set $D \subseteq V$ in a graph G is said to be a *dominating target*, if $D \cup S$ is a dominating set in G for every set $S \subseteq V$ for which the subgraph of G induced by $D \cup S$ is connected. It is shown in [7] that any graph G has a dominating target D with $D \leq \mathsf{an}(G)$. Furthermore, every repulsive asteroidal set of G is such a dominating target of G.

In the journal version of the paper, we prove the following stronger version of the result above. Let $D \subseteq V$ be a repulsive asteroidal set of a graph $G = (V, E)$.

Lemma 4. *For every $x, y \in V$, there exist $a, b \in D$ such that $x, y \in N[P]$ for any path P of G between a and b.*

Consider two arbitrary vertices a, b of D and a shortest path $P(a, b) := (a = x_0, x_1, \ldots, x_l = b)$ connecting a and b in G. We can build two spanning trees $T_1(a, b)$ and $T_2(a, b)$ for G, both containing the edges of $P(a, b)$, in such a way that for any two vertices $x, y \in N[P(a, b)]$, $d_{T_1(a,b)}(x, y) \leq d_G(x, y) + 4$ and $\min\{d_{T_1(a,b)}(x, y), d_{T_2(a,b)}(x, y)\} \leq d_G(x, y) + 3$.

Our trees $T_1(a, b)$ and $T_2(a, b)$ are very similar to the trees constructed for AT-free graphs. The tree $T_1(a, b) = (V, E_1)$ is constructed as follows. Add to initially empty set E_1 all edges of path $P(a, b)$. Then, for each vertex $z \in N(P(a, b))$ choose an arbitrary neighbor w in $P(a, b)$ and add edge zw to E_1. The obtained subtree of G (which covers so far only vertices from $N[P(a, b)]$) extends to a spanning tree $T_1(a, b)$ arbitrarily. The tree $T_1(a, b)$ is an analog of the caterpillar-tree constructed for an AT-free graph. The second tree is an analog of the cactus-tree considered for an AT-free graph. The tree $T_2(a, b) = (V, E_2)$ is a special breadth-first-search-tree T_a with vertex a as the root. The detailed algorithm for constructing $T_2(a, b)$ and the proof of the following lemma will appear in the journal version.

Lemma 5. *For any two vertices $x, y \in N[P(a, b)]$:*

1. $d_{T_1(a,b)}(x, y) \leq d_G(x, y) + 4$;
2. *there is a tree $T' \in \{T_1(a, b), T_2(a, b)\}$ such that $d_{T'}(x, y) \leq d_G(x, y) + 3$.*

If we construct trees $T_1(a, b)$ and $T_2(a, b)$ for each pair of vertices $a, b \in D$, from Lemma 4 and Lemma 5, we get (recall that $|D| \leq \mathsf{an}(G)$):

Theorem 4. *Any graph G with asteroidal number $\mathsf{an}(G)$ admits a system of $\mathsf{an}(G)(\mathsf{an}(G) - 1)/2$ collective additive tree 4-spanners and a system of $\mathsf{an}(G)(\mathsf{an}(G) - 1)$ collective additive tree 3-spanners.*

Corollary 1. *Any graph G with asteroidal number bounded by a constant admits a system of a constant number of collective additive tree 3-spanners. Moreover, given a repulsive asteroidal set of G, all trees are constructable in total linear time.*

3 Routing Labeling Schemes in AT-free Graphs

In this section, we use the results obtained above to design compact and efficient routing labeling schemes for AT-free graphs. For DSP-graphs and graphs with bounded asteroidal number, corresponding routing labeling schemes are described in the journal version of the paper. We will show that any AT-free graph with diameter $\mathcal{D} := diam(G)$ and maximum vertex degree Δ admits a $(3\log_2 \mathcal{D} + 6\log_2 \Delta + O(1))$-bit routing labeling scheme of deviation at most 2. Moreover, the scheme is computable in linear time, and the routing decision is made in constant time per vertex.

It is worth mentioning that any AT-free graph admits a $(\log_2 \mathcal{D} + 1)$-bit distance labeling scheme of deviation at most 2 (see [5]). That is, there is a function L labeling the vertices of each AT-free graph G with (not necessarily distinct) labels of up to $\log_2 \mathcal{D}+1$ bits such that given two labels $L(v), L(u)$ of two vertices v, u of G, it is possible to compute in constant time, by merely inspecting the labels of u and v, a value $\widehat{d}(u,v)$ such that $0 \leq \widehat{d}(u,v) - d_G(u,v) \leq 2$. To the best of our knowledge, the method of [5] cannot be used (at least directly) to design a routing labeling scheme for AT-free graphs.

Labels. In subsection 2.1, we showed that any AT-free graph $G = (V, E)$ admits a system of two collective additive tree 2-spanners. During the construction of the cactus-tree T_2 for G, each vertex $z \in V$ received a deviation number $dev(z)$ which is the distance in T_2 between z and the stem $P := (x_0, x_1, \ldots, x_l)$ of T_2. To simplify the routing decision, it will be useful to construct one more spanning tree $T' = (V, E')$ for G. Let $P := (x_0, x_1, \ldots, x_l)$ be the dominating path of G described in Lemma 1.

set $E' := \{$all edges of the path $P := (x_0, x_1, \ldots, x_l)\}$;
set $dev'(x_i) := 0$ for each x_i of the path P and $dev'(z) := l + 1$ for any $z \in V \setminus P$;
for each $z \in N_l(x_0)$ which is adjacent to x_l, set $dev'(z) := 1$ and add edge zx_l to E';
for $i = l - 1$ down to 1 do
  for each vertex $z \in N_i(x_0) \setminus \{x_i\}$ do
    if z is adjacent to x_i in G then add edge zx_i to E' and set $dev'(z) := 1$;
    else if z has neighbors in $N_{i+1}(x_0)$ then
      among all neighbors of z in $N_{i+1}(x_0)$, choose a neighbor w with
        minimum deviation $dev'(w)$ (break ties arbitrarily);
      if $dev'(w) < l + 1$ then add edge zw to E' and set $dev'(z) := dev'(w) + 1$;
  enddo
enddo
for each vertex z with $dev'(z)$ still $l + 1$ do
  let $z \in N_i(x_0)$;
  add edge zx_{i-1} to E';
enddo.

We name tree T' the *willow-tree* of G. As a result of its construction, each vertex $z \in V$ received a second deviation number $dev'(z)$, which is either $l + 1$ or the distance in T' between z and the path $P := (x_0, x_1, \ldots, x_l)$ of T'.

Now we are ready to describe the routing labels of the vertices of G. For each vertex $x_i \in P$ $(i = 0, 1, \ldots, l)$, we have

$$Label(x_i) := (b(x_i), level(x_i), port_{up}(x_i), port_{down}(x_i)),$$

where

- $b(x_i) := 1$, a bit indicating that x_i belongs to P;
- $level(x_i) (= i)$ is the index of x_i in P, i.e., the distance $d_G(x_i, x_0)$;
- $port_{up}(x_i)$ is the port number at vertex x_i of the edge $x_i x_{i+1}$ (if $i = l$, $port_{up}(x_i) :=$ nil);
- $port_{down}(x_i)$ is the port number at vertex x_i of the edge $x_i x_{i-1}$ (if $i = 0$, $port_{down}(x_i) :=$ nil).

For each vertex $z \in V \setminus P$, we have

$$Label(z) := (b(z), level(z), a_v(z), port_{v-in}(z), port_{v-out}(z), a_h(z), port_{h-in}(z),$$

$$port_{h-out}(z), dev(z), port_{down}(z), dev'(z), port_{up}(z)),$$

where

- $b(z) := 0$, a bit indicating that z does not belong to P;
- $level(z)$ is the distance $d_G(z, x_0)$;
- $a_v(z)$ is a bit indicating whether z is adjacent to $x_{label(z)-1}$;
- $port_{v-in}(z)$ is the port number at vertex $x_{label(z)-1}$ of the edge $x_{label(z)-1}z$ (if z and $x_{label(z)-1}$ are not adjacent in G, then $port_{v-in}(z) :=$ nil);
- $port_{v-out}(z)$ is the port number at vertex z of the edge $zx_{label(z)-1}$ (if z and $x_{label(z)-1}$ are not adjacent in G, then $port_{v-out}(z) :=$ nil);
- $a_h(z)$ is a bit indicating whether z is adjacent to $x_{label(z)}$;
- $port_{h-in}(z)$ is the port number at vertex $x_{label(z)}$ of the edge $x_{label(z)}z$ (if z and $x_{label(z)}$ are not adjacent in G, then $port_{h-in}(z) :=$ nil);
- $port_{h-out}(z)$ is the port number at vertex z of the edge $zx_{label(z)}$ (if z and $x_{label(z)}$ are not adjacent in G, then $port_{h-out}(z) :=$ nil);
- $dev(z)$ is the deviation of z in tree T_2;
- $port_{down}(z)$ is the port number at vertex z of the edge zw, where w is the father of z in T_2;
- $dev'(z)$ is the deviation of z in tree T';
- $port_{up}(z)$ is the port number at vertex z of the edge zw, where w is the father of z in T' (if $dev'(z) = l + 1$, $port_{up}(z) :=$ nil).

Clearly, the label size of each vertex of G is at most $3\lceil \log_2 l \rceil + 6 \lceil \log_2 \Delta \rceil + 3 \leq 3 \log_2 \mathcal{D} + 6 \log_2 \Delta + O(1)$ bits.

Routing Decision. The routing decision algorithm is obvious. Suppose that a packet with the header (address of destination) $Label(y)$ arrives at vertex x.

The vertex x can use the following constant time algorithm to decide where to submit the packet. Note that each vertex v of G is uniquely identified by its label $Label(v)$.

function routing_decision_AT-free($Label(x), Label(y)$)

if $Label(x) = Label(y)$ then return "packet reached its destination";
else do case
  case $(b(x) = 1)$
  /* x belongs to P and routing is performed on the caterpillar-tree T_1 of G */
    do case
      case $(level(x) > level(y))$
        send packet via $port_{down}(x)$;
      case $(level(x) < level(y))$
        if $b(y) = 1$ then send packet via $port_{up}(x)$;
        else if $level(y) = level(x) + 1$ and $a_v(y) = 1$ then
            send packet via $port_{v-in}(y)$;
          else send packet via $port_{up}(x)$;
      case $(level(x) = level(y))$
        if $a_h(y) = 1$ then send packet via $port_{h-in}(y)$;
        else send packet via $port_{down}(x)$;
    endcase;
  /* now x does not belong to P */
  case $(level(x) > level(y))$
    do case
      case $(a_v(x) = 1)$
        send packet via $port_{v-out}(x)$; /* routing is performed on T_1 */
      case $(b(y) = 1$ or $b(y) = 0$ and $a_h(y) = 1)$
        send packet via $port_{h-out}(x)$; /* routing is performed on T_1 */
      otherwise /* here we have $d_{T_1}(x, y) = level(x) - level(y) + 3$ */
        if $dev(x) \leq level(x) - level(y) + 1$ then send packet via $port_{down}(x)$;
            /* the cactus-tree T_2 of G is used for routing */
        else send packet via $port_{h-out}(x)$; /* routing is performed on T_1 */
    endcase;
  case $(level(x) < level(y))$
    do case
      case $(a_h(x) = 1)$
        send packet via $port_{h-out}(x)$; /* routing is performed on T_1 */
      case $(b(y) = 1$ or $b(y) = 0$ and $a_v(y) = 1)$
        send packet via $port_{v-out}(x)$; /* routing is performed on T_1 */
      otherwise /* here we have $d_{T_1}(x, y) = level(y) - level(x) + 3$ */
        if $dev'(x) \leq level(y) - level(x) + 1$ then send packet via $port_{up}(x)$;
            /* the willow-tree T' of G is used for routing */
        else send packet via $port_{v-out}(x)$; /* routing is performed on T_1 */
    endcase;
  case $(level(x) = level(y))$ /* routing is performed on T_1 */
    if $a_h(x) = 1$ then send packet via $port_{h-out}(x)$;
    else send packet via $port_{v-out}(x)$;
endcase.

Thus, we have the following result.

Theorem 5. *Every AT-free graph of diameter $\mathcal{D} := diam(G)$ and of maximum vertex degree Δ admits a $(3\log_2 \mathcal{D} + 6\log_2 \Delta + O(1))$-bit routing labeling scheme of deviation at most 2. Moreover, the scheme is computable in linear time, and the routing decision is made in constant time per vertex.*

Acknowledgements

DGC wishes to thank the Natural Sciences and Engineering Research Council of Canada for financial assistance and the Department of Computer Science at Kent State University for hosting his visit there.

References

1. L.P. CHEW, There are planar graphs almost as good as the complete graph, *J. of Computer and System Sciences,* 39 (1989), 205–219.
2. D.G. CORNEIL, S. OLARIU and L. STEWART, Asteroidal Triple–free Graphs, *SIAM J. Discrete Math.,* 10 (1997), 399–430.
3. D.G. CORNEIL, S. OLARIU and L. STEWART, Linear time algorithms for dominating pairs in asteroidal triple–free graphs, *SIAM J. on Computing,* 28 (1999), 1284–1297.
4. F.F. DRAGAN, C. YAN and I. LOMONOSOV, Collective tree spanners of graphs, *Algorithm Theory - SWAT 2004, 9th Scandinavian Workshop on Algorithm Theory,* Humlebaek, Denmark, July 8-10, 2004, *Lecture Notes in Computer Science,* Vol. 3111, pp. 64 - 76.
5. C. GAVOILLE, M. KATZ, N.A. KATZ, C. PAUL, and D. PELEG, Approximate Distance Labeling Schemes, in *Proceedings of the 9th Annual European Symposium on Algorithms" (ESA'01),* Aarhus, Denmark, August 28-31, 2001, *Lecture Notes in Computer Science 2161,* Springer, 2001, pp. 476-487.
6. T. KLOKS, D. KRATSCH and H. MÜLLER, Approximating the bandwidth for asteroidal triple-free graphs, *J. Algorithms,* 32 (1999), 41–57.
7. T. KLOKS, D. KRATSCH and H. MÜLLER, On the structure of graphs with bounded asteroidal number, *Graphs and Combinatorics,* 17 (2001), 295–306.
8. A.L. LIESTMAN AND T. SHERMER, Additive graph spanners, *Networks,* 23 (1993), 343–364.
9. T.H. MA and J.P. SPINRAD, On the two-chain subgraph cover and related problems, *J. of Algorithms,* 17 (1994), 251–268.
10. M.S. MADANLAL, G. VENKATESAN, and C. PANDU RANGAN, Tree 3-spanners on interval, permutation and regular bipartite graphs, *Inform. Process. Lett.,* 59 (1996), 97–102.
11. R.M. McCONNELL and J.P. SPINRAD, Linear-time transitive orientation, In *Proceedings of the Eighth Annual ACM-SIAM Symposium on Discrete Algorithms,* New Orleans, Louisiana, 5-7 January 1997, pp. 19-25.
12. D. PELEG, Distributed Computing: A Locality-Sensitive Approach, *SIAM Monographs on Discrete Math. Appl.,* SIAM, Philadelphia, 2000.
13. D. PELEG, and A.A. SCHÄFFER, Graph Spanners, *J. Graph Theory,* 13 (1989), 99-116.
14. D. PELEG AND J.D. ULLMAN, An optimal synchronizer for the hypercube, *in Proc. 6th ACM Symposium on Principles of Distributed Computing,* Vancouver, 1987, 77–85.
15. E. PRISNER, D. KRATSCH, H.-O. LE, H. MÜLLER, and D. WAGNER, Additive tree spanners, *SIAM Journal on Discrete Mathematics,* 17 (2003), 332–340.

On the Maximum Cardinality Search
Lower Bound for Treewidth[*]

Hans L. Bodlaender[1] and Arie M.C.A. Koster[2]

[1] Institute of Information and Computing Sciences, Utrecht University,
P.O. Box 80.089, 3508 TB Utrecht, the Netherlands
`hansb@cs.uu.nl`
[2] Konrad-Zuse-Zentrum für Informationstechnik Berlin,
Takustraße 7,
D-14194 Berlin, Germany
`koster@zib.de`

Abstract. The Maximum Cardinality Search algorithm visits the vertices of a graph in an order such that at any point, a vertex is visited that has the largest number of visited neighbours. An MCS-ordering of a graph is an ordering of the vertices that can be generated by the Maximum Cardinality Search algorithm. The visited degree of a vertex v in an MCS-ordering is the number of neighbours of v that are before v in the ordering. The MCSLB of an MCS-ordering ψ of G is the maximum visited degree over all vertices v in ψ. Lucena [10] showed that the treewidth of a graph G is at least the MCSLB of any MCS-ordering of G.
In this paper, we analyse the maximum MCSLB over all possible MCS-orderings of given graphs G. We give upper and lower bounds for this number for planar graphs. Given a graph G, it is NP-complete to determine if G has an MCS-ordering with MCSLB at least k, for any fixed $k \geq 7$. Also, this problem does not have a polynomial time approximation algorithm with constant ratio, unless P=NP. Variants of the problem are also shown to be NP-complete.
We also propose and experimentally analysed some heuristics for the problem. Several tiebreakers for the MCS algorithm are proposed and evaluated. We also give heuristics that give upper bounds on the maximum MCSLB that an MCS-ordering can obtain which appear to give results close to optimal on several graphs from real life applications.

1 Introduction

Recent research has shown that the notion of treewidth is not only of theoretical interest, but can also be used to solve problems arising from real life applications in practice (see e.g., [8,9].) One important issue when using treewidth in implementations is the problem to find tree decompositions of given graphs of optimal

[*] This research has been partially supported by the DFG research group "Algorithms, Structure, Randomness" (Grant number GR 883/9-3, GR 883/9-4, and partially by the Netherlands Organisation for Scientific Research NWO (project *Treewidth and Combinatorial Optimisation*).

J. Hromkovič, M. Nagl, and B. Westfechtel (Eds.): WG 2004, LNCS 3353, pp. 81–92, 2004.

or close to optimal width. Many of the theoretical solutions to this problem seem not to be applicable in practice, e.g., some have very large constant factors hidden in the O-notation (like the algorithm from [1], see [12].) Thus, there is a need for practical algorithms for determining the treewidth and finding tree decompositions. Recent investigations brought us preprocessing methods (e.g., [2, 3, 15]), heuristics that often give close to optimal results (e.g., [5–7]), lower bound methods [4, 5, 10, 11], and some exact methods. Still, in many cases, exact methods are too slow, and there are large gaps between the bounds given by upper bound and lower bound heuristics. This paper concentrates on the study of a lower bound on the treewidth that is due to Lucena [10]. We analyse this bound both theoretically and experimentally.

The lower bound on treewidth of Lucena [10] is based on the Maximum Cardinality Search (or, in short: MCS) algorithm. This algorithm that visits all vertices of a given graph in order was first proposed in 1984 by Tarjan and Yannakakis for the recognition of chordal graphs [14]. The order in which the Maximum Cardinality Search algorithm visits the vertices of a graph must fulfil the following property: at each step, a vertex must be visited that has the largest *current visited degree*, where we define the current visited degree of an unvisited vertex at a certain step as the number of its visited neighbours. We call any ordering of the vertices of the graph $G = (V, E)$ that fulfils this property an *MCS-ordering* of G. The *visited degree* of a vertex v (with respect to a given MCS-ordering) is its current visited degree at the moment it is visited, i.e., the number of neighbours that are before v in the MCS-ordering. The MCSLB of an MCS-ordering is the maximum visited degree of the vertices. Lucena [10] showed that for every graph G and MCS-ordering ψ of G, the MCSLB of ψ is at most the treewidth of G.

Thus, Maximum Cardinality Search provides us with a lower bound heuristic for the treewidth of a given graph. A graph with more than one vertex has more than one MCS-ordering: we can start at any vertex, and often, the MCS has the choice between more than one unvisited vertex with maximum current visited degree.

In this paper, we give a theoretical, and an experimental evaluation of this lower bound for treewidth. Amongst others, we compare MCSLB with other treewidth lower bounds, look for the complexity of determining what is the best MCSLB that can be obtained for a given graph amongst all possible MCS-orderings, give heuristics that provide upper bounds on the MCSLB, and report on experiments with different tiebreakers for constructing MCS-orderings.

It is interesting to note that Maximum Cardinality Search also has been used as a heuristic for obtaining upper bounds on the treewidth; an experimental evaluation has been reported in [7].

1.1   Definitions

Some definitions are given above. Many other notions in this paper follow standard conventions from graph theory and graph algorithms.

A *tree decomposition* of a graph $G = (V, E)$ is a pair $(\{X_i \mid i \in I\}, T = (I, F))$, with $\{X_i \mid i \in I\}$ a family of subsets of V and T a tree, such that $\bigcup_{i \in I} X_i = V$, for all $\{v, w\} \in E$, there is an $i \in I$ with $v, w \in X_i$, and for all $i_0, i_1, i_2 \in I$: if i_1 is on the path from i_0 to i_2 in T, then $X_{i_0} \cap X_{i_2} \subseteq X_{i_1}$. The *width* of tree decomposition $(\{X_i \mid i \in I\}, T = (I, F))$ is $\max_{i \in I} |X_i| - 1$. The treewidth of a graph G is the minimum width among all tree decompositions of G.

Define $mcslb_{\max}(G, v)$ as the maximum over all MCS-orderings ψ of G of the visited degree of v in ψ. Define $mcslb'_{\max}(G, v)$ as the maximum over all MCS-orderings ψ of G where v has at least one neighbour that is visited after v of the visited degree of v in ψ. Denote the degree of v in G by $d_G(v)$. Clearly, we have $mcslb'_{\max}(G, v) = \min\{mcslb_{\max}(G, v), d_G(v) - 1\}$. Define $mcslb_{\max}(G, v, w)$, with w a neighbour of v in G as the maximum over all MCS-orderings ψ of G where w is visited after v of the visited degree of v in ψ.

2   MCSLB for General Graphs

The following lemma is of great help for getting upper bounds on the MCSLB and the maximum visited degree of a vertex amongst all possible MCS-orderings.

Lemma 1. *Let ψ be an MCS-ordering of G and suppose v has visited degree k in ψ. Then v has distinct neighbours $w_1, \ldots, w_k$, such that the visited degree of w_i is at least $i - 1$, and each w_i is visited before v.*

Proof. Let w_i be the ith visited neighbour of v. Each w_i, $i \leq k$ is visited before v. Just before w_i is visited, v has current visited degree exactly $i - 1$. As the MCS visits w_i instead of v at that point, w_i has visited degree at least $i - 1$. $\square$

A direct consequence of this lemma is that the visited degree of a vertex cannot be larger than its degree, or than the maximum degree of its neighbours.

The *degeneracy* of a graph is the maximum over all induced subgraphs of the minimum degree of a vertex [13]. As the degeneracy of a graph can be computed very quickly, and the treewidth of a graph is never smaller than its degeneracy, this notion provides us with a successful lower bound for treewidth. (An improvement on the lower bound is the *contraction degeneracy*, proposed and studied in [4].) While computing an MCS-ordering with corresponding MCSLB is somewhat slower than computing the degeneracy, the MCSLB gives a lower bound that is at least as good as the degeneracy.

Lemma 2. *Let ψ be an MCS-ordering of G. The MCSLB of ψ is at least the degeneracy of G.*

Proof. Let the degeneracy of G be δ. Let H be a subgraph of G such that every vertex in H has degree at least δ. Let v be the last vertex in H that is visited by ψ. v has visited degree at least δ, as each neighbour of v that belongs to H is visited before v. $\square$

It is interesting to note that the maximum and minimum MCSLB over all possible MCS-orderings of a graph are not closed under taking subgraphs, induced subgraphs, or minors. I.e., there is a graph H that is a subgraph (induced subgraph, minor) of a graph G, such that H has an MCS-ordering with MCSLB larger than the maximum MCSLB over all MCS-orderings of G. The example with minors is easy to give: take a clique with $k \geq 4$ vertices (each MCS-ordering has MCSLB $k - 1$), and then subdivide each edge (each MCS-ordering now has MCSLB two). The examples for subgraphs and induced subgraphs are more involved and can e.g., be created by adding additional vertices and edges to the graph given in the next section with Theorem 1. Thus, it is also interesting to study the maximum MCSLB over all orderings over all minors (or subgraphs, induced subgraphs) [4].

3   MCSLB for Planar Graphs

In this section, we show that there are planar graphs with maximum MCSLB $\Omega(\log n / \log \log n)$ and that the maximum MCSLB is bounded by $O(\log n)$ for planar graphs. As planar graphs can have treewidth $\Theta(\sqrt{n})$ (e.g., an r by r grid has treewidth exactly r), this shows that the MCSLB can be far from the treewidth on planar graphs. On the other hand, the MCSLB can be much larger than the degeneracy, as the degeneracy of a planar graph is never larger than 5. (Every planar graph has a vertex of degree at most five.)

We first show that the starting vertex of an MCS-ordering can have a dramatic impact on the resulting MCSLB, and provide an example that shows that planar graphs can have arbitrary large MCSLB.

Theorem 1. *For every k, there is a planar graph G_k, such that*

1. *The treewidth of G_k is k.*
2. *There is an MCS-ordering ψ with maximum visited degree k.*
3. *There is a vertex v in G_k such that every MCS-ordering that starts in v has maximum visited degree 2.*
4. *G_k has $O((k - 1)!)$ vertices.*

Proof. (Sketch.) An example of the construction is shown in Figure 1. Basically, we have a tree with $k + 1$ levels, with all vertices on the same level of the tree connected by a path. The root node has two children, and vertices at distance $k - i \geq 1$ from the root have i children.

We can visit the vertices with MCS level by level, starting at the bottom level. Vertices in the ith level from below have $i - 1$ visited neighbours in the

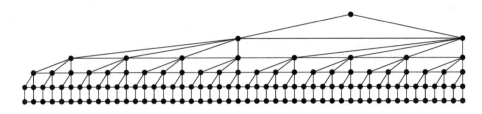

Fig. 1. The construction for Theorem 1 for $k = 5$.

level below, and possibly one visited neighbour in the same level. If we have $k+1$ levels, the second vertex in the highest-but-one level has visited degree k.

On the other hand, if we start an MCS with the top vertex, then we also must visit the vertices level by level, but now from top to bottom. It is not hard to see that in this case, each vertex receives visited degree at most two.

It is not hard to construct a tree decomposition of G_k with treewidth k.   □

We now show an upper bound of $5 \log n + 4$ on the MCSLB possible on planar graphs. (A more detailed argument can give a bound of $4 \log n + O(1)$.) We first introduce the notion of special tree. A *special tree* is a rooted tree with each vertex labelled with a nonnegative integer, such that a vertex labelled with i has at least $\min\{0, i - 4\}$ children, and we can order its children $w_1, \ldots, w_k$ ($k \geq \min\{0, i-4\}$) such that the label of w_j is at least j. Let $n(i)$ be the minimum number of vertices in a special tree whose root is labelled i. The following lemma can be easily shown with induction.

Lemma 3. *For all i, $n(i) \geq 2^{\lfloor i/5 \rfloor}$.*

Theorem 2. *If G is a planar graph with n vertices, and π an MCS-ordering of G with MCSLB k, then $k \leq 5 \log n + 4$.*

Proof. Let G, n, π, k be as stated above. Let v be the first vertex visited by π with visited degree k. Suppose G does not contain a proper subgraph that has an MCS-ordering with MCSLB k. Hence, G is connected.

For i, $1 \leq i \leq n$, let v_i be the ith vertex visited by π. Denote $G_{>i}$ as the subgraph, induced by the vertices $\{v_{i+1}, \ldots, v_n\}$, and $G_{\leq i}$ as the subgraph, induced by vertices $\{v_1, \ldots, v_i\}$.

For each i, we may assume that $G_{>i}$ is connected. If not, take any connected component of $G_{>i}$ that does not contain v. Let G' be the graph obtained by removing that component from G. Restricting π to the vertices in G' still gives an MCS-ordering where v has visited degree k, contracting the minimality of G. $G_{\leq i}$ is also connected: if G is connected, then at any point in the MCS, the set of visited vertices is connected.

Now, for each i, $1 \leq i \leq n$, v_i has at most three neighbours in $G_{\leq i}$ that are adjacent to vertices in $G_{>i}$. Suppose v_i has four such neighbours. These belong together with v_i to one face of $G_{\leq i}$. Number these, with respect to the order in which they come after v_i on this face as w_1, w_2, w_3, and w_4. W.l.o.g., suppose w_2 is visited before w_3. $\{v_i, w_2\}$ and $\{v_i, w_3\}$ are separators of the graph $G_{\leq i}$. After w_3 is visited, the component of $G_{\leq i} - \{v_i, w_3\}$ that contains w_4 has only unvisited vertices, and before v_i and the first vertex in this component is visited, these vertices can only have w_3 as visited neighbour, so cannot be visited before v_i, contradiction.

Now, mark an edge $\{v_j, v_i\}$, $j < i$ to be *special* when v_j is not adjacent to a vertex in $G_{>i}$. Directing special edges towards the higher numbered vertex gives a forest. Taking from this forest the subtree with root v and labelling each vertex with its visited degree gives a special tree: by Lemma 1, a vertex w with visited degree α has earlier visited neighbours with visited degree at least $0, 1, 2, \ldots,$

$\alpha - 1$, and at most three of these do not have a special edge to w. So, G contains a subtree with at least $n(k) \geq 2^{\lfloor k/5 \rfloor}$ vertices. The theorem now follows. $\square$

4   Complexity of MCSLB

In this section, we give some results on the complexity of the problem of determining the maximum MCSLB that can be obtained on a given graph. Unfortunately, the proof of the main result is much too long to give here.

MAX MCSLB
Instance: Graph $G = (V, E)$, integer $k \leq |V|$.
Question: Is there an MCS-ordering for G with MCSLB at least k?

Lemma 4. *Let* $k \geq 7$ *be a constant. Let* C *be a collection of clauses over a set* U *of Boolean variables, each clause of size three. There is a graph* $G_{C,k}$ *with* $O(|C| + k!)$ *vertices, such that if* C *is satisfiable, then* $G_{C,k}$ *has an MCS-ordering with MCSLB at least* k, *and if* C *is not satisfiable, then every MCS-ordering of* $G_{C,k}$ *has MCSLB at most 6.*

The complicated construction puts together several parts. Around a modification of the planar graph of Theorem 1, structures are build that guarantee that this subgraph can be visited from bottom to top (and not in the other direction), if and only if the set of clauses is satisfiable. The 18 pages long proof is omitted here. From the construction, we then can obtain the following result.

Theorem 3. *(i)* MAX MCSLB *is NP-complete, even for fixed* $k \geq 7$.
(ii) If $P \neq NP$, *then every polynomial time approximation algorithm for* MAX MCSLB *has approximation ratio* $\Omega(\log n / \log \log n)$.

With similar techniques, the problem to determine whether there is an MCS-ordering with MCSLB at most k can be shown to be NP-complete, for fixed constants $k \geq 7$. Variants where the starting vertex is fixed are also NP-complete, when $k \geq 6$; the proofs are slightly easier. A similar bound for the approximation ratio holds also in these cases. A few cases can be seen to be easy: the case $k = 2$, and the case that G is chordal.

Proposition 1. *(i)* G *is a forest, if and only if every MCS-ordering of* G *has MCSLB 1, if and only if there exists an MCS-ordering of* G *with MCSLB 1.*
(ii) If G *is a chordal graph, then every MCS-ordering of* G *has MCSLB equal to the maximum clique size in* G *minus 1.*

5   Upper Bounds on MCSLB

In this section we propose some heuristics for obtaining upper bounds on the maximum MCSLB over all possible MCS-orderings of a given graph G.

Our heuristics maintain for each vertex upper bounds on $mcslb_{\max}(G, v)$, $mcslb'_{\max}(G, v)$, and $mcslb_{\max}(G, v, w)$, respectively, and improve these upper bounds stepwise, until no improvements can be found.

The first heuristic takes for each vertex v a variable $u(v)$, and maintains as an invariant that for each $v \in V$: $mcslb_{\max}(G, v) \leq u(v)$. A local improvement step for a vertex v has the following form:

PROCEDURE ImproveMCSLBMAXv (**Graph** G, **Vertex** v)
    Compute $UN(v) = \{u(w) \mid \{v, w\} \in E\}$, and sort $UN(v)$.
    Suppose $UN(v) = \{u_1, u_2, \ldots, u_d\}$, with $u_1 \leq u_2 \leq \cdots \leq u_d$.
    count $= 0$;
    for $j = 1$ **to** $u(v)$
    do
        if $(u_{d-u(v)+j} \geq$ count$)$
        then count $++$.
    if (count $< u(v)$)
    then $u(v) =$ count; return **true**
    else return **false**

So, we start by sorting the values $u(w)$ for all neighbours of v. Next, we select as many vertices as possible that satisfy the condition stated in Lemma 1. This number then defines the new $u(v)$ and we return true on improvement.

We start by setting $u(v)$ to $d_G(v)$ for all vertices. Then, we repeatedly run ImproveMCSLBMAXv on the different vertices v until it gives no improvement on any of the vertices in the graph. We implemented this by using a set S which initially contains all vertices. Repeatedly, a vertex v is taken out of S, ImproveMCSLBMAXv is run on v, and when this causes a decrease of $u(v)$, all neighbours of v are added (again) to S when they are not already in S. From Lemma 1, we can conclude:

Lemma 5. *The procedure ImproveMCSLBMAXv maintains as invariant that for all vertices $x \in V$: $mcslb_{\max}(G, x) \leq u(x)$.*

The bound can be somewhat improved by working with upper bounds on $mcslb'_{\max}(G, x)$ instead. In the second heuristic, each vertex $v \in V$ has a variable $u'(v)$ and as invariant, we maintain that for all $v \in V$: $mcslb'_{\max}(G, v) \leq u'(v)$. Initially, we set $u'(v) = d_G(v) - 1$. An improvement step and schedule similar to ImproveMCSLBMAXv is used, but now we work with values $u'(v)$. Finally, with a step very similar to ImproveMCSLBMAXv (let the for-loop run from 1 to $u'(v) + 1$), we can compute an upper bound on $mcslb_{\max}(G, v)$ for each v.

Our experiments show that this improvement sometimes gives better results. A small example that also shows this is when we consider again the graph, obtained by subdividing each edge of a K_4. Here, we get values $u(v) = 2$ for subdivision vertices and $u(v) = 3$ for clique vertices, giving an upper bound of 3, while the second heuristic gives $u'(v) = 1$ for subdivision vertices and $u'(v) = 2$ for clique vertices, yielding an upper bound of 2 on the maximum value of MCSLB.

The third heuristic gives a further refinement by looking at which neighbour of v is visited after v. We maintain upper bounds $u(v, w)$ on $mcslb_{\max}(G, v, w)$, and refine these again stepwise, until no improvements are possible, with the following procedure.

PROCEDURE ImproveMCSLBMAXe (**Graph** G, **Vertex** v, **Vertex** w)
  Compute $UN(v, w) = \{u(x, v) \mid \{v, x\} \in E, x \neq w\}$, and sort $UN(v, w)$.
  Suppose $UN(v, w) = \{u_1, u_2, \ldots, u_d\}$, with $u_1 \leq u_2 \leq \cdots \leq u_d$.
  count $= 0$;
  for $j = 1$ **to** $u(v, w)$
  do
      if $(u_{d-u(v,w)+j} \geq \text{count})$
      then count $++$.
    if (count $< u(v, w)$)
    then $u(v, w) = \text{count}$; return **true**
    else return **false**

Lemma 6. *The procedure ImproveMCSLBMAXe maintains as invariant that for all pairs of adjacent vertices vertices* $x, y \in V$: $mcslb_{\max}(G, x, y) \leq u(x, y)$.

When ImproveMCSLBMAXe cannot decrease a value $u(v, w)$ for any pair of adjacent vertices v, w, then as a final step, upper bounds $u(v)$ on $mcslb_{\max}(G, v)$ are computed for all $v \in V$; the code of this final step is as in ImproveMCSLB-MAXv, except that we take as $UN(v)$ the set $\{u(x, v) \mid \{v, w\} \in E\}$, taking for the old value of $u(v)$ the degree of v. Our experiments show that this third heuristic gives some additional improvements on the upper bounds on the maximum value of MCSLB.

6   Computational Results

In this section, we perform an experimental evaluation of the MCSLB. For this purpose, we selected some graphs from frequency assignment [8] and probabilistic networks [9]. These instances have been preprocessed by the methods described in [3] and have been used for other experiments as well [2]. All algorithms have been coded in C++ and the computations have been carried out on a Linux operated PC with Intel Pentium 4 processor with 3.0 GHz CPU. All reported CPU times are in seconds.

Our experiments are divided in two parts. First, we examine the value of MCSLB obtained by different start vertices and tiebreaking rules. Second, we report on upper bounds on MCSLB (cf. Section 5).

6.1   Start Vertices and Tiebreakers

Each MCS-ordering ψ provides a lower bound for treewidth. The start vertex of an MCS-ordering influences the final ordering directly. Computational experiments however have shown that the outcome varies only marginally depending on the start vertex. Typically, an overwhelming majority of the start vertices results in the same MCSLB, with a few exceptions to lower and/or higher values.

During the ordering process, multiple vertices can have the highest visited degree, e.g., after the start vertex is fixed all neighbours have the same visited degree and can be ordered next. To select the next vertex various tiebreakers can be applied.

Table 1. MMD and MCSLB for test graphs with different tiebreakers.

instance	$\|V\|$	$\|E\|$	MMD		default		max-degree		min-degree		MCS-UB	
			LB	CPU	LB	CPU	LB	CPU	LB	CPU	UB	CPU
barley-pp	26	78	5	0.00	6	0.00	5	0.01	6	0.00	7	0.01
diabetes-pp	116	276	4	0.00	4	0.07	4	0.12	4	0.12	5	0.11
link-pp	308	1158	6	0.01	6	0.60	6	1.27	6	1.20	27	9.76
munin1-pp	66	188	4	0.00	5	0.02	5	0.04	5	0.05	17	0.31
munin2-pp	167	455	4	0.00	5	0.15	5	0.30	5	0.29	8	0.47
munin3-pp	96	313	4	0.00	5	0.05	4	0.10	5	0.10	17	1.42
munin4-pp	217	646	5	0.00	5	0.26	5	0.61	5	0.60	12	1.10
munin-kgo-pp	16	41	4	0.00	**5**	0.00	4	0.00	**5**	0.00	**5**	0.00
oesoca+-pp	14	75	9	0.00	10	0.00	9	0.00	10	0.00	11	0.00
oow-trad-pp	23	54	4	0.00	4	0.00	4	0.00	4	0.01	6	0.01
oow-solo-pp	27	63	4	0.00	5	0.00	4	0.01	4	0.00	6	0.01
pathfinder-pp	12	43	5	0.00	**6**	0.00	5	0.00	**6**	0.00	**6**	0.00
pignet2-pp	1024	3774	5	0.01	6	7.59	6	17.42	6	18.02	239	2600.27
pigs-pp	48	137	4	0.00	5	0.01	4	0.02	5	0.02	13	0.23
ship-ship-pp	30	77	4	0.00	4	0.00	4	0.01	4	0.01	8	0.05
water-pp	22	96	6	0.00	8	0.00	7	0.00	8	0.01	10	0.02
celar01-pp	157	804	8	0.00	10	0.16	9	0.39	10	0.39	18	1.21
celar02-pp	19	115	9	0.00	**10**	0.00	9	0.01	**10**	0.00	**10**	0.01
celar03-pp	81	413	9	0.00	10	0.04	10	0.10	10	0.10	16	0.36
celar04-pp	114	524	9	0.00	11	0.08	10	0.18	10	0.17	18	0.98
celar05-pp	80	426	9	0.00	9	0.04	9	0.10	9	0.09	17	0.50
celar06-pp	16	101	**11**	0.00	**11**	0.00	**11**	0.01	**11**	0.00	**11**	0.00
celar07-pp	92	521	11	0.00	12	0.11	11	0.15	12	0.14	18	0.57
celar08-pp	189	1016	11	0.01	12	0.25	11	0.67	12	0.61	19	1.74
celar09-pp	133	646	11	0.01	12	0.12	11	0.28	12	0.26	18	1.20
celar10-pp	133	646	11	0.00	12	0.12	11	0.31	12	0.36	18	1.23
celar11-pp	96	470	9	0.00	10	0.06	9	0.13	10	0.14	17	0.65

In Table 1 we compare three different tiebreakers for selecting the next vertex among all vertices of highest visited degree. For each tiebreaker we report the maximum MCSLB taken over all possible start vertices. The columns 'default' present the results without a tiebreaker. The 'max-degree' tiebreaker selects the vertex with maximum degree among the vertices with highest visited degree, whereas the 'min-degree' tiebreaker selects the vertex with minimum degree. The idea behind the maximum degree strategy is to push the visited degree for as much vertices as possible. On the other hand, the minimum degree strategy tries to keep a vertex of high degree as long as possible unvisited such that more and more neighbours are visited before it, and thus, its visited degree increases.

The figures in Table 1 show that typically the 'default' and 'min-degree' tiebreakers perform best with respectively 27 and 25 times the best value (out of 27 instances). The 'max-degree' tiebreaker obtains only 10 times this value.

For comparison, the degeneracy (or MMD lower bound) is also included in the table as well as the upper bound computed by the MCS heuristic [7]. As proved in Lemma 2, the MCSLB is always at least as good as the MMD. The experiments show that in more than half the cases the best MCSLB is typically one better than the MMD. In four cases the MCSLB equals the MCS upper bound (**bold** values) and thus the reported value is the treewidth of those graphs. In other cases the gap between lower and upper bound is still large, in particular for instance pignet2-pp. The computation times are larger than those for MMD, but still very small.

6.2   Upper Bounds on MCSLB

In Section 5 we have reported on three ways to compute an upper bound on the maximum MCSLB value. All three methods as well as the maximum degree, the actual best value achieved (LB; cf. Table 1) and the MCS upper bound (UB) are reported in Table 2. The maximum degree of each graph is reported since the algorithm to compute $u(v)$ is initialised with the degree $d_G(v)$. Table 2 shows that in several cases the final maximum of $u(v)$ over all vertices is significantly smaller than the maximum degree. Only in cases where the maximum degree is

Table 2. Upper bounds on the MCSLB for test graphs.

instance	$|V|$	$|E|$	$\Delta(G)$	$mcslb_{\max}(v)$ value	CPU	$mcslb'_{\max}(v)$ value	CPU	$mcslb_{\max}(e)$ value	CPU	best LB	UB
barley-pp	26	78	11	10	0.00	9	0.00	8	0.00	6	7
diabetes-pp	116	276	48	7	0.00	6	0.00	6	0.01	4	5
link-pp	308	1158	30	10	0.00	10	0.00	10	0.03	6	27
munin1-pp	66	188	17	10	0.00	9	0.00	8	0.00	5	17
munin2-pp	167	455	25	10	0.00	9	0.01	8	0.01	5	8
munin3-pp	96	313	46	11	0.00	10	0.01	9	0.01	5	17
munin4-pp	217	646	62	12	0.00	11	0.00	10	0.02	5	12
munin-kgo-pp	16	41	7	7	0.00	6	0.00	6	0.00	5	5
oesoca+-pp	14	75	13	13	0.00	13	0.00	13	0.00	10	11
oow-trad-pp	23	54	6	6	0.00	6	0.00	6	0.00	4	6
oow-solo-pp	27	63	6	6	0.00	6	0.00	6	0.00	5	6
pathfinder-pp	12	43	11	9	0.00	8	0.00	8	0.00	6	6
pignet2-pp	1024	3774	172	26	0.01	25	0.02	23	1.82	6	239
pigs-pp	48	137	28	12	0.00	11	0.00	10	0.00	5	13
ship-ship-pp	30	77	10	8	0.00	8	0.00	8	0.00	4	8
water-pp	22	96	13	13	0.00	12	0.00	12	0.00	8	10
celar01-pp	157	804	35	22	0.00	21	0.00	21	0.02	10	18
celar02-pp	19	115	16	16	0.00	16	0.00	15	0.00	10	10
celar03-pp	81	413	31	23	0.00	22	0.00	22	0.01	10	16
celar04-pp	114	524	34	23	0.00	22	0.01	21	0.01	11	18
celar05-pp	80	426	31	22	0.00	21	0.01	21	0.01	9	17
celar06-pp	16	101	15	15	0.00	15	0.00	15	0.00	11	11
celar07-pp	92	521	36	26	0.00	25	0.01	24	0.01	12	18
celar08-pp	189	1016	37	25	0.00	24	0.01	24	0.02	12	19
celar09-pp	133	646	37	25	0.00	24	0.00	24	0.02	12	18
celar10-pp	133	646	37	25	0.00	24	0.00	24	0.02	12	18
celar11-pp	96	470	32	22	0.01	21	0.00	20	0.01	10	17

close to the treewidth, only minor improvement could be achieved. The successive improvement steps lower the bounds with at most one per step.

Compared to the actually computed MCSLB, there is either space for increasing the MCSLB or the upper bounds are not tight. For some instances, the latter is supported by the upper bounds for treewidth computed by MCS. For six probabilistic networks and all frequency assignment graphs this bound is better than the specifically for MCSLB computed upper bounds. Computation of the MCS upper bound however is more time consuming.

Regardless whether or not these upper bounds for MCSLB are tight, the results show that they have limited explanatory power. For those probabilistic networks where the gap between lower and upper bound is large, it cannot be closed by computing the best MCSLB over all MCS-orderings. For the frequency assignment graphs this could be the case but the values are in fact useless since they are larger than the treewidth upper bound.

7   Conclusions

In this paper, we analysed the lower bound on the treewidth, introduced by Lucena in [10], based on Maximum Cardinality Search. While computing the MCS-ordering with a maximum MCSLB is NP-hard, we see that in practice, an arbitrary MCS-ordering gives reasonable results. A method to obtain upper bounds on the maximum MCSLB shows that in several cases, an arbitrary MCS-ordering gives an MCSLB that is not far from that of the best MCS-ordering.

Comparing the MCSLB lower bound with other lower bounds for treewidth, we see that it gives bounds that are at least as good as the degeneracy (termed MMD in some papers), while it still can be computed very fast. Combining the method with contracting edges can give a further improvement to the bounds [4]. Still, on many graphs, there are large differences between the lower bounds that can be obtained in this way and the actual treewidth: for instance, on planar graphs, the treewidth can be $\Omega(\sqrt{n})$ while an MCS-ordering has MCSLB bounded by $O(\log n)$. So, the search for further lower bound heuristics for treewidth remains important and interesting.

Several interesting theoretical questions are left open in this paper. We mention a few. What is the complexity of MAX MCSLB when k is 3, 4, 5, or 6? (We conjecture NP-completeness when $k = 4$, $k = 5$, and $k = 6$, and polynomial time solvability when $k = 3$.) Can we find an approximation algorithm for MAX MCSLB with performance ratio $O(\log n)$? Can we solve the MAX MCSLB problem exactly on interesting graph classes, like planar graphs or permutation graphs?

References

1. H. L. Bodlaender. A linear time algorithm for finding tree-decompositions of small treewidth. *SIAM J. Comput.*, 25:1305–1317, 1996.
2. H. L. Bodlaender and A. M. C. A. Koster. Safe separators for treewidth. In *Proceedings 6th Workshop on Algorithm Engineering and Experiments ALENEX04*, pages 70–78, 2004.

3. H. L. Bodlaender, A. M. C. A. Koster, F. van den Eijkhof, and L. C. van der Gaag. Pre-processing for triangulation of probabilistic networks. In J. Breese and D. Koller, editors, *Proceedings of the 17th Conference on Uncertainty in Artificial Intelligence*, pages 32–39, San Francisco, 2001. Morgan Kaufmann.
4. H. L. Bodlaender, A. M. C. A. Koster, and T. Wolle. Contraction and treewidth lower bounds. Technical report UU-CS-2004-034, Institute of Information and Computing Sciences, Utrecht University, 2004. To appear in proceedings ESA 2004.
5. F. Clautiaux, J. Carlier, A. Moukrim, and S. Négre. New lower and upper bounds for graph treewidth. In J. D. P. Rolim, editor, *Proceedings International Workshop on Experimental and Efficient Algorithms, WEA 2003*, pages 70–80. Springer Verlag, Lecture Notes in Computer Science, vol. 2647, 2003.
6. V. Gogate and R. Dechter. A complete anytime algorithm for treewidth. To appear in proceedings UAI'04, Uncertainty in Artificial Intelligence, 2004.
7. A. M. C. A. Koster, H. L. Bodlaender, and S. P. M. van Hoesel. Treewidth: Computational experiments. In H. Broersma, U. Faigle, J. Hurink, and S. Pickl, editors, *Electronic Notes in Discrete Mathematics*, volume 8. Elsevier Science Publishers, 2001.
8. A. M. C. A. Koster, S. P. M. van Hoesel, and A. W. J. Kolen. Solving partial constraint satisfaction problems with tree decomposition. *Networks*, 40:170–180, 2002.
9. S. J. Lauritzen and D. J. Spiegelhalter. Local computations with probabilities on graphical structures and their application to expert systems. *The Journal of the Royal Statistical Society. Series B (Methodological)*, 50:157–224, 1988.
10. B. Lucena. A new lower bound for tree-width using maximum cardinality search. *SIAM J. Disc. Math.*, 16:345–353, 2003.
11. S. Ramachandramurthi. The structure and number of obstructions to treewidth. *SIAM J. Disc. Math.*, 10:146–157, 1997.
12. H. Röhrig. Tree decomposition: A feasibility study. Master's thesis, Max-Planck-Institut für Informatik, Saarbrücken, Germany, 1998.
13. G. Szekeres and H. S. Wilf. An inequality for the chromatic number of a graph. *J. Comb. Theory*, 4:1–3, 1968.
14. R. E. Tarjan and M. Yannakakis. Simple linear time algorithms to test chordiality of graphs, test acyclicity of graphs, and selectively reduce acyclic hypergraphs. *SIAM J. Comput.*, 13:566–579, 1984.
15. F. van den Eijkhof and H. L. Bodlaender. Safe reduction rules for weighted treewidth. In L. Kučera, editor, *Proceedings 28th Int. Workshop on Graph Theoretic Concepts in Computer Science, WG'02*, pages 176–185. Springer Verlag, Lecture Notes in Computer Science, vol. 2573, 2002.

Fully-Dynamic Recognition Algorithm and Certificate for Directed Cographs

Christophe Crespelle and Christophe Paul

CNRS – LIRMM, 161 rue Ada, 34392 Montpellier Cedex 5, France
{crespell,paul}@lirmm.fr

Abstract. This paper presents an optimal fully-dynamic recognition algorithm for directed cographs. Given the modular decomposition tree of a directed cograph G, the algorithm supports arc and vertex modification (insertion or deletion) in $\mathcal{O}(d)$ time where d is the number of arcs involved in the operation. Moreover, if the modified graph remains a directed cograph, the modular tree decomposition is updated; otherwise, a certificate is returned within the same complexity.

1 Introduction

Directed cographs is the family of digraphs recursively defined from the single vertex under the closure of the operations of *disjoint union*, *series* and *order*. Let $G_1, \ldots, G_k$ be a set of k disjoint digraphs. The *disjoint union* (or *parallel composition*) of the G_i's is the digraph whose connected components are precisely the G_i's. The *series* composition of the G_i's is the union of these k graphs plus all possible arcs between vertices of different G_i's. The *order* composition of the G_i's is the union of these k graphs plus all possible arcs from G_i towards G_j, with $1 \leqslant i < j \leqslant k$. These operations define a unique tree representation of the directed cograph referred which corresponds to its modular decomposition tree [12]. The leaves are mapped to the vertices of the graph and the inner nodes are labeled by the different composition operations (see Figure 1). Notice that by definition of the composition operations, the complement of a directed cograph is a directed cograph. Indeed, the term *cograph* [3] stands for *complement reducible graph*. Moreover the directed cograph family is *hereditary*: any induced subgraph of a directed cograph is also a directed cograph. It should also be noticed that directed cographs can be characterized by forbidden subgraphs (see Theorem 2 and Figure 2).

Restricted to posets, directed cographs are the *series-parallel orders* [11] for which the recognition problem has been solved in linear time [14]. In the case of undirected graphs, the series composition and the order composition are equivalent. The family of undirected graphs defined from the single vertex graph by the closure of the series and the disjoint composition is the *cographs*. The modular decomposition tree of a cograph is called a *cotree*. A number of linear time cograph recognition algorithms is now known: the first one was presented in [4] and the most recent one in [1].

J. Hromkovič, M. Nagl, and B. Westfechtel (Eds.): WG 2004, LNCS 3353, pp. 93–104, 2004.

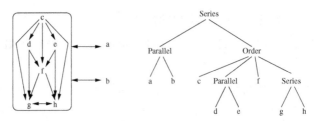

Fig. 1. A directed cograph and its modular decomposition tree. Since set $\{a, b\}$ is in series composition with the rest of the vertices, for any $x \notin \{a, b\}$ and $y \in \{a, b\}$, both arcs xy and yx exist.

The *dynamic recognition and representation problem* for a family $\mathcal{F}$ of graphs aims to maintain a characteristic representation of dynamically changing graphs as long as the modified graph belongs to $\mathcal{F}$. The input of the problem is a graph $G \in \mathcal{F}$ with its representation and a series of modifications. Any modification is of the following: adding a vertex (along with the arcs incident to it), deleting a vertex (and its incident arcs), adding or deleting an arc or two symmetric arcs (notice that the insertion/deletion of only one of these arcs may not result in a graph of $\mathcal{F}$, while the insertion/deletion of both would). Moreover, as pointed out by [10], if the property of belonging to $\mathcal{F}$ is no longer satisfied, providing a certificate would be highly desirable in practice (eg. for debugging features). This paper considers that problem for the family of directed cographs. The representation we maintain is the modular decomposition tree.

Related Works. Several authors have considered the dynamic recognition and representation problem for various graphs families. [9] devised a fully dynamic recognition algorithm for chordal graphs which handles edge operations in $\mathcal{O}(n)$ time. For proper interval graphs [8], each update can be supported in $\mathcal{O}(d + \log n)$ where d is the number of edges involved in the operation. Concerning cographs, a constant time algorithm for edge modification (insertion or deletion) has been designed in [13]. The undirected cograph recognition algorithm of [4] is incremental: given a cograph G, its cotree T and a vertex x, it modifies T iff $G + x$ is a cograph. Merging the results of [4] and [13] provides a fully dynamic recognition algorithm for cographs with $\mathcal{O}(d)$ worst case time complexity per operation. Pushing further Algorithm of [4], if $G + x$ is not a cograph, it is possible, within the same complexity, to extract a certificate (namely a P_4, an induced path of 4 vertices).

The work of [4] has recently been extended for bipartite graphs. A new decomposition dedicated to bipartite graphs has been proposed in [6] and the family of bipartite graphs totally decomposable, as are the cographs for the modular decomposition, are defined: the weak-bisplit graphs. In [7], a linear time recognition algorithm of weak-bisplit graphs is given. It turns out that the incidence bipartite graph of a directed cograph is a weak-bisplit graph. As for cographs, the decomposition tree is built by adding the vertices one by one. But unfortunately, to get linear time complexity, the vertices have to be ordered with respect to their degree. It follows that the incremental aspect cannot be guaranteed.

Our Results. We present an optimal algorithm for the dynamic recognition and representation problem for the family of directed cographs. If needed, our algorithm is also able to find a certificate. Therefore, it extends the algorithms of [4,13]. Moreover, unlike the algorithm of [7] restricted to directed cographs, our algorithm supports arc modification and the dynamical aspect is guaranteed (that is the updates can be handled in arbitrary order).

Theorem 1. *The dynamic recognition and representation problem for directed cographs is solvable in $\mathcal{O}(d)$ worst case time per update, where d is the number of edges involved in the updating operation. Moreover, if needed, a certificate that the modified graph is not a directed cograph, is provided within the same time complexity.*

Note that the results of [4] and [13] for undirected cographs cannot solve the directed case since there is no way, to our knowledge, to determine if an orientation of an undirected cograph is a directed cograph or not.

2   Preliminaries

Any graph $G = (V, E)$ considered here will be finite, loopless and directed, with $n = |V|$ and $m = |E|$. The complement of a graph G is denoted by $\overline{G}$. If X is a subset of vertices, then $G[X]$ is the subgraph of G induced by X. Since the graphs are directed, the arc xy differs from yx. Let x be a vertex, then $N^+(x) = \{z \in V, xz \in E\}$, $N^-(x) = \{y \in V, yx \in E\}$ and $N(x) = N^-(x) \cup N^+(x)$ stand respectively for the *out-neighborhood* of x, its *in-neighborhood* and its neighborhood. The non-neighborhood of x will be designed by $\overline{N}(x)$. The degree $d(x)$ of a vertex x is the sum of its in-degree, $d^-(x) = |N^-(x)|$, and its out-degree, $d^+(x) = |N^+(x)|$. Let $G = (V, E)$ be a digraph, $x \notin V$ be a vertex and $N^-(x) \subseteq V$, $N^+(x) \subseteq V$ be its in and out-neighborhoods. Then $G + x$ denotes the digraph $G' = (V \cup \{x\}, E \cup \{xz, z \in N^+(x)\} \cup \{yx, y \in N^-(x)\})$. If $xy \in E$, $G - xy$ will be the digraph $G' = (V, E \setminus \{xy\})$. $G - x$ and $G + xy$ are similarly defined.

As for the cographs family, directed cographs can be characterized by forbidden subgraphs. Unfortunately, such a characterization does not help for an efficient recognition algorithm (even for a non-dynamical one). Nevertheless, these subgraphs will be useful to provide a certificate if the referred graph is not a directed cograph. This characterization can be retrieved from a result of [5].

Theorem 2. *A digraph is a directed cograph iff it does not contain any graph of Figure 2 as induced subgraph.*

A *module* M is a set of vertices such that for any $x \notin M$ and $y \in M$, $xy \in E$ iff $\forall z \in M$, $xz \in E$ and $yx \in E$ iff $\forall z \in M$, $zx \in E$. The modules of a graph are a potentially exponential-sized family. However, the sub-family of *strong* modules, the modules that overlap[1] no other module, has size $O(n)$. The inclusion order of this family defines the *modular decomposition tree*, which

---

[1] *A overlaps B if $A \cap B \neq \emptyset$, $A \setminus B \neq \emptyset$ and $B \setminus A \neq \emptyset$.*

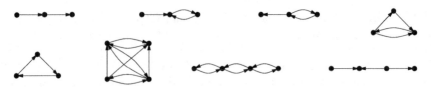

Fig. 2. The set of forbidden subgraphs for the directed cographs family. Notice that this set is closed under complementation.

is enough to store the module family of a graph [12]. The root of this tree is the trivial module V and its n leaves are the trivial modules $\{x\}, x \in V$. In the case of directed cographs, the internal nodes are labeled by one of the three composition operations: *parallel (disjoint union)*, *series* or *order* (see Figure 1). Let us call the modular decomposition tree of a directed cograph, the *di-cotree*.

Any node p of the di-cotree corresponds to a set of vertices $M(p)$. To shorten the notations, the set $M(p)$ will be denoted by P. A set $S \subseteq V$ of vertices is *uniform* wrt. $x \notin S$ in G if S is a module of the graph $G[S \cup \{x\}]$. If S is not uniform, then it is *mixed*. We say that p is uniform (resp. mixed) wrt. x if P is. Finally, a set S of vertices (resp. a node p of the di-cotree) is *linked* to a vertex $x \notin S$ in G, if there exists $y \in S$ (resp. $y \in P$) st. $xy \in E$ or $yx \in E$. In the following, if no confusion is possible, we will omit to mention the graph in which the above notions are applied. The subtree of the di-cotree T rooted at a node q will be denoted by T_q. The path between any node p and the root r of T will be denoted by P_q^r. Finally, M_{xy} stands for the minimum (wrt. the inclusion order) module that contains vertices x and y. Since M_{xy} is not necessarily strong, it is a subset of $M(p_{xy})$ where p_{xy} is the least common ancestor of the leaves corresponding to x and y. A *factorizing permutation* [2] τ is a permutation of the vertices such that any strong module M is a factor of τ (the vertices of M occur consecutively). A DFS of the modular decomposition tree orders the leaves as a factorizing permutation. Maintaining factorizing permutation is helpful to find a certificate.

3   Dynamic Vertex Operations

This section deals with Theorem 1 in the case of vertex modification (insertion or deletion). Vertex deletion is first considered. Then a theorem characterizes the cases where the insertion of a vertex is possible. This theorem is the basis of an insertion algorithm that either updates the di-cotree or finds a certificate that G is not a directed cograph. For sake of simplicity, the certificate consists in a set of 4 vertices that induces a subgraph containing a forbidden subgraph. Pushing further the algorithm, an exact forbidden subgraph can be found.

3.1   Deleting a Vertex

As already noticed, the family of directed cographs is hereditary. It follows that deleting a vertex of a directed cograph G only requires to update its di-cotree T.

It can be done in $\mathcal{O}(d(x))$ as follows (see [13] for a similar algorithm). The case where x is the only vertex is trivial. Otherwise, let q be the parent node of x in T.

1. If x has at least 2 siblings, then x is removed from T.
2. Otherwise, let p be the sibling of x.
   (a) If q is the root of T or the label of $parent(q) = \tilde{q}$ is different from the one of p, x is removed from T and q replaced by p.
   (b) If $label(\tilde{q}) = label(p)$, nodes x, q and p are removed from T and q is replaced by the children of p, respecting their relative order if p is an order node.

For complexity issues, the case where $p, \tilde{q}$ have the same label, has to be handled carefully: only nodes containing neighbors of x can be touched. If q is a parallel node, its siblings are disconnected from $\tilde{q}$ and reconnected as new children of p (at their right place if $\tilde{q}$ is an order node). Finally p replaces $\tilde{q}$. If q is not a parallel node, the children of p can be moved similarly.

3.2   Adding a Vertex

The main difficulty consists in maintaining a di-cotree under vertex insertion. Theorem 3 characterizes the cases where given a directed cograph G, a vertex x and its neighborhoods, the augmented graph $G + x$ remains a directed cograph. As in [4], the algorithm first proceeds a marking step of the di-cotree T of G. Then it tests whether the marks satisfy Theorem 3. In the positive, the di-cotree is updated; otherwise a certificate that $G + x$ is not a directed cograph is given.

Theorem 3. *Let $G = (V, E)$ be a directed cograph and T be its di-cotree. Let $x \notin V$ be a vertex and $N^-(x)$, $N^+(x)$ be its in and out-neighborhoods. $G' = G+x$ is a directed cograph iff for any node p of T one of the following conditions holds:*

1. *P is uniform wrt. x;*
2. *P is mixed and has a unique mixed child f such that $F \cup \{x\}$ is a module of $G'[P \cup \{x\}]$;*
3. *P is mixed, has no mixed child and either*
   (a) *there exists a unique non-empty set $S \subset C(p)$ of children of p such that $S = \bigcup_{k \in S} K$ is uniform wrt. x and $S \cup \{x\}$ is a module of $G'[P \cup \{x\}]$,*
   (b) *or there exists a non-empty set $S \subset C(p)$ of children of p such that $S \cup \{x\}$, $(P \setminus S) \cup \{x\}$ are both modules of $G'[P \cup \{x\}]$.*

Corollary 1 shows that the mixed nodes cannot be spread anywhere in T.

Corollary 1. *Assume $G + x$ is a directed cograph. The set of mixed nodes induces a path between the root and a certain node p of the di-cotree. Node p is the only mixed node without mixed child.*

Before describing the marking process, let us rephrase Theorem 3. Hereafter the only mixed node without mixed child will be called the *terminal mixed node* of T. A *single mixed node* will be a node satisfying condition 2 of Theorem 3. As illustrated by Figure 3, it is worth to notice that case 3.a and 3.b of Theorem 3 are exclusive. Indeed, in case 3.a, x is not a maximal (wrt. inclusion order) strong module of $G[P \cup \{x\}]$: therefore x will be inserted as a grand-child of the terminal mixed node p. While in case 3.b, x is a maximal strong module of $G[P \cup \{x\}]$ and should be inserted as a child of p. Moreover in the last case, p is an order node.

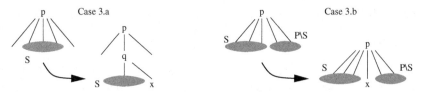

Fig. 3. Modifications of the modular decomposition tree according to cases 3 of Theorem 3. In case 3.b, the node p is an order node. In case 3.a, depending on the cardinality of $\mathcal{S}$, the label of the nodes of $\mathcal{S}$ and their adjacency with x, intermediate nodes may be inserted between q and S (see Subsection *Inserting a vertex*).

The Marking Process. The first step of our algorithm colors nodes of the modular decomposition tree T according to the neighborhood of the vertex to be inserted. This preliminary step can be seen as an extension of the marking process of [4].

Initially each leaf $l = \{y\}$, such that $y \in N(x)$, is colored red. Depending on the adjacency relationship between y and x, these leaves are given a type: $type(l) = InOut$ if $xy \in E$ and $yx \in E$; $type(l) = In$ if $yx \in E$; or $type(l) = Out$ if $xy \in E$. The process is a bottom-up search: each red node p forwards its type to its parent node q and depending on the different types received by q, a color is given to q. The first time a node receives a forwarded type from one of its children, it is colored black. A node q becomes red if all its children are of the same type (ie. the corresponding set of vertices Q is uniformly linked to x). A

Type$(G, T, \mathcal{R}$ a set of typed red leaves)
1. **While** some red node p exists **Do**
2. $color(p) \leftarrow grey$
3.   **If** p is not the root of T **Then**
4.   Let q be the parent node of p
5.   Add p to the list $greyChild(q)$
6.   Increase $\#type(q, type(p))$ by one
7.   **If** $\#type(q, type(p)) = \#child(q)$ **Then**
8. $color(q) \leftarrow red$ and $type(q) \leftarrow type(p)$
9.   **Else** $color(q) \leftarrow black$
10. **End of while**

Fig. 4. Marking process.

red node receives the type of its children. Once a red node has been processed it becomes grey. In order to prepare the possible insertion of x, a list of the grey children is maintained for each node.

For sake of simplicity, let us say that the default color is *white*. Also notice that the absence of type can be considered as a non-adjacency type, we will use the notation $type(p) = None$. However, the marking algorithm will never manage neither the white nodes nor the $None$ type.

Each node stores the list of its grey children and a few counters: eg. $\#child(q)$ indicates the number of children of q, $\#type(q, In)$ the number of children of q whose type is In. It is straightforward to see that the running time of Routine **Type** is $\mathcal{O}(d(x))$. Let T^c be the resulting colored di-cotree. The number of grey nodes and of black nodes are both bounded by $\mathcal{O}(d(x))$.

Lemma 1. *If there exists a black node q such that:*

1. *any black node belongs to P_q^r,*
2. *any black node of $P_{parent(q)}^r$ is series or order and*
3. *any white node of $P_{parent(q)}^r$ is parallel,*

then the white nodes of P_q^r are single mixed nodes.

The set of black nodes will be denoted $\mathcal{B}$. By definition, a black node is a mixed node. A white node p can be mixed if T_p contains a black node, otherwise it is uniform. Lemma 1 implies that the set of white mixed nodes is exactly the set $\mathcal{W}$ of white nodes of the path P_q^r mentioned in Lemma 1. Therefore $\mathcal{W} \cup \mathcal{B}$ is the set of mixed nodes of T^c.

Testing the Insertion. Assume the following conditions are satisfied: there exists a terminal mixed node $q \in \mathcal{B}$ such that any node of $\mathcal{B}$ belongs to P_q^r and any nodes of $P_{parent(q)}^r$ is a single mixed node. Then by Lemma 1, any node of T satisfies the hypothesis of Theorem 3. It implies that x can be inserted. Therefore Routine **Check** (see Figure 5) only has to test these conditions. If one of them is not satisfied, then a call to Routine **Find-Certificate** enables us to find a set Z of 3 vertices such that $G'[Z \cup \{x\}]$, with $G' = G + x$, contains one of the forbidden subgraphs of Figure 2. The insertion of vertex x, if possible, is handled by Routine **Insert**.

Let p be the current node in Routine **Check**. If p has already been visited (test Line 6), then by Corollary 1 G' is not a directed cograph. The tests of Line 7 and 8 check whether p is a single mixed node. As shown by Lemma 2, depending on their color, the label of single mixed nodes are constrained.

Lemma 2. *Let p be a single mixed node of T^c. If p is black, then p is either a series or an order node. Otherwise it is a white parallel node.*

Let p be a black (series or order) node. For p to be a single mixed node, all but one of its children have to be colored grey. If p is a series node, the children distinct from the only non-grey child q should be typed $InOut$. If p is an order node, the children that occur before (resp. after) q have to be typed In (resp. Out).

Check$(G, T^c, \mathcal{B}, x)$
1. $bottom \leftarrow r$, where r is the root of T
2. **While** some node q in $\mathcal{B}$ exists **Do**
3. $p \leftarrow q$ and remove q from $\mathcal{B}$
4.   **While** $p \neq bottom$ **Do**
5. $p \leftarrow parent(p)$
6.   **If** p has been visited **Then** Find-Certificate(p)
7.   **If** p is a white non-parallel node **Then** Find-Certificate(p)
8.   **If** $p \in \mathcal{B}$ is not a single mixed node **Then** Find-Certificate(p)
9.   **If** $p \in \mathcal{B}$ **Then** Remove p from $\mathcal{B}$
10.   Mark p as visited
11.   **End of while**
12. $bottom \leftarrow q$
13. **End of while**
14. **If** q is a terminal mixed node **Then** Insert(x, q, T)
15. **Else** Find-Certificate(q)

Fig. 5. Testing the insertion.

Finally let q be the last node considered by routine Check (*bottom*). There is no constraint on the label of q. By Theorem 3, it has to be terminal mixed, which can be tested as follows:

- if q is a parallel node: check that $\#type(In) = \#grey$ or $\#type(Out) = \#grey$ or $\#type(InOut) = \#grey$ (since in that case, any node of $\mathcal{S}$ is a grey node);
- if q is a series node: check that $\#type(In) + \#type(InOut) = |\mathcal{C}(q)|$ or $\#type(Out) + \#type(InOut) = |\mathcal{C}(q)|$ or $\#type(In) = \#type(Out) = 0$;
- if q is an order node: first, test if either $\#type(InOut) + \#type(In) + \#type(Out) = |\mathcal{C}(q)|$ or $\#type(InOut) = 0$. Then check whether the first (wrt. the relative order of q) $\#type(In)$ children of q are typed In and the last $\#type(Out)$ are typed Out.

If each node p of the di-cotree stores its number of children $|\mathcal{C}|$, these tests can be done by a simple search in the grey children. Since the number of grey nodes and black nodes is $\mathcal{O}(d(x))$, Routine Check runs in $\mathcal{O}(d(x))$ time.

Inserting a Vertex. As illustrated by Figure 3, the modification occurs in the di-cotree T_q where q is the only terminal mixed node (the *bottom* node in Routine Check). We know that any child of q is uniform and since q is mixed, it has at least two children of different types (remind that the absence of type can be considered as a non-adjacency type).

Assume q is a series node (the case where q is parallel is similar). As already noticed, since q is not an order node, x has to be inserted as a grand-child of q. By theorem 3, a set $\mathcal{S}$ of children such that $S = \bigcup_{k \in \mathcal{S}} K$ is uniform wrt. x and

$S \cup \{x\}$ is a module of $G'[Q \cup \{x\}]$, where $G' = G + x$, exists. Since q is a series node, a child p belongs to S iff $type(p) \neq InOut$. Remark that the uniformity of S implies that the nodes belonging to S all have the same type. To update the di-cotree, three cases should be considered. First, if $S = \{p\}$ and the label of node p coincides with its type, $type(p)$, then x is added as a leaf of p. Otherwise, a new node p', labeled by the composition operation corresponding to the type of nodes of S, is inserted as a child of q instead of nodes of S. If $S = \{p\}$, then p is made a child of this new node p'. If $|S| \geq 2$, a node q', whose children are the nodes of S, is made a child of p. q' get the same label than q (ie. series). In both cases, x is added as a leaf of p'.

Assume q is an order node. The difference with the previous cases, is that three subsets of $\mathcal{C}(q)$ can be identified: $\mathcal{S}_{In}$ (resp. $\mathcal{S}_{Out}$) the children with type In (resp. Out) and $\mathcal{S}$ the other children. The nodes of $\mathcal{S}_{In}$ appears before those of $\mathcal{S}$ that appears before those of $\mathcal{S}_{Out}$ in the order defined by q. Notice that one of these three sets could be empty. If $\mathcal{S} = \emptyset$, x is inserted as a child of q between $\mathcal{S}_{In}$ and $\mathcal{S}_{Out}$. Otherwise, a new child p of q has to be inserted between q and $\mathcal{S}$, and x is made a child of p.

As done for the vertex deletion, to update the di-cotree, we have to carefully handle the moving of non-neighborhood of x. Any insertion costs $\mathcal{O}(d(x))$ time.

Finding a Certificate. Routine `Find-Certificate`(p) looks for one of the forbidden induced subgraphs of Figure 2. Assume this routine is also given the parameters P^r_{bottom} and P^p_q where $bottom$ and q are the nodes respectively defined at Line 12 and 2 of Routine `Check`. Thanks to the lists of grey children for each node of T^c and the factorizing permutation, the search is processed in $\mathcal{O}(d(x))$ time. The call to `Find-Certificate` at Line 6 occurs if the current node p has already been visited before. At Lines 7 and 8, node p should have been a single mixed node but is not. At Line 15, the last visited node q is not terminal mixed. In each case, a subgraph of Figure 2 can be found in $\mathcal{O}(d(x))$ time. Though Routine `Find-Certificate` returns the exact subgraph, for sake of simplicity, we just give some hints of the following:

Lemma 3. *If* $G' = G + x$ *is not a directed cograph, a set* Z *of 3 vertices can be found in* $\mathcal{O}(d(x))$ *time such that* $G'[Z \cup \{x\}]$ *contains one of the graphs of Figure 2.*

Let us detail the former call of Line 6. Notice that p has to be a parallel node (otherwise, `Check` would have found out that p is not a single mixed node). Indeed, p has at least 2 black mixed children: namely h, the child of p on the path P^r_{bottom}, and h', the child of p on the path P^p_q. Since h and h' are black, they both received a type from a grey child, say k and k' respectively. Let a be a vertex of $K = M(k)$ and b be a vertex of $K' = M(k')$. Finally, since h' is mixed and k' is uniform, a vertex $c \in H' \setminus K'$ (with $H' = M(h')$) such that $type(c) \neq type(b)$ exists. A simple case by case analysis of the different possible types for nodes k, k' and vertices a, b, c proves that $G'[\{a, b, c, x\}]$, with $G' = G + x$, contains a certificate (one of the graphs of Figure 2). Figure 6 illustrates two different cases.

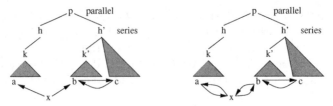

Fig. 6. Since p is a parallel node, h' is either a series or an order node. Assume that h' is a series nodes, therefore bc and cb exist. In the first example, the certificate is induced by $\{b, c, x\}$, in the second by $\{a, b, c, x\}$.

4   Dynamic Arc Operations

This section deals with Theorem 1 in the case of arc modification. For lack of space, we only present how to handle arc deletion. Since the family of directed cographs is closed under complementation, the graph $G + xy$ is a directed cograph iff the graph $\overline{G} - xy$ is. Similarly, a certificate that $\overline{G} - xy$ is not a directed cograph, is a certificate for $G + xy$. As remarked by [13], since the di-cotree of the complement of a directed cograph G can easily be deduced from the di-cotree of G a recognition algorithm that supports edge deletion can be extended to support edge insertion within the same complexity. Finally, we shall assume that if an arc deletion query is asked for, then the arc involved exists.

Deleting an Arc. Two types of arc based modifications should be distinguished. The first one concerns the simultaneous removal of two symmetric arcs, say xy and yx. This modification can be compared to the deletion of an edge in an undirected cograph, see [13]. The proof of [13] can be generalized to the case of directed cographs. Let q_x (resp. q_y) be the child of p_{xy} containing x (resp. y)[2].

Theorem 4. *[13] The graph $G' = G - \{xy, yx\}$ is a directed cograph iff $|Q_x| = 1$ and $Q_y \setminus \{y\} \subseteq \overline{N}(y)$ or $|Q_y| = 1$ and $Q_x \setminus \{x\} \subseteq \overline{N}(x)$.*

Theorem 5 extends Theorem 4 so that any valid arc modification of a directed cograph can be characterized.

Theorem 5. *The graph $G' = G - xy$ is a directed cograph iff*

1. *p_{xy} is an order node, $M_{xy} = Q_x \cup Q_y$ and:[3]*
   (a) either $|Q_x| = 1$ and $Q_y \setminus \{y\} \subseteq \overline{N}(y)$,
   (b) or $|Q_y| = 1$ and $Q_x \setminus \{x\} \subseteq \overline{N}(x)$.
2. *p_{xy} is a series node and:*
   (a) either $|Q_x| = 1$ and $Q_y \setminus \{y\} \subseteq N^+(y) \setminus N^-(y)$,
   (b) or $|Q_y| = 1$ and $Q_x \setminus \{x\} \subseteq N^-(x) \setminus N^+(x)$.

It is straightforward from Theorem 4 and 5 that the deletion test can be done in $\mathcal{O}(1)$. Indeed, x and y has to be the child and the grand-child of p_{xy}.

---

[2] p_{xy} is the lca of x and y in T.
[3] M_{xy} is defined as the minimum module containing x and y. Therefore $M_{xy} \subseteq P_{xy}$.

Wlog. assume x is the child. Then, it suffices to check the label of p_{xy} and of its child q which is the parent of y. If the deletion is possible, the modifications of the di-cotree are carried out in constant time. The different cases are depicted in Figure 7.

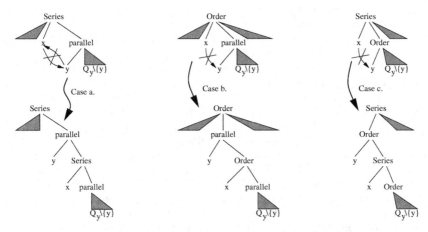

Fig. 7. Case **a.** illustrates the modification implied by the simultaneous removal of two symmetric arcs (see Theorem 4); cases **b.** and **c.** illustrate the removal of the arc xy described in Theorem 5. Depending on the number of siblings of y, the resulting di-cotrees may contain less node than above.

Finding a Certificate. Assume the test of the xy deletion fails. As done for the vertex certificate, our algorithm returns a small subgraph containing one of the graphs of Figure 2. Thanks to the factorizing permutation, the vertices of this subgraph can be found in constant time. If an exact certificate is wished, it can be found in $\mathcal{O}(min(d(x), d(y)))$.

Lemma 4. *If $G' = G - xy$ is not a directed cograph, a set Z of at most 4 vertices can be found in $\mathcal{O}(1)$ such that $G'[Z \cup \{x, y\}]$ contains one of the graphs of Fig. 2.*

Let us describe how the set Z is defined. Let p_x (resp. p_y) be the parent of x (resp. y) in T. If $p_x \neq r$ (resp. $p_y \neq r$), let q_x (resp. q_y) be the parent of p_x (resp. p_y) in T. If $q_x \neq r$ (resp. $q_y \neq r$), let k_x (resp. k_y) be the parent of q_x (resp. q_y) in T. Let us define 6 vertices, namely a_x, b_x, c_x and a_y, b_y, c_y. Vertex a_x belongs to $P_x \setminus \{x\}$ and if p_x is an order node and $P_x \cap N^+(x) \neq \emptyset$, then choose for a_x an out-neighbor of x. Vertices b_x and c_x belongs respectively to $Q_x \setminus P_x$ and $K_x \setminus Q_x$ if these sets exist. The last 3 vertices a_y, b_y, c_y are similarly defined wrt. y. If possible, a_y should be picked in $N^-(y)$. Note that, even if they exist, these vertices may not be all distinct. Finally a case by case analysis of the labels of parents and grand-parents of x and y enables us to select at most 4 vertices among $a_x, a_y, b_x, b_y, c_x, c_y$.

References

1. A. Bretscher, D.G. Corneil, M. Habib, and C. Paul. A simple linear time lexbfs cograph recognition algorithm. In H. Bodlaender, editor, *29th International Workshop on Graph Theoretical Concepts in Computer Science (WG'03)*, number 2880 in Lecture Notes in Computer Science, pages 119–130, 2003.
2. C. Capelle and M. Habib. Graph decompositions and factorizing permutations. In *Fifth Israel Symposium on the Theory of Computing Systems (ISTCS'97)*, IEEE Computer Society, pages 132–143, 1997.
3. D.G. Corneil, H. Lerchs, and L. Stewart Burlingham. Complement reducible graphs. *Discrete Applied Mathematics*, 3(1):163–174, 1981.
4. D.G. Corneil, Y. Perl, and L.K. Stewart. A linear time recognition algorithm for cographs. *SIAM Journal on Computing*, 14(4):926–934, 1985.
5. A. Ehrenfeucht and G. Rozenberg. Primitivity is hereditary for 2-structures. *Theoretical Computer Science*, 70(3):343–359, 1990.
6. J.-L. Fouquet, V. Giakoumakis, and J.-M. Vanherpe. Bipartite graphs totally decomposable by canonical decomposition. *International Journal of Foundation of Computer Science*, 10(4):513–533, 1999.
7. V. Giakoumakis and J.-M. Vanherpe. Linear time recognition and optimizations for weak-bisplit graphs, bi-cographs and bipartite p_6-free graphs. *International Journal of Foundation of Computer Science*, 14(1):107–136, 2003.
8. P. Hell, R. Shamir, and R. Sharan. A fully dynamic algorithm for recognizing and representing proper interval graphs. *SIAM Journal on Computing*, 31(1):289–305, 2002.
9. L. Ibarra. Fully dynamic algorithms for chordal graphs. In *10th ACM-SIAM Annual Symposium on Discrete Algorithm (SODA'03)*, pages 923–924, 1999.
10. D. Kratsch, R.M. McConnell, K. Mehlhorn, and J.P. Spinrad. Certifying algorithm for recognition of interval graphs and permutation graphs. In *14th ACM-SIAM Annual Symposium on Discrete Algorithm (SODA'03)*, pages 153–167, 2003.
11. E.L. Lawler. Graphical algorithm and their complexity. *Mathematical center tracts*, 81:3–32, 1976.
12. R.H. Möhring and F. J. Radermacher. Substitution decomposition for discrete structures and connections with combinatorial optimization. *Annals of Discrete Mathematics*, 19:257–356, 1984.
13. R. Shamir and R. Sharan. A fully dynamic algorithm for modular decomposition and representation of cographs. *Discrete Applied Mathematics*, 136(2-3):329–340, 2004.
14. J. Valdes, R.E. Tarjan, and E.L. Lawler. The recognition of series parallel digraphs. *SIAM Journal on Computing*, 11:298–313, 1982.

Recognizing HHD-free
and Welsh-Powell Opposition Graphs*

Stavros D. Nikolopoulos and Leonidas Palios

Department of Computer Science, University of Ioannina
GR-45110 Ioannina, Greece
{stavros,palios}@cs.uoi.gr

Abstract. In this paper, we consider the recognition problem on two classes of perfectly orderable graphs, namely, the HHD-free and the Welsh-Powell opposition graphs (or WPO-graphs). In particular, we prove properties of the chordal completion of a graph and show that a modified version of the classic linear-time algorithm for testing for a perfect elimination ordering can be efficiently used to determine in $O(\min\{nm\alpha(n),\ nm + n^2 \log n\})$ time whether a given graph G on n vertices and m edges contains a house or a hole; this leads to an $O(\min\{nm\alpha(n),\ nm + n^2 \log n\})$-time and $O(n+m)$-space algorithm for recognizing HHD-free graphs. We also show that determining whether the complement $\overline{G}$ of the graph G contains a house or a hole can be efficiently resolved in $O(nm)$ time using $O(n^2)$ space; this in turn leads to an $O(nm)$-time and $O(n^2)$-space algorithm for recognizing WPO-graphs. The previously best algorithms for recognizing HHD-free and WPO-graphs required $O(n^3)$ time and $O(n^2)$ space.

1 Introduction

A linear order $\prec$ on the vertices of a graph G is *perfect* if the ordered graph $(G, \prec)$ contains no induced P_4 $abcd$ with $a \prec b$ and $d \prec c$ (such a P_4 is called an *obstruction*). In the early 1980s, Chvátal [2] defined the class of graphs that admit a perfect order and called them *perfectly orderable* graphs.

The perfectly orderable graphs are perfect; thus, many interesting problems in graph theory, which are NP-complete in general graphs, have polynomial-time solutions in graphs that admit a perfect order [1, 5]; unfortunately, it is NP-complete to decide whether a graph admits a perfect order [12]. Since the recognition of perfectly orderable graphs is NP-complete, we are interested in characterizing graphs which form polynomially recognizable subclasses of perfectly orderable graphs. Many such classes of graphs, with very interesting structural and algorithmic properties, have been defined so far and shown to admit polynomial-time recognitions (see [1, 5]); note however that not all subclasses of perfectly orderable graphs admit polynomial-time recognition [7].

---

* Research partially funded by the European Union and the Hellenic Ministry of Education through EPEAEK II.

J. Hromkovič, M. Nagl, and B. Westfechtel (Eds.): WG 2004, LNCS 3353, pp. 105–116, 2004.
© Springer-Verlag Berlin Heidelberg 2004

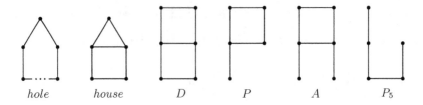

Fig. 1. Some simple graphs.

In this paper, we consider two classes of perfectly orderable graphs, namely, the HHD-free and the Welsh-Powell opposition graphs. A graph is *HHD-free* if it contains no hole (i.e., a chordless cycle on ≥ 5 vertices), no house, and no domino (D) as induced subgraphs (see Figure 1). In [8], Hoàng and Khouzam proved that the HHD-free graphs admit a perfect order, and thus are perfectly orderable. It is important to note that the HHD-free graphs properly generalize the class of triangulated (or chordal) graphs, i.e., graphs with no induced chordless cycles of length greater than or equal to four [5]. A subclass of HHD-free graphs, which also properly generalizes the class of triangulated graphs, is the class of HH-free graphs; a graph is HH-free if it contains no hole and no house as induced subgraphs (see Figure 1). Chvátal conjectured and later Hayward [6] proved that the complement $\overline{G}$ of an HH-free graph G is also perfectly orderable.

A graph is called an *Opposition graph* if it admits a linear order $\prec$ on its vertices such that there is no P_4 $abcd$ with $a \prec b$ and $c \prec d$. Opposition graphs belong to the class of **bip*** graphs (see [1]), and hence are perfect graphs [14]. The complexity of recognizing opposition graphs is unknown. It is also open whether there is an opposition graph that is not perfectly orderable [1]. The class of opposition graphs contains several known classes of perfectly orderable graphs. For example, bipolarizable graphs are, by definition, opposition graphs; a graph is bipolarizable if it admits a linear order $\prec$ on its vertices such that every P_4 $abcd$ has $b \prec a$ and $c \prec d$ [15]. Another subclass of opposition graphs, which we study in this paper, are the Welsh-Powell opposition graphs. A graph is defined to be a *Welsh-Powell Opposition graph* (or WPO-graph for short), if it is an opposition graph for every Welsh-Powell ordering; a Welsh-Powell ordering for a graph is an ordering of its vertices in nondecreasing degree [18].

Hoàng and Khouzam [8], while studying the class of brittle graphs (a well-known class of perfectly orderable graphs which contains the HHD-free graphs), showed that HHD-free graphs can be recognized in $O(n^4)$ time, where n denotes the number of vertices of the input graph. An improved result was obtained by Hoàng and Sritharan [9] who presented an $O(n^3)$-time algorithm for recognizing HH-free graphs and showed that HHD-free graphs can be recognized in $O(n^3)$ time as well; one of the key ingredients in their algorithms is the reduction to the recognition of triangulated graphs. Recently, Eschen *et al.* [4] described recognition algorithms for several classes of perfectly orderable graphs, among which a recognition algorithm for HHP-free graphs; a graph is HHP-free if it contains no hole, no house, and no "P" as induced subgraphs (see Figure 1).

Their algorithm is based on the property that every HHP-free graph is HHDA-free graph (a graph with no induced hole, house, domino D, or "A"), and thus a graph G is HHP-free graph if and only if G is a HHDA-free and contains no "P" as an induced subgraph. The characterization of HHDA-free graphs due to Olariu (a graph G is HHDA-free if and only if every induced subgraph of G either is triangulated or contains a non-trivial module [15]) and the use of modular decomposition [11] allowed Eschen *et al.* to present an $O(nm)$-time recognition algorithm for HHP-free graphs.

For the class of WPO-graphs, Olariu and Randall [16] gave the following characterization: a graph G is WPO-graph if and only if G contains no induced C_5 (i.e., a hole on 5 vertices), house, P_5, or "P" (see Figure 1). It follows that G is a WPO-graph if and only if G is HHP-free and $\overline{G}$ is HH-free. Eschen *et al.* [4] combined their $O(nm)$-time recognition algorithm for HHP-free graphs with the $O(n^3)$-time recognition algorithm for HH-free graphs proposed in [9], and showed that WPO-graphs can be recognized in $O(n^3)$ time.

In this paper, we present efficient algorithms for recognizing HHD-free graphs and WPO-graphs. We show that a variant of the classic linear-time algorithm for testing whether an ordering of the vertices of a graph is a perfect elimination ordering can be used to determine whether a vertex of a graph G belongs to a hole or is the top of a house or a building in G. We take advantage of properties characterizing the chordal completion of a graph and show how to efficiently compute for each vertex v the leftmost among v's neighbors in the chordal completion which are to the right of v without explicitly computing the chordal completion. As a result, we obtain an $O(\min\{nm\alpha(n), nm + n^2 \log n\})$-time and $O(n + m)$-space algorithm for determining whether a graph on n vertices and m edges is HH-free. This result along with results by Jamison and Olariu [10], and by Hoàng and Khouzam [8] enable us to describe an algorithm for recognizing HHD-free graphs which runs in $O(\min\{nm\alpha(n), nm + n^2 \log n\})$ time and requires $O(n + m)$ space.

Additionally, for a graph G on n vertices and m edges, we show that we can detect whether the complement $\overline{G}$ of G contains a hole or a house in $O(nm)$ time using $O(n^2)$ space. In light of the characterization of WPO-graphs due to Olariu and Randall [16] which implies that a graph G is a WPO-graph if and only if G is HHP-free and its complement $\overline{G}$ is HH-free, and the $O(nm)$-time recognition algorithm for HHP-free graphs of Eschen *et al.* [4], our result yields an $O(nm)$-time and $O(n^2)$-space algorithm for recognizing WPO-graphs.

2 Preliminaries

We consider finite undirected graphs with no loops or multiple edges. Let G be such a graph; then, $V(G)$ and $E(G)$ denote the set of vertices and of edges of G respectively. The subgraph of a graph G induced by a subset S of G's vertices is denoted by $G[S]$. A subset $B \subseteq V(G)$ of vertices is a *module* if $2 \leq |B| < |V(G)|$ and each vertex $x \in V(G) - B$ is adjacent to either all vertices or no vertex in B. The *neighborhood* $N(x)$ of a vertex $x \in V(G)$ is the set of all the vertices of G which are adjacent to x. The *closed neighborhood* of x is

Algorithm PEO(G, σ)

---

1. **for** each vertex $u \in V(G)$ **do**
        $A(u) \leftarrow \emptyset$;
2. **for** $i \leftarrow 1$ to $n - 1$ **do**
3. $u \leftarrow \sigma(i)$;
4. $X \leftarrow \{x \in N(u) \mid \sigma^{-1}(u) < \sigma^{-1}(x)\}$;
5.   **if** $X \neq \emptyset$ **then**
6. $w \leftarrow \sigma(min\{\sigma^{-1}(x) \mid x \in X\})$;
7.   concatenate $X - \{w\}$ to $A(w)$;
8.   **if** $A(u) - N(u) \neq \emptyset$ **then return** "false";
9. **return** "true";

Fig. 2. The perfect elimination ordering testing algorithm.

defined as $N[x] := N(x) \cup \{x\}$. We use $M(x)$ to denote the set $V(G) - N[x]$ of non-neighbors of x. Furthermore, for a vertex $v \in M(x)$, we use $n(v, x)$ to denote the number of vertices in the set $N(v) \cap N(x)$, i.e., the set of common neighbors of v and x. The *degree* of a vertex x in a graph G, denoted $deg(x)$, is the number of edges incident on x; thus, $deg(x) = |N(x)|$.

Let G be a graph and let x, y be a pair of vertices. If G contains a path from vertex x to vertex y, we say that x *is connected to* y. The graph G is *connected* if x is connected to y for every pair of vertices $x, y \in V(G)$. The *connected components* (or *components*) of G are the equivalence classes of the "is connected to" relation on the vertex set $V(G)$. The *co-connected components* (or *co-components*) of G are the connected components of the complement $\overline{G}$ of the graph G.

A graph G has a *perfect elimination ordering* if its vertices can be linearly ordered $(v_1, v_2, \ldots, v_n)$ such that each vertex v_i is simplicial in the graph G_i induced by the vertex set $\{v_i, \ldots, v_n\}$, $1 \leq i \leq n$; a vertex of a graph is *simplicial* if its neighborhood induces a complete subgraph. It is well-known that a graph is triangulated if and only if it has a perfect elimination ordering [1, 5, 17]. The notion of a simplicial vertex was generalized by Jamison and Olariu [10] who defined the notion of a semi-simplicial vertex: a vertex of a graph G is *semi-simplicial* if it is not a midpoint of any P_4 of G. A graph G has a *semi-perfect elimination ordering* if its vertices can be linearly ordered $(v_1, v_2, \ldots, v_n)$ such that each vertex v_i is semi-simplicial in the graph $G_i = G[\{v_i, \ldots, v_n\}]$, $1 \leq i \leq n$. A graph is a *semi-simplicial graph* if and only if it has a semi-perfect elimination ordering (see [4]).

Let $\sigma = (v_1, v_2, \ldots, v_n)$ be an ordering of the vertices of a graph G; $\sigma(i)$ is the i-th vertex in σ, i.e., $\sigma(i) = v_i$, while $\sigma^{-1}(v_i)$ denotes the position of vertex v_i in σ, i.e., $\sigma^{-1}(v_i) = i$, $1 \leq i \leq n$. In Figure 2, we include the classic algorithm PEO(G, σ) for testing whether the ordering σ is a perfect elimination ordering; if the graph G has n vertices and m edges, the algorithm runs in $O(n + m)$ time and requires $O(n + m)$ space [5, 17]. Note that, in Step 4 of the Algorithm PEO(G, σ), the set X is assigned the neighbors of the vertex u which have larger $\sigma^{-1}(\ )$-values; that is, $X = N(u) \cap \{\sigma(i+1), \ldots, \sigma(n)\}$. Thus, in Step 6, the vertex w is the neighbor of u in G which is first met among the vertices to the right of u along the ordering σ. Since neither the graph G nor the ordering σ changes during the execution of the Algorithm PEO, we can without error replace Step 6 by

6. $w \leftarrow Next\_Neighbor_{G,\sigma}[u];$

where $Next\_Neighbor_{G,\sigma}[\ ]$ is an array whose values have been precomputed in accordance with the assignment in Step 6 of the Algorithm PEO.

Note: Due to lack of space we have omitted the proofs of several Lemmata and Theorems of this paper; all the proofs can be found in [13].

3   Recognizing HH-free Graphs

The most important ingredient (and the bottleneck too) of the HHD-free graph recognition algorithm of Hoàng and Sritharan [9] is an algorithm to determine whether a simplicial vertex v of a graph G is *high*, i.e., it is the top of a house or a building[1] (or belongs to a hole) in G, which involves the following steps:

▷ They compute an ordering of the set $M(v)$ of non-neighbors of v in G where, for two vertices $y, y' \in M(v)$, y precedes y' whenever $n(y, v) \leq n(y', v)$; recall that, $n(y, v)$ is the number of common neighbors of y and v, or, equivalently, the degree of the vertex $y \in M(v)$ in the graph induced by the set $N(v) \cup \{y\}$. As we will be using this ordering in the description of our approach, we call it a *DegMN-ordering* of $M(v)$.

▷ They perform chordal completion on $G[M(v)]$ with respect to a DegMN-ordering of $M(v)$.

▷ The vertex v is high if and only if the graph G'_v resulting from G after the chordal completion on $G[M(v)]$ is triangulated.

As we mentioned in the introduction, the algorithm of Hoàng and Sritharan runs in $O(n^3)$ time, where n is the number of vertices of the input graph. In order to be able to beat this, we need to avoid the chordal completion step. Indeed, we show how we can take advantage of the Algorithm PEO and of properties of the chordal completion in order to compute all necessary information without actually performing the chordal completion. In particular, we prove that the following results hold:

Lemma 3.1. *Let G be a graph, v a vertex of G, and $(y_1, y_2, \ldots, y_k)$ a DegMN-ordering of the non-neighbors $M(v)$ of v in G. Moreover, let G'_v be the graph resulting from G after the chordal completion on $G[M(v)]$ with respect to the DegMN-ordering $(y_1, y_2, \ldots, y_k)$ and let $\sigma = (y_1, y_2, \ldots, y_k, x_1, x_2, \ldots, x_{deg(v)}, v)$ where $x_1, x_2, \ldots, x_{deg(v)}$ is an arbitrary ordering of the neighbors of v in G. If Algorithm $PEO(G'_v, \sigma)$ returns "false" while processing vertex $y_i \in M(v)$, then $A(y_i) - N(y_i) \subseteq N(v)$.*

---

[1] A building is a graph on vertices $v_1, v_2, \ldots, v_p$, where $p \geq 6$, and edges $v_1 v_p$, $v_2 v_p$, and $v_i v_{i+1}$ for $i = 1, 2, \ldots, p - 1$; the vertex v_1 is called the *top* of the building.

Algorithm Not-in-HHB(G, v)

---

1. Compute a DegMN-ordering $\sigma = (y_1, y_2, \ldots, y_k)$ of the non-neighbors of v in the graph G;
   compute the array $Next\_Neighbor_{G'_v, \sigma}[\,]$ for the non-neighbors of v;
   for each non-neighbor u of v **do**
   $\quad A(u) \leftarrow \emptyset$;
2. **for** $i \leftarrow 1$ to k **do**
3. $\quad u \leftarrow \sigma(i)$;
4. $\quad X \leftarrow N(u) \cap N(v)$;   {*note:* $\forall x \in X,\ \sigma^{-1}(u) < \sigma^{-1}(x)$}
5. $\quad$ **if** $X \neq \emptyset$ **then**
6. $\qquad w \leftarrow Next\_Neighbor_{G'_v, \sigma}[u]$;
7. $\qquad$ **if** $w \in M(v)$ **then** concatenate X to $A(w)$;   {*note:* $w \notin X$}
8. $\quad$ **if** $A(u) - N(u) \neq \emptyset$ **then return** "false";
9. **return** "true";

Fig. 3. Algorithm for determining whether a vertex v belongs to a hole or is the top of a house or a building.

Proof: Since the Algorithm PEO returns "false" while processing vertex $y_i \in M(v)$, then $A(y_i) - N(y_i) \neq \emptyset$. Suppose that there exists a vertex $y_j \in M(v)$ belonging to $A(y_i) - N(y_i)$. The vertex y_j was added to $A(y_i)$ at Step 7 of a prior iteration of the for-loop, say, while processing vertex y_ℓ. It follows that $\sigma^{-1}(y_\ell) < \sigma^{-1}(y_i) < \sigma^{-1}(y_j)$, and $y_i, y_j \in N(y_\ell)$. Since $y_j \notin N(y_i)$, we have that y_ℓ is not simplicial in $G'_v[\{y_\ell, y_{\ell+1}, \ldots, y_k\}]$; a contradiction. ∎

Lemma 3.2. *Let G'_v and σ be as in the statement of Lemma 3.1. The vertex v belongs to a C_5 or is the top of a house in the graph G'_v if and only if Algorithm $PEO(G'_v, \sigma)$ returns "false" while processing vertex z, where $z \in M(v)$.*

Lemma 3.1 implies that, while running Algorithm $PEO(G'_v, \sigma)$, it suffices to collect in the set X (Step 4) only the common neighbors of u and v; in turn, Lemma 3.2 implies that it suffices to execute the for-loop of Steps 2-8 only for the non-neighbors of v. The above can be used to yield the Algorithm Not-in-HHB, presented in Figure 3, which takes as input a graph G and a vertex v of G, and returns "true" if and only if the vertex v does not belong to a hole, and it is not the top of a house or a building in G. That is, we can show the following result.

Theorem 3.1. *Algorithm Not-in-HHB(G, v) returns "false" if and only if the vertex v belongs to a hole or is the top of a house or a building in G.*

3.1   Computation of the Values of $Next\_Neighbor_{G'_v, \sigma}[\,]$

In order to avoid computing the graph G'_v, we take advantage of the following property of the chordal completion of a graph:

Lemma 3.3. *Let G be a graph, let $(v_1, v_2, \ldots, v_k)$ be an ordering of its vertices, and let G' be the graph resulting from G after the addition of edges so that, for all $i = 1, 2, \ldots, k$, vertex v_i is simplicial in the subgraph induced by the vertices*

Algorithm Compute-Next_Neighbor(G, σ, v)

---

1. {*let $y_1, y_2, \ldots, y_k$ be the non-neighbors of v in the order they appear in σ*}
    make a set containing the vertex y_1;
2. **for** $j = 2, 3, \ldots, k$ **do**
3.   make a set containing the vertex y_j;
4.   **for** each edge $y_i y_j$ of G, where $i < j$, **do**
5. $y_r \leftarrow$ the rightmost (w.r.t. σ) vertex in the set to which y_i belongs;
6.   **if** $y_r \neq y_j$ {*y_i and y_j belong to different sets*}
7.   **then**
8. $Next\_Neighbor_{G'_v, \sigma}[y_r] \leftarrow y_j$;
9.   union the sets to which y_i and y_j belong;

Fig. 4. Algorithm for computing the contents of the array $Next\_Neighbor_{G'_v, \sigma}[\ ]$.

$v_i, v_{i+1}, \ldots, v_k$. Then, the graph G' contains the edge $v_r v_j$, where $r < j$, if and only if there exists an edge $v_i v_j$ in G such that $i \leq r$ and the vertices v_i, v_r belong to the same connected component of the subgraph of G induced by the vertices $v_1, v_2, \ldots, v_i, \ldots, v_r$.

We note that the above lemma implies Lemma 2 of [9] as a corollary. Lemma 3.3 implies that for the computation of the value $Next\_Neighbor_{G'_v, \sigma}[y_r]$, where $\sigma = (y_1, y_2, \ldots, y_k)$, it suffices to find the leftmost (w.r.t. σ) vertex among $y_{r+1}, y_{r+2}, \ldots, y_k$ which is adjacent in G to a vertex in the connected component of $G[\{y_1, y_2, \ldots, y_r\}]$ to which y_r belongs. This can be efficiently done by processing the vertices in the order they appear in σ. In detail, the algorithm to compute the contents of the array $Next\_Neighbor_{G'_v, \sigma}[\ ]$ is presented in Figure 4.

It is important to observe that, at the completion of the processing of vertex y_j, the sets of vertices maintained by the algorithm are in a bijection with the connected components of $G[\{y_1, y_2, \ldots, y_j\}]$; while processing y_j, we consider the edges $y_i y_j$ where $i < j$, and we union the set containing y_j (which has vertex y_j as its rightmost vertex with respect to σ) to another set iff y_j is adjacent to a vertex in that set. The correctness of the algorithm is established in the following lemma.

Lemma 3.4. *The Algorithm Compute-Next_Neighbor correctly computes the values of $Next\_Neighbor_{G'_v, \sigma}[y_i]$ for all the vertices $y_i \in M(v)$ (i.e., all the vertices that are not adjacent to v in G).*

3.2   Time and Space Complexity

Let us assume that the graph G has n vertices and m edges and that vertex v of G has k non-neighbors in G. The execution of the Algorithm Not-in-HHB(G, σ, v) for vertex v takes $O(n+m)$ time and space plus the time and space needed for the computation of the entries of the array $Next\_Neighbor_{G'_v, \sigma}[\ ]$. So, let us now turn to the time and space complexity of the Algorithm Compute-Next_Neighbor(G, σ, v). If we ignore the operations to process sets (i.e., make a set, union sets, or find the rightmost (w.r.t. σ) vertex in a set), then the rest of the execution of the Algorithm Compute-Next_Neighbor takes $O(n + m)$ time. The sets are maintained by our algorithm in a fashion amenable for Union-Find operations, where additionally the representative of each set also contains a field storing the rightmost (w.r.t. σ) vertex in the set. Then,

- making a set which contains a single vertex y_i requires building the set and setting the rightmost (w.r.t. σ) vertex in the set to y_i;
- finding the rightmost (w.r.t. σ) vertex in the set to which a vertex y_j belongs requires performing a Find operation to locate the representative of the set from which the rightmost vertex is obtained in constant time per Find operation;
- unioning two sets requires constructing a single set out of the elements of the two sets, and updating the rightmost (w.r.t. σ) vertex information; since we always union a set with the set containing y_j, where y_j is the rightmost vertex in any of the sets, then the rightmost vertex of the resulting set is y_j, and this assignment can be done in constant time per union.

As the Algorithm Compute-Next_Neighbor creates one set for each one of the vertices $y_1, y_2, \ldots, y_k$, it executes k make-set operations; this also implies that the number of union operations is less than k. The number of times to find the rightmost (w.r.t. σ) vertex in a set is $O(m)$ since the algorithm executes one such operation for each edge connecting two non-neighbors of v. Hence, if we use disjoint-set forests to maintain the sets, the time to execute the above operations is $O(m\alpha(k))$ [3], where $\alpha(\ )$ is a very slowly growing function; if instead we use the linked-list representation, then the time is $O(m + k \log k)$ [3]. In either case, the space required (in addition to the space needed to store the graph G) is $O(k)$. Thus, the computation of the values of the array $Next\_Neighbor_{G'_v, \sigma}[\ ]$ for the k non-neighbors of the vertex v takes a total of $O(n + \min\{m\alpha(k), m + k \log k\})$ time and $O(k)$ space. Therefore, we have:

Theorem 3.2. *Let G be a graph on n vertices and m edges. Determining whether a vertex v of G belongs to a hole or is the top of a house or a building can be done in $O(n + \min\{m\alpha(k), m + k \log k\})$ time and $O(n + m)$ space, where k is the number of non-neighbors of v in G.*

Applying the Algorithm Non-in-HHB on every vertex of a graph and observing that a building contains a hole, we obtain the following corollary:

Corollary 3.1. *Determining whether a graph G on n vertices and m edges contains a hole or a house (i.e., is not HH-free) can be done in $O(\min\{nm\alpha(n), nm + n^2 \log n\})$ time and $O(n + m)$ space.*

4   Recognition of HHD-free Graphs

Our HHD-free graph recognition algorithm is motivated by the corresponding algorithm of Hoàng and Sritharan [9], which in turn is motivated by the work of Hoàng and Khouzam [8] and relies on the following characterization of HHD-free graphs proved by Jamison and Olariu:

Theorem 4.1. (Jamison and Olariu [10]) *The following two statements are equivalent:*

(i) The graph G is HHD-free;
(ii) For every induced subgraph H of the graph G, every ordering of vertices of H produced by LexBFS is a semi-perfect elimination.

In fact, we could use the Algorithm Not-in-HHB(G, v) in Hoàng and Sritharan's HHD-free graph recognition algorithm in order to determine if vertex v is high, and we would achieve the improved time and space complexities stated in this paper. However, we can get the much simpler algorithm which we give below.

Algorithm Rec-HHD-free

Input: an undirected graph G on n vertices and m edges.

Output: "true," if G is an HHD-free graph; otherwise, "false."

1. **if** the graph G is not HH-free
    then **return**("false");
2. Run LexBFS on G starting at an arbitrary vertex w, and let $(v_1, v_2, \ldots, v_n)$ be the resulting ordering, where $v_n = w$.
3. **for** $i = 1, 2, \ldots, n - 5$ **do**
       if v_i is not semi-simplicial in $G[\{v_i, v_{i+1}, \ldots, v_n\}]$
       then **return**("false");
4. **return**("true").

Note that, after Step 1, we need only check whether the input graph G contains a domino; this is why, we only process the $n - 5$ vertices $v_1, v_2, \ldots, v_{n-5}$ in Step 3. Additionally, it is important to observe that, for all $i = 1, 2, \ldots, n$, the ordering $(v_i, v_{i+1}, \ldots, v_n)$ is an ordering which can be produced by running LexBFS on the subgraph $G[\{v_i, v_{i+1}, \ldots, v_n\}]$ starting at vertex v_n. The correctness of the algorithm follows from Theorem 4.1 and the fact that if the currently processed vertex v_i in Step 3 is semi-simplicial then clearly it cannot participate in a domino (note that none of the vertices of a domino is semi-simplicial in any graph containing the domino as induced subgraph).

4.1   Time and Space Complexity

According to Corollary 3.1, Step 1 takes $O(\min\{nm\alpha(n), nm + n^2 \log n\})$ time and $O(n+m)$ space. Step 2 takes $O(n+m)$ time and space [5, 17]. The construction of the subgraphs $G[\{v_i, v_{i+1}, \ldots, v_n\}]$ in Step 3 can be done in a systematic fashion by observing that $G[\{v_1, \ldots, v_n\}] = G$ and that $G[\{v_{i+1}, \ldots, v_n\}]$ can be obtained from $G[\{v_i, \ldots, v_n\}]$ by removing vertex v_i and all its incident edges; if the graph G is stored using a (doubly-connected) adjacency-list representation with pointers for every edge ab connecting the record storing b in the adjacency list of a to the record storing a in the adjacency list of b and back, then obtaining $G[\{v_{i+1}, \ldots, v_n\}]$ from $G[\{v_i, \ldots, v_n\}]$ takes time proportional to the degree of v_i in $G[\{v_i, \ldots, v_n\}]$ and hence $O(deg(v_i))$ time, where $deg(v_i)$ denotes the degree of vertex v_i in G. Additionally, in order to check whether a vertex is semi-simplicial, we take advantage of the following result of Hoàng and Khouzam (which was also used in [9]):

Theorem 4.2. (Hoàng and Khouzam [8]) *Let G be a graph and x be a semi-simplicial vertex of G. If x is not simplicial, then each big co-component of the subgraph $G[N(x)]$ is a module of G.*

(A connected component or co-component of a graph is called *big* if it has at least two vertices; we also note that if a vertex x is simplicial then none of the co-components of the subgraph $G[N(x)]$ is big.) Since computing the subgraph induced by the neighbors of vertex v_i in $G[\{v_i, \ldots, v_n\}]$, computing its co-components, and testing whether a vertex set is a module in $G[\{v_i, \ldots, v_n\}]$ can all be done in time and space linear in the size of $G[\{v_i, \ldots, v_n\}]$, Step 3 takes a total of $O\left(\sum_i \left(n + m + deg(v_i)\right)\right) = O(nm)$ time and $O(n + m)$ space. Finally, Step 4 takes constant time. Therefore, we obtain the following theorem.

Theorem 4.3. *Let G be an undirected graph on n vertices and m edges. Then, it can be determined whether G is an HHD-free graph in $O(\min\{nm\alpha(n), nm + n^2 \log n\})$ time and $O(n + m)$ space.*

5 Recognition of WPO-graphs

Our algorithm for recognizing WPO-graphs relies on the fact that a graph G is a WPO-graph if and only if G is HHP-free and its complement $\overline{G}$ is HH-free, which follows from the following characterization due to Olariu and Randall [16].

Theorem 5.1. (Olariu and Randall [16]) *A graph G is a WPO-graph if and only if G contains no induced C_5, P_5, house, or "P".*

Eschen *et al.* [4] described an $O(nm)$-time algorithm for recognizing whether a graph G on n vertices and m edges is HHP-free by using the modular decomposition tree of G and Theorem 4.2 due to Hoàng and Khouzam [8]. We next show that we can detect whether the complement $\overline{G}$ of G contains a hole or a house in $O(nm)$ time. Combining these two algorithms, we get an $O(nm)$-time algorithm for recognizing WPO-graphs.

Let G be a graph and let v be an arbitrary vertex of G. We construct the graph $\widehat{G}_v$ from G as follows:

- $V(\widehat{G}_v) = V(G)$
- $E(\widehat{G}_v) = \{\, vy \mid y \in M(v)\,\}$
  $\cup \;\{\, xy \mid x \in N(v),\; y \in M(v),\; \text{and}\; xy \notin E(G)\,\}$
  $\cup \;\{\, xx' \mid x, x' \in N(v)\; \text{and}\; xx' \notin E(G)\,\}$

Note that in $\overline{G}$ the neighbors of v are the vertices in $M(v)$, i.e., the non-neighbors of v in G, and the non-neighbors are the vertices in $N(v)$. Thus, the graph $\widehat{G}_v$ is precisely $\overline{G}$ with any edges between vertices in $M(v)$ removed. Then, it is not difficult to see that the following result holds.

Lemma 5.1. *The vertex v belongs to a hole or is the top of a house or a building in $\overline{G}$ if and only if v belongs to a hole in $\widehat{G}_v$.*

Because in $\widehat{G}_v$ there are no edges between vertices adjacent to v, the vertex v cannot be the top of a house or a building. Thus, we can run the Algorithm Not-in-HHB($\widehat{G}_v, v$) and the vertex v belongs to a hole in $\widehat{G}_v$ if and only if the algorithm returns "false." Assuming that the graph G has n vertices and m edges, the graph $\widehat{G}_v$ has n vertices and $O(n\,deg(v) + deg^2(v)) = O(n\,deg(v))$ edges, where $deg(v)$ is the degree of the vertex v in G; then, the construction of $\widehat{G}_v$ takes $O(m + n\,deg(v))$ time and $O(n\,deg(v))$ space, and the execution of Not-in-HHB($\widehat{G}_v, v$) runs in $O(n + n\,deg(v) + deg(v)\log deg(v)) = O(n\,deg(v))$ time (Theorem 3.2; note that $k = deg(v)$). Thus, we can determine whether the vertex v belongs to a hole in $\widehat{G}_v$ in $O(m + n\,deg(v))$ time and $O(n\,deg(v))$ space.

Therefore, in light of Lemma 5.1, we have the following result.

Theorem 5.2. *Let G be an undirected graph on n vertices and m edges. Then, it can be determined whether the complement $\overline{G}$ is an HH-free graph in $O(nm)$ time and $O(n^2)$ space.*

From Theorem 5.2 and the result of Eschen *et al.* [4] (i.e., HHP-free graphs can be recognized in $O(nm)$ time and $O(n + m)$ space), we obtain the following theorem.

Theorem 5.3. *Let G be an undirected graph on n vertices and m edges. Then, it can be determined whether G is a WPO-graph in $O(nm)$ time and $O(n^2)$ space.*

6   Concluding Remarks

We have presented recognition algorithms for the classes of HHD-free graphs and WPO-graphs running in $O(\min\{nm\alpha(n), nm + n^2\log n\})$ and $O(nm)$ time, respectively, where n is the number of vertices and m is the number of edges of the input graph. Our proposed algorithms are simple, use simple data structures, and require $O(n + m)$ and $O(n^2)$ space, respectively. Moreover, our HH-free and HHD-free graph recognition algorithms can be easily augmented to yield a certificate (a hole, a house, or a domino) whenever they decide that the input graph is not HH-free or HHD-free [13].

We leave as an open problem the designing of $O(nm)$-time algorithms for recognizing HHD-free graphs. In light of the $O(nm)$-time recognition of P_4-comparability, P_4-simplicial, bipolarizable, and WPO-graphs, it would be worth investigating whether the recognition of brittle and semi-simplicial graphs is inherently more difficult.

References

1. A. Brandstädt, V.B. Le, and J.P. Spinrad, *Graph classes: A survey*, SIAM Monographs on Discrete Mathematics and Applications, 1999.
2. V. Chvátal, Perfectly ordered graphs, *Annals of Discrete Math.* **21**, 63–65, 1984.

3. T.H. Cormen, C.E. Leiserson, R.L. Rivest, and C. Stein, *Introduction to Algorithms* (2nd edition), MIT Press, Inc., 2001.
4. E.M. Eschen, J.L. Johnson, J.P. Spinrad, and R. Sritharan, Recognition of some perfectly orderable graph classes, *Discrete Appl. Math.* **128**, 355–373, 2003.
5. M.C. Golumbic, *Algorithmic Graph Theory and Perfect Graphs*, Academic Press, Inc., 1980.
6. R. Hayward, Meyniel weakly triangulated graphs I: co-perfect orderability, *Discrete Appl. Math.* **73**, 199–210, 1997.
7. C.T. Hoàng, On the complexity of recognizing a class of perfectly orderable graphs, *Discrete Appl. Math.* **66**, 219–226, 1996.
8. C.T. Hoàng and N. Khouzam, On brittle graphs, *J. Graph Theory* **12**, 391–404, 1988.
9. C.T. Hoàng and R. Sritharan, Finding houses and holes in graphs, *Theoret. Comput. Sci.* **259**, 233–244, 2001.
10. B. Jamison and S. Olariu, On the semi-perfect elimination, *Adv. Appl. Math.* **9**, 364–376, 1988.
11. R.M. McConnell and J. Spinrad, Linear-time modular decomposition and efficient transitive orientation, *Proc. 5th Annual ACM-SIAM Symp. on Discrete Algorithms (SODA'94)*, 536–545, 1994.
12. M. Middendorf and F. Pfeiffer, On the complexity of recognizing perfectly orderable graphs, *Discrete Math.* **80**, 327–333, 1990.
13. S.D. Nikolopoulos and L. Palios, Recognizing HHD-free and Welsh-Powell opposition graphs, Technical Report TR-16-04, Dept. of Computer Science, University of Ioannina, 2004.
14. S. Olariu, All variations on perfectly orderable graphs, *J. Combin. Theory* Ser. B **45**, 150–159, 1988.
15. S. Olariu, Weak bipolarizable graphs, *Discrete Math.* **74**, 159–171, 1989.
16. S. Olariu and J. Randall, Welsh-Powell opposition graphs, *Inform. Process. Lett.* **31**, 43–46, 1989.
17. D.J. Rose, R.E. Tarjan, and G.S. Lueker, Algorithmic aspects of vertex elimination on graphs, *SIAM J. Comput.* **5**, 266–283, 1976.
18. D.J.A. Welsh and M.B. Powell, An upper bound on the chromatic number of a graph and its applications to timetabling problems, *Comput. J.* **10**, 85–87, 1967.

Bimodular Decomposition of Bipartite Graphs

Jean-Luc Fouquet[1], Michel Habib[2],
Fabien de Montgolfier[3], and Jean-Marie Vanherpe[1]

[1] Université du Maine
{Jean-Luc.Fouquet,Jean-Marie.Vanherpe}@univ-lemans.fr
[2] LIRMM, université Montpellier II
habib@lirmm.fr
[3] LIAFA, université Paris 7
fm@liafa.jussieu.fr

Abstract. This paper gives a decomposition theory for bipartite graphs. It uses *bimodules*, a special case of *2-modules* (also known as *homogeneous pairs*, an extension of both modules and splits). It is shown how a unique decomposition tree represents the bimodular decomposition of a bipartite graph, with strong analogs with modular decomposition of graphs. An $O(mn^3)$ algorithm for this decomposition is provided. At least a classification of the 2-modules of a bipartite graph is given.

1 Introduction

There are many ways to decompose a graph. Among the most popular is the *modular decomposition*, less known is the *split decomposition*. Both produce a *decomposition tree* of a given graph, providing a better understanding of its structure, and a way to solve a wide class of optimization problems using a divide-and-conquer method, when the graph is decomposable. An indecomposable graph is said to be *prime*. Unfortunately, "most" graphs are prime with respect to the modular or split decomposition. This leads to search for more powerful decomposition tools.

The modular decomposition of bipartite graphs is poor, giving only connected components and twins (vertices with the same neighborhood and the same color). A decomposition theory was already given by Fouquet, Giakoumakis and Vanherpe in [1], namely *canonical decomposition* for bipartite graphs. This theory uses three decomposition operation while modular decomposition knows four of them, the three first very similar to canonical decomposition. But the fourth case was missing. In order to extend this decomposition we consider a particular family of 2-modules in the class of bipartite graphs, namely the *bimodules*. We present a way to represent all the bimodules of a bipartite graph. For this we use a *bimodular* decomposition which is a generalization of the canonical decomposition.

Some Set Theory Background. Let us now give a summary of a known and powerful decomposition frame [2–4] that has applications in graph decomposition. Two sets A and B *overlap* whenever $A \cap B \neq \emptyset$, $A \setminus B \neq \emptyset$, and $B \setminus A \neq \emptyset$.

J. Hromkovič, M. Nagl, and B. Westfechtel (Eds.): WG 2004, LNCS 3353, pp. 117–128, 2004.
© Springer-Verlag Berlin Heidelberg 2004

Definition 1. *A* point-partitive hypergraph *[3] is a couple* $(V, \mathcal{F})$, *where* V *is a finite set and* $\mathcal{F}$ *is a family of subsets of* V *such that*
1. $\emptyset \in \mathcal{F}, V \in \mathcal{F}$, *and* $\{v\} \in \mathcal{F}$ *for each* $v \in V$
2. *If* $A, B \in \mathcal{F}$ *are overlapping then* $A \cup B \in \mathcal{F}$ *and* $A \cap B \in \mathcal{F}$ *and* $A \setminus B \in \mathcal{F}$.

If $A \in \mathcal{F}$ does not overlap with any other member of $\mathcal{F}$, A is said *strong*. Let $\mathcal{T}$ be the inclusion tree of the strong subsets. Its root is V and its leaves are $\{v\}, v \in V$. Obviously it has $O(|V|)$ nodes.

Theorem 1 ([3]). *The nodes of* $\mathcal{T}$ *can be marked* **complete**, **linear** *or* **prime** *in such a way that*
1) The union of an arbitrary choice of sons of a complete *node belongs to* $\mathcal{F}$.
2) There is a unique (up to reversal) ordering on the sons of a linear *node such that every union of sons of a* linear *node that are consecutive in this ordering is a member of* $\mathcal{F}$.
3) A prime *node is a member of* $\mathcal{F}$.
4) There is no other member in $\mathcal{F}$ *than those defined above.*

Modular Decomposition of Graphs. One of the most known point-partitive hypergraph is the family of modules of a graph. Here a graph $G = (V, E)$ is finite, loopless, and directed. Let $A \subset V$ be a set of vertices. $x \notin A$ *distinguishes* A if A overlaps the in-neighborhood of x, or if it overlaps x out-neighborhood.

Definition 2. *A* module *in a graph is a subset of vertices that no vertex distinguishes.*

The sets $\emptyset$, V and the singletons $\{v\}, v \in V$ are *trivial* modules. The number of modules may be exponential (2^n in a complete graph), but the family of modules can be stored in the $O(n)$-sized tree of Theorem 1. In this case the tree $\mathcal{T}$ is called the *modular decomposition tree*. In the case of the undirected graphs there is no *linear* nodes ; the *complete* nodes correspond to the *series* and *parallel* composition, and the *prime* nodes to a prime quotient graph. In the case of directed graphs, the *linear* nodes correspond to the *total order* composition. The modular decomposition tree can be computed in $O(n + m)$ time [5–8].

Motivation of Our Article. The modular decomposition of bipartite graphs is poor, giving only connected components and twins. A decomposition theory was already given by Fouquet, Giakoumakis and Vanherpe in [1], namely *canonical decomposition* for bipartite graphs. But internal nodes in the associated decomposition tree were only of two types : *complete* or *linear* and the analog of the *prime* case was missing. The class of totally decomposable bipartite graphs, called *weak-bisplit graphs*, is thus an analog of *cographs* [9] with respect to modular decomposition of graph. In this paper, we extend the canonical decomposition with a fourth decomposition case. We show that this extended canonical decomposition can be changed into a bimodular decomposition, that describes all bimodules of a graph.

2   Canonical Decomposition of Bipartite Graphs

In the remaining of the paper, we focus on *bipartite* (without odd cycles) graphs. Let us (re)define some graph concepts to this special case. A bipartite graph is written as the triple $(\mathcal{B}, \mathcal{W}, E)$ where $\mathcal{B} \cup \mathcal{W} = V$. The sets $\mathcal{B}$ (black) and $\mathcal{W}$ (white) are the *color classes* of G. A subset of vertices which is either contained in $\mathcal{B}$ or in $\mathcal{W}$ is *monochromatic* (*bichromatic* otherwise). $G[A]$ denotes the graph induced by A. The *bipartite complement* of $G = (\mathcal{B}, \mathcal{W}, E)$ is defined by $\overline{G}^{bip} = (\mathcal{B}, \mathcal{W}, (\mathcal{B} \times \mathcal{W}) \setminus E)$. Let $A \subset V$ be a set of vertices. x is *isolated* if it has no neighbor and *universal* if its neighborhood is $\mathcal{B}$ or $\mathcal{W}$.

2.1   The Canonical Decomposition [1]

In [1], Fouquet, Giakoumakis and Vanherpe introduced the notion of K+S-decomposition. An ordered partition $(V_1, V_2, \ldots, V_k)$ of the vertex set in a bipartite graph is a K+S-partition whenever the following is verified: *if i and j are such that $1 \leq i < j \leq k$, the black vertices of V_i are all adjacent to the white vertices of V_j while there is no edge connecting a white vertex of V_i to a black vertex of V_j.* It is shown in [1] that the K+S-partition that maximizes k is unique, and called the **K+S decomposition** of G. Set V_i will be said a *K+S-component* of the K+S-decomposition.

For a bipartite graph G, *canonical decomposition* also defined in [1] recursively applies K+S-decomposition or *parallel* decomposition (following the connected components of G) or *series* decomposition (following the connected components of $\overline{G}^{bip}$).

A graph is **c-decomposable** (*c* for "components") if it admits a nontrivial (into several components) parallel, series or K+S decomposition. As Proposition 3 below shows (in a more general cases) the decompositions are mutually exclusives under the conditions of Algorithm 1.

2.2   Bimodules

When dealing with bipartite graphs, a black (resp. white) vertex $x \notin A$ is said to *distinguish* A if x has a white (resp. black) neighbor and a white (resp. black) non-neighbor in A. This notion of distinction is more convenient for bipartite graphs and allows to define:

Definition 3. *M is a* bimodule *if no vertex distinguishes M.*

Unfortunately the bimodules are not a point-partitive hypergraph: if M_1 and M_2 are overlapping bimodules, $M_1 \cup M_2$ or $M_1 \setminus M_2$ can fail to be a bimodule (see Figure 2 for an example). Theorem 1 does not apply. The canonical decomposition can be extended, using a fourth decomposition operation. The goal of this article is to show that the extended decomposition tree represents the structure of all bimodules of a graph.

Fig. 1. K+S components of a graph. The circled vertices form a bimodule that falls in last case of Lemma 4 (see below)

Algorithm 1: Canonical decomposition process of a bipartite graph.

Data: A bipartite graph G
if G has at least two K+S components **then**
    decompose G following its K+S-components
else if G has at least 5 vertices and is not connected **then**
    /* G has a *Parallel* decomposition */
    decompose G following its connected components
else if $\overline{G}^{bip}$ has at least 5 vertices and is not connected **then**
    /* G has a *Series* decomposition */
    decompose G following the connected components of $\overline{G}^{bip}$
else /* G is c-indecomposable */
    stop.

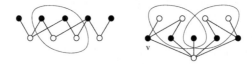

Fig. 2. Left: example of bimodule. Right: the two bimodules overlap but their union is distinguished by vertex v: it is a **conflict** case.

2.3   Relationships Between Bimodules and Canonical Decomposition

One can check that every component (following any decomposition) of a graph is a bimodule. Furthermore every union of connected components of G or of $\overline{G}^{bip}$ is a bimodule, and any union of consecutive K+S components $V_i \cup V_{i+1} \cup \ldots V_{j-1} \cup V_j$ also is a bimodule, while a union of non-consecutive K+S components is not a bimodule. The canonical decomposition tree can thus be seen as the tree of a point-partitive hypergraph, that has only *complete* and *linear* nodes. But not all bimodules are found in that way.

In the following we will consider only *twin-free* bipartite graphs, where twins are vertices with the same color and the same neighborhood. This is not a strong constraint because twin classes can easily be factorized. Trivially, in a twin-free graph a bimodule induces a subgraph which is also a twin-free graph. Recursivity can be used to identify bimodules, because one can check that:

Lemma 1. *Let M be a bimodule and $M' \subset M$. M' is a bimodule of G iff M' is a bimodule of $G[M]$.*

Proposition 1. *Every bimodule of a c-decomposable graph either is a union of components, or is included in one component, or is included in two or three K+S components that follow consecutively. In the last case only one component may have more than one vertex.*

This proposition allows to reduce the problem of finding the bimodules of a c-decomposable graph to the problem of finding the bimodules in the graphs induced by every connected component, if the graph has no K+S decomposition.

For K+S decomposable graphs, the recursive search misses the bimodules that fall into the last case. Section 3 focuses on such bimodules.

The proof of the proposition is the straightforward consequence of the following three lemmas:

Lemma 2. *Let G be a twin-free bipartite graph with no isolated vertex, and $C_1 \ldots C_k$ its connected components. A bimodule of G is*

- *either a union of components*
- *or included (or equal) in a component C_i*
- *or trivial (one black vertex, one white vertex)*

Proof. If G fulfills the conditions of the lemma, then every connected components is bichromatic. Let M be a bimodule. If M is neither trivial nor included in a component, then M contains two vertices with the same color (say white) $w \in C_i$ and $w' \in C_j$, $i \neq j$. If there exists a black vertex in C_i that does not belong to M, as C_i is connected there is a vertex $b \in C_i \setminus M$ adjacent to a white vertex w'' of $C_i \cap M$ and not to w': b distinguishes M, impossible. M thus contains all black vertices of C_i, and we can show it also contains black and white vertices of C_i and C_j, thus is a union of connected components.

Lemma 3. *M is a bimodule of a bipartite graph G iff it is a bimodule of its bipartite complement $\overline{G}^{bip}$*

This is an immediate consequence of definitions. If $\overline{G}^{bip}$ is not connected, then Lemma 2 applies to $\overline{G}^{bip}$, and reduces to the *parallel* case.

Lemma 4. *Let G be a twin-free bipartite graph, $C_1 \ldots C_k$ its ordered K+S components and M a bimodule of G.*
- *Either M is a union of consecutive components,*
- *or M is included (or equal) in a K+S component C_i,*
- *or M is trivial (one black vertex, one white vertex),*
- *or all the vertices of M belong to a component C_i, excepted a vertex $c \in M$ such $C_{i-1} = \{c\}$ and/or a vertex $c' \in M$ such $C_{i+1} = \{c'\}$.*

Proof. Let C be a K+S component and M a bimodule that overlaps C. First, notice that, as the K+S components are not K+S decomposable, $G[C]$ either is a singleton, or has at least four vertices and has no universal nor isolated vertex. As M is a bimodule, $C \setminus M$ contains at least two vertices, and is bichromatic because G is twin-free.

Let us suppose $C \cap M = \{x\}$ and, w.l.o.g., that x is black. If M contains another black vertex in a component that follows C, no white vertex of C can be adjacent to M so x is isolated in $G[C]$, a contradiction. And universal if there exists a black vertex that precedes C. If M has only one white vertex, it has at least two white ones: swap colors. So $M \cap C$ is bichromatic.

Let us suppose that M contains a black vertex b in a component C' that follows C. The white vertices of C are not neighbor of b, so are not adjacent to $C \cap M$. If M also contains a white vertex in a component that follows C, then C

has a K+S partition into $C \cap M$ and $C \setminus M$, a contradiction. According to what has been said, C' can not overlap M (the intersection would be bichromatic) so $C' \subset M$ and, as C' has no white vertices and no twins, $C' = \{b\}$. If there is a component between C and C', it must contains a white vertex that distinguishes M: C' immediately follows C. Same proof can be done for the other color or the other side of C.

2.4   The Prime Decomposition

Let us now try to decompose c-indecomposable graphs. A bimodule M is **maximal** if $M \neq V$ and the only bimodule that contains M is V. M is **nontrivial** if it has at least three vertices. Two maximal bimodule may overlap. A vertex is **shared** if it belong to more than one nontrivial maximal bimodule.

Proposition 2. *Let G be a twin-free c-indecomposable bipartite graph.*

1. *The intersection of two maximal bimodules of G contains at most one vertex*
2. *If $|M| \geq 4$, a shared vertex of M is universal or isolated in $G[M]$*
3. *Each maximal bimodule of G contains at most two shared vertices*

Before the proof, let us define the family of *strong components* of a c-indecomposable graph G. A vertex x is *single* if it belong to no nontrivial maximal bimodule. Two vertices are *G-equivalent* if they belong to exactly the same nontrivial maximal bimodules of G. But say that a *single* vertex is G-equivalent to itself only (and not to other *singles*).

Definition 4. *The* strong components *of a c-indecomposable graph G are the equivalence classes of the G-equivalence.*

A strong component C, according to Proposition 2, is either $\{x\}$ where x is *single*, or $\{x\}$ where x is shared, or a non-trivial bimodule M without its shared vertices. If $|M| \geq 4$ then, according to Point 2, C is a bimodule. If $|M| = 3$ then C is a trivial bimodule. $|M \setminus C| \leq 2$, therefore the strong components are "almost" the maximal nontrivial bimodules. Not let us prove three lemmas, then Proposition 2.

Lemma 5. *Let $M_1 \neq V$ and $M_2 \neq V$ be two overlapping bimodules such that $M_1 \cup M_2 = V$. G is c-decomposable.*

Proof. Since M_2 is a bimodule, the vertices of $(M_1 \setminus M_2) \cap \mathcal{B}$ do not distinguish those of $M_2 \cap \mathcal{W}$ and the vertices of $(M_1 \setminus M_2) \cap \mathcal{W}$ do not distinguish those of $M_2 \cap \mathcal{B}$. According to the cases, $M_1 \setminus M_2$ is either a union of connected components of G, a union of connected components of $\overline{G}^{bip}$ or a union of K+S-components of G.

Lemma 6. *Let M and M' be two overlapping bimodules. If $M \cap M'$ is bichromatic, then $M \cup M'$ is a bimodule.*

Proof. Immediate: if a white vertex not in $M \cup M'$ is adjacent to a black vertex of M, it is adjacent to all vertices of M, thus to a black vertex of $M \cap M'$, thus to all vertices of M'. Same for black vertices.

Lemma 7. *A twin-free bipartite graph contains at most two universal or isolated vertices.*

Proof. It has no twins, so at most four such vertices; furthermore cases *black universal* and *white isolated* are mutually exclusive, and so are the two others cases.

Now we can demonstrate Proposition 2. Point 1. If a maximal bimodule has less than three vertices, it intersects other ones on at most one vertex. So let us consider two overlapping bimodules M and M' with at least three vertices each. First, Lemma 5 shows that $M \cup M' \neq V$. $M \cup M'$ is therefore not a bimodule. According to Lemma 6, $M \cap M'$ is monochromatic. Black vertices from $V \setminus M$ and from $V \setminus M'$ can not distinguish $M \cap M'$, therefore $M \cap M'$ vertices are twins. As M is twin-free, $M \cap M' = \{m\}$.

Point 2. Let us prove that m is universal or isolated in $G[M]$ and in $G[M']$. Let us suppose w.l.o.g. that m is white. If M' contain another white vertex w', let b_1 and b_2 be two (possibly equal) black vertices of M. If b_1 is neighbor of m, then it is neighbor of w' (bimodularity of M'), w' is neighbor of b_2 (bimodularity of M) and then b_2 is neighbor of m: m is universal in $G[M]$. And if M' has no other white vertex than w, M' is a triple $\{w, b, b'\}$ with w neighbor of b and non-neighbor of b', thus if M has another white vertex w'' it must distinguish M', contradiction. So M also is a triple $\{w, b'', b'''\}$

Point 3. Lemma 7 ends the proof.

2.5   The Extended Canonical Decomposition

We can now extend the canonical decomposition: in Algorithm 1 just replace the last line "stop" by "decompose G into its strong components".

The decomposition sets (K+S, parallel, series or strong components according to the case) are bimodules of the graph, so according to Lemma 1, this algorithm defines a family of non-overlapping bimodules, the **strong canonical bimodules** of the graph. Their inclusion tree is called the **bimodular decomposition tree** and is denoted $T(G)$ in the following. Internal nodes are labeled either *Parallel* or *Series* or *K+S* or *Prime* according to the type of decomposition applied. The **canonical bimodules** are the strong ones plus any union of sons of a parallel or series node, plus any union of consecutive sons of a K+S node. Clearly we construct a point-partitive hypergraph:

Theorem 2. *Let G be a twins-free bipartite graph, let $\mathcal{F}$ be the family of canonical bimodules of G. The couple $(V, \mathcal{F})$ is a point-partitive hypergraph.*

$T(G)$ is precisely the tree associated with the hypergraph. According to Theorem 1, Parallel and Series nodes can be marked **complete**, K+S-nodes can be marked **linear**, the remaining internal nodes being marked **prime**. Furthermore:

Proposition 3. *The four decomposition cases are mutually exclusive*

Proof. First, notice that the only twins-free bipartite graph that is not connected, with a not connected bipartite complement, and no isolated nor universal vertex is $2K_2$ (four vertices, two unconnected edges), that is thus the smallest c-indecomposable graph. And if a graph has a K+S decomposition, the K+S structure connects all vertices excepted the white of first components and the black of last components. If they are not connected to their component, they are isolated and the graph has no parallel decomposition. And no series decomposition, because the bipartite complement of a K+S graph is a K+S graph.

3   Bimodular Decomposition of Bipartite Graphs

Let n be the number of vertices of a graph. Any singleton and bichromatic pair of vertices is a bimodule: there are $O(n^2)$ of them. In a twin-free bipartite graph, a bimodule with three vertices is a triple $\{u, u', v\}$ where u and u' have the same color and their neighborhoods differ only on v. Let us consider the bipartite graph with b black vertices, identified to bits from 0 to $b-1$, and 2^b numbered white vertices, where a white vertex is adjacent to the bits set in its number (binary written). It is twin-free and has $b2^{b-1} = \Theta(n \log n)$ bimodule with size 3 (all bimodules $\{w, w', i\}$ where w and w' numbers differ on bit i, $i \in \mathcal{B}$). All the canonical bimodules of this graph are trivial. This worst case shows that an $\Omega(n \log n)$ coding of bimodules is needed; [10] shows that the bottleneck indeed is the three-vertices bimodules that can be stored in $O(n \log n)$ size. In this section we give an $O(n)$ representation of all the bimodules with at least four vertices of a twin-free graph, using the decomposition tree.

The canonical bimodules are already represented by the canonical decomposition tree. It has $O(n)$ leaves and no degree-1 node, so its size is $O(n)$. Let M be a bimodule that is not canonical and S be the smallest strong canonical bimodule that contains M. M also is a bimodule of $G[S]$. Now we use Propositions 1 and 2: $G[S]$ has

- either a K+S decomposition and M is almost contained in one of its K+S components C, in the sense that $|M \setminus C| \leq 2$. The vertices of $M \setminus C$ are the singleton components that immediately precedes or follows C
- or $G[S]$ is c-indecomposable and M is almost contained in one of its strong components C, $|M \setminus C| \leq 2$. The vertices of $M \setminus C$ are shared.

In both case the vertices that are not in the components are isolated or universal in $G[M]$. As the graph has no twin, given a canonical bimodule, there are only three way to *extend* it to a non-canonical 2-module: adding one of these vertices, the other one, or both. Conversely all canonical bimodules can be generated like this.

Definition 5. *Let C be a canonical bimodule and $v \notin C$. v is* augmenting *C if $\{v\} \cup C$ is a non-canonical bimodule. v is* minimally augmenting *C if no canonical bimodule containing C is augmented by v.*

Is is easy to see that for each bimodule included in M, v is augmenting. The canonical bimodules that admit minimally augmenting vertices are the strong ones; Lemma 7 then apply: a canonical bimodule has at most two vertices that

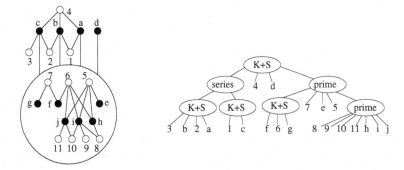

Fig. 3. Example of extended canonical decomposition tree of a graph. Vertices a..d and 7..11 are neighbors. The non-canonical bimodules are $\{a, 2, 4\}$ $\{a, b, 2, 4\}$ $\{a, b, 2, 3, 4\}$ $\{c, 1, 4\}$ $\{h, 8, 9\}$ $\{i, 10, 11\}$ $\{j, 9, 10\}$ and bichromatic couples.

minimally augment him. At most two pointers, for each node of $T(G)$, store the augmenting vertices and thus all the bimodule, canonical or not, are *stored* in $O(n)$ size by this pointers-labeled canonical decomposition tree, then called *bimodular decomposition tree* (this means that the full list of bimodules with a least four vertices can be output from the tree, in time linear in the list size). In fact for K+S decomposition nodes, such pointers may be omitted, the order of the components is enough to identify augmenting vertices.

4   Decomposition Algorithm

The bimodular decomposition tree can be computed in polynomial time, recursively. The connected components of G and $\overline{G}^{bip}$ are easily computed. Finding the K+S components can be achieved in $O(n)$ time only when degrees sequences are known [1]. The strong components can be found as follow.

Given two same-colored vertices u and v, the unique *smallest* (with respect to inclusion) bimodule containing u and v, $sb(u, v)$, can be found greedily, by adding distinguishers until convergence, in $O(m)$ time (where n is the number of edges and n the number of vertices).

Lemma 8. *Let N be the least common ancestor of u and v in $T(G)$, and C_i (resp. C_j) the son of N in $T(G)$ containing u (resp. v) If N is*

1. *series or parallel then* $sb(u, v) = C_i \cup C_j$
2. *K+S then* $sb(u, v) = C_i \cup C_{i+1} \cup \ldots \cup C_{j-1} \cup C_j$
3. *prime then* $sb(u, v) = N$

Consider a c-indecomposable graph G, and compute $sb(u, v)$ for all possible monochromatic pairs. Discard the cases $sb(u, v) = V$ and perform a union algorithm on the remaining sets (union of sets that overlap): according to the previous lemma you find the maximal bimodules. It is then easy to remove shared vertices to get the strong components. And then to find the minimally augmenting vertices, in $O(n)$ per bimodule. All this is done in $O(mn^2)$, thus, recursively:

Theorem 3. *The bimodular decomposition tree can be computed in $O(mn^3)$ time.*

5   Classification of the 2-Modules of a Bipartite Graph

The *2-modules*, also called *homogeneous pair*, introduced in [11], are a generalization of both modules and *splits* in undirected graphs. In fact, in a given graph there are more 2-modules than modules and splits, but this makes the structure of the 2-modules very difficult to understand. The theory of modular decomposition and of split decomposition have been extensively studied and are well understood. Their algorithmic aspect is also very nice, because both have fast decomposition algorithms. The "2-modular" decomposition seems harder, and – as far as we know – the only work about it is a paper from Everett, Klein and Reed [12] that finds one 2-module in a graph, if it exists, in $O(mn^3)$ time.

Definition 6. *A homogeneous pair [11], also know as 2-module of an undirected graph G is a pair $M = \{M_a, M_b\}$ of disjoint subsets of V such that all vertices that distinguish M_a are in M_b and all vertices that distinguish M_b are in M_a.*

A bimodule M is a 2-module $\{M \cap \mathcal{W}, M \cap \mathcal{B}\}$. A module M is a 2-module $\{M, \emptyset\}$.

Definition 7. *A split [13] is a partition (V_1, V_2) of V such that there exists four disjoint subsets of V, namely A, B, C and D, verifying $V_1 = A \cup B$, $V_2 = C \cup D$, B and C are totally adjacent while there is no edge connecting A to C, A to D and B to D.*

In the following a split will be denoted (A, B, C, D). $\{A, B\}$ and $\{C, D\}$ are then two 2-modules.

Theorem 4. *Let $G = (\mathcal{B}, \mathcal{W}, E)$ be a bipartite graph and $M = \{M_a, M_b\}$ be a 2-module of G then*

- *either M is monochromatic and all the vertices of M are twins,*
- *or $(M_b, M_a, N(M_a) \setminus M_b, \overline{N(M_a)} \setminus M_b)$ is a split,*
- *or $(M_a, M_b, N(M_b) \setminus M_a, \overline{N(M_b)} \setminus M_a)$ is a split,*
- *or $M_a \cup M_b$ is disconnected from the rest of the graph,*
- *or M is a bimodule.*

Proof. Let A be $(N(M_a) \cap \overline{N(M_b)}) \setminus M$, B be $(N(M_a) \cap N(M_b)) \setminus M$, C be $(\overline{N(M_a)} \cap N(M_b)) \setminus M$ and D be $(\overline{N(M_a)} \cap \overline{N(M_b)}) \setminus M$. Two vertices having a common neighbor have the same color.

If $B \neq \emptyset$ all the vertices of $M_a \cup M_b$ have a common neighbor, thus $M_a \cup M_b$ is monochromatic and there is no edge between M_a and M_b. Therefore, all the vertices of $M_a \cup M_b$ are twins.

Suppose $B = \emptyset$. If $A \neq \emptyset$ and $C \neq \emptyset$ all the vertices of M_a have the same color, and all the vertices of M_b have the same color. If this color is the same then M is a monochromatic set of twin vertices else $\{M_a, M_b\}$ is a bimodule.

If $A \neq \emptyset$ and $C = \emptyset$ then (M_b, M_a, A, D) is a split. If $A = \emptyset$ and $C \neq \emptyset$ then(M_a, M_b, C, D) is a split. And when $A = B = C = \emptyset$ the set $M_a \cup M_b$ is disconnected from the rest D of the graph.

Monochromatic 2-modules are given by modular decomposition (see [5–7] for linear time decomposition algorithms). Moreover, 2-modules which are splits are given by the split decomposition of the bipartite graph. In [13], Cunningham explains how this family can be uniquely represented, using an unrooted tree whose leaves are the vertices of the graph and whose edges are the *strong splits*. In [14], Dahlhaus gives an $O(n+m)$ time algorithm for the split decomposition. In this article we have shown that the last family of 2-modules, the bimodules, also is organized into a tree computable in polynomial time. The 2-modular decomposition of a bipartite graph is therefore divided into two "orthogonal" families, splits and bimodules, with two decomposition trees having no clear relationships. Both generalize modules, in different ways.

6   Conclusion and Perspectives

Many more things could be said about bimodular decomposition. For instance, the decomposition is invariant under bipartite complement (that swaps series and parallel, and reverse the K+S orders). Or the *substitution* operation: one can substitute a bipartite graph H to two vertices b and w of a bipartite graph G, building a larger graph. There is information loss, because the result is the same whenever b and w are neighbor or not. On the other hand, one can *quotient* a graph from a bimodule: all black vertices of the bimodule are replaced by a new one, and so for white vertices. We cannot decide if new vertices are to be adjacent or not: they must be linked by a *special edge*. The graph quotiented by M plus the graph induced by M is a smaller representation of the initial graph.

Our decomposition promises wide uses. For instance [15] uses a variant of it and proves that the maximum induced matching problem, NP-complete for arbitrary bipartite graphs, can be solved in linear time for the class of $Star_{123}$-free graphs, using a top-down computation on the decomposition tree. Or the cliquewidth may be computed from the decomposition tree, knowing the cliquewidth of all quotients. It is challenging to find new classes of bipartite graphs, defined both by their decomposition tree and another property (like [1] that show that weak-bisplit graphs, that have no prime node, are characterized by two forbidden induced subgraphs) and to use this for new efficient algorithms design...

References

1. Fouquet, J., Giakoumakis, V., Vanherpe, J.: Bipartite graphs totally decomposable by canonical decomposition. International Journal of Fundations of Computer Science **10** (1999) 513–534
2. Cunningham, W., Edmonds, J.: A combinatorial decomposition theory. Canadian Journal of Mathematics **32** (1980) 734–765

3. Chein, M., Habib, M., Maurer, M.C.: Partitive hypergraphs. Discrete Mathematics **37** (1981) 35–50
4. Mohring, R., Radermacher, F.: Substitution decomposition for discrete structures and connections with combinatorial optimization. Annals of Discrete Mathematics **19** (1984) 257–356
5. Cournier, A., Habib, M.: A new linear Algorithm for Modular Decomposition. Lectures notes in Computer Science (1994) 68–84
6. Dahlhaus, E., Gustedt, J., McConnell, R.: Efficient and practical modular decomposition. SODA 97 (1997) 26–35
7. McConnell, R., Spinrad, J.: Modular decomposition and transitive orientation. Discrete Mathematics **201** (1999) 189–241
8. McConnell, R.M., de Montgolfier, F.: Linear-time modular decomposition of directed graphs. To appear in *Discrete Applied Mathematics* (2003)
9. Sumner, D.P.: Graphs indecomposable with respect to the X-join. Discrete Mathematics **6** (1973) 281–298
10. de Montgolfier, F.: Décomposition modulaire des graphes. Théorie, extensions et algorithmes. PhD thesis, Université Montpellier II (2003) In French, available at `http://www.lirmm.fr/~montgolfier`.
11. Chvátal, V., Sbihi, N.: Bull-free berge graphs are perfect. Graphs Combinatorics **3** (1987) 127–139
12. Everett, H., Klein, S., Reed, B.: An algorithm for finding homogeneous pairs. Discrete Applied Mathematics **72** (1997) 209–218
13. Cunningham, W.: Decomposition of directed graphs. SIAM Journal of algebraic and discrete methods **3** (1982) 214–228
14. Dahlhaus, E.: Parallel algorithms for hierarchical clustering, and applications to split decomposition and parity graph recognition. Journal of Algorithms **36** (2000) 205–240
15. Lozin, V.: On maximum induced matchings in bipartite graphs. Information Processing Letters **81** (2002) 7–11

Coloring a Graph Using Split Decomposition

Michaël Rao

Université de Metz, Laboratoire d'Informatique Théorique et Appliquée
57045 Metz Cedex 01, France
rao@sciences.univ-metz.fr

Abstract. We show how to use split decomposition to compute the weighted clique number and the chromatic number of a graph and we apply these results to some classes of graphs. In particular we present an $O(n^2 m)$ algorithm to compute the chromatic number for all those graphs having a split decomposition in which every prime graph is an induced subgraph of either a C_k or a $\overline{C_k}$ for some $k \geq 3$.

1 Introduction

Decompositions play an important role in graph theory. The central role of decompositions in the recent proof of one of the major open conjectures in Graph Theory, the Strong Perfect Graph Conjecture of C. Berge, is an exciting example [7]. Furthermore various decompositions of graphs such as decomposition by clique separators, tree decomposition and clique decomposition are often used to design efficient graph algorithms. There are even beautiful general results stating that a variety of NP-complete graph problems can be solved in linear time for graphs of bounded treewidth and bounded clique-width, respectively [1, 9].

The typical approach to design efficient algorithms using graph decompositions works as follows. The algorithm recursively decomposes the graph into smaller graphs, until the obtained graphs cannot be decomposed further. Such graphs are called prime. Then the algorithm solves the problem on the prime graphs, and combines the solutions recursively to find eventually the solution for the original graph. In order to obtain an efficient algorithm by this approach, the input graphs have to be restricted to a graph class nicely decomposable with respect to the decomposition.

Several decompositions have been studied in this direction. Tarjan has given some NP-complete problems which can be solved using decomposition by clique separators [24]. Modular decomposition for discrete structures is known and studied for a long time. A nice survey on this topic has been written by Möhring and Radermacher [21]. Recently, a lot of work has been done studying modular decomposition on graphs. This includes linear time modular decomposition algorithms [10, 20], the study of modular decomposition for certain graphs classes [6], and efficient algorithms for some NP-complete graph problems using modular decomposition [4, 3, 5]. Many NP-complete problems can be solved by polynomial time (or even linear time) algorithms using tree decomposition or clique decomposition, if the treewidth or the clique-width of the graph is bounded (and the

J. Hromkovič, M. Nagl, and B. Westfechtel (Eds.): WG 2004, LNCS 3353, pp. 129–141, 2004.
© Springer-Verlag Berlin Heidelberg 2004

graph is given with a clique decomposition in the case of the clique-width) [1, 9, 13, 18].

In this paper, we consider the split decomposition (also called join decomposition) which can be seen as an extension of the modular decomposition. There are only few papers presenting algorithms for NP-complete problems using split decomposition. Cunningham has given an algorithm for the independent set problem [11]. Cicerone and Di Stefano showed how to apply this algorithm to parity graphs, which are exactly those graphs for which all prime graphs with respect to the split decomposition are bipartite or complete [8]. They also presented an algorithm unfortunately uncorrect for the clique problem, and applications to parity graphs [8]. Split decomposition is also used for circle graph recognition [14, 23], and parity graph recognition [8, 12].

This paper is organized as follows: Section 2 gives several preliminaries and Section 3 introduces the split decomposition. In Section 4 we discuss known and new results in the approach for obtaining efficients algorithms for NP-complete problems using the split decomposition. In Section 5 we present an algorithm for the coloring problem using split decomposition. Finally, in Section 6 we present some polynomial time coloring algorithms for nicely decomposable graph classes based on the algorithm of Section 5.

2 Preliminaries

Let $G = (V, E)$ be an undirected, simple and finite graph. We denote by $N_G(v)$ $= \{u \in V : \{u, v\} \in E\}$ the neighborhood of v in G and by $N_G[v] = N_G(v) \cup \{v\}$ its closed neighborhood. We shall write $N(v)$ and $N[v]$ if there is no ambiguity. Let $V' \subseteq V$. We denote by $G[V'] = (V', \{\{u, v\} \in E : u, v \in V'\})$ the subgraph of G induced by V'. We denote by $G - V' = G[V \setminus V']$ the subgraph induced by $V \setminus V'$ and, if $v \in V$, we write $G - v$ instead of $G - \{v\}$. A *module* is a set $M \subseteq V$ such that for all $v \in V \setminus M$ either $N(v) \cap M = \emptyset$ or $M \subseteq N(v)$.

Let $w : V \to \mathbb{N}$ be a weight function. The weight of a subset $V' \subseteq V$ is $w(V') = \sum_{v \in V'} w(v)$. A *stable set* of G is a subset S of V such that for all $u, v \in S$, $\{u, v\} \notin E$. The *weighted stability number*, denoted by $\alpha_w(G)$, of a weighted graph (G, w) is the maximum weight of a stable set of G. A *clique* of G is a subset C of V such that for all $u, v \in C$, $\{u, v\} \in E$. The *weighted clique number*, denoted by $\omega_w(G)$, of a weighted graph (G, w) is the maximum weight of a clique of G.

The *chromatic number* of a graph $G = (V, E)$, denoted by $\chi(G)$, is the smallest integer k such that there is a function $f : V \to \{1, 2, \dots, k\}$ with for all $u, v \in V$, $\{u, v\} \in E$ implies $f(u) \neq f(v)$. Clearly a *coloring* C of G can be seen as a partition of V into stable sets.

A *multiset* may contain multiple instances of the same element. We denote it by $\langle e_1, \dots e_k \rangle$. A *weighted coloring* of the weighted graph (G, w) (or a *w-weighted coloring* of G) is a multiset C of stable sets of G such that for all $v \in V$, $|\langle S \in C : v \in S \rangle| \geq w(v)$. The *weighted chromatic number* of a weighted graph (G, w), denoted by $\chi_w(G)$ is the minimum cardinality of a w-weighted coloring of G. It is well known that $\chi_w(G) \geq \omega_w(G)$.

Throughout the paper we use the following notation in order to simplify the presentation. Let $w : V \to \mathbb{N}$ be a function and let $a \in \mathbb{N}$. We denote $w|_{v \to a}$ the function of domain $V \cup \{v\}$ such that $w|_{v \to a}(v) = a$ and $w|_{v \to a}(u) = w(u)$ for all $u \in V \setminus \{v\}$.

3   Reviewing Split Decomposition

A *split* of a graph $G = (V, E)$ is a partition of V into two sets V_1 and V_2 such that $|V_1| \geq 2$, $|V_2| \geq 2$, and every vertex in V_1 with a neighbor in V_2 has the same neighborhood in V_2. Following this definition, we can define the *simple decomposition* of $G = (V, E)$ by the split V_1, V_2. G is decomposed into G_1 and G_2, where, for $i \in \{1, 2\}$, G_i is the subgraph of G induced by V_i with an additional vertex v, called a *marker*, such that the neighborhood of v in G_i is the set of those vertices in V_i which are adjacent in G to a vertex outside of V_i. A graph is *prime* if it does not have a split, and we say that G is decomposable into G_1 and G_2 if there is a split V_1, V_2 such that G is decomposable into G_1 and G_2 with the split V_1, V_2.

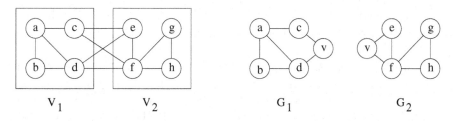

Fig. 1. A graph with a split V_1, V_2, and the two graphs G_1 and G_2 obtained by the simple decomposition.

Additionally, we define a related composition. Let $G_1 = (V_1, E_1)$ and $G_2 = (V_2, E_2)$ be two graphs, such that $V_1 \cap V_2 = \{v\}$. Then $G_1 * G_2$ is the graph with vertex set $(V_1 \cup V_2) \setminus \{v\}$, and edge set $\{\{x, y\} \in E_1 : x \neq v$ and $y \neq v\} \cup \{\{x, y\} \in E_2 : x \neq v$ and $y \neq v\} \cup \{\{x, y\} : x \in N_{G_1}(v)$ and $y \in N_{G_2}(v)\}$. Obviously, if G is decomposable into G_1 and G_2, then $G = G_1 * G_2$. We write $G_1 * \ldots * G_k$ instead of $((G_1 * G_2) \cdots) * G_k$.

The *split decomposition* of a graph is the recursive decomposition of the graph using simple decomposition until none of the obtained graphs can be decomposed further. The *split decomposition tree* of the graph G is the tree T in which each node h corresponds to a prime graph denoted by G_h^* obtained by the split decomposition. Furthermore two nodes h and h' of T are adjacent iff the corresponding graphs G_h^* and $G_{h'}^*$ have a common marker (see figure 2).

Remark 1. The split decomposition of a graph is not necessarily unique. Cunningham [11] showed that every graph has a unique decomposition by splits into prime graphs, stars and complete graphs, with a minimum number of non

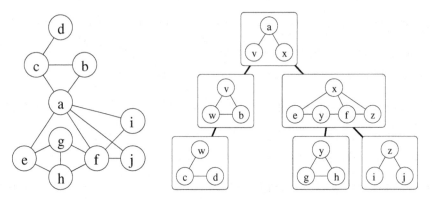

Fig. 2. A graph and its split decomposition tree. The markers are v, w, x, y and z.

decomposed graphs. This decomposition, which we call a Cunningham decomposition, is not necessarily a split decomposition, since stars and complete graphs can be decomposable. Notice that it is easy to obtain a split decomposition from a Cunningham decomposition. Furthermore a split in a graph is either a split in the Cunningham decomposition or a split in a star or a complete graph in the Cunningham decomposition [11]. Thus each split decomposition can be obtained from the Cunningham decomposition, and the set of all prime graphs is the same for every split decomposition, up to isomorphism.

A simple induction shows that a split decomposition tree of a graph G with n vertices and m edges has at most $n - 2$ nodes. The sum of the number of vertices of all prime graphs is at most $3n - 4$ since each vertex in a prime graph is either a vertex of G, or a marker. The sum of the number of edges of all prime graphs is at most $m + n - 3$, since each simple decomposition adds at most one edge to the overall number of edges.

All known algorithms to compute a split decomposition or a split decomposition tree compute in fact a Cunningham decomposition. The first algorithm was given by Cunningham and has running time $O(n^3)$ [11]. This has been improved to $O(nm)$ in [14], and to $O(n^2)$ in [19]. Finally Dahlhaus has given a linear time algorithm in [12].

4   Split Decomposition, Graph Classes and the Clique Problem

Some graphs classes are nicely decomposable by split decompositions. Distance hereditary graphs are completely decomposable by split decomposition (i.e. all primes graphs have at most 3 vertices) [16]. A graph is a circle graph if and only if every prime graph in its split decomposition is a circle graph [14]. A graph is a parity graph if and only if every prime graph in its split decomposition is bipartite or complete [8]. The best known recognition algorithms for circle graphs [23] and parity graphs [8, 12] are based on split decomposition and these forementioned properties.

We already know that a graph is perfect iff every prime graph in the split decomposition is perfect [2]. A simple decomposition cannot destroy a C_k (induced cycle on k vertices) or its complement $\overline{C_k}$ for some $k \geq 5$. Thus if every prime graph in a split decomposition of G is weakly chordal then G is weakly chordal. (Notice that C_4 and $\overline{C_4}$ are decomposable by a simple decomposition.)

Cunningham has given an algorithm to compute the weighted stability number using split decomposition [11]. In [8] the authors apply this algorithm to parity graphs in the split decomposition of G. This algorithm has running time $O(n^{2.5})$. They also present algorithms to compute the clique number and the weighted clique number of a parity graph using split decomposition. Unfortunately theses algorithms are flawed. Figure 3 gives a counter-example for the clique number algorithm (and also a counter-example for the weighted clique number when taking all vertex weights to be 1).

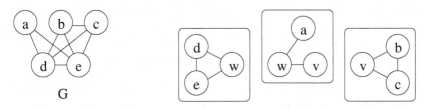

Fig. 3. A parity graph for which the algorithm for the clique number in [8] fails, and its unique split decomposition, which is also its Cunningham decomposition. The markers are v and w. $\omega(G) = 4$ and the output of the algorithm is 3.

Lemma 1 recalls Cunningham's algorithm. Lemma 2 provides a correct algorithm to compute the weighted clique number using split decomposition.

Let $G = (V, E)$ be a graph such that $G = G_1 * G_2$, $G_1 = (V_1, E_1)$, $G_2 = (V_2, E_2)$ and $V_1 \cap V_2 = \{v\}$. Let $w : V \to \mathbb{N}$ be a weight function.

Lemma 1 ([11]). Let $a = \alpha_w(G_2 - N_{G_2}[v])$ and $a' = \alpha_w(G_2 - v)$. Then

$$\alpha_w(G) = \alpha_{w|_{v \to a'-a}}(G_1) + a.$$

Lemma 2. Let $a = \omega_w(G_2[N_{G_2}(v)])$. Then

$$\omega_w(G) = \max(\omega_w(G_2 - v), \omega_{w|_{v \to a}}(G_1)).$$

Proof. Obviously, $\omega_w(G) \geq \omega_w(G_2 - v)$. Furthermore $\omega_w(G) \geq \omega_w(G[(V_1 \setminus \{v\}) \cup N_{G_2}(v)])$. Now $\omega_w(G[(V_1 \setminus \{v\}) \cup N_{G_2}(v)]) = \omega_{w|_{v \to a}}(G_1)$, since $N_{G_2}(v)$ is a module in $G[(V_1 \setminus \{v\}) \cup N_{G_2}(v)]$. Consequently $\omega_w(G) \geq \max(\omega_w(G_2 - v), \omega_{w|_{v \to a}}(G_1))$.

Furthermore $\omega_w(G) = \max(\omega_w(G_2 - v), \omega_{w|_{v \to a}}(G_1))$ since if C is a clique of G then either $C \subseteq V_2 \setminus \{v\}$ and $w(C) \leq \omega_w(G_2 - v)$ or $C \subseteq (V_1 \setminus \{v\}) \cup N_{G_2}(v)$ and $w(C) \leq \omega_{w|_{v \to a}}(G_1)$. □

Lemma 1 and Lemma 2 respectively can be used to calculate $\alpha_w(G)$ and $\omega_w(G)$ respectively of a graph G using a split decomposition tree T of G (see [8]).

We know that we can use the split decomposition to compute a optimal coloring of a perfect graph [2]. But in the general case, the problem seems to be more complex than the problems in this section, since if $G = G_1 * G_2$, $\chi(G)$ cannot easily be obtained from solutions of subproblems for G_1 and G_2.

5   Coloring a Graph Using a Split Decomposition Tree

In this section we present an algorithm to compute the chromatic number of a graph $G = (V, E)$ using a split decomposition tree of G. Typically our algorithm has to compute the weighted chromatic number for a variety of induced subgraphs of G corresponding to nodes of the split decomposition tree.

Let T be a split decomposition tree of the graph G computed using the linear time algorithm of Dahlhaus [12]. Our coloring algorithm runs recursively on the split decomposition tree T for which we choose any vertex to be its root r.

We recall that for a node h of T, G_h^* is the prime graph of the split decomposition corresponding to the node h. For each node $h \neq r$ of T and its parent h' in T, let v_h be the unique marker belonging to G_h^* and $G_{h'}^*$. We define for every leaf h of T, $G_h = G_h^*$. We define recursively for each internal node h of T with children $h_1, h_2, \ldots, h_k$, $G_h = G_h^* * G_{h_1} * \ldots * G_{h_k}$. We call G_h the graph corresponding to the subtree of T rooted at h. Notice that $G_r = G$.

We present our algorithm in Figure 4. Its correctness proof is based on Lemma 3 and 4. The following notation is used throughout the proofs of these Lemmas. Let C be a w-weighted coloring of $G = (V, E)$ and $V' \subseteq V$. We denote by $C_{V'}$ the collection of all stables sets of the coloring C that contain at least one vertex of V', i.e. $C_{V'} = \langle S \in C : S \cap V' \neq \emptyset \rangle$. We denote by $\mathcal{D}(G, w, V')$ the set of all pairs $(a, b) \in \mathbb{N} \times \mathbb{N}$ such that there is a w-weighted coloring C of G with $a + b$ colors and $|C_{V'}| = a$. Obviously, if $(a, b) \in \mathcal{D}(G, w, V')$, then $(a', b) \in \mathcal{D}(G, w, V')$ for all $a' > a$ and $(a, b') \in \mathcal{D}(G, w, V')$ for all $b' > b$. We call a pair $(a, b) \in \mathcal{D}(G, w, V')$ minimal if $(a - 1, b) \notin \mathcal{D}(G, w, V')$ and $(a, b - 1) \notin \mathcal{D}(G, w, V')$. Note that for every minimal pair (a, b) of $\mathcal{D}(G, w, V')$, both $a \leq \chi_w(G)$ and $b \leq \chi_w(G)$.

From now on, for every node $h \neq r$, we call $\mathcal{D}(h) = \mathcal{D}(G_h - v_h, w, N_{G_h}(v_h))$ the $\mathcal{D}$-set of h, where for all $v \in V_h \setminus \{v_h\}$, $w(v) = 1$. Our algorithm uses a dynamic programming and computes in a bottom-up fashion for all nodes $h \neq r$ of T, the $\mathcal{D}$-set of h from the $\mathcal{D}$-sets of all its children. Finally it computes $\chi(G)$ from the $\mathcal{D}$-sets of the children of the root r of T.

The following lemma provides the operation to be executed for each node of the split decomposition tree when computing its $\mathcal{D}$-set and its chromatic number using the $\mathcal{D}$-sets of its children. Let h be a node of T. Let $G_h^* = (V_h^*, E_h^*)$ be the prime graph corresponding to h, and let $G_h = (V_h, E_h)$ be the graph corresponding to the subtree of T rooted at h. Let $h_1, h_2, \ldots, h_k$ be the children of h in T. Then G_{h_i} is the graph corresponding to the subtree in T rooted at h_i and v_{h_i} is the unique marker belonging to G_h^* and G_{h_i}. For simplifying the notations in Lemma 3 and its proof, we note $G_i = G_{h_i} = (V_i, E_i)$, $v_i = v_{h_i}$ and $G^* = G_h^* = (V^*, E^*)$. As discussed above, $G_h = G^* * G_1 * \ldots * G_k$. Let $w : V_h \to \mathbb{N}$ be a weight function.

Function CHROMATICNUMBER(T)
Input: split decomposition tree T of G, Output: chromatic number of G
begin
   for each node h of T, in bottom-up fashion, where k is the number of children of h
         and $h_1, h_2, \ldots, h_k$ are its children
    if $h \neq r$ **then**
      $D_h \leftarrow \emptyset$
      for b **from** 0 **to** n
        $D_h \leftarrow D_h \cup \{(\text{CHROMSPLIT}(h, b, k, D_{h_1}, D_{h_2}, \ldots, D_{h_k}) - b, b)\}$
      endfor
    else return CHROMSPLIT($r, 0, k, D_{h_1}, D_{h_2}, \ldots, D_{h_k}$)
    endif
   endfor
end {ChromaticNumber}

Function CHROMSPLIT($h, b, k, D_{h_1}, D_{h_2}, \ldots, D_{h_k}$)
begin
   if $k = 0$ **then**
    let $w^* : V_h^* \to \mathbb{N}$ such that for all $v \in V_h^* \setminus \{v_h\}$, $w^*(v) = 1$ and, if $h \neq r$, $w(v_h) = b$
    return $\chi_{w^*}(G_h^*)$
   else
    let $m \leftarrow +\infty$
    for c **from** 0 **to** $2n$ **do**
      for i **from** 1 **to** k **do**
        let $a_i \leftarrow \min\{a' : \exists b'$ such that $(a', b') \in D_{h_i}$ and $a' + b' \leq c\}$
      endfor
      let $w^* : V_h^* \to \mathbb{N}$ such that for all $i \in \{1, 2, \ldots, k\}$, $w^*(v_{h_i}) = a_i$,
        for all $v \in V_h^* \setminus \{v_h, v_{h_1}, \ldots v_{h_k}\}$, $w^*(v) = 1$ and, if $h \neq r$, $w(v_h) = b$
      $m \leftarrow \min(m, \max(c, \chi_{w^*}(G_h^*)))$
    endfor
    return m
   endif
end {ChromSplit}

Fig. 4. Algorithm to compute the chromatic number.

Lemma 3. *For all $i \in \{1, \ldots k\}$, let $(a_i, b_i) \in \mathcal{D}(G_i - v_i, w, N_{G_i}(v_i))$. Let $\mathfrak{C}$ be the set of all w-weighted colorings C of G_h with $|C_{V_i \setminus \{v_i\}}| = a_i + b_i$ and $|C_{N_{G_i}(v_i)}| = a_i$. Then*

$$\min_{C \in \mathfrak{C}} |C| = \max(\chi_{w^*}(G^*), \max_{i \in \{1, 2, \ldots, k\}} (a_i + b_i))$$

where $w^* : V^* \to \mathbb{N}$ such that for all $i \in \{1, 2, \ldots, k\}, w^*(v_i) = a_i$, and for all $v \in V^* \setminus \{v_1, v_2, \ldots, v_k\}$, $w^*(v) = w(v)$.

Proof. Recall that if C is a weighted coloring of G and $V' \subseteq V$ then $C_{V'}$ is defined to be the multiset of all stable sets belonging to C having a non-empty intersection with $V' \subseteq V$. For all $i \in \{1, 2, \ldots, k\}$, let $A_i = N_{G_i}(v_i)$ and let $B_i = V_i \setminus N_{G_i}[v_i]$.

Firstly, we prove that every coloring $C \in \mathfrak{C}$ contains at least $\max(\chi_{w^*}(G^*),$ $\max_{i \in \{1,2,\ldots,k\}}(a_i + b_i))$ stable sets. For all $i \in \{1, \ldots k\}$, $a_i + b_i = |C_{V_i \setminus \{v_i\}}| \leq |C|$. It remains to show that $\chi_{w^*}(G^*) \leq |C|$. To do so we construct a w^*-weighted coloring C^* of G^* from $C_{V_h \setminus \bigcup_{i=1}^{k} B_i}$. For each $S \in C_{V_h \setminus \bigcup_{i=1}^{k} B_i}$, let $S^* \subseteq V^*$ be the set satisfying $S^* \cap (V^* \setminus \{v_1, \ldots v_k\}) = S \cap (V^* \setminus \{v_1, \ldots v_k\})$, and for all $i \in \{1, 2, \ldots, k\}$, $v_i \in S^*$ iff $S \cap A_i \neq \emptyset$. Notice that S^* is a stable set of G^*. Let C^* be the multiset of all sets S^* Clearly $|C^*| \leq |C|$. Furthermore C^* is a w^*-weighted coloring of G^* since for all i, $|C^*_{\{v_i\}}| = |C_{A_i}| = a_i$. Hence $\min_{C \in \mathfrak{C}} |C| \geq \max(\chi_{w^*}(G^*), \max_{i \in \{1,2,\ldots,k\}}(a_i + b_i))$.

It remains to prove equality. To do so we construct a w-weighted coloring C of G_h such that $C \in \mathfrak{C}$ and $|C| = \max(\chi_{w^*}(G^*), \max_{i \in \{1,2,\ldots,k\}}(a_i + b_i))$. Let C^* be a minimum w^*-weighted coloring of G^*, and for all i, let C^i be a coloring of $G_i - v_i$ such that $|C^i| = a_i + b_i$ and $|C^i_{A_i}| = a_i$.

We construct C as follows. In the first stage, we pick a stable set $S^* \in C^*$ and for all $i \in \{1, 2, \ldots, k\}$, if $v_i \in S^*$ we pick a stable set S^i in $C^i_{A_i}$, otherwise we pick S^i in $C^i \setminus C^i_{A_i}$. If there is no remaining stable set, we take $S^i = \emptyset$. We add to C the set $S = S^* \setminus \{v_1, \ldots v_k\} \cup \bigcup_{i=1}^{k} S^i$. Notice that S is a stable set. We repeat this operation until C^* is empty. Since $w^*(v_i) = a_i = |C^i_{A_i}|$, there is no remaining set in $S^i \in C^i$ such that $S_i \cap A_i \neq \emptyset$ at the end of the first stage.

In the second stage, as long as there is an $i \in \{1, 2, \ldots, k\}$ such that $C^i \neq \emptyset$, we pick for all $i \in \{1, 2, \ldots, k\}$ a stable set $S^i \in C^i$ if $C^i \neq \emptyset$, otherwise we take $S^i = \emptyset$. We add to C the set $S = \bigcup_{i=1}^{k} S^i$. Notice that S is a stable set. It is not hard to see that C is a w-weighted coloring of G_h, and $C \in \mathfrak{C}$.

At the end of the first stage, C has $\chi_{w^*}(G^*)$ stable sets, and for each $i \in \{1, 2, \ldots, k\}$, C^i has $\max(0, |C^i \setminus C^i_{A_i}| - |C^* \setminus C^*_{v_i}|) = \max(0, b_i - \chi_{w^*}(G^*) + a_i)$ remaining stable sets. Then we add in the second stage $\max(0, -\chi_{w^*}(G^*) + \max_{i \in \{1,2,\ldots,k\}}(a_i + b_i))$ stable sets to C. Thus, at the end $|C| = \max(\chi_{w^*}(G^*), \max_{i \in \{1,2,\ldots,k\}}(a_i + b_i))$. $\qquad\square$

Then $\chi_w(G_h)$ is the minimum of $\max(\chi_{w^*}(G_h^*), \max_{i \in \{1,2,\ldots,k\}}(a_i + b_i))$ over all possible choice of the k-tuple $((a_1, b_1), (a_2, b_2), \ldots, (a_k, b_k))$ into $\mathcal{D}(h_1) \times \mathcal{D}(h_2) \times \ldots \times \mathcal{D}(h_k)$. For all $c \in \mathbb{N}$, let t_c be the k-tuple such that each for all $i \in \{1, 2, \ldots k\}$, (a_i, b_i) is a pair of $\{(a, b) \in D(h_i) : a + b \leq c\}$ with the smallest a. Obviously, if t is a k-tuple $((a_1, b_1), (a_2, b_2), \ldots, (a_k, b_k))$ and $c = \max_i(a_i + b_i)$, then the choice t_c at least as good as t. The algorithm computes $\chi_w(G_h)$ by taking the minimum over all k-tuple t_c, $c \in \{0, 1, \ldots, 2n\}$.

The previous lemma has shown how to compute the weighted chromatic number of G_h from the $\mathcal{D}$-set of its children. If h is a node different from r, then we have to compute the $\mathcal{D}$-set of h. The following lemma shows how this can be done.

Lemma 4. *Let $G = (V, E)$ be a graph, $v \in V$ and $w : V \setminus \{v\} \to \mathbb{N}$ a weight function. Let $b \in \mathbb{N}$ and $w' : V \to \mathbb{N}$ such that $w'(v) = w(v)$ for all $v \in V \setminus \{v\}$, and $w'(v) = b$. Then $(\chi_{w'}(G) - b, b) \in \mathcal{D}(G - v, w, N(v))$, and $(\chi_{w'}(G) - b - 1, b) \notin \mathcal{D}(G - v, w, N(v))$.*

Proof. Let C be a minimum w'-weighted coloring of G. Then $C' = \langle S \setminus \{v\} :$ $S \in C \rangle$ is a w-weighted coloring of G, and has at least b stable sets not having a vertex in common with $N(v)$. Thus $(\chi_{w'}(G) - b, b) \in \mathcal{D}(G - v, w, N(v))$.

If there would be a a w-weighted coloring $\widehat{C}$ of $G - v$ such that $|\widehat{C}| < \chi_{w'}(G)$ having at least b stable sets having no vertex in common with $N(v)$, then the multiset C obtained by adding v to all stable sets having no vertex in common with $N(v)$, would be a w'-weighted coloring of G, contradiction. □

Now the minimals pairs of the $\mathcal{D}$-set of h can be computed using $\chi_{w|_{v_h \to b}}(G_h)$, $b = 0, 1, \ldots, n$. Now we are ready to summarize the correctness proof.

Theorem 1. *The algorithm* CHROMATICNUMBER *takes as input a graph G and it split decomposition tree T and computes the chromatic number of G.*

Proof. The function CHROMSPLIT$(h, b, \ldots)$ returns $\chi_w(G_h)$, with $w(v) = 1$ for all v in $V_h \setminus \{v_h\}$ and, if $h \neq r$, $w(v_h) = b$, using Lemma 3 and the remark following the proof of Lemma 3. For all $h \neq r$ the main loop computes D_h which contains all the minimal pairs of the $\mathcal{D}$-set of h, using the Lemma 4. Finally, the algorithm computes the chromatic number of $G_r = G$. □

Remark 2. Our algorithm computes the chromatic number of the input graph. We mention that it is not hard to modify the algorithm such that it computes a minimum coloring if we use as sub-function an algorithm which computes a minimal weighted coloring of a prime graph.

Theorem 2. *If the algorithm for weighted chromatic number for the prime graphs in the input split decomposition tree has running time $f(n, m)$ then the total running time of the algorithm* CHROMATICNUMBER *is $O(n^3 \cdot f(n, m))$.*

Proof. This algorithm executes $O(n^2)$ times the algorithm for the weighted chromatic number for each prime graph in the split decomposition, and the split decomposition tree has $O(n)$ nodes. The remainder can be done in time $O(n^3)$ by pre-calculating $\min\{a' : \exists b'$ such that $(a', b') \in D_h$ and $a' + b' \leq c\}$ for all node h of T and for all $c \in \{0, 1, \ldots, 2n\}$. Consequently the overall running time is $O(n^3 \cdot f(n, m))$. □

Remark 3. This running time can be improved to $O(n^2 \cdot f(n, m))$ for functions f such that the running time to execute the algorithm for weighted chromatic number on each prime graph is $O(f(n, m))$.

6   Polynomial Time Algorithms

6.1   Some Classes of Perfect Graphs

In order to compute the chromatic number (respectively a minimum coloring) using the split decomposition, we have to compute the weighted chromatic number (respectively a minimum weighted coloring) of the prime graphs. Hoàng has given an $O(nm)$ algorithm for minimum weighted coloring for perfectly orderable graphs if a perfect order is given with the graph, and $O(n^2)$ algorithms for comparability graphs and for chordal graphs [17]. Grötschel, Lovász and Schrijver have given a weakly polynomial algorithm for this problem for perfect graphs (i.e. the algorithm is polynomial on the sum of the vertex weights) [15], which is applicable in our algorithm since the sum of the weights is bounded by n.

Using Hoàng's $O(n^2)$ algorithm for the minimum weighted coloring for the comparability graphs and the algorithm in Section 5, we obtain an $O(n^4)$ algorithm for the weighted coloring on the class of graphs for which every prime graph in the split decomposition is a comparability graph. Raschle and Simon have given $O(m^2)$ algorithms for orienting P_4-comparability graphs and P_4-indifference graphs, which are subclasses of perfectly orderable graphs [22]. So using these algorithms, Hoàng's $O(nm)$ algorithm for the minimum weighted coloring for perfectly orderable graphs, and the algorithm in Section 5, we obtain an $O(n^2m^2)$ algorithm for the weighted coloring on the class of graphs for which every prime graph in the split decomposition is a P_4-comparability graph or a P_4-indifference graph.

6.2   Every Prime Graph Has Bounded Size

Let $\mathcal{G}_k$, $k \geq 3$ be the class of graph for which every prime graph in a split decomposition has at most k vertices. Notice that the $\mathcal{G}_3$ is the class of all distance hereditary graphs. For graphs of bounded size the weighted stability number and the weighted clique number can be computed in constant time. The weighted chromatic number of graphs of bounded size can be computed in constant time [3].

Corollary 1. *For any fixed $k \geq 3$, there is an $O(n)$ algorithm to compute the weighted stability number and the weighted clique number, respectively, and an $O(n^3)$ algorithm to compute the chromatic number of graph in the class $\mathcal{G}_k$, if a split decomposition tree is given with the graph.*

6.3   Every Prime Graph Is an Induced Subgraph of C_k or $\overline{C_k}$ for Some $k \geq 3$

Let $\mathcal{G}_c$ be the class of all graphs for which every prime graph in the split decomposition is an induced subgraph of C_k or $\overline{C_k}$, $k \geq 3$. For all these graphs, computing the weighted stability number and the weighted clique number is easy and can be done in linear time. Thus using results of Section 4, we obtain

Corollary 2. *There are $O(n + m)$ algorithms to compute the weighted stability number and the weighted clique number, respectively, on the graph class $\mathcal{G}_c$.*

A more challenging problem is to compute the chromatic number for $\mathcal{G}_c$. The following lemma generalizes a theorem of Vanherpe for the graph C_5 [25].

Lemma 5. *Let $G = (V, E)$ with $V = \{v_1, v_2, \ldots, v_n\}$ be a cycle of length $n \geq 3$, and let $w : V \to \mathbb{N}$ be a weight function of G. Then*

$$\chi_w(G) = \max \left(\omega_w(G), \left\lceil \frac{\sum_{i=1}^n w(v_i)}{\lfloor \frac{n}{2} \rfloor} \right\rceil \right).$$

Proof. We already mentioned that $\chi_w(G) \geq \omega_w(G)$. Furthermore each color class has at most $\lfloor \frac{n}{2} \rfloor$ vertices, thus $\chi_w(G) \geq \left\lceil \frac{\sum_{i=1}^n w(v_i)}{\lfloor \frac{n}{2} \rfloor} \right\rceil$. If n is even or $n = 3$, then G is perfect. Thus $\chi_w(G) = \omega_w(G)$ [15].

Now we assume that n is odd and $n \neq 3$. We show the claimed equality by induction on the sum of the weights of G. If there is a $v \in V$ such that $w(v) = 0$, then $\chi_w(G) = \chi_w(G - v) = \omega_w(G)$, since $G - v$ is perfect. Otherwise, let E' be the set of all edges of G for which the sum of the weights of its two endpoints is $\omega_w(G)$.

Case 1: $E' = E$. Then all weights are equal (since n is odd), so $\left\lceil \frac{\sum_{i=1}^n w(v_i)}{\lfloor \frac{n}{2} \rfloor} \right\rceil > \omega_w(G)$. Let S be a stable set of $\lfloor \frac{n}{2} \rfloor$ vertices, and w' be the weight function defined by $w'(v) = w(v) - 1$ for all $v \in S$ and $w'(v) = w(v)$ for all $v \in V \setminus S$. Then

$$\chi_w(G) \leq 1 + \chi_{w'}(G) = 1 + \max\left(\omega_{w'}(G), \left\lceil \frac{\sum_{i=1}^n w'(v_i)}{\lfloor \frac{n}{2} \rfloor} \right\rceil\right)$$

$$\leq \left\lceil \frac{\sum_{i=1}^n w(v_i)}{\lfloor \frac{n}{2} \rfloor} \right\rceil$$

since $\left\lceil \frac{\sum_{i=1}^n w(v_i)}{\lfloor \frac{n}{2} \rfloor} \right\rceil > \omega_w(G) = \omega_{w'}(G)$ and $\left\lceil \frac{\sum_{i=1}^n w'(v_i)}{\lfloor \frac{n}{2} \rfloor} \right\rceil = \left\lceil \frac{\sum_{i=1}^n w(v_i)}{\lfloor \frac{n}{2} \rfloor} \right\rceil - 1$.

Case 2: $E' \neq E$. Then there is a stable set S of $\lfloor \frac{n}{2} \rfloor$ vertices containing an endpoint for each edge of E'. Let w' be the weight function defined by $w'(v) = w(v) - 1$ for all $v \in S$ and $w'(v) = w(v)$ for all $v \in V \setminus S$. Then

$$\chi_w(G) \leq 1 + \chi_{w'}(G) = 1 + \max\left(\omega_{w'}(G), \left\lceil \frac{\sum_{i=1}^n w'(v_i)}{\lfloor \frac{n}{2} \rfloor} \right\rceil\right)$$

$$\leq \max\left(\omega_w(G), \left\lceil \frac{\sum_{i=1}^n w(v_i)}{\lfloor \frac{n}{2} \rfloor} \right\rceil\right)$$

since $\left\lceil \frac{\sum_{i=1}^n w'(v_i)}{\lfloor \frac{n}{2} \rfloor} \right\rceil = \left\lceil \frac{\sum_{i=1}^n w(v_i)}{\lfloor \frac{n}{2} \rfloor} \right\rceil - 1$ and $\omega_{w'}(G) = \omega_w(G) - 1$. $\square$

The *weighted clique partition number* of a weighted graph (G, w) is $\kappa_w(G) = \chi_w(\overline{G})$. The following lemma gives the weighted clique partition number of a graph C_n for all $n \geq 3$. Thus it gives the weighted chromatic number of a graph $\overline{C_n}$ for all $n \geq 3$. The proof is similar to the proof of Lemma 5 and omitted.

Lemma 6. *Let $G = (V, E)$ with $V = \{v_1, v_2, \ldots, v_n\}$ be a cycle of length $n \geq 3$, and let $w : V \to \mathbb{N}$ be a weight function of G. Then*

$$\kappa_w(G) = \max\left(\alpha_w(G), \left\lceil \frac{\sum_{i=1}^n w(v_i)}{2} \right\rceil\right).$$

By Lemma 5 and 6 the weighted chromatic number of a graph C_n or $\overline{C_n}$, $n \geq 3$ can be computed in linear time. Using Theorem 2 and it's the remark we obtain

Corollary 3. *There is an $O(n^2 m)$ algorithm to compute the chromatic number for all graphs in the class $\mathcal{G}_c$.*

Acknowledgement

I am grateful to my supervisor D. Kratsch and anonymous referees for helpful comments.

References

1. S. ARNBORG, J. LAGERGREN, D. SEESE, Easy problems for tree-decomposable graphs, *Journal of Algorithms* 12 (1991) 308-340
2. R. BIXBY, A composition for perfect graphs, *Annals of Discrete Math.* 21 (1984) 221-224
3. H. L. BODLAENDER, A. BRANDSTÄDT, D. KRATSCH, M. RAO, J. SPINRAD, Linear time algorithms for some NP-complete problems on (P_5,gem)-free graphs, *Proceedings of FCT'2003*, LNCS 2751, 61-72, Springer-Verlag, Berlin, 2003
4. H.L. BODLAENDER, K. JANSEN, On the complexity of the maximum cut problem, *Nord. J. Comput.* 7 (2000) 14-31
5. H. L. BODLAENDER, U. ROTICS, Computing the treewidth and the minimum fill-in with the modular decomposition, *Algorithmica* 36 (2003) 375-408
6. A. BRANDSTÄDT, D. KRATSCH, On the structure of (P_5,gem)-free graphs, *Manuscript 2002*
   http://www.informatik.uni-rostock.de/(en)/~ab/ps-files/p5gemdam.ps
7. M. CHUDNOVSKY, N. ROBERTSON, P.D.SEYMOUR, R.THOMAS, The strong perfect graph theorem, *Manuscript 2002*
   http://www.math.gatech.edu/~thomas/spgc.ps.gz
8. S. CICERONE, D. DI STEFANO, On the extension of bipartite graphs to parity graphs, *Discrete Applied Math.* 95 (1999) 181-195
9. B. COURCELLE, J. A. MAKOWSKY, U. ROTICS, Linear time solvable optimization problems on graphs of bounded clique-width, *Theory of Computing Systems* 33 (2000) 125-150
10. A. COURNIER, M. HABIB, A new linear algorithm for modular decomposition, *Trees in Algebra and Programming – CAAP '94*, LNCS 787 (1994) 68-84
11. W. CUNNINGHAM, Decomposition of directed graphs, *SIAM Journal on Algebraic and Discrete Methods* 3 (1982) 214-228
12. E. DAHLHAUS, Parallel algorithms for hierarchical clustering and applications to split decomposition and parity graph recognition, *Journal of Algorithms* 36 (2000) 205-240
13. W. ESPELAGE, F. GURSKI, E. WANKE, How to solve NP-hard graph problems on clique-width bounded graphs in polynomial time, *Proceedings of WG 2001*, LNCS 2204, 117-128, Springer-Verlag, Berlin, 2001
14. C. P. GABOR, W. L. HSU, K. J. SUPOWIT, Recognizing circle graphs in polynomial time, *Journal of the ACM* 36 (1989) 435-473

15. M. GRÖTSCHEL, L. LOVÁSZ, A. SCHRIJVER, Polynomial algorithms for perfect graphs, *Annals of Discrete Math.* 21 (1984) 325-356

16. P. HAMMER, F. MAFFRAY, Completely separable graphs, *Discrete Applied Math.* 27 (1990) 85-99

17. C. T. HOÀNG, Efficient algorithms for minimum weighted colouring of some classes of perfect graphs, *Discrete Applied Math.* 55 (1994) 133-143

18. D. KOBLER, U. ROTICS, Edge dominating set and colorings on graphs with fixed clique-width, *Discrete Applied Math.* 126 (2003) 197-221

19. T.-H. MA, J. SPINRAD, An $O(n^2)$ algorithm for undirected split decomposition, *Journal of Algorithms* 16 (1994) 154-160

20. R.M. MCCONNELL, J. SPINRAD, Modular decomposition and transitive orientation, *Discrete Math.* 201 (1999) 189-241

21. R.H MÖHRING, F.J. RADERMACHER, Substitution decomposition for discrete structures and connections with combinatorial optimization, *Annals of Discrete Math.* 19 (1984) 257-356

22. T. RASCHLE, K. SIMON, On the P_4-components of graphs, *Discrete Applied Math.* 100 (2000) 215-235

23. J. SPINRAD, Recognition of circle graphs, *Journal of Algorithms* 16 (1994) 264-282

24. R. E. TARJAN, Decomposition by clique separators, *Discrete Math.* 55 (1985) 221-232

25. J.-M. VANHERPE, Décomposition et algorithmes efficaces sur les graphes, *Ph.D. thesis*, Université de Picardie, LaRIA, 1999

Decremental Clique Problem*

Fabrizio Grandoni and Giuseppe F. Italiano

Dipartimento di Informatica, Sistemi e Produzione
Università di Roma "Tor Vergata"
Via del Politecnico 1
00133 Roma, Italy
{grandoni,italiano}@disp.uniroma2.it

Abstract. The clique problem consists in determining whether an undirected graph G of order n contains a clique of order ℓ. In this paper we are concerned with the decremental version of clique problem, where the property of containing an ℓ-clique is dynamically checked during deletions of nodes. We provide an improved dynamic algorithm for this problem for every fixed value of $\ell \geq 3$. Our algorithm naturally applies to filtering for the constraint satisfaction problem. In particular, we show how to speed up the filtering based on an important local consistency property: the inverse consistency.

1 Introduction

There is a wealth of research on *dynamic graph problems*, which consist in checking a given property on graphs subject to dynamic changes, such as deletions or insertions of nodes or edges [4–6, 9, 10, 13, 14, 18, 19]. If only deletions or insertions are allowed, the dynamic problem is also called *decremental* or *incremental* respectively.

A *clique* is an undirected graph such that its nodes are pairwise adjacent. The *decremental clique problem* consists in dynamically determining whether a graph G of n nodes contains an ℓ-clique (a clique of ℓ nodes), during deletions of nodes.

To the best of our knowledge, no non-trivial algorithm is known for this problem, while several non-trivial results are available for its static version. Itai and Rodeh [12] showed how to detect a *triangle* (clique of three nodes) in G in $O(n^\omega)$ steps, where the complexity of multiplying two $n \times n$ matrices is $O(n^\omega)$, $\omega < 2.376$ [1]. Nešetřil and Poljak [17] generalized the algorithm of Itai and Rodeh to the detection of cliques of arbitrary cardinality ℓ. Their algorithm has a $O(n^{\alpha(\ell)})$ time complexity, where $\alpha(\ell) = \omega\lfloor \ell/3 \rfloor + \ell \pmod 3$.

Recently, Eisenbrand and Grandoni [7] developed a faster algorithm for the same task. Their algorithm has a $O(n^{\beta(\ell)})$ time complexity, where $\beta(\ell) = \omega(\lfloor \ell/3 \rfloor,$

---

* This work has been partially supported by the IST Programme of the EU under contract n. IST-1999-14.186 (ALCOM-FT), by the Italian Ministry of University and Research (Project "ALINWEB: Algorithmics for Internet and the Web").

J. Hromkovič, M. Nagl, and B. Westfechtel (Eds.): WG 2004, LNCS 3353, pp. 142–153, 2004.

$\lceil(\ell-1)/3\rceil, \lceil\ell/3\rceil)$, and the time complexity of multiplying a $n^r \times n^s$ matrix by a $n^s \times n^t$ matrix is denoted by $O(n^{\omega(r,s,t)})$.

All the algorithms above can be easily adapted so as to count the number of ℓ-cliques in which each node is contained.

In this paper we present a dynamic algorithm for the decremental clique problem. In particular, we show how to efficiently update the number of ℓ-cliques in which each node of a graph is contained, during deletions of nodes. Our algorithm, which builds up on the algorithm of Eisenbrand and Grandoni, performs updates in $O(n^{\beta(\ell)-0.8})$ time for every fixed ℓ, that is, roughly, $n^{0.8}$ times faster than recomputing everything from scratch.

1.1 An Application to the Constraint Satisfaction Problem

The *constraint satisfaction problem* consists in determining whether a set of k variables, defined on domains of size at most d, admits an instantiation which satisfies a given set of constraints. Any such instantiation is a *solution* for the *constraint network*. Without loss of generality [16], we can assume that all the constraints are *binary* (a constraint is binary if it involves only a pair of variables).

An assignment (i, a) of a value a to a variable i is *consistent* if there is a solution which assigns a to i, and *inconsistent* otherwise. Inconsistent assignments can be removed from the constraint network without loosing any solution (by removing an assignment (i, a), we mean removing a from the domain of i).

Detecting inconsistent assignments is a NP-hard problem [16]. For this reason, many heuristic filtering techniques have been developed, which allow to efficiently detect (and remove) part of the inconsistent assignments. Most of them are based on some kind of *local consistency property* $\mathcal{P}$, which all the consistent assignments *need* to satisfy. The assignments which do not satisfy $\mathcal{P}$ are iteratively filtered out.

Note that an assignment which initially satisfies $\mathcal{P}$, may not satisfy $\mathcal{P}$ any more after some deletions. Thus the same assignment may be checked for consistency many times along the filtering process. This suggests the idea of performing such repeated checks dynamically, instead of doing it each time from scratch. In fact, this approach is used by most of the fastest filtering algorithms.

Maybe the simplest and most studied local consistency property is *arc consistency* [15]. An assignment (i, a) is arc consistent if, for every other variable j, there is at least one assignment (j, b) compatible with (i, a). Clearly, if a node is not arc consistent, it cannot be consistent (unless i is the unique variable in the network).

Arc consistency can be easily generalized. An assignment (i, a) is *path-inverse consistent* [8] if, for every other two variables j and h, there are assignments (j, b) and (h, c) which are mutually compatible and compatible with (i, a). The ℓ-inverse consistency [8] is the natural generalization of arc-consistency ($\ell = 2$) and path-inverse consistency ($\ell = 3$) to arbitrary (fixed) values of $\ell \leq k$.

The currently fastest filtering algorithm based on ℓ-inverse consistency is the $O(k^\ell d^\ell)$ algorithm of Debruyne [3]. This algorithm is based on a very simple dynamic strategy to check whether an assignment is ℓ-inverse consistent.

We show how to reduce the problem of dynamically checking ℓ-inverse consistency to the decremental clique problem on graphs of $O(d)$ nodes. By applying this reduction and our $O(d^{\beta(\ell)-0.8})$ decremental algorithm, we reduce the complexity of ℓ-inverse consistency based filtering to $O(k^\ell d^{\beta(\ell)+0.2})$. This implies an improvement for every $\ell \geq 3$.

The remainder of this paper is organized as follows. In Section 2 we introduce some preliminaries. In Section 3 we describe the algorithm of Eisenbrand and Grandoni, upon which our decremental algorithm builds up. In Section 4 we present our decremental algorithm to maintain the number of ℓ-cliques. Eventually, in Section 5 we show how to speed up the filtering based on inverse consistency.

2 Preliminaries

We use standard graph notation as contained for instance in [2]. An *undirected graph* G is a pair (V, E), where V is a finite set of *nodes* and the *edge* set E consists of unordered pairs of nodes. Without loss of generality, we can assume $V = \{1, 2 \ldots |V|\}$, where $|V|$ is the *cardinality* of G. Two nodes v and w are *adjacent* if $\{v, w\} \in E$. A graph is *complete* if each pair of distinct nodes is adjacent. An ℓ-*clique* is a complete graph of ℓ nodes. The 3-cliques are also called *triangles*. The graph $G[V']$ *induced* on G by a subset V' of nodes is the graph obtained from G by removing all the nodes not in V' and the edges incident on them.

The adjacency matrix A of G is a 0-1 matrix such that, for each pair of nodes v and w, $A[v, w] = 1$ if and only if v and w are adjacent (in particular A is symmetric and the main diagonal is set to zero).

A k-*partite graph* $G = (\{V_1, V_2 \ldots V_k\}, E)$ is a graph where the set of nodes is $V = \bigcup_i V_i$, the set of edges is E, the subsets V_i (*partitions*) are disjoint, and the nodes in the same partition are not adjacent.

A binary constraint network of k variables can be naturally represented via a k-partite graph $G = (\{V_1, V_2 \ldots V_k\}, E)$, the *consistency graph*, which has a node for each possible assignment (i, a) and an edge between all the pairs of assignments which are compatible according to the constraints. In particular, partition V_i is formed by all the assignments corresponding to variable i (two values cannot be assigned to the same variable). An example of consistency graph is given in Figure 1.

It is not hard to show that a k-clique in G corresponds to each solution of the binary constraint network. In other words, the binary constraint satisfaction problem is equivalent to the problem of determining whether the consistency graph contains a k-clique.

The definitions concerning the assignments can be naturally extended to the nodes of the consistency graph. In particular, a node (i, a) is *consistent* if it belongs to at least one k-clique, and *inconsistent* otherwise. Node (i, a) is ℓ-*inverse consistent*, $\ell \leq k$, if, taken ℓ partitions $V_{j_1}, V_{j_2} \ldots V_{j_\ell}$ including V_i, node (i, a) is contained in at least one ℓ-clique of the graph $G[\cup_k V_{j_k}]$ induced on G by such partitions. Clearly, if a node is not ℓ-inverse consistent, it cannot be consistent.

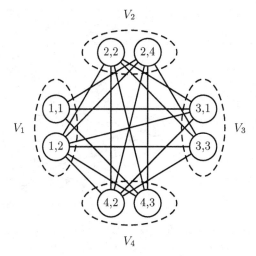

$$V_2$$

$$V_1 \qquad V_3$$

$$V_4$$

Fig. 1. Example of consistency graph. The partitions corresponding to each variable are included into dashed ellipses. A solution is given by the assignments $(1, 2)$, $(2, 2)$, $(3, 1)$ and $(4, 2)$. The assignment $(1, 1)$ is arc consistent, while it is not path-inverse consistent.

3   Static Algorithm

In this section we describe a static algorithm to count the number of ℓ-cliques in which each node of an undirected graph $G = (V, E)$, with n nodes, is contained.

We adapt the algorithm of Eisenbrand and Grandoni for the clique problem to this purpose. We first recall their algorithm. They compute the following 3-partite auxiliary graph $\widetilde{G}_\ell = (\{W_1, W_2, W_3\}, F) = \widetilde{G}$. Let ℓ_1, ℓ_2 and ℓ_3 be equal to $\lfloor \ell/3 \rfloor$, $\lceil (\ell-1)/3 \rceil$ and $\lceil \ell/3 \rceil$ respectively (notice that $\ell = \ell_1 + \ell_2 + \ell_3$). Partition W_i, $i \in \{1, 2, 3\}$, is formed by the ℓ_i-cliques of G. A node $w_i \in W_i$ is adjacent to a node $w_j \in W_j$, if $i \neq j$ and the nodes of w_i and w_j induce a clique of order $(\ell_i + \ell_j)$ in G. Then the algorithm return *yes* if and only if $\widetilde{G}$ contains a triangle. In Figure 2 an example of graph with the corresponding auxiliary graph in the case $\ell = 4$ is depicted.

Lemma 1. *For every fixed $\ell \geq 3$, the algorithm above determines whether an undirected graph G of n nodes contains a clique of ℓ nodes in time $\mathrm{O}(n^{\beta(\ell)}) = \mathrm{O}(n^{\omega(\lfloor \ell/3 \rfloor, \lceil (\ell-1)/3 \rceil, \lceil \ell/3 \rceil)})$ time.*

Proof. Let $\widetilde{G}$ be the auxiliary graph defined above. We show that G contains a ℓ-clique if and only if $\widetilde{G}$ contains a triangle. Assume that G contains a ℓ-clique $\{v_1, v_2 \ldots v_\ell\}$. Thus the partitions W_1, W_2 and W_3 of $\widetilde{G}$ contain the nodes $w_1 = \{v_1, v_2 \ldots v_{\ell_1}\}$, $w_2 = \{v_{\ell_1+1}, v_{\ell_1+2} \ldots v_{\ell_1+\ell_2}\}$ and $w_3 = \{v_{\ell_1+\ell_2+1}, v_{\ell_1+\ell_2+2} \ldots v_\ell\}$ respectively. Moreover w_1, w_2 and w_3 are pairwise adjacent. Thus $\widetilde{G}$ contains a triangle.

Assume now that $\widetilde{G}$ contains a triangle $\{w_1, w_2, w_3\}$. Let $T = \bigcup_i w_i$. Since the graph is 3-partite, the nodes w_i must belong to distinct partitions. Moreover

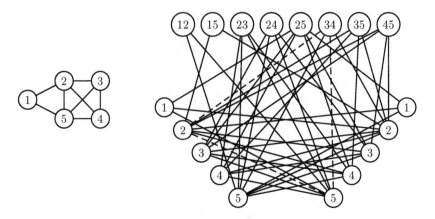

Fig. 2. An example of graph G (on the left) with the corresponding auxiliary graph $\widetilde{G}$ in the case $\ell = 4$. The nodes of $\widetilde{G}$ are labelled with the corresponding subset of nodes of G. One of the $\binom{4}{1,1,2} = 12$ triangles of $\widetilde{G}$ corresponding to the clique $\{2,3,4,5\}$ of G is pointed out via dashed lines.

two nodes of $\widetilde{G}$ which contain the same node v of G cannot be adjacent. Thus $|T| = \ell_1 + \ell_2 + \ell_3 = \ell$. Every two distinct nodes of T are adjacent in G. Thus T is a subset of ℓ pairwise adjacent nodes of G, that is a ℓ-clique of G.

Consider now the time complexity of the algorithm. Partition W_i contains $O(n^{\ell_i})$ nodes, $i \in \{1,2,3\}$. A triangle of $\widetilde{G}$ can be detected in the following way. For each pair of nodes $\{w_1, w_3\}$, $w_1 \in W_1$ and $w_3 \in W_3$, one computes the number $P_{1,2,3}(w_1, w_3)$ of 2-length paths of the kind (w_1, w_2, w_3), where $w_2 \in W_2$. The value of $P_{1,2,3}$ is obtained by multiplying the $n^{\ell_1} \times n^{\ell_2}$ adjacency matrix of the nodes in W_1 with the nodes in W_2 by the $n^{\ell_2} \times n^{\ell_3}$ adjacency matrix of the nodes in W_2 with the nodes in W_3. Graph $\widetilde{G}$ contains a triangle if and only if there is a pair of adjacent nodes $\{w_1, w_3\}$, $w_1 \in W_1$ and $w_3 \in W_3$, such that $P_{1,2,3}(w_1, w_3) > 0$. Computing $P_{1,2,3}$ costs $= O(n^{\omega(\ell_1,\ell_2,\ell_3)}) = O(n^{\beta(\ell)})$. This is also an upper bound on the complexity of the algorithm.

A rectangular matrix multiplication can be executed through a straightforward decomposition into square blocks and fast square matrix multiplication. In other words:

$$\omega(r,s,t) \leq r + s + t + (\omega - 3)\min\{r,s,t\}.$$

More sophisticated fast rectangular matrix multiplication algorithms are available. In particular, for every $0 \leq r \leq 1$, the following bound holds [1,11]:

$$\omega(1,1,r) \leq \begin{cases} 2 + o(1) & \text{if } 0 \leq r \leq \alpha = 0.294; \\ \omega - (1-r)\frac{\omega-2}{1-\alpha} & \text{if } \alpha < r \leq 1. \end{cases} \tag{1}$$

With this bound at hand, one obtains:

$$\beta(\ell) \leq \begin{cases} \lfloor \frac{\ell}{3} \rfloor \omega & \text{if } \ell \pmod 3 = 0; \\ \lfloor \frac{\ell}{3} \rfloor \omega + 1 & \text{if } \ell \pmod 3 = 1; \\ \lfloor \frac{\ell}{3} \rfloor \omega + 2 - \frac{\alpha(\omega-2)}{1-\alpha} & \text{if } \ell \pmod 3 = 2. \end{cases}$$

Better bounds are available for $\omega(r,s,t)$ [11]. These bounds lead to a tighter bound on $\beta(\ell)$ in the case $\ell \pmod 3 \neq 0$. For simplicity we will not consider these tighter bounds, since they are not expressed via a closed formula.

3.1   Counting Cliques

Consider now the problem of counting the number $K_\ell(v)$ of ℓ-cliques in which each node v of G is contained. The algorithm of Eisenbrand and Grandoni can be easily adapted to count, in $O(n^{\beta(\ell)})$ time, the number $\widetilde{K}_3(w)$ of triangles in which each node w of $\widetilde{G}$ is contained. Note that many triangles in $\widetilde{G}$ may correspond to the same ℓ-clique of G.

More precisely, the number of distinct triangles of $\widetilde{G}$ which correspond to the same ℓ-clique of G is equal to the number of ways in which one can partition a set of cardinality ℓ in three subsets of cardinality ℓ_1, ℓ_2 and ℓ_3 respectively, that is $\binom{\ell}{\ell_1, \ell_2, \ell_3}$.

Let $W_i(v)$, for each $i \in \{1, 2, 3\}$, be the set of nodes of W_i which contain node v. It is not hard to show that the sum of $\widetilde{K}_3(w)$ over $W_1(v)$ is equal to $K_\ell(v)$, multiplied by the number of ways in which one can partition a set of cardinality $(\ell - 1)$ in three subsets of cardinality $(\ell_1 - 1)$, ℓ_2 and ℓ_3 respectively:

$$\sum_{w \in W_1(v)} \widetilde{K}_3(w) = \binom{\ell - 1}{\ell_1 - 1, \ell_2, \ell_3} K_\ell(v). \tag{2}$$

Then we can compute $K_\ell(v)$, for each $v \in V$, in $O(n^{\beta(\ell)})$ steps.

Corollary 1. *The algorithm above counts the number of cliques of ℓ nodes in which each node of an undirected graph G of n nodes is contained, in time $O(n^{\beta(\ell)})$.*

4   Decremental Algorithm

In this section we consider the problem of decrementally updating the number $K_\ell(v)$ of cliques of cardinality ℓ in which each node v of the undirected graph G is contained, during deletions of nodes.

The idea is to update the value of $\widetilde{K}_3(w)$, for each w in W_1, and then update $K_\ell(v)$, for each v in w, following Equation (2). Consider the deletion of a node u. The deletion of u corresponds to the deletion of the subsets of nodes $W_1(u)$, $W_2(u)$ and $W_3(u)$ in W_1, W_2 and W_3 respectively.

As two nodes of $\widetilde{G}$ which contain the same node u of G cannot belong to the same triangle, one can safely consider the effects of the deletion of each node in $W_i(u)$, $i \in \{1, 2, 3\}$, separately. First of all, for each $w \in W_1(u)$, one sets $\widetilde{K}_3(w)$ to zero (in linear time).

Consider now the deletion of the nodes in $W_2(u)$ (the deletion of the nodes in $W_3(u)$ is completely analogous). For each $w_1 \in W_1$ and for each deleted node $w_2 \in W_2(u)$, the value of $\widetilde{K}_3(w_1)$ has to be decreased by the number of triangles in which both nodes w_1 and w_2 are contained (at the same time). This quantity is zero if w_1 and w_2 are not adjacent, and it is equal to the number $P_{1,3,2}(w_1, w_2)$ of 2-length paths from w_1 to w_2 through a node in W_3 otherwise. Then one needs to compute $P_{1,3,2}(w_1, w_2)$, for each $w_1 \in W_1$ and for each $w_2 \in W_2(u)$. A simple-minded approach is to compute the number of such paths from scratch.

A better time bound can be obtained as follows. One maintains a *lazy* value $P'_{1,3,2}(w_1, w_2)$ of $P_{1,3,2}(w_1, w_2)$, for each $w_1 \in W_1$ and $w_2 \in W_2$. Whenever a node w_3 is removed from W_3, instead of updating $P'_{1,3,2}$, one stores w_3 in a set D_3. When the cardinality of D_3 reaches a given threshold, one updates $P'_{1,3,2}$ and empties D_3. Clearly the current value of $P_{1,3,2}(w_1, w_2)$ depends on both $P'_{1,3,2}(w_1, w_2)$ and D_3.

In more details, one initially sets $D_3 = \emptyset$ and $P'_{1,3,2} = P_{1,3,2}$. Let $\mu_3 \in [0, 1]$ be a parameter to be fixed later. When one removes a node w_3 from W_3, w_3 is added to D_3 and, if $|D_3| > n^{\ell_3 - 1 + \mu_3}$, one executes the following steps:

- $P'_{1,3,2}$ is updated by subtracting from $P'_{1,3,2}(w_1, w_2)$ the number $\Delta P_{1,3,2}(w_1, w_2)$ of 2-length paths from w_1 to w_2 through a node in D_3.
- Set D_3 is emptied.

The current value of $P_{1,3,2}(w_1, w_2)$, for every $w_1 \in W_1$ and $w_2 \in W_2$, is given by:

$$P_{1,3,2}(w_1, w_2) = P'_{1,3,2}(w_1, w_2) - \Delta P_{1,3,2}(w_1, w_2). \tag{3}$$

Let

$$\widetilde{\beta}(\ell) = \min_{\mu_2, \mu_3 \in [0,1]} \max\{\omega(\ell_1, \ell_3 - 1 + \mu_3, \ell_2) - \mu_3, \omega(\ell_1, \ell_3 - 1 + \mu_3, \ell_2 - 1), \tag{4}$$

$$\omega(\ell_1, \ell_2 - 1 + \mu_2, \ell_3) - \mu_2, \omega(\ell_1, \ell_2 - 1 + \mu_2, \ell_3 - 1)\}.$$

Theorem 1. *The algorithm above maintains the number of cliques of fixed cardinality ℓ in which each node of a graph of n nodes is contained, during deletion of nodes. The preprocessing time of the algorithm is $O(n^{\beta(\ell)})$ and its amortized update time per deletion is $O(n^{\widetilde{\beta}(\ell)})$.*

Proof. The correctness of the algorithm is a consequence of Equations (2) and (3).

Consider now the time complexity of the algorithm. Set $W_i(u)$, $i \in \{1, 2, 3\}$, contains $O(n^{\ell_i - 1})$ nodes. The number $\Delta P_{1,3,2}(w_1, w_2)$ of 2-length paths from w_1 to w_2 through a node in D_3 can be obtained by multiplying the $n^{\ell_1} \times n^{\ell_3 - 1 + \mu_3}$ adjacency matrix of the nodes in W_1 with the nodes in D_3 by the $n^{\ell_3 - 1 + \mu_3} \times n^{\ell_2}$ adjacency matrix of the nodes in D_3 with the nodes in W_2. This costs $O(n^{\omega(\ell_1, \ell_3 - 1 + \mu_3, \ell_2)})$. The value of $P'_{1,3,2}$ can be updated within the same time bound. Note that the deletion of $O(n^{\ell_3 - 1})$ nodes in W_3 corresponds to each deletion of one node in G. This means that one updates $P'_{1,3,2}$ every $\Omega(n^{\mu_3})$ deletions in G. Then the amortized update cost per deletion is $O(n^{\omega(\ell_1, \ell_3 - 1 + \mu_3, \ell_2) - \mu_3})$.

Table 1. Optimal values of μ_2 and μ_3 for varying ℓ.

ℓ	μ_2 and μ_3
3	$\mu_2 = \mu_3 = \frac{1+\alpha-\alpha\omega}{4-2\alpha-\omega}$
4	$\mu_2 = \frac{1-\alpha(3-\omega)}{\omega-1-\alpha}$; $\mu_3 = \omega - 2$
$3k+2 \geq 5$	$\mu_2 = \mu_3 = \frac{1-\alpha}{1+\alpha(\omega-3)}$
$3k \geq 6$	$\mu_2 = \mu_3 = \frac{1+\alpha-\alpha\omega}{4-\alpha\omega-\omega}$
$3k+1 \geq 7$	$\mu_2 = 1$; $\mu_3 = \alpha\frac{\omega-2}{1-\alpha}$

Table 2. Values of $\widetilde{\beta}(\ell)$ for varying ℓ.

ℓ	$\widetilde{\beta}(\ell)$
3	$\frac{5-\alpha-\omega(\alpha+1)}{4-2\alpha-\omega}$
$3k+1 \geq 4$	$k\omega$
$3k+2 \geq 5$	$k\omega + \frac{1-\alpha}{1+\alpha(\omega-3)}$
$3k \geq 6$	$k\omega - \omega + 1 + (\frac{1+\alpha-\alpha\omega}{4-\alpha\omega-\omega})(1 - \alpha\frac{\omega-2}{1-\alpha})$

Following Equation (3), the current value of $P_{1,3,2}(w_1, w_2)$, for every $w_1 \in W_1$ and $w_2 \in W_2(u)$, can be computed in $O(n^{\omega(\ell_1,\ell_3-1+\mu_3,\ell_2-1)})$ time. This is the time required to multiply the $n^{\ell_1} \times n^{\ell_3-1+\mu_3}$ adjacency matrix of the nodes in W_1 with the nodes in D_3 by the $n^{\ell_3-1+\mu_3} \times n^{\ell_2-1}$ adjacency matrix of the nodes in D_3 with the nodes in $W_2(u)$.

Then the cost of updating $\widetilde{K}_3(w)$, for each $w \in W_1$, after the deletion of the nodes in $W_2(u)$, is $O(n^{\omega(\ell_1,\ell_3-1+\mu_3,\ell_2)-\mu_3} + n^{\omega(\ell_1,\ell_3-1+\mu_3,\ell_2-1)})$. Analogously, the cost of updating $\widetilde{K}_3(w)$ after the deletion of the nodes in $W_3(u)$ is $O(n^{\omega(\ell_1,\ell_2-1+\mu_2,\ell_3)-\mu_2} + n^{\omega(\ell_1,\ell_2-1+\mu_2,\ell_3-1)})$, where $\mu_2 \in [0,1]$. The claim follows by fixing μ_2 and μ_3 as suggested by Equation (4).

In next section we will show how to choose μ_2 and μ_3 so as to minimize the complexity of the algorithm.

4.1 Bounds on the Complexity

In this subsection we want to estimate the complexity of the decremental algorithm described above according to the current best bounds on rectangular matrix multiplication. For this purpose, we will use the bounds on $\omega(r, s, t)$ given by Equation (1) and by the following equation [11]. For any $0 \leq t \leq 1 \leq r$:

$$\omega(t, 1, r) \leq \begin{cases} r+1+o(1) & 0 \leq t \leq \alpha; \\ r+1+(t-\alpha)\frac{\omega-2}{1-\alpha}+o(1) & \alpha < t \leq 1. \end{cases} \tag{5}$$

With simple calculations involving Equations (1) and (5) one obtains the results of Table 1 and 2.

Summarizing these results, the update cost is $O(n^{\widetilde{\beta}(\ell)}) = O(n^{\beta(\ell)-\delta(\ell)})$, where:

$$\delta(\ell) \geq \begin{cases} 0.800 & \text{if } \ell = 3; \\ 0.832 & \text{if } \ell \pmod 3 = 0, \ell \geq 6; \\ 1.000 & \text{if } \ell \pmod 3 = 1; \\ 0.978 & \text{if } \ell \pmod 3 = 2. \end{cases}$$

Thus our algorithm performs updates roughly $n^{0.8}$ times faster than recomputing everything from scratch.

In Table 3, the complexity of the decremental algorithm is compared with the complexity of the static algorithm for $3 \leq \ell \leq 8$.

Table 3. Running time comparison of the static and dynamic algorithms for counting the cliques of cardinality ℓ.

ℓ	Static [7]	Dynamic
3	$O(n^{2.376})$	$O(n^{1.575})$
4	$O(n^{3.376})$	$O(n^{2.376})$
5	$O(n^{4.220})$	$O(n^{3.241})$
6	$O(n^{4.751})$	$O(n^{3.919})$
7	$O(n^{5.751})$	$O(n^{4.751})$
8	$O(n^{6.595})$	$O(n^{5.616})$

5 Fast Inverse Consistency

Let G be a consistency graph with k partitions of size at most d. Consider the problem of computing the largest (for number of nodes) induced subgraph $G_{\ell\text{-IC}}$ of G such that all its partitions are non-empty and all its nodes are ℓ-inverse consistent, or determine that such graph does not exist. This problem is well defined:

Lemma 2. Let $\mathcal{G}_{\ell\text{-IC}}$ be the set of all the induced subgraphs of G such that all their partitions are non-empty and all their nodes are ℓ-inverse consistent. If $\mathcal{G}_{\ell\text{-IC}}$ is not empty, it contains a unique graph $G_{\ell\text{-IC}}$ of maximum cardinality.

Proof. Suppose that there exist two distinct graphs $G_1 = G[V_1]$ and $G_2 = G[V_2]$ in $\mathcal{G}_{\ell\text{-IC}}$ of maximum cardinality. Then $G' = G[V_1 \cup V_2]$ is an induced subgraph of G, of cardinality strictly greater than G_1 and G_2, whose nodes are ℓ-inverse consistent. This is a contradiction.

The fastest algorithm known to solve this problem [3] has a $O(k^\ell d^\ell)$ time complexity. In this section we present a faster algorithm for the same problem, which is based on the decremental algorithm of Section 4. Its time complexity is $O(k^\ell d^{\tilde{\beta}(\ell)+1}) = O(k^\ell d^{\beta(\ell)+0.2})$. This improves on the $O(k^\ell d^\ell)$ bound for any $\ell \geq 3$.

Our algorithm works as follows. Nodes which are not ℓ-inverse consistent are removed from G one by one. The procedure ends when all the nodes in G are

ℓ-inverse consistent or a partition becomes empty. In the first case, at the end of the procedure G is equal to $G_{\ell\text{-IC}}$. In the second case, $G_{\ell\text{-IC}}$ is empty.

We have to show how nodes which are not ℓ-inverse consistent are detected along the way. First of all, one checks all the nodes and removes the nodes which are not ℓ-inverse consistent. Then one has to propagate efficiently the effects of deletions. In fact, the deletion of one node can induce as a side effect the deletion of other nodes (which were previously recognized as ℓ-inverse consistent).

Consider the deletion of a node $v \in V_i$. Let $G_\ell(i)$ be the set of graphs induced on G by the nodes of ℓ distinct partitions, including partition V_i. For each graph G' in $G_\ell(i)$ and for each node w of G', one has to check whether w belongs to at least one ℓ-clique of G'.

In more details, a set DelSet of integers is used to keep trace of the partitions into which a deletion occurred: whenever a node v in a partition V_i is removed, i is stored in DelSet. We can distinguish two main steps in the algorithm: an initialization step and a propagation step. In the initialization step, for each node v in each partition V_i, one checks for each G' in $G_\ell(i)$ whether v is contained in at least one ℓ-clique of G'. If this is not true, v is removed from G. In the propagation step, until DelSet is not empty, one extracts an integer j from DelSet and executes the following steps. For each G' in $G_\ell(j)$ and for each node v in G', one checks whether v is contained in at least one ℓ-clique of G'. If not, v is removed from G.

We have to show how to check whether a node of a graph G' is contained in at least one ℓ-clique. The idea is to use the algorithm of previous section. For each graph G' induced by ℓ distinct partitions, we maintain the number of ℓ-cliques in which each one of its nodes is contained. Whenever a node v in a partition V_i is removed, one updates consequently these quantities for each graph G' in $G_\ell(i)$.

Theorem 2. *The algorithm above computes $G_{\ell\text{-IC}}$ or determines that it does not exist in time* $O(k^\ell d^{\tilde{\beta}(\ell)+1})$.

Proof. The number of iterations of the propagation step is bounded by the number of nodes. Then the algorithm halts.

An ℓ-inverse consistent node is clearly never removed. Consider the nontrivial case that the algorithm halts when no partition is empty. To show correctness, we prove that all the remaining nodes in G are ℓ-inverse consistent. Assume by contradiction that, when the algorithm halts, G contains a node $v \in V_i$ which is not ℓ-inverse consistent.

Since all the nodes which are not ℓ-inverse consistent in the original graph are removed during the initialization step, v must be not ℓ-inverse consistent because of the deletions which occurred during the initialization and/or the propagation step.

Consider the sequence $v^{(1)}, v^{(2)} \ldots v^{(p)}$ in which nodes are removed from the graph. Let q, $q \in \{1, 2 \ldots p\}$, be the smallest index such that v is not ℓ-inverse consistent in the graph $G[V^{(q)}]$, where $V^{(q)} = V \backslash \{v^{(1)}, v^{(2)} \ldots v^{(q)}\}$. Let V_j be the partition of $v^{(q)}$. Notice that $v^{(q)}$ must belong to a graph G' which contains partition V_i, and thus node v.

Table 4. Time complexity comparison of our algorithm to enforce ℓ-inverse consistency and the previous best.

ℓ	Previous best [3]	This paper
3	$O(k^3 d^3)$	$O(k^3 d^{2.575})$
4	$O(k^4 d^4)$	$O(k^4 d^{3.376})$
5	$O(k^5 d^5)$	$O(k^5 d^{4.241})$
6	$O(k^6 d^6)$	$O(k^6 d^{4.919})$
7	$O(k^7 d^7)$	$O(k^7 d^{5.751})$
8	$O(k^8 d^8)$	$O(k^8 d^{6.616})$

After the deletion of node $v^{(q)}$, j is inserted in `DelSet`. In one of the following steps, j is extracted from `DelSet` and all the nodes in any graph G' of $\mathcal{G}_\ell(j)$ are checked. In particular, node v is checked. Since in that iteration the set of nodes still in G is a subset of $V^{(q)}$, the node v is recognized as a node which is not ℓ-inverse consistent and it is thus removed, which is a contradiction.

The time complexity of the algorithm is bounded by the cost of maintaining the number of ℓ-cliques in which each node of each graph G' is contained. The number of such graphs is $O(k^\ell)$ (that is the number of ways one can select ℓ from k partitions), and each graph contains $O(d)$ nodes. Then the total initialization cost is $O(k^\ell d^{\beta(\ell)})$. Since each graph G' is interested by at most $O(d)$ deletions, the total update cost is $O(k^\ell d^{\widetilde{\beta}(\ell)+1})$. Thus the time complexity of the algorithm is $O(k^\ell(d^{\beta(\ell)} + d^{\widetilde{\beta}(\ell)+1})) = O(k^\ell d^{\widetilde{\beta}(\ell)+1})$

The performance of our algorithm and of the previous best are compared in Table 4 for $3 \leq \ell \leq 8$. In particular, our algorithm reduces the time complexity to *enforce* path-inverse consistency from $O(k^3 d^3)$ to $O(k^3 d^{2.575})$.

References

1. D. Coppersmith and S. Winograd. Matrix multiplication via arithmetic progressions. *Journal of Symbolic Computation*, 9(3):251–280, 1990.
2. T. H. Cormen, C. E. Leiserson, R. L. Rivest, and C. Stein. *Introduction to Algorithms*. The MIT Press/McGraw-Hill Book Company, 2 edition, 2001.
3. R. Debruyne. A property of path inverse consistency leading to an optimal PIC algorithm. In *European Conference on Artificial Intelligence*, pages 88–92, 2000.
4. C. Demetrescu and G. F. Italiano. Fully dynamic transitive closure: Breaking through the $O(n^2)$ barrier. In *IEEE Symposium on Foundations of Computer Science*, pages 381–389, 2000.
5. C. Demetrescu and G. F. Italiano. Fully dynamic all pairs shortest paths with real edge weights. In *IEEE Symposium on Foundations of Computer Science*, pages 260–267, 2001.
6. C. Demetrescu and G. F. Italiano. A new approach to dynamic all pairs shortest paths. In *ACM Symposium on the Theory of Computing*, pages 159–166, 2003.
7. F. Eisenbrand and F. Grandoni. On the complexity of fixed parameter clique and dominating set. 2003. To appear in "Theoretical Computer Science".

8. C. D. Elfe and E. C. Freuder. Neighborhood inverse consistency preprocessing. In *National Conference on Artificial Intelligence/Innovative Applications of Artificial Intelligence*, volume 1, pages 202–208, 1996.

9. D. Frigioni, A. Marchetti-Spaccamela, and U. Nanni. Fully dynamic shortest paths and negative cycles detection on digraphs with arbitrary arc weights. In *European Symposium on Algorithms*, pages 320–331, 1998.

10. J. Holm, K. de Lichtenberg, and M. Thorup. Poly-logarithmic deterministic fully-dynamic algorithms for connectivity, minimum spanning tree, 2-edge, and biconnectivity. *Journal of the Association for Computing Machinery*, 48(4):723–760, 2001.

11. X. Huang and V. Pan. Fast rectangular matrix multiplication and applications. *Journal of Complexity*, 14(2):257–299, 1998.

12. A. Itai and M. Rodeh. Finding a minimum circuit in a graph. *SIAM Journal on Computing*, 7(4):413–423, 1978.

13. V. King. Fully dynamic algorithms for maintaining all-pairs shortest paths and transitive closure in digraphs. In *IEEE Symposium on Foundations of Computer Science*, pages 81–91, 1999.

14. V. King and G. Sagert. A fully dynamic algorithm for maintaining the transitive closure. In *ACM Symposium on the Theory of Computing*, pages 492–498, 1999.

15. A. K. Mackworth. Consistency in networks of relations. *Artificial Intelligence*, 8:99–118, 1977.

16. U. Montanari. Networks of constraints: Fundamental properties and applications to picture processing. *Information Sciences*, 7:95–132, 1974.

17. J. Nešetřil and S. Poljak. On the complexity of the subgraph problem. *Commentationes Mathematicae Universitatis Carolinae*, 26(2):415–419, 1985.

18. A. Shoshan and U. Zwick. All pairs shortest paths in undirected graphs with integer weights. In *IEEE Symposium on Foundations of Computer Science*, pages 605–615, 1999.

19. U. Zwick. All pairs shortest paths in weighted directed graphs - exact and almost exact algorithms. In *IEEE Symposium on Foundations of Computer Science*, pages 310–319, 1998.

A Symbolic Approach
to the All-Pairs Shortest-Paths Problem

Daniel Sawitzki[*]

University of Dortmund, Computer Science 2
D-44221 Dortmund, Germany
daniel.sawitzki@cs.uni-dortmund.de
http://ls2-www.cs.uni-dortmund.de/~sawitzki/

Abstract. Graphs can be represented symbolically by the Ordered Binary Decision Diagram (OBDD) of their characteristic function. To solve problems in such implicitly given graphs, specialized symbolic algorithms are needed which are restricted to the use of functional operations offered by the OBDD data structure. In this paper, a symbolic algorithm for the all-pairs shortest-paths (APSP) problem in loopless directed graphs with strictly positive integral edge weights is presented. It requires $\Theta\big(\log^2(NB)\big)$ OBDD-operations to obtain the lengths and edges of all shortest paths in graphs with N nodes and maximum edge weight B. It is proved that runtime and space usage are polylogarithmic w.r.t. N and B on graph sequences with characteristic bounded-width functions. This convenient property is closed under certain graph composition operations. Moreover, an alternative symbolic approach for general integral edge weights is sketched which does not behave efficiently on general graph sequences with bounded-width functions. Finally, two variants of the APSP problem are briefly discussed.

1 Introduction

Algorithms on graphs G with node set V and edge set $E \subseteq V^2$ typically work on adjacency lists of size $\Theta(|V| + |E|)$ or on adjacency matrices of size $\Theta(|V|^2)$. These representations are called *explicit*. However, there are application areas in which problems on graphs of such large size have to be solved that an explicit representation on today's computers is not possible. In the area of logic synthesis and verification, state-transition graphs with for example 10^{27} nodes and 10^{36} edges occur. Other applications produce graphs which are representable in explicit form, but for which even runtimes of efficient polynomial algorithms are not practicable anymore. Modeling of the WWW, street, or social networks are examples of this problem scenario.

Yet, we expect the large graphs occurring in application areas to contain regularities. If we consider graphs as Boolean functions, we can represent them by *Ordered Binary Decision Diagrams (OBDDs)* [3, 4, 24]. This data structure

---

[*] Supported by the Deutsche Forschungsgemeinschaft (DFG) as part of the Research Cluster "Algorithms on Large and Complex Networks" (1126).

J. Hromkovič, M. Nagl, and B. Westfechtel (Eds.): WG 2004, LNCS 3353, pp. 154–167, 2004.

is well established in verification and synthesis of sequential circuits [11, 12, 14, 15, 24] due to its good compression of regular structures. In order to represent a graph $G = (V, E)$ by an OBDD, its edge set E is considered as a *characteristic Boolean function* χ_E, which maps binary encodings of E's elements to 1 and all others to 0. This representation is called *implicit* or *symbolic*, and is not essentially larger than explicit ones. Nevertheless, we hope that advantageous properties of G lead to small, that is sublinear OBDD-sizes [23, 25].

Having such an OBDD-representation of a graph, we are interested in solving problems on it without extracting too much explicit information from it. Algorithms that are mainly restricted to the use of functional operations are called *implicit* or *symbolic algorithms* [9, 10, 13, 19, 20, 24, 26, 27]. They are considered as heuristics to save time and/or space when large structured input graphs do not fit into the internal memory anymore. Then, we hope that each OBDD-operation processes many edges in parallel. The runtime of such methods depends on the number of executed operations as well as on the efficiency of each single one. The latter in turn depends on the size of the operand OBDDs.

Bahar et al. [1] presented a symbolic shortest-path algorithm for graphs represented by Algebraic Decision Diagrams (ADDs), which are difficult to analyze and useful only for a small number of different weight values. In [19], the algorithms of Dijkstra and Bellman-Ford are transformed into symbolic methods and evaluated in experiments. Although they perform efficiently on a variety of instances, their runtime is always at least linear in the depth of the shortest-paths tree. In this paper, we present a symbolic OBDD-algorithm for the all-pairs shortest-paths problem (called APSP-algorithm) that enables polylogarithmic runtime independent of the input graph's diameter. Given a symbolically represented loopless directed graph $G = (V, E, c)$ with strictly positive integral edge weights, it computes the length of shortest paths from node u to node v (called u–v-path in the following) for every connected pair $(u, v) \in V^2$, as well as the edges of such paths. The algorithm performs $\Theta(\log^2(NB))$ OBDD-operations on graphs with N nodes and maximum edge weight B.

The paper is organized as follows: Sections 2 and 3 introduce the principles of symbolic graph representation and preliminaries before presenting the symbolic APSP-algorithm in Sect. 4. Section 5 investigates its runtime and space usage on bounded-width functions as well as graph composition operations preserving the bounded-width property. In Sect. 6, we consider an alternative symbolic approach for general integral edge weights and point to a major disadvantage. Adaptations to two variants of the APSP problem are briefly presented in Sect. 7. Finally, Sect. 8 gives conclusions on the work.

2 Symbolic Graph Representation

We denote the class of Boolean functions $f \colon \{0, 1\}^n \to \{0, 1\}$ by B_n. The ith character of a binary number $x \in \{0, 1\}^n$ is denoted by x_i and $|x| := \sum_{i=0}^{n-1} x_i 2^i$ identifies its value.

Consider a directed graph $G = (V, E)$ with node set $V = \{v_0, \ldots, v_{N-1}\}$ and edge set $E \subseteq V^2$. G can be represented by a *characteristic* Boolean function $\chi_E \in B_{2n}$ which maps pairs $(x, y) \in \{0, 1\}^{2n}$ of binary node numbers of length $n :=$ $\lceil \log N \rceil$ to 1 iff $(v_{|x|}, v_{|y|}) \in E$. We can capture more complex graph properties by adding further arguments to characteristic functions. An additional weight function $c\colon E \to \{0, \ldots, 2^m - 1\}$ is modeled by $\chi_C \in B_{2n+m}$ which maps triples (x, y, d) to 1 iff $(v_{|x|}, v_{|y|}) \in E$ and $c(v_{|x|}, v_{|y|}) = |d|$.

A Boolean function $f \in B_n$ defined on variables $x_0, \ldots, x_{n-1}$ can be represented by an *Ordered Binary Decision Diagram (OBDD)* [3, 4, 24]. An OBDD $\mathcal{G}$ is a directed acyclic graph consisting of *internal nodes* and *sink nodes*. Each internal node is labeled with a Boolean variable x_i, while each sink node is labeled with a Boolean constant. Each internal node is left by two edges one labeled by 0 and the other by 1. A *function pointer* p marks a special node that represents f. Moreover, a permutation $\pi \in \Sigma_n$ called *variable order* must be respected by the internal nodes' labels on every path from p to a sink. For a given variable assignment $a \in \{0, 1\}^n$, we compute the function value $f(a)$ by traversing $\mathcal{G}$ from p to a sink labeled with $f(a)$ while leaving a node x_i via its a_i-edge.

An OBDD $\mathcal{G}$ with variable order π is called π-OBDD. Its size $\text{size}(\mathcal{G})$ is measured by the number of its nodes. The minimal-size π-OBDD for a function $f \in B_n$ is known to be canonical and will be denoted by $\pi\mathcal{G}[f]$ in this paper. We adopt the usual assumption that all OBDDs occurring in symbolic algorithms have minimal size, since all essential OBDD-operations produce minimized diagrams. Figure 1(a) shows the minimal OBDD for an example function. There is an upper bound of $(2 + o(1))2^n/n$ for the OBDD-size of every $f \in B_n$; hence, an edge set $E \subseteq V^2$ has worst-case OBDD-size $\mathcal{O}(V^2/\log|V|)$.

The satisfiability of f can be decided in time $\mathcal{O}(1)$. The negation $\overline{f}$ as well as the replacement of a function variable x_i by a constant a_i (i.e., $f_{|x_i=a_i}$) is obtained in time $\mathcal{O}(\text{size}(\pi\mathcal{G}[f]))$ without enlarging the OBDD. Whether two functions f and g are equivalent (i.e., $f = g$) can be decided in time $\mathcal{O}(\text{size}(\pi\mathcal{G}[f]) + \text{size}(\pi\mathcal{G}[g]))$. These operations are called *cheap*. Further essential operations are the *binary synthesis* $f \otimes g$ for $f, g \in B_n$, $\otimes \in B_2$ (e.g., "$\wedge$" and "$\vee$"), and the *quantification* $(\mathcal{Q}x_i)f$ for a quantifier $\mathcal{Q} \in \{\exists, \forall\}$. In general, the result $\pi\mathcal{G}[f \otimes g]$ has size $\mathcal{O}(\text{size}(\pi\mathcal{G}[f]) \cdot \text{size}(\pi\mathcal{G}[g]))$, which is also the general runtime of this operation. The computation of $\pi\mathcal{G}[(\mathcal{Q}x_i)f]$ can be realized by two cheap operations and one binary synthesis in time and space $\mathcal{O}(\text{size}^2(\pi\mathcal{G}[f]))$.

3   Preliminaries

The characteristic functions used for symbolic representation are typically defined on a number of k subsets of Boolean variables, each representing a different argument (e.g., $C(x, y, d)$ is defined on nodes x, y and weight d). We assume w.l.o.g. that all arguments consist of the same number of n Boolean variables. If there is no confusion, both a function $\chi_S \in B_{kn}$ defined on $x^{(1)}, \ldots, x^{(k)} \in \{0, 1\}^n$ as well as its OBDD-representation $\pi\mathcal{G}[\chi_S]$ will be denoted by $S(x^{(1)}, \ldots, x^{(k)})$ in this paper. Quantifications $(\mathcal{Q}x_0^{(i)}, \ldots, x_{n-1}^{(i)})$ over all n variables of argument i will be denoted by $(\mathcal{Q}x^{(i)})$.

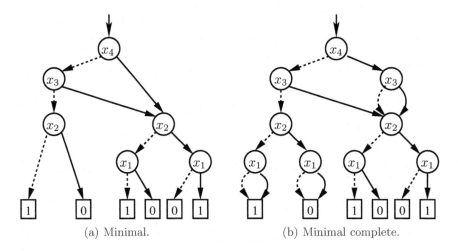

(a) Minimal.   (b) Minimal complete.

Fig. 1. Minimal (left) and minimal complete (right) π-OBDD for $f(x_1, \ldots, x_4) :=$ $\bar{x}_1 \bar{x}_2 + \bar{x}_2 \bar{x}_3 \bar{x}_4 + x_1 x_2 x_3 + x_1 x_2 x_4$ with $\pi := (4, 3, 2, 1)$

Interleaved Variable Orders. Assume that each of the k function arguments $x^{(1)}, \ldots, x^{(k)} \in \{0, 1\}^n$ has its own variable order $\tau_i \in \Sigma_n$. The global order π is called k-*interleaved* if it respects each τ_i while reading variables $x_j^{(i)}$ with same bit index j en bloc, that is, $\pi := (x_{\tau_1(0)}^{(1)}, x_{\tau_2(0)}^{(2)}, \ldots, x_{\tau_k(0)}^{(k)}, x_{\tau_1(1)}^{(1)}, \ldots, x_{\tau_k(n-1)}^{(k)})$.

Definition 1. *Let* $\rho \in \Sigma_k$ *and* $f \in B_{kn}$ *be defined on variables* $x^{(1)}, \ldots, x^{(k)} \in \{0, 1\}^n$. *The* argument reordering $\mathcal{R}_\rho : B_{kn} \to B_{kn}$ *is defined by* $\mathcal{R}_\rho(f(x^{(1)}, \ldots, \ldots, x^{(k)})) := f(x^{(\rho(1))}, \ldots, x^{(\rho(k))})$.

When using a k-interleaved variable order π, the resulting OBDD $\pi\mathcal{G}[\mathcal{R}_\rho(f)]$ has worst-case size $k^3 3^k \cdot \text{size}(\pi\mathcal{G}[f])$ and can be computed in time and space $\mathcal{O}(k^2 3^k \cdot \text{size}(\pi\mathcal{G}[f]))$ (see [23]). Because k is independent of f, this is considered as linear in $\text{size}(\pi\mathcal{G}[f])$. For example, argument reordering is used in (2) to replace the original arguments of $C(x, y, d)$ by temporary ones $x^{(1)}$, $x^{(2)}$, and $d^{(2)}$.

Multivariate Threshold Functions. The APSP-algorithm contains comparisons like $F(x, y, z) := (|x| + |y| = |z|)$, which can be realized by *multivariate threshold functions*.

Definition 2 (Woelfel [26]). *Let* $f \in B_{kn}$ *be defined on variables* $x^{(1)}, \ldots, x^{(k)} \in \{0, 1\}^n$. *Moreover, let* $W, T \in \mathbb{Z}$, *and* $w_1, \ldots, w_k \in \{-W, \ldots, W\}$. f *is called* k-variate threshold function *iff*

$$f(x^{(1)}, \ldots, x^{(k)}) = \left(\sum_{i=1}^{k} w_i \cdot |x^{(i)}| \geq T \right) .$$

W *is called the* maximum absolute weight *of* f. *The class of* k-variate threshold functions $f \in B_{kn}$ *with maximum absolute weight* W *is denoted by* $\mathbb{T}_{k,n}^{W}$.

Obviously, F can be expressed as $(|x| + |y| - |z| \geq 0) \wedge (|z| - |x| - |y| \geq 0)$. Analogue, the relations $>$, $\leq$, and $<$ can be composed of multivariate threshold functions, too. For constant W and k, such comparisons have π-OBDDs of size $\mathcal{O}(n)$ using a k-interleaved variable order π with increasing bit significance (i. e., $\tau_i = \mathrm{id}$) [26].

4   The Symbolic APSP-Algorithm

We now describe the APSP-algorithm for symbolically represented loopless directed graphs $G = (V, E, c)$ with node set $V = \{v_0, \ldots, v_{N-1}\}$, edge set $E \subseteq V^2$, edge weight function $c \colon E \to \mathbb{N}_{>0}$, and $B := \max\{c(e) \mid e \in E\}$. The maximum path length in G is $B(N - 1) =: L$. Let $n := \lceil \log(L + 1) \rceil = \Theta(\log N + \log B)$ the number of bits encoding one node number or distance value. The algorithm receives the input graph G as an OBDD for the characteristic function $C(x, y, d)$ with

$$C(x, y, d) = 1 :\Leftrightarrow \left[(v_{|x|}, v_{|y|}) \in E \right] \wedge \left[c(v_{|x|}, v_{|y|}) = |d| \right] .$$

4.1   Computing the Shortest Paths' Lengths

At first, we are interested in the distance function $\mathrm{dist} \colon V^2 \to \mathbb{N}_0 \cup \{\infty\}$ which maps node pairs $(u, v) \in V^2$ to the length $\|\bar{p}\|$ of a shortest path $\bar{p} = (u, \ldots, v)$ with $\|\bar{p}\| := \sum_{e \in \bar{p}} c(e)$. The algorithm computes dist's OBDD $S(x, y, d)$ with

$$S(x, y, d) = 1 :\Leftrightarrow \mathrm{dist}(v_{|x|}, v_{|y|}) = |d| .$$

We use functions $S_i(x, y, d)$ to represent shortest paths of maximal length $2^i - 1$, i.e., $S_i(x, y, d) = S(x, y, d) \wedge (|d| < 2^i)$. These are computed iteratively for $i \in \{1, \ldots, n\}$ until the output of $S(x, y, d) = S_n(x, y, d)$.

We consider $S_1(x, y, d)$. Due to $c(e) \in \mathbb{N}_{>0}$, paths of length $2^1 - 1 = 1$ correspond to edges contained in $C(x, y, d)$, whereas $|d| = 0$ implies $x = y$. Hence, $S_1(x, y, d)$ is defined by

$$S_1(x, y, d) := \left[(|d| = 1) \wedge C(x, y, d) \right] \vee \left[(|d| = 0) \wedge (x = y) \right] . \tag{1}$$

In general, we compute $S_{i+1}(x, y, d)$ from $C(x, y, d)$ and $S_i(x, y, d)$ using the following lemma.

Lemma 1. *For every path $\bar{p} = (p_1, \ldots, p_K)$ in G with $K \geq 1$, $\|\bar{p}\| < 2^{i+1}$, $i \in \mathbb{N}_0$, there is an edge $e := (p_j, p_{j+1}) \in \bar{p}$ such that $\bar{p}_1 := (p_1, \ldots, p_j)$ and $\bar{p}_2 := (p_{j+1}, \ldots, p_K)$ have length $\|\bar{p}_1\|, \|\bar{p}_2\| < 2^i$.*

Proof. We choose edge $e = (p_j, p_{j+1})$ with the smallest index j such that $\|\bar{p}_2\| < 2^i$. If $j = 1$ then $\|\bar{p}_1\| = \|(p_1, p_1)\| = 0$. If $j > 1$ we conclude $\|\bar{p}_1\| = \|(p_1, \ldots, p_j)\| < 2^i$ from $\|(p_j, \ldots, p_K)\| \geq 2^i$ and $\|\bar{p}\| < 2^{i+1}$. $\qquad\square$

In order to obtain a superset of all paths $\bar{p}$ with $2^i \leq \|\bar{p}\| < 2^{i+1}$, we compute the OBDD $H_{i+1}(x, y, d)$ of all connected pairs $(v_{|x|}, v_{|y|}) \in V^2$ having $v_{|x|}$–$v_{|y|}$-paths of length $|d|$ that can be partitioned into parts $\bar{p}_1$ and $\bar{p}_2$ by means of Lemma 1.

$$H_{i+1}(x, y, d) := (\exists x^{(1)}, x^{(2)}, d^{(1)}, d^{(2)}, d^{(3)}) \left[(|d^{(1)}| + |d^{(2)}| + |d^{(3)}| = |d|) \right.$$
$$\left. \wedge S_i(x, x^{(1)}, d^{(1)}) \wedge C(x^{(1)}, x^{(2)}, d^{(2)}) \wedge S_i(x^{(2)}, y, d^{(3)}) \right] \quad (2)$$

Then, we restrict $H_{i+1}(x, y, d)$ to those triples (x, y, d) with $|d| < 2^{i+1}$ (3) and $\text{dist}(v_{|x|}, v_{|y|}) = |d|$ (4); i.e., there is no shorter $|d^{(1)}|$ fulfilling $H_{i+1}(x, y, d^{(1)})$.

$$H'_{i+1}(x, y, d) := H_{i+1}(x, y, d) \wedge (|d| < 2^{i+1}) \quad (3)$$
$$\wedge \overline{(\exists d^{(1)}) \left[(|d^{(1)}| < |d|) \wedge H_{i+1}(x, y, d^{(1)}) \right]} \quad (4)$$

Finally, we cover paths shorter than 2^i by adding the previously computed $S_i(x, y, d)$ and obtain $S_{i+1}(x, y, d)$.

$$S_{i+1}(x, y, d) := S_i(x, y, d) \vee H'_{i+1}(x, y, d) \quad (5)$$

The output $S_n(x, y, d) = S(x, y, d)$ represents the OBDD for the all-pairs shortest-paths function dist.

4.2   Computing the Shortest Paths' Edges

Having computed $S(x, y, d)$, we are interested in the edges being part of shortest paths. We represent these by the OBDD $P(w, x, y, z)$ with

$$P(w, x, y, z) = 1 :\Leftrightarrow (v_{|w|}, v_{|x|}) \in E \text{ is part of a shortest } v_{|y|}\text{–}v_{|z|}\text{-path} .$$

Edge $(v_{|w|}, v_{|x|})$ lies on a shortest $v_{|y|}$–$v_{|z|}$-path iff $\text{dist}(v_{|y|}, v_{|w|}) + c(v_{|w|}, v_{|x|}) + \text{dist}(v_{|x|}, v_{|z|}) = \text{dist}(v_{|y|}, v_{|z|})$. This is expressed as

$$P(w, x, y, z) := (\exists d, d^{(1)}, d^{(2)}, d^{(3)}) \left[(|d^{(1)}| + |d^{(2)}| + |d^{(3)}| = |d|) \right.$$
$$\left. \wedge S(y, w, d^{(1)}) \wedge C(w, x, d^{(2)}) \wedge S(x, z, d^{(3)}) \wedge S(y, z, d) \right] . \quad (6)$$

We now consider the number of OBDD-operations the APSP-algorithm requires.

Theorem 1. *The symbolic APSP-algorithm computes the functions $S(x, y, d)$ and $P(w, x, y, z)$ by $\Theta(n^2) = \Theta\left(\log^2(NB)\right)$ OBDD-operations.*

Proof. The computation of each OBDD $S_i(x, y, d)$, $i \in \{1, \ldots, n\}$, as well as $P(w, x, y, z)$ consists of a constant number of cheap operations, argument re-orderings, binary syntheses, and quantifications over node numbers or distance values. Each such quantification involves $\Theta(n)$ cheap operations and binary syntheses. Due to $n = \Theta\left(\log(NB)\right)$, a number of $\Theta(n^2) = \Theta\left(\log^2(NB)\right)$ OBDD-operations is executed. $\square$

Remark 1. Heuristic methods may behave much worse than the best known general methods in the worst case. Most papers on symbolic methods do not contain any worst-case bounds, because these are not considered as representative. Taking into account that the OBDD-size of any function $f \in B_n$ is bounded by $(2 + o(1))2^n/n$, the pseudopolynomial bounds of $\mathcal{O}(N^{16}B^8 \log^2(NB))$ on runtime and $\mathcal{O}(N^{16}B^8)$ on space are obtained for the symbolic APSP-algorithm.

4.3   Computing Concrete Shortest Paths

In order to obtain the edges of a concrete shortest $v_{|y^*|}$–$v_{|z^*|}$-path for fixed y^* and z^*, different methods can be used. A straight-forward method to construct the path nodes $v_{|y^*|} = p_1, \ldots, p_K = v_{|z^*|}$ in time $\mathcal{O}(K \cdot \text{size}(P))$ is to replace the y- and z-variables in $P(w, x, y, z)$ by the corresponding Boolean constants y^* and z^*. Then, we maintain a current node number w^* (starting with $w^* := y^*$) which replaces the argument w. The resulting OBDD $P(x)$ depends only on the target node of an edge $(v_{|w^*|}, \cdot)$ being part of a shortest path $(p_1, \ldots, v_{|w^*|}, \ldots, p_K)$. In time $\mathcal{O}(\text{size}(P))$ we obtain an arbitrary satisfying assignment of x, which becomes the new actual w^*. This is repeated until $w^* = z^*$.

Alternatively, the shortest $v_{|y^*|}$–$v_{|z^*|}$-path can be computed by the symbolic blocking-flow construction method presented in [20]. This performs only $\mathcal{O}(\log^2 N)$ OBDD-operations independent from $K = \mathcal{O}(N)$, while it may cause an exponential blow-up of the OBDD-sizes.

5   Bounded-Width Functions

Symbolic algorithms are well established in logic synthesis because they often behave better than explicit methods on interesting instances [11–15, 24]. To be efficient w.r.t. the size of an input graph (i.e., the number of nodes and edges), this graph must have a compact OBDD-representation. The latter in turn is a property of the input and does not depend on the algorithm itself. Therefore, it is reasonable to investigate the behavior of symbolic methods w.r.t. the input's and output's OBDD-size.

Unfortunately, a number of $\Theta(n)$ quantification operations applied on a characteristic function $f \in B_n$ may suffice to cause an exponential blow-up of its OBDD-size, which makes it difficult to analyze symbolic algorithms. Moreover, Feigenbaum et al. [8] proved that even the basic problem of reachability analysis on OBDD-represented graphs is PSPACE-complete. So in in most papers the usability of symbolic algorithms is just proved by experiments on benchmark inputs from special application areas [13–15, 17, 27]. In other works considering more general graph problems, mostly the number of OBDD-operations (often referred to as "symbolic steps") is bounded as a hint on the actual runtime [2, 9, 10, 18].

Therefore, we propose to consider a class of characteristic functions that enables statements on the over-all runtime and space usage of the symbolic APSP-algorithm, and which has also been successfully used in the analysis of symbolic topological sorting [26] and maximum flow algorithms [20–22].

Definition 3. *A π-OBDD for a function $f \in B_n$ is called* complete *if every path from its function pointer to a sink has length n.*

That is, complete OBDDs are not allowed to skip variable tests. The minimal-size complete π-OBDD for $f \in B_n$ is also known to be canonical [24] and will be denoted by $\pi\mathcal{G}_c[f]$ in the following. Figure 1(b) shows the minimal complete OBDD for an example function.

Definition 4. *Let $F := (f_n)_{n \in \mathbb{N}}$ be a sequence of functions $f_n \in B_{\mathcal{N}(n)}$, $\mathcal{N}: \mathbb{N} \to \mathbb{N}$, defined on variables $x_0, \ldots, x_{\mathcal{N}(n)-1}$. Moreover, let $\Pi := (\pi_n)_{n \in \mathbb{N}}$ be a sequence of variable orders $\pi_n \in \Sigma_{\mathcal{N}(n)}$. F has* bounded width b w.r.t. Π *(F is b-bounded by Π) iff for all $n \in \mathbb{N}$ the OBDD $\pi_n\mathcal{G}_c[f_n]$ contains no more than b nodes labeled with the same variable x_i for $i \in \{0, \ldots, \mathcal{N}(n) - 1\}$.*

Note that $\pi_n\mathcal{G}[f_n] \le \pi_n\mathcal{G}_c[f_n] = \mathcal{O}(\mathcal{N}(n)b)$.

Theorem 2 (Sawitzki [23]). *Let $F^{(1)} := \left(f_n^{(1)}\right)_{n \in \mathbb{N}}$ and $F^{(2)} := \left(f_n^{(2)}\right)_{n \in \mathbb{N}}$ be sequences of functions $f_n^{(1)}, f_n^{(2)} \in B_{k\mathcal{N}(n)}$, $k \in \mathbb{N}$, $\mathcal{N}: \mathbb{N} \to \mathbb{N}$, defined on variables $x^{(1)}, \ldots, x^{(k)} \in \{0,1\}^{\mathcal{N}(n)}$. Assume that $F^{(1)}$ and $F^{(2)}$ have bounded width b_1 resp. b_2 w.r.t. variable orders $\Pi := (\pi_n)_{n \in \mathbb{N}}$, $\pi_n \in \Sigma_{k\mathcal{N}(n)}$.*

1. *(Binary Synthesis)*
   For all $n \in \mathbb{N}$, the OBDD $\pi_n\mathcal{G}[f_n^{(1)} \otimes f_n^{(2)}]$, $\otimes \in B_2$, can be computed in time and space $\mathcal{O}(k\mathcal{N}(n)b_1 b_2)$. The resulting sequence $\left(f_n^{(1)} \otimes f_n^{(2)}\right)_{n \in \mathbb{N}}$ is $b_1 b_2$-bounded by Π.

2. *(Quantification)*
   Let $X := (X_n)_{n \in \mathbb{N}}$ be a sequence of variable sets $X_n \subseteq \{x_i^{(j)} | i \in \{0, \ldots, \mathcal{N}(n)-1\}, j \in \{1, \ldots, k\}\}$. For all $n \in \mathbb{N}$, the OBDD $\pi_n\mathcal{G}[(\mathcal{Q}X_n)f_n^{(1)}]$, $\mathcal{Q} \in \{\exists, \forall\}$, can be computed in time and space $\mathcal{O}(|X_n|k\mathcal{N}(n)2^{2b_1})$. The resulting sequence $\left((\mathcal{Q}X_n)f_n^{(1)}\right)_{n \in \mathbb{N}}$ is 2^{b_1}-bounded by Π.

3. *(Argument Reordering)*
   Let $\rho \in \Sigma_k$ and assume that Π is k-interleaved. For all $n \in \mathbb{N}$, the OBDD $\pi_n\mathcal{G}[\mathcal{R}_\rho(f_n^{(1)})]$ can be computed in time and space $\mathcal{O}(\mathcal{N}(n)b_1 k^3 2^k)$. The resulting sequence $\left(\mathcal{R}_\rho(f_n^{(1)})\right)_{n \in \mathbb{N}}$ is $b_1 2^k$-bounded by Π.

The resulting width bounds are worst cases. However, because b_1, b_2, and k are independent of n, each operation takes linear time and space w.r.t. the number $\mathcal{N}(n)$ of variables. We conclude that bounded-width functions are closed under all operations used by the symbolic APSP-algorithm.

Theorem 3 (Woelfel [26]). *Let $F := (f_n)_{n \in \mathbb{N}}$ be a sequence of functions $f_n \in B_{k\mathcal{N}(n)}$, $k \in \mathbb{N}$, $\mathcal{N}: \mathbb{N} \to \mathbb{N}$, and $\Pi := (\pi_n)_{n \in \mathbb{N}}$ k-interleaved variable orders $\pi_n \in \Sigma_{k\mathcal{N}(n)}$ with increasing bit significance. If for all $n \in \mathbb{N}$ it is $f_n \in \mathbb{T}_{k,\mathcal{N}(n)}^W$ then F is $\mathcal{O}(k^2 W)$-bounded by Π.*

Theorem 3 implies that the comparison functions introduced in Sect. 3 are bounded-width functions.

5.1   Analysis on Graphs
with Characteristic Bounded-Width Functions

Consider a sequence $G = (G_n)_{n \in \mathbb{N}}$ of valid input graphs $G_n = (V_n, E_n, c_n)$ for the symbolic APSP-algorithm. Assume that G_n has $N_n := |V_n|$ nodes and maximum edge weight $B_n \in \mathbb{N}_{>0}$. Let $C = (C(x, y, d)_n)_{n \in \mathbb{N}}$ be the sequence of G's characteristic functions and $S = (S(x, y, d)_n)_{n \in \mathbb{N}}$ be the characteristic functions of G's shortest path distances $(\mathrm{dist}_n)_{n \in \mathbb{N}}$. Let $\mathcal{N}(n) = \Theta(\log(N_n B_n))$ be the number of bits encoding one node number $|x|$ or distance value $|d| \leq B_n(N_n - 1)$ of G_n. Moreover, assume a sequence $\Pi := (\pi_n)_{n \in \mathbb{N}}$ of interleaved variable orders which read bits of distance values with increasing significance.

Theorem 4. *If both C and S are b-bounded by Π, the symbolic APSP-algorithm computes $S(x, y, d)_n$ from $C(x, y, d)_n$ in time $\mathcal{O}(\log^3(N_n B_n) \cdot \alpha(b))$ and space $\mathcal{O}(\log(N_n B_N) \cdot \alpha(b))$ for all $n \in \mathbb{N}$ and*

$$\alpha(b) := 2^{2^{\mathcal{O}(b^3)}} .$$

Proof. All characteristic functions are defined on a constant number of binary node and distance numbers. Hence, the over-all number of Boolean variables is $\Theta(\log(N_n B_n))$ and reordering causes only a linear width growth to bounded-with functions. We show that all occurring functions are $\alpha(b)$-bounded by Π. (A similar analysis technique has been used in [16].)

Using Π enables to realize the comparison $(|d| < 2^i)$ by multivariate threshold functions of $\mathbb{T}_{1, \mathcal{N}(n)}^{\mathcal{O}(1)}$ with width bound $\mathcal{O}(1)$ (see Theorem 3). From $S_i(x, y, d)_n = S(x, y, d)_n \wedge (|d| < 2^i)$, $i \in \{1, \dots, \mathcal{N}(n)\}$, and the width bound b of S we conclude that S_i is $\mathcal{O}(b)$-bounded (see Theorem 2).

It remains to show that each intermediate result is $\alpha(b)$-bounded. In (1), $S_1(x, y, d)_n$ is initialized by three syntheses involving comparisons and the input OBDD $C(x, y, d)_n$. Analogue to S_i, each occurring OBDD has bounded width $\mathcal{O}(b)$.

In (2), three binary syntheses are performed before the existential quantifications. Being a composition of multivariate threshold functions, the comparison $(|d^{(1)}| + |d^{(2)}| + |d^{(3)}| = |d|)$ is $\mathcal{O}(1)$-bounded. Due to Theorem 2, each intermediate conjunction result is $\mathcal{O}(b^3)$-bounded, whereas each quantification result (including H_{i+1}) is $2^{\mathcal{O}(b^3)}$-bounded.

At next, $H'_{i+1}(x, y, d)_n$ is obtained by restricting $H_{i+1}(x, y, d)_n$ in (3). The conjunctions with $(|d| < 2^{i+1})$ resp. $(|d^{(1)}| < |d|)$ do not change the asymptotical width bound $2^{\mathcal{O}(b^3)}$. Finally, the quantification $(\exists d^{(1)})$ causes one further exponentiation and the new width bound is $2^{ab^3} \cdot 2^{2^{ab^3}} = \alpha(b)$ for an appropriate constant a. This still holds after disjunction with $S_i(x, y, d)_n$ due to S_i being $\mathcal{O}(b)$-bounded.

Hence, all occurring characteristic functions are $\alpha(b)$-bounded and have π_n-OBDD-size $\mathcal{O}(\log(N_n B_n) \cdot \alpha(b))$. Due to Theorem 2, each of the $\Theta(\log^2(N_n B_n))$ executed OBDD-operations takes time and space $\mathcal{O}(\log(N_n B_n) \cdot \alpha(b))$, which implies an over-all runtime bound of $\mathcal{O}(\log^3(N_n B_n) \cdot \alpha(b))$. Because only a constant number of OBDDs has to be stored at any time, the over-all space usage is of the same magnitude as each single OBDD-size. □

Analogue to this proof, a width bound of $2^{\mathcal{O}(b^4)}$ can be obtained for $P(w,x,y,z)$. Moreover, a less elegant formulation of (4) improves $\alpha(b)$ to $2^{\mathcal{O}(b^3)}$.

Corollary 1. *If both C and S are b-bounded by Π, the symbolic APSP-algorithm computes $P(w,x,y,z)_n$ from $S(x,y,d)_n$ in time $\mathcal{O}\big(\log^2(N_nB_n) \cdot 2^{\mathcal{O}(b^4)}\big)$ and space $\mathcal{O}\big(\log(N_nB_n) \cdot 2^{\mathcal{O}(b^4)}\big)$.*

How to classify this result? It is desirable that symbolic algorithms behave efficiently on "small" input OBDDs, which could be defined most general by being polynomial in the number $\mathcal{N}(n) = \Theta\big(\log(N_nB_n)\big)$ of Boolean variables. Theorem 4 can be considered as showing this convenient property for the more restricted case of bounded-width functions, whose π_n-OBDD-size is even linear in $\mathcal{N}(n)$; here, "efficiently" means polylogarithmic w.r.t. N_n and B_n.

Hence, the symbolic APSP-algorithm can be considered as being *fixed-parameter tractable* [7] for the parameter b of characteristic b-bounded functions.

5.2 Composition of Graphs with Characteristic Bounded-Width Functions

Having the results on bounded-with functions, we ask what kinds of graphs can be represented by them. Obviously, sequences $G = (G_n)_{n \in \mathbb{N}}$ consisting of a single graph $G_1 = \cdots = G_n$, $n \in \mathbb{N}$, have characteristic bounded-width functions. We already know multivariate threshold functions to have bounded width. These in turn can be used to build many simple sequences like empty, complete, complete bipartite, and grid graphs [23]. From the closedness under OBDD-operations we now conclude the closedness under four graph composition operations.

Let $G^{(i)} := \big(G_n^{(i)} = (V_n^{(i)}, E_n^{(i)}, c_n^{(i)})\big)_{n \in \mathbb{N}}$, $i \in \{1,2,3\}$, be sequences of valid input graphs for the symbolic APSP-algorithm with same notation as G in Sect. 5.1. Assume $V_n^{(1)} \cap V_n^{(2)} = \emptyset$ for all $n \in \mathbb{N}$.

Definition 5. *Graph Composition Operations.*

1. *$G^{(3)}$ is called the cojoin of $G^{(1)}$ and $G^{(2)}$ iff for all $n \in \mathbb{N}$ it is $V_n^{(3)} = V_n^{(1)} \cup V_n^{(2)}$, $E_n^{(3)} = E_n^{(1)} \cup E_n^{(2)}$, and $c_n^{(3)}(e) = c_n^{(i)}(e)$ for $e \in E_n^{(i)}$.*
2. *$G^{(3)}$ is called the $\mathcal{A}$-join of $G^{(1)}$ and $G^{(2)}$, $\mathcal{A}: \mathbb{N} \to \mathbb{N}_{>0}$, iff for all $n \in \mathbb{N}$ it is $V_n^{(3)} = V_n^{(1)} \cup V_n^{(2)}$, $E_n^{(3)} = E_n^{(1)} \cup E_n^{(2)} \cup (V_n^{(1)} \times V_n^{(2)})$, and $c_n^{(3)}(e) = c_n^{(i)}(e)$ for $e \in E_n^{(i)}$ resp. $c_n^{(3)}(e) = \mathcal{A}(n)$ for $e \in V_n^{(1)} \times V_n^{(2)}$.*
3. *$G^{(3)}$ is called the node substitution of $G^{(1)}$ in $G^{(2)}$ iff for all $n \in \mathbb{N}$ it is $V_n^{(3)} = V_n^{(1)} \times V_n^{(2)}$,*

   $$E_n^{(3)} = \Big\{ \big((t,u),(v,w)\big) \mid \big((t,v) \in E_n^{(1)} \wedge (u=w)\big) \vee (u,w) \in E_n^{(2)} \Big\},$$

   and $c_n^{(3)}$ weights edge $\big((t,u),(v,w)\big)$ with $c_n^{(1)}(t,v)$ if $u=w$ resp. $c_n^{(2)}(u,w)$ if $(u,w) \in E_n^{(2)}$.

4. $G^{(3)}$ *is called the* product *of $G^{(1)}$ and $G^{(2)}$ iff for all $n \in \mathbb{N}$ it is $V_n^{(3)} = V_n^{(1)} \times V_n^{(2)}$,*

$$E_n^{(3)} = \Big\{ ((t, u), (v, w)) \mid ((t, v) \in E_n^{(1)} \wedge (u = w))$$
$$\vee \, ((t = v) \wedge (u, w) \in E_n^{(2)}) \Big\} \,,$$

and $c_n^{(3)}$ weights edge $((t, u), (v, w))$ with $c_n^{(1)}(t, v)$ if $u = w$ resp. $c_n^{(2)}(u, w)$ if $t = v$.

Theorem 5. *Let $G^{(3)}$ be the product of $G^{(1)}$ and $G^{(2)}$. If $C^{(i)}$ and $S^{(i)}$ are b-bounded by Π for $i \in \{1, 2\}$, then $C^{(3)}$ is $\mathcal{O}(b^2)$-bounded by Π and $S^{(3)}$ is $2^{\mathcal{O}(b^2)}$-bounded by Π.*

Proof. Let $(x^{(1)}, x^{(2)})$ denote the binary node number of $(v_{|x^{(1)}|}, v_{|x^{(2)}|}) \in V_n^{(3)}$ corresponding to argument x of a characteristic function. A shortest path in the product $G_n^{(3)}$ is composed of shortest paths in $G_n^{(1)}$ and $G_n^{(2)}$. We express $C^{(3)}$ and $S^{(3)}$ in terms of $C^{(1)}$, $C^{(2)}$, $S^{(1)}$, and $S^{(2)}$:

$$C^{(3)}(x^{(1)}, x^{(2)}, y^{(1)}, y^{(2)}, d)_n = \big[C^{(1)}(x^{(1)}, y^{(1)}, d)_n \wedge (x^{(2)} = y^{(2)}) \big]$$
$$\vee \big[(x^{(1)} = y^{(1)}) \wedge C^{(2)}(x^{(2)}, y^{(2)}, d)_n \big] \,,$$

$$S^{(3)}(x^{(1)}, x^{(2)}, y^{(1)}, y^{(2)}, d)_n = (\exists d^{(1)}, d^{(2)}) \big[(|d^{(1)}| + |d^{(2)}| = |d|)$$
$$\wedge S^{(1)}(x^{(1)}, y^{(1)}, d^{(1)})_n \wedge S^{(2)}(x^{(2)}, y^{(2)}, d^{(2)})_n \big] \,.$$

Analogue to S_i in Theorem 4, we conclude $C^{(3)}$ to be $\mathcal{O}(b^2)$-bounded. For $S^{(3)}$, the quantifiers are applied to an intermediate result of width $\mathcal{O}(b^2)$ and cause an exponentiation leading to the final bound of $2^{\mathcal{O}(b^2)}$. $\square$

The same width bounds can be obtained for the case of node substitution, whereas both $C^{(3)}$ and $S^{(3)}$ are $\mathcal{O}(b^2)$-bounded if a cojoin or $\mathcal{A}$-join has been applied. That is, the efficiency results of Theorem 4 also hold for complex graphs builded from basic ones having characteristic bounded-width functions.

For the complete proofs, further composition operations, and a more comprehensive discussion of graphs with characteristic bounded-width functions, the reader is referred to [23].

6   A Reason for Restricting to Positive Edge Weights

The proof of Theorem 4 makes use of the fact that the intermediate results $S_i(x, y, d)$ can be expressed in terms of the final result $S(x, y, d)$ and the $\mathcal{O}(1)$-bounded comparison $(|d| < 2^i)$ by $S_i(x, y, d) = S(x, y, d) \wedge (|d| < 2^i)$. For the correctness of the APSP-algorithm it is essential that only empty paths have length 0 due to the strictly positive edge weights.

Instead of computing $S(x, y, d)$ by iterating over the number $i \in \{1, \ldots, n\}$ of considered distance bits, we could iteratively double the number of edges of considered paths. A corresponding recursion would be

$$H_{i+1}(x, y, d) := (\exists x^{(1)}, d^{(1)}, d^{(2)}) \left[(|d^{(1)}| + |d^{(2)}| = |d|) \right.$$
$$\left. \wedge\, S_i(x, x^{(1)}, d^{(1)}) \wedge S_i(x^{(1)}, y, d^{(2)}) \right] ,$$
$$S_{i+1}(x, y, d) := H_{i+1}(x, y, d) \wedge \overline{(\exists d^{(1)}) \left[(|d^{(1)}| < |d|) \wedge H_{i+1}(x, y, d^{(1)}) \right]} ,$$

where $S_i(u, v, d)$ now represents all shortest paths consisting of no more than 2^i edges. The resulting algorithm is able to handle graphs with general integral edge weights that contain no negative cycles. Nevertheless, it does not provide a counterpart to Theorem 4: There are graphs $G^* := (G_n^*)_{n \in \mathbb{N}}$ fulfilling the bounded-width conditions, but whose intermediate functions S_i have not bounded width in general.

The construction of G^* makes use of the fact that shortest paths may consist of many edges. If the shortest paths have more then 2^i edges, the functions S_i have to represent longer ones, which can be chosen such that the bounded width of S_i would also imply bounded width for the multiplication function MUL_n. This contradicts exponential lower bounds on the π-OBDD-size of MUL_n for every variable order π. For a detailed discussion, the reader is referred to [23].

There is no symbolic algorithm known to the author which is able to handle general integral edge weights and that has properties comparable to Theorem 4.

7   Related Problems

The symbolic APSP-algorithm can be easily adapted to compute the edges of almost shortest-paths of small stretch [5, 6]. This is done by replacing $P(w, x, y, z)$ (see (6)) by a function $P^{a,b}(w, x, y, z)$ representing edges $(v_{|w|}, v_{|x|})$ on $v_{|y|} - v_{|z|}$-paths $\bar{p}$ of length $\|\bar{p}\| \le a \cdot \operatorname{dist}(v_{|y|}, v_{|z|}) + b$.

$$P^{a,b}(w, x, y, z) := (\exists d, d^{(1)}, d^{(2)}, d^{(3)}) \left[(|d^{(1)}| + |d^{(2)}| + |d^{(3)}| \le a \cdot |d| + b) \right.$$
$$\left. \wedge\, S(y, w, d^{(1)}) \wedge C(w, x, d^{(2)}) \wedge S(x, z, d^{(3)}) \wedge S(y, z, d) \right] .$$

Due to the results on the bounded width of multivariate threshold functions, Corollary 1 holds for $P^{a,b}$, too.

Finally, we consider a dynamic scenario: After the computation of $S(x, y, d)$, the weights of an edge set $E' \subseteq E$ are decreased resulting in the new symbolic graph $C'(x, y, d)$. If every graph path contains at most one updated edge, the new distances $S'(x, y, d)$ can be computed by $\Theta(\log(NB))$ OBDD-operations. $v_{|x|} - v_{|y|}$-paths of length $|d|$ containing a decreased edge $(v_{x^{(1)}}, v_{x^{(2)}})$ are expressed as

$$F(x, y, d) := (\exists x^{(1)}, x^{(2)}, d^{(1)}, d^{(2)}, d^{(3)}) \left[(|d^{(1)}| + |d^{(2)}| + |d^{(3)}| = |d|) \right.$$
$$\left. \wedge\, S(x, x^{(1)}, d^{(1)}) \wedge C'(x^{(1)}, x^{(2)}, d^{(2)}) \wedge S(x^{(2)}, y, d^{(3)}) \right] .$$

To obtain $S'(x, y, d)$, we just have to select the smallest $|d|$ with $S(x, y, d) \vee F(x, y, d) = 1$ similar to (4). Again, bounded width of C' and S imply time and space bounds as in Theorem 4.

8   Conclusions

We presented a symbolic algorithm for the all-pairs shortest-paths problem. The algorithm works on OBDD-representations of loopless directed graphs $G = (V, E, c)$ with strictly positive integral edge weights. It computes the lengths and edges of shortest paths by performing a polylogarithmic number of OBDD-operations w. r. t. $N := |V|$ and $B := \max\{c(e) \mid e \in E\}$.

In order to investigate runtime and space usage, bounded-width functions have been introduced, which have small OBDDs and allow efficient OBDD-operations. The algorithm is proved to have polylogarithmic runtime and space usage w. r. t. N and B on graphs whose characteristic functions have bounded width. This property is closed under important graph composition operations. In contrast, the bounded-width of input and output does not guarantee efficiency for symbolic algorithms which iterate over the number of edges of paths instead of their length.

Finally, adaptations of the symbolic APSP-algorithm to dynamic edge weights and almost shortest paths of small stretch have been briefly discussed.

Acknowledgments

Thanks to Thomas Franke and Ingo Wegener for proofreading and helpful discussions.

The author's technical reports can be obtained via his homepage.

References

1. R.I. Bahar, E.A. Frohm, C.M. Gaona, G.D. Hachtel, E. Macii, A. Pardo, and F. Somenzi. Algebraic decision diagrams and their applications. In *ICCAD'93*, pages 188–191. IEEE Press, 1993.
2. R. Bloem, H.N. Gabow, and F. Somenzi. An algorithm for strongly connected component analysis in $n \log n$ symbolic steps. In *FMCAD'00*, volume 1954 of *LNCS*, pages 37–54. Springer, 2000.
3. R.E. Bryant. Symbolic manipulation of Boolean functions using a graphical representation. In *DAC'85*, pages 688–694. ACM Press, 1985.
4. R.E. Bryant. Graph-based algorithms for Boolean function manipulation. *IEEE Transactions on Computers*, 35:677–691, 1986.
5. E. Cohen and U. Zwick. All-pairs small-stretch paths. *Journal of Algorithms*, 38:335–353, 2001.
6. D. Dor, S. Halperin, and U. Zwick. All-pairs almost shortest paths. *SIAM Journal on Computing*, 29:1740–1759, 2000.
7. R.G. Downey and M.R. Fellows. *Parameterized Complexity*. Springer, Berlin Heidelberg New-York, 1999.

8. J. Feigenbaum, S. Kannan, M.Y. Vardi, and M. Viswanathan. Complexity of problems on graphs represented as OBDDs. *Chicago Journal of Theoretical Computer Science*, 1999, 1999.
9. R. Gentilini, C. Piazza, and A. Policriti. Computing strongly connected components in a linear number of symbolic steps. In *SODA'03*, pages 573–582. ACM Press, 2003.
10. R. Gentilini and A. Policriti. Biconnectivity on symbolically represented graphs: A linear solution. In *ISAAC'03*, volume 2906 of *LNCS*, pages 554–564. Springer, 2003.
11. G.D. Hachtel, M. Hermida, A. Pardo, M. Poncino, and F. Somenzi. Re-Encoding sequential circuits to reduce power dissipation. In *ICCAD'94*, pages 70–73. IEEE Press, 1994.
12. G.D. Hachtel and F. Somenzi. *Logic Synthesis and Verification Algorithms*. Kluwer Academic Publishers, Boston, 1996.
13. G.D. Hachtel and F. Somenzi. A symbolic algorithm for maximum flow in 0–1 networks. *Formal Methods in System Design*, 10:207–219, 1997.
14. R. Hojati, H. Touati, R.P. Kurshan, and R.K. Brayton. Efficient ω-regular language containment. In *CAV'93*, volume 663 of *LNCS*, pages 396–409. Springer, 1993.
15. H. Jin, A. Kuehlmann, and F. Somenzi. Fine-grain conjunction scheduling for symbolic reachability analysis. In *TACAS'02*, volume 2280 of *LNCS*, pages 312–326. Springer, 2002.
16. M. Krause. BDD-Based cryptanalysis of keystream generators. In *EURO-CRYPT'02*, volume 2332 of *LNCS*, pages 222–237. Springer, 2002.
17. I. Moon, J.H. Kukula, K. Ravi, and F. Somenzi. To split or to conjoin: The question in image computation. In *DAC'00*, pages 23–28. ACM Press, 2000.
18. K. Ravi, R. Bloem, and F. Somenzi. A comparative study of symbolic algorithms for the computation of fair cycles. In *FMCAD'00*, volume 1954 of *LNCS*, pages 143–160. Springer, 2000.
19. D. Sawitzki. Experimental studies of symbolic shortest-path algorithms. In *WEA'04*, volume 3059 of *LNCS*, pages 482–497. Springer, 2004.
20. D. Sawitzki. Implicit flow maximization by iterative squaring. In *SOFSEM'04*, volume 2932 of *LNCS*, pages 301–313. Springer, 2004.
21. D. Sawitzki. Implicit flow maximization on grid networks. Technical report, Universität Dortmund, 2004.
22. D. Sawitzki. Implicit maximization of flows over time. Technical report, Universität Dortmund, 2004.
23. D. Sawitzki. On graphs with characteristic bounded-width functions. Technical report, Universität Dortmund, 2004.
24. I. Wegener. *Branching Programs and Binary Decision Diagrams*. SIAM, Philadelphia, 2000.
25. P. Woelfel. The OBDD-size of cographs. Internal report, Universität Dortmund, 2003.
26. P. Woelfel. Symbolic topological sorting with OBDDs. In *MFCS'03*, volume 2747 of *LNCS*, pages 671–680. Springer, 2003.
27. A. Xie and P.A. Beerel. Implicit enumeration of strongly connected components. In *ICCAD'99*, pages 37–40. ACM Press, 1999.

Minimal de Bruijn Sequence
in a Language with Forbidden Substrings[*]

Eduardo Moreno[1,3] and Martín Matamala[1,2]

[1] Departamento de Ingeniería Matemática
Facultad de Ciencias Físicas y Matemáticas
Universidad de Chile
Casilla 170-3, Correo 3, Santiago, Chile
{emoreno,mmatamal}@dim.uchile.cl
[2] Centro de Modelamiento Matemático, UMR 2071, UCHILE-CNRS
Casilla 170-3, Correo 3, Santiago, Chile
[3] Institut Gaspard Monge, Université de Marne-la-Vallée
Champs-sur-Marne, 77454 Marne-la-Vallée cedex 2, France

Abstract. Let be the following strategy to construct a walk in a labeled digraph: at each vertex, we follow the unvisited arc of minimum label. In this work we study for which languages, applying the previous strategy over the corresponding de Bruijn graph, we finish with an Eulerian cycle, in order to obtain the minimal de Bruijn sequence of the language.

1   Introduction

Given a language, a de Bruijn sequence of span n is a periodic sequence such that every n-tuple in the language (and no other n-tuple) occurs exactly once. Its first known description appears as a Sanskrit word *yamátárájabhánasalagám* which was a memory aid for Indian drummers, where the accented/unaccented syllables represent long/shorts beats, so all possible triplets of short and long beats are included in the word. De Bruijn sequences are also known as "shift register sequences" and was originally studied by N. G. De Bruijn for the binary alphabet [1]. These sequences have many different applications, such as memory wheels in computers and other technological device, network models, DNA algorithms, pseudo-random number generation, modern public-key cryptographic schemes, to mention a few (see [2],[3],[4]). Historically, de Bruijn sequence was studied in an arbitrary alphabet considering the language of all the n-tuples. There is a large number of de Bruijn sequence in this case, but only a few can be generated efficiently, see [5] for a survey about this subject. In 1978, Fredricksen and Maiorana [6] give an algorithm to generate a de Bruijn sequence of span n based in the Lyndon words of the language, which resulted to be the minimal one in the lexicographic order, and this algorithm was proved to be efficient [7]. Recently, the study of these concepts was extended to languages with forbidden

---

[*] Partially supported by ECOS C00E03 (French-Chilean Cooperation), Programa Iniciativa Científica Milenio P01-005, and CONICYT Ph.D. Fellowship.

J. Hromkovič, M. Nagl, and B. Westfechtel (Eds.): WG 2004, LNCS 3353, pp. 168–176, 2004.
© Springer-Verlag Berlin Heidelberg 2004

substrings: in [8] it was given efficient algorithms to generate all the words in a language with one forbidden substring, in [9] the concept of de Bruijn sequences was generalized to restricted languages with a finite set of forbidden substrings and it was proved the existence of these sequences and presented an algorithm to generate one of them, however, to find the minimal sequence is a non-trivial problem in this more general case. This problem is closely related to the "shortest common super-string problem" which is a important problem in the areas of DNA sequencing and data compression.

In this work we study the de Bruijn sequence of minimal lexicographical label. In section 2 we present some definitions and previous results on de Bruijn sequences and the BEST Theorem, necessary to understand the main problem, and we prove a result related with the BEST Theorem which will be useful in the following sections. In section 3 we study the main problem, giving some results on the structure of the de Bruijn graph. Finally, in section 4 we present some remarks and extensions of this work.

2   De Bruijn Sequence of Restricted Languages

2.1   Definitions

Let A be a finite set with a linear order $<$. A *word* on the alphabet A is a finite sequence of elements of A, whose length is denoted by $|w|$.

A word p is said to be a *factor* of a word w if there exist words $u, v \in A^*$ such that $w = upv$. If u is the empty word ε then p is called a *prefix* of w, and if v is empty then is called a *suffix* of w. If $p \neq w$ then p is a *proper factor, proper prefix* or *proper suffix*, respectively.

The set A^* of all the words on the alphabet A is linearly ordered by the alphabetic order induced by the order $<$ on A. By definition, $x < y$ either if x is a prefix of y or if $x = uav$, $y = ubw$ with $u, v, w \in A^*$, $a, b \in A$ and $a < b$. A basic property of the alphabetic order is the following: if $x < y$ and if x is not a prefix of y, then for any pair of words u, v, $xu < yv$.

Given an alphabet A, a full shift $A^{\mathbb{Z}}$ is the collection of all bi-infinite sequences of symbols from A. Let $\mathcal{F}$ be a set of words over A^*. A *subshift of finite type* (SFT) is the subset of sequences in $A^{\mathbb{Z}}$ which does not contain any factor in $\mathcal{F}$. We will refer to $\mathcal{F}$ as the set of *forbidden blocks* or *forbidden factors*.

Given a set $\mathcal{F}$ of forbidden blocks, in this work we will say that a word w is in the language if the periodical word w^∞, composed by infinite repetitions of w, is in the language of the SFT defined by $\mathcal{F}$. The set of all the words of length n in the language defined by $\mathcal{F}$ will be denoted by $\mathcal{W}^{\mathcal{F}}(n)$.

A SFT is *irreducible* if for every ordered pair of blocks u, v in the language there is a block w in the language so that uwv is a block of the language.

A de Bruijn sequence of span n in a restricted language is a circular string $B^{\mathcal{F},n}$ of length $|\mathcal{W}^{\mathcal{F}}(n)|$ such that all the words in the language of length n are factors of $B^{\mathcal{F},n}$. In other words,

$$\{(B^{\mathcal{F},n})_i \dots (B^{\mathcal{F},n})_{i+n-1 \bmod n} | i = 0 \dots n - 1\} = \mathcal{W}^{\mathcal{F}}(n)$$

These concepts are studied in [9], extending the known results on subshifts of finite type to this context. In particular two results are relevant in this work, the first one is a bound in the number of words of length n in the language:

$$\left|\mathcal{W}^{\mathcal{F}}(n)\right| = \Theta\left(\lambda^n\right)$$

where $\log(\lambda)$ is the *entropy* of the system (see [10]). The second result proves the existence of a de Bruijn sequence:

Theorem 1. *For any set of forbidden substrings $\mathcal{F}$ defining an irreducible subshift of finite type, there exists a de Bruijn sequence of span n.*

This last theorem is a direct consequence of the fact that the de Bruijn graph of span n is an Eulerian graph. The *de Bruijn graph* of span n, denoted by $G^{\mathcal{F},n}$, is the largest strongly connected component of the directed graph with $|A|^n$ vertices, labeled by the words in A^n, and the set of arcs

$$E = \left\{(as, sb) | a, b \in A, s \in A^{n-1}, asb \in \mathcal{W}^{\mathcal{F}}(n+1)\right\}$$

where the label of the arc $e = (as, sb)$ is $l(e) = b$. Note that if the SFT is irreducible, this graph has only one strongly connected component of size greater than 1, so there is no ambiguity in the definition.

There are not two vertices with the same label, hence from now we identify a vertex by its label. If $W = e_1 \dots e_k$ is a walk over $G^{\mathcal{F},n}$, we denote the label of W by $l(W) = l(e_1) \dots l(e_k)$, and by $l(W)^j$ the concatenation of of j times $l(W)$.

There exists a bijection between the arcs of $G^{\mathcal{F},n}$ and the words in $\mathcal{W}^{\mathcal{F}}(n+1)$, because to each arc with label $a \in A$ with tail at $w' \in A^n$ we can associate the word $w'a$ which is, by definition, a word in $\mathcal{W}^{\mathcal{F}}(n+1)$. Equally if $w'a$ is a word of $\mathcal{W}^{\mathcal{F}}(n+1)$, with $a \in A$, then there exists a vertex w' and an arc with tail at this vertex with label a.

Furthermore, if a word w is a label of a walk from u to v then v is a suffix of length n of uw. In the same way, if $w \in \mathcal{W}^{\mathcal{F}}(n+1)$ then there is a cycle C in $G^{\mathcal{F},n}$ with label $l(C)$ such that $l(C)^{\frac{n+1}{|C|}} = w$.

With all these properties it is easy to see that a de Bruijn sequence of span $n+1$ is exactly the label of an Eulerian cycle over $G^{\mathcal{F},n}$.

2.2   The BEST Theorem

BEST is an acronym of N. G. de Bruijn, T. van Aardenne-Ehrenfest, C. A. B. Smith and W. T. Tutte, the BEST Theorem (see [11]) gives a correspondence between Eulerian cycles in a digraph and its rooted trees converging to the root vertex.

Let r be a vertex of an Eulerian digraph $G = (V, E)$, a spanning tree converging to the root r is a spanning tree such that there exists a directed path from each vertex to the root.

Given an Eulerian cycle starting at the root of an Eulerian digraph, if for every vertex of G we take the last arc with tail at this vertex in the cycle then

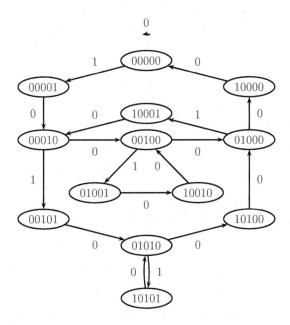

Fig. 1. De Bruijn digraph of span 5 for the Golden Mean ($\mathcal{F} = \{11\}$).

we obtain a spanning tree converging to the root. Conversely, given a spanning tree converging to the root, a walk over G starting at the root and using the arc in the tree only if all the arcs with tail at this vertex has been used, is an Eulerian cycle. A walk over the graph of this kind will be called a walk "avoiding the tree".

The BEST Theorem proves that for every different spanning tree we have a different Eulerian cycle. Therefore it also allows us to calculate the exact number of Eulerian cycles on a digraph, which is given by

$$C_{\mathcal{F}} = M_T \cdot \prod_{i=1}^{|V|} (d^+(v_i) - 1)!$$

where M_T is the number of rooted spanning trees converging to a given vertex. We bound the second term by $((\bar{d}^+ - 1)!)^{|V|}$ where $\bar{d}^+$ is the mean of the outgoing degrees over all the vertices, so we have a lower bound to the number of de Bruijn sequences

$$C_{\mathcal{F}} = \Omega\left(\lfloor\lambda - 1\rfloor!^{\lambda^{n-1}}\right)$$

in particular, for a system with $\lambda \geq 3$ the number of the Bruijn sequences of span n is exponential in the number of words in the language of length $n \doteq 1$. In the systems with $3 > \lambda > 1$ this bound is generally also true, because the underestimated term M_T is generally exponential, for example, in the system without restrictions of alphabet $\{0, 1\}$, this term is equal to $2^{2^{n-1}}$.

Now, we define formally a walk "avoiding a subgraph". Let r be any vertex. For each vertex $v \neq r$ in $G^{\mathcal{F},n}$ let e_v be any arc starting at v. Let H be the spanning subgraph of $G^{\mathcal{F},n}$ with arc set $\{e_v : v \in V(G^{\mathcal{F},n}) \setminus \{r\}\}$.

Is easy to see that H is composed by cycles, subtrees converging to a cycle, and one subtree converging to r. For a vertex not in a cycle of H, we define H_v as the directed subtree converging to v in H.

We define recursively a walk in $G^{\mathcal{F},n}$ which *avoid* H. It starts at the root vertex r. Let $v_0 e_0 \cdots v_i$ be the current walk. If there is an unvisited arc $e_i = (v_i, v_{i+1})$ not in H we extend the walk by $e_i v_{i+1}$. Otherwise we use the arc e_{v_i} in H.

We say that a walk over the graph *exhausts* a vertex if the walk use all the arc having the vertex as head or tail.

The next lemma studies in which order the vertices are exhausted in a walk avoiding H

Lemma 1. *Let W be a walk starting at vertex r avoiding H, let v be a vertex and let Wv the subpath of W starting at vertex r and finishing when it exhausts the vertex v. Then for each vertex u in H_v, u is exhausted in Wv.*

Proof. By induction in the depth of the subtree with root v. If v is a leaf of H then $H_v = \{v\}$. If v is not a leaf and Wv exhaust v, then Wv visit all arc $(v, w) \in E$, and therefore all the arcs $(u, v) \in E$, applying induction hypothesis to all vertices u such that $(u, v) \in E$ we prove the result.   □

3   Minimal de Bruijn Sequence

Let $m = m_1, \ldots m_n$ be the vertex of $G^{\mathcal{F},n}$ of maximum label in the lexicographic order. We are interested in to obtain the Eulerian cycle of minimum label starting at m. In order to obtain this cycle, we define the following walk: Starting at m, at each vertex we continue by the arc with the lowest label between the unvisited arcs with tail at this vertex. A walk constructed by this way will be called a *minimal walk*. By definition, there is no walk with a lexicographically lower label, except its subwalks. In this section we characterize when a minimal walk starting at m is an Eulerian cycle, obtaining the minimal de Bruijn sequence.

For each vertex v let $e(v)$ be the arc with tail at the vertex v and with maximum label. Let T be the spanning subgraph of $G^{\mathcal{F},n}$ composed by the set of arcs $e(v)$, for $v \in V(G^{\mathcal{F},n})$, $v \neq m$. The label of $e(v)$ will be denoted by $\gamma(v)$.

Is easy to see that a minimal walk is a walk avoiding T, hence we can study a minimal walk analyzing the structure of T.

Theorem 2. *A minimal walk is an Eulerian cycle if and only if T is a tree.*

Proof. A minimal walk W exhaust m, if T is a tree then by Lemma 1 all vertices of T are exhausted by W, hence W is an Eulerian cycle. Conversely, if W is an Eulerian cycle, by the BEST Theorem the subgraph composed by the last arc visited at each vertex is a tree, but this subgraph is T, concluding that T is a tree.   □

In the unrestricted case (when $\mathcal{W}^{\mathcal{F}}(n) = A^n$), the subgraph T is a regular tree of depth n where each non-leaf vertex has $|A|$ sons, therefore the minimal walk is an Eulerian cycle.

In the restricted case, we do not obtain necessarily an Eulerian cycle, because T is not necessarily a spanning tree converging to the root due to the existence of cycles.

We will study the structure of the graph $G^{\mathcal{F},n}$ and the subgraph T, specially the cycles in T. The main theorem of this section characterizes the label of cycles in T, allowing us to characterize the languages where the minimal walk is an Eulerian cycle.

First of all, we will prove some properties of the de Bruijn graph to understand the structure of the arcs and cycles in T.

Lemma 2. *Let $k \geq n + 2$. Let $W = v_0 e_0 v_1 e_1 \cdots e_{k-1} v_k$ be a walk in T. Then $l(e_0) \leq l(e_{n+1})$.*

Proof. Since $v_n = l(e_0) \cdots l(e_{n-1})$ we have that $l(e_1) \cdots l(e_{n-1}) l(e_n) l(e_0) \in \mathcal{W}^{\mathcal{F}}(n + 1)$. Hence there exists an arc (v_{n+1}, u) with label $l(e_0)$, where $v_{n+1} = l(e_1) \cdots l(e_{n-1}) l(e_n)$. By the definition of T, $l(e_0) \leq \gamma(v_{n+1}) = l(e_{n+1})$. □

Corollary 1. *Let C be a cycle in T. Then $|C|$ divides $n + 1$. Moreover for every vertex u in C, $u\gamma(u) = l(C)^{\frac{n+1}{|C|}}$.*

Proof. Let consider the walk $W = v_0 e_0 \cdots e_{|C|-1} v_{|C|} = v_0 e_0 \cdots e_{(n+1)|C|-1} v_0 e_0 v_1$ as $n + 1$ repetitions of the cycle C. From Lemma 2 we have $l(e_0) \leq l(e_{n+1}) \leq l(e_{2(n+1)}) \leq l(e_{(n+1)|C|}) = l(e_0)$. Since we can start the cycle in any vertex we conclude that $l(e_i) = l(e_{(n+1)+i})$ for every $i = 0, \ldots, |C| - 1$. Hence $|C|$ divides $n + 1$. The second conclusion comes from the fact that the label of any walk of length at most n ending in a vertex u is a suffix of u. □

Let $u \neq m$ be a vertex. Among all the words which are prefix of m and suffix of u, let $g(u)$ be the longest one (notice that $g(u)$ could be the empty word ε and $|g(u)| < n$). Let $\alpha(u) = m_{|g(u)|+1}$ be the letter following the end of $g(u)$ in m.

Notice that in the unrestricted case, $|g(u)|$ is the distance over the graph from the vertex u to m. This function will be essential in the study of T. The next lemma give us a bound over the label of the arcs in terms of the function $g(\cdot)$.

Lemma 3. *For all pairs of adjacent vertices u and v, $l(uv) \leq \alpha(u)$. Moreover, if $l(uv) < \alpha(u)$ then $g(v) = \varepsilon$ and if $l(uv) = \alpha(u)$ then $g(v) = g(u)l(uv)$.*

Proof. $g(u)$ is a suffix of u, and $ul(uv) \in \mathcal{W}^{\mathcal{F}}(n + 1)$, so $g(u)l(uv)$ is a prefix of a word in $\mathcal{W}^{\mathcal{F}}(n + 1)$. Since m is the maximal word and $g(u)$ is a prefix of m we get $l(uv) \leq \alpha(u)$.

If $l(uv) = \alpha(u)$ then $g(u)l(uv)$ is a prefix of m and a suffix of v. Hence $g(u)l(uv)$ is a suffix of $g(v)$. Since by removing the last letter of a suffix of v we obtain a suffix of u we conclude $g(v) = g(u)l(uv)$.

We show that if $g(v) \neq \varepsilon$ then $\alpha(u) \geq l(uv)$. Let $g(v) = g'(v)l(uv)$, then $g'(v)$ is a suffix of u and a prefix of m. Hence $g'(v)$ is a suffix of $g(u)$. Therefore $g'(v)\alpha(u)$ is a factor of m. By the definition of $g(v)$ and the maximality of m $g(v)$ is greater or equal (lexicographically) than $g'(v)\alpha(u)$. We conclude that $\alpha(u) \geq l(uv)$.   □

In the unrestricted case, where T is a tree of depth n, all the arcs not in T go to a leaf. In the general case we can define an analog to the leaves.

We say that a vertex u is a *floor* vertex if $g(u) = \varepsilon$. Notice that in the unrestricted case the leaves of T are the floor vertices. We say that a vertex u is a *restricted* vertex if $\gamma(u) < \alpha(u)$.

Corollary 2. *If a cycle in T contains l restricted vertices, then it has exactly l floor vertices.*

Proof. From Lemma 3 we know that if a vertex u is restricted then for every arc (u, v) the vertex v is a floor vertex. To conclude it is enough to see that in T an arc (u, v) with u unrestricted has label $\alpha(u)$. Then v is not a floor vertex.   □

Corollary 3. *Let P be a path in T starting in a floor vertex, ending in a vertex v and with unrestricted inner vertices. Then $l(P) = g(v)$.*

Proof. We apply induction on the length of P. The case where the length of P is zero is direct since v is a floor vertex. Let us consider the case where P has length at least 1. Since v is not a restricted vertex, from Lemma 3 we know that $g(v) = g(u)l(uv)$, where u is its neighbor in P. By the induction assumption $g(u) = l(P')$ where P' is the path obtained from P removing the arc (u, v). Hence $g(v) = l(P')l(uv) = l(P)$.   □

We will use these results to characterize the label of cycles in T, specially we will characterize the restricted vertices of a cycle.

Theorem 3. *Let C be a cycle in T, let $u^0, \ldots, u^{k-1}$ be the restricted vertices in C ordered according to the order of C. Then $u^i = g(u^{i+1})\gamma(u^{i+1}) \cdots \gamma(u^{i-1})g(u^i)$ for $i = 0, \ldots, k-1$, where $i+1, \ldots, i-1$ are computed $\mod k$.*

Proof. From Corollary 3 the label of C is $g(u^0)\gamma(u^0) \cdots g(u^{k-1}) \gamma(u^{k-1})$, and by definition of $G^{\mathcal{F},n}$, u^i is the label of any walk over $G^{\mathcal{F},n}$ of length n finishing in u^i, so by Corollary 1 we can take the walk C^k composed by $k = (n+1)/|C|$ repetitions of C finishing in u^i, concluding that $u^i = g(u^{i+1})\gamma(u^{i+1}) \cdots \gamma(u^l)$ $l(C^{k-1})g(u_1) \cdots \gamma(u^{i-1}) g(u^i)$.   □

Now we are able to give a characterization of the languages where a minimal walk produces an Eulerian cycle.

Let $\mathcal{H}$ be the subset of $\mathcal{W}^{\mathcal{F}}(n+1)$ where $w \in \mathcal{H}$ if and only if w can be decomposed by $w = h^0\beta_1 \ldots h^{k-1}\beta_{k-1}$ where each $h^i \in A^*$ and $\beta_i \in A$ satisfy the following conditions:

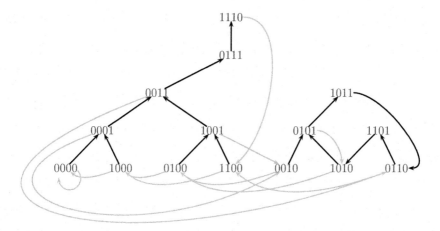

Fig. 2. Example of the subgraph T for $n = 4$ and $\mathcal{F} = \{01111\}$ in a binary alphabet.

1. $h^i = m_1 \ldots m_{|h^i|}$ (a prefix of m)
2. $\beta_i < m_{|h^i|+1}$
3. $\forall \beta' > \beta_i,\ h^{i+1}\beta_{i+1} \ldots \beta_{i-1}h^i\beta' \notin \mathcal{W}^{\mathcal{F}}(n+1)$

Now, we are able to characterize the languages where a minimal walk is an Eulerian cycle.

Theorem 4. *A minimal walk is an Eulerian cycle if and only if $\mathcal{H} = \emptyset$.*

Proof. From Theorem 2, we have to prove that T is a tree if and only if $\mathcal{H} = \emptyset$.

If T is not a tree then T has a cycle C. Let $u^0 \ldots u^{k-1}$ be the restricted vertices of the cycle. By Theorem 3 $l(C) = g(u^0)\gamma(u^0) \ldots g(u^{k-1})\gamma(u^{k-1})$ and by Corollary 1 $|C|$ divides $n + 1$. Therefore there exists a word w in $\mathcal{W}^{\mathcal{F}}(n+1)$ composed by $(n+1)/|C|$ repetitions of C. By definition of $\mathcal{H}$ we conclude that $w \in \mathcal{H}$.

Conversely, let us assume that T has no cycles and $\mathcal{H} \neq \emptyset$. Let w be a word in $\mathcal{H}$. By definition of $G^{\mathcal{F},n}$, there is a cycle C in $G^{\mathcal{F},n}$ of length dividing $n + 1$ such that C (or repetitions of C) has label w. We shall prove that C is also a cycle in T.

Let v be a vertex of C, with $v = \ldots \beta_{i-1}(h^i)_1 \ldots (h^i)_j$ where $j = 0 \ldots |h^i|$. If $0 < j < |h^i|$, then $m_1 \ldots m_j$ is a suffix of v, so $\alpha(v) = m_{j+1} = (h^i)_{j+1}$ hence the arc of C with tail at v is in T. If $j = 0$ then $\gamma(v) = m_1$ therefore the arc in C is in T. Finally, let consider the case $j = |h^i|$. If (v, v') is the arc in C then $l(vv') = \beta_i$. Since $w \in \mathcal{H}$, no arc in $G^{\mathcal{F},n}$ with tail at v has a label greater than β_i. Then $(v, v') \in T$. We conclude that C is a cycle in T which leads to a contradiction. □

4   Some Remarks

The previous analysis considers only the minimal walk starting at the root vertex. This case does not necessarily produce the minimal label over all Eulerian cycles, because there can be Eulerian cycles starting at a non root vertex with a lexicographically lower label.

It is also possible to construct an algorithm which modifies T in order to destroy cycles in T, and obtain the minimal de Bruijn sequence for any irreducible subshift of finite type. However further research in this subject allow us to construct an algorithm to obtain the minimal Eulerian cycle for any edge-labeled digraph (see [12]), but this result escapes to the scope of this work.

References

1. de Bruijn, N.G.: A combinatorial problem. Nederl. Akad. Wetensch., Proc. **49** (1946) 758–764
2. Stein, S.K.: The mathematician as an explorer. Sci. Amer. **204** (1961) 148–158
3. Bermond, J.C., Dawes, R.W., Ergincan, F.Ö.: De Bruijn and Kautz bus networks. Networks **30** (1997) 205–218
4. Chung, F., Diaconis, P., Graham, R.: Universal cycles for combinatorial structures. Discrete Math. **110** (1992) 43–59
5. Fredricksen, H.: A survey of full length nonlinear shift register cycle algorithms. SIAM Rev. **24** (1982) 195–221
6. Fredricksen, H., Maiorana, J.: Necklaces of beads in k colors and k-ary de Bruijn sequences. Discrete Math. **23** (1978) 207–210
7. Ruskey, F., Savage, C., Wang, T.M.: Generating necklaces. J. Algorithms **13** (1992) 414–430
8. Ruskey, F., Sawada, J.: Generating necklaces and strings with forbidden substrings. Lect. Notes Comput. Sci. **1858** (2000) 330–339
9. Moreno, E.: Lyndon words and de bruijn sequences in a subshift of finite type. In Harju, T., Karhumäki, J., eds.: Proceedings of WORDS'03. Number 27 in TUCS General Publications, Turku, Finland, Turku Centre for Computer Science (2003) 400–410
10. Lind, D., Marcus, B.: Symbolic Dynamics and Codings. Cambridge University Press (1995)
11. Tutte, W.T.: Graph theory. Volume 21 of Encyclopedia of Mathematics and its Applications. Addison-Wesley Publishing Company Advanced Book Program, Reading, MA (1984)
12. Matamala, M., Moreno, E.: Minimal Eulerian cycle in a labeled digraph. Technical Report CMM-B-04/08-108, DIM-CMM, Universidad de Chile (2004)

A Graph-Theoretic Generalization of the Least Common Subsumer and the Most Specific Concept in the Description Logic $\mathcal{EL}$

Franz Baader*

Theoretical Computer Science, TU Dresden, D-01062 Dresden, Germany
baader@tcs.inf.tu-dresden.de

Abstract. In two previous papers we have investigates the problem of computing the least common subsumer (lcs) and the most specific concept (msc) for the description logic $\mathcal{EL}$ in the presence of terminological cycles that are interpreted with descriptive semantics, which is the usual first-order semantics for description logics. In this setting, neither the lcs nor the msc needs to exist. We were able to characterize the cases in which the lcs/msc exists, but it was not clear whether this characterization yields decidability of the existence problem.

In the present paper, we develop a common graph-theoretic generalization of these characterizations, and show that the resulting property is indeed decidable, thus yielding decidability of the existence of the lcs and the msc. This is achieved by expressing the property in monadic second-order logic on infinite trees. We also show that, if it exists, then the lcs/msc can be computed in polynomial time.

1 Introduction

Description Logics (DLs) [6] are a class of knowledge representation formalisms in the tradition of semantic networks and frames, which can be used to represent the terminological knowledge of an application domain in a structured and formally well-understood way. DL systems provide their users with standard inference services (like subsumption and instance checking) that deduce implicit knowledge from the explicitly represented knowledge. More recently, non-standard inferences [8] were introduced to support building and maintaining large DL knowledge bases. For example, computing the most specific concept (msc) of an individual and the least common subsumer (lcs) of concepts can be used in the bottom-up construction of description logic knowledge bases. Instead of defining the relevant concepts of an application domain from scratch, this methodology allows the user to give typical examples of individuals belonging to the concept to be defined. These individuals are then generalized to a concept by first computing the most specific concept of each individual (i.e., the least concept description in the available description language that has this individual as an instance), and then computing the least common subsumer of

---

* Partially supported by DFG (BA 1122/4-3) and by National ICT Australia Limited.

J. Hromkovič, M. Nagl, and B. Westfechtel (Eds.): WG 2004, LNCS 3353, pp. 177–188, 2004.
© Springer-Verlag Berlin Heidelberg 2004

these concepts (i.e., the least concept description in the available description language that subsumes all these concepts). The knowledge engineer can then use the computed concept as a starting point for the concept definition.

The motivation for the graph-theoretic problem solved in the present paper comes from non-standard inferences in the DL $\mathcal{EL}$, which is rather inexpressive, but nevertheless has significant applications. For example, SNOMED, the Systematized Nomenclature of Medicine [11, 10] employs $\mathcal{EL}$. Unfortunately, the most specific concept of a given individual need not exist in $\mathcal{EL}$. For other DLs, this problem had been overcome by allowing for cyclic concept definitions [7]. In order to adapt this approach also to $\mathcal{EL}$, the impact on both standard and non-standard inferences of cyclic definitions in this DL had to be investigated first. This investigation was carried out in a series of papers [4, 3, 1, 2] that gives an almost complete picture of the computational properties of the above mentioned standard and non-standard inferences in $\mathcal{EL}$ with cyclic concept definitions[1]. Regarding standard inferences, the subsumption and the instance problem turned out to be polynomial for both types of semantics. Regarding non-standard inferences, w.r.t. gfp-semantics the lcs and the msc always exist and can be computed in polynomial time. Descriptive semantics is less well-behaved. In [1] it was shown that, in general, the lcs need not exist. The paper gave a characterization for the existence of the lcs, but the question of how to decide this condition remained open. In [2], analogous results were shown for the msc.

The present paper introduces a common graph-theoretic generalization of these open problems: the problem whether a so-called two-level graph is of bounded cycle depth. Then it shows that this problem is decidable by reducing it to monadic second-order logic on infinite trees [9]. Finally, it shows that, if a two-level graph is of bounded cycle depth, then its cycle depth is polynomially bounded by the size of the graph. This implies that the lcs/msc can be computed in polynomial time, provided that it exists.

Because of the space constraints, we concentrate on the graph-theoretic problems. The reader is referred to [6] for more information on DLs in general, to [4, 3, 1, 2] for previous results on $\mathcal{EL}$ with cyclic definitions, and to [5] for a long version of this paper containing full proofs and the connection to the lcs/msc.

2   The Cycle Depth of Two-Level Graphs

In this section, we define the relevant graph-theoretic notions, and relate them to the problem of computing the lcs and the msc in $\mathcal{EL}$.

For the purpose of this paper, a *graph* is of the form (V, E, L), where V is a finite set of nodes, $E \subseteq V \times N_e \times V$ is a set of edges labeled by elements of the finite set N_e, and L is a labelling function that assigns to every node $v \in V$ a subset $L(v)$ of the finite set N_n.

Simulations are binary relations on the nodes of a graph that respect node labels and edges in the sense defined below.

---

[1] Cyclic definitions in $\mathcal{EL}$ can either be interpreted with greatest fixpoint (gfp) or with descriptive semantics, which is the usual first-order semantics for DLs.

$$\begin{array}{lccccccccc} p_1: & u = u_0 & \stackrel{r_1}{\rightarrow} & u_1 & \stackrel{r_2}{\rightarrow} & u_2 & \stackrel{r_3}{\rightarrow} & u_3 & \stackrel{r_4}{\rightarrow} & \cdots \\ & Z\downarrow & & Z\downarrow & & Z\downarrow & & Z\downarrow & & \\ p_2: & v = v_0 & \stackrel{r_1}{\rightarrow} & v_1 & \stackrel{r_2}{\rightarrow} & v_2 & \stackrel{r_3}{\rightarrow} & v_3 & \stackrel{r_4}{\rightarrow} & \cdots \end{array}$$

Fig. 1. An infinite (u, v)-simulation chain.

$$\begin{array}{ccccccccc} u = u_0 & \stackrel{r_1}{\rightarrow} & u_1 & \stackrel{r_2}{\rightarrow} & \cdots & \stackrel{r_{n-1}}{\rightarrow} & u_{n-1} & \stackrel{r_n}{\rightarrow} & u_n \\ Z\downarrow & & Z\downarrow & & & & Z\downarrow & & \\ v = v_0 & \stackrel{r_1}{\rightarrow} & v_1 & \stackrel{r_2}{\rightarrow} & \cdots & \stackrel{r_{n-1}}{\rightarrow} & v_{n-1} & & \end{array}$$

Fig. 2. A partial (u, v)-simulation chain.

Definition 1. *Let $\mathcal{G} = (V, E, L)$ be a graph. The binary relation $Z \subseteq V \times V$ is a* simulation *on $\mathcal{G}$ iff*

(S1) *$(v_1, v_2) \in Z$ implies $L(v_1) \subseteq L(v_2)$; and*
(S2) *if $(v_1, v_2) \in Z$ and $(v_1, r, v_1') \in E$, then there exists a node $v_2' \in V$ such that $(v_1', v_2') \in Z$ and $(v_2, r, v_2') \in E$.*

Here, we are not interested in arbitrary simulations containing a given pair of nodes, but in ones that are synchronized in the sense defined below. If $(u, v) \in Z$, then any infinite path p_1 starting with u can be simulated by an infinite path p_2 starting with v. We call the pair p_1, p_2 a *(u, v)-simulation chain* (see Figure 1). Given an infinite path p_1 starting with u, we construct a simulating path p_2 step by step. The main point is, however, that the decision which node v_n to take in step n should depend only on the partial simulation chain already constructed, and *not* on the parts of the path p_1 not yet considered.

Definition 2. *Let $\mathcal{G}$ be a graph, Z a simulation on $\mathcal{G}$, and $(u, v) \in Z$.*

(1) A partial (u, v)-simulation chain *is of the form depicted in Figure 2. A* selection function *S for u, v and Z assigns to each partial (u, v)-simulation chain of this form a node v_n such that (v_{n-1}, r_n, v_n) is an edge in $\mathcal{G}$ and $(u_n, v_n) \in Z$.*

*(2) Given an infinite path $u = u_0 \stackrel{r_1}{\rightarrow} u_1 \stackrel{r_2}{\rightarrow} u_2 \stackrel{r_3}{\rightarrow} u_3 \stackrel{r_4}{\rightarrow} \cdots$, one can use the selection function S to construct a simulating path. In this case we say that the resulting infinite (u, v)-simulation chain is S-*selected.

(3) The simulation Z is called (u, v)-synchronized *iff there exists a selection function S for Z such that the following holds: for every infinite S-selected (u, v)-simulation chain of the form depicted in Figure 1 there exists an $i \geq 0$ such that $u_i = v_i$.*

As shown in [4, 2], the subsumption and the instance problem in $\mathcal{EL}$ can be reduced to the problem of deciding whether there exists a synchronized simulation on a given graph (which is a problem decidable in polynomial time [4]).

To define the main graph-theoretic problem addressed in this paper, we must first introduce two-level graphs.

Definition 3. *The graph $\mathcal{G} = (V, E, L)$ is called* two-level graph *iff V can be partitioned into disjoint sets $V = V_1 \cup V_2$ such that $(v, r, v') \in E$ implies $v \in V_1$ or $v' \in V_2$. To make this partition explicit, we write two-level graphs as $\mathcal{G} = (V_1 \cup V_2, E, L)$.*

Intuitively, a two-level graph $\mathcal{G} = (V_1 \cup V_2, E, L)$ consists of a subgraph $\mathcal{G}_1$ on V_1, a subgraph $\mathcal{G}_2$ on V_2, and possibly additional edges from nodes of $\mathcal{G}_1$ to nodes of $\mathcal{G}_2$. Next, we consider graphs obtained from $\mathcal{G}$ by unraveling cycles in $\mathcal{G}_1$ up to a certain length.

Definition 4. *Let $\mathcal{G} = (V_1 \cup V_2, E, L)$ be a two-level graph and $u \in V_1$. The k-unraveling of $\mathcal{G}$ w.r.t. u is the two-level graph $\mathcal{G}_u^{(k)} := (V_1^{(k)} \cup V_2, E^{(k)}, L^{(k)})$, where*

$$V_1^{(k)} := \{u_0^{(k)}\} \cup \{v_i^{(k)} \mid v \in V_1 \text{ and } 1 \leq i \leq k\};$$

$$E^{(k)} := \{(v, r, w) \mid (v, r, w) \in E \text{ and } v, w \in V_2\} \cup$$
$$\{(v_i^{(k)}, r, w_{i+1}^{(k)}) \mid (v, r, w) \in E \text{ and } v_i^{(k)}, w_{i+1}^{(k)} \in V_1^{(k)}\} \cup$$
$$\{(v_i^{(k)}, r, w) \mid (v, r, w) \in E \text{ and } v_i^{(k)} \in V_1^{(k)}, w \in V_2\};$$

$$L^{(k)}(v) := L(v) \quad \text{if } v \in V_2,$$
$$L^{(k)}(v_i^{(k)}) := L(v) \quad \text{if } v_i^{(k)} \in V_1^{(k)}.$$

Given two different such unravelings $\mathcal{G}_u^{(k)} = (V_1^{(k)} \cup V_2, E^{(k)}, L^{(k)})$ and $\mathcal{G}_u^{(\ell)} = (V_1^{(\ell)} \cup V_2, E^{(\ell)}, L^{(\ell)})$ of $\mathcal{G} = (V_1 \cup V_2, E, L)$, their union $\mathcal{G}_u^{(k)} \cup \mathcal{G}_u^{(\ell)}$ is defined in the obvious way by building the union of the node sets, the edge sets, and the labeling functions[2].

Definition 5. *Let $\mathcal{G} = (V_1 \cup V_2, E, L)$ be a two-level graph, $u \in V_1$, and $k \neq \ell$. We say that $\mathcal{G}_u^{(\ell)}$ subsumes $\mathcal{G}_u^{(k)}$ ($\mathcal{G}_u^{(k)} \sqsubseteq \mathcal{G}_u^{(\ell)}$) iff there is a $(u_0^{(\ell)}, u_0^{(k)})$-synchronized simulation Z on $\mathcal{G}_u^{(k)} \cup \mathcal{G}_u^{(\ell)}$ such that $(u_0^{(\ell)}, u_0^{(k)}) \in Z$.*

It is easy to see that $\ell > k$ implies $\mathcal{G}_u^{(\ell)} \sqsubseteq \mathcal{G}_u^{(k)}$ (see also Lemma 3 in [2]). Given a node $u \in V_1$ of a two-level graph $\mathcal{G} = (V_1 \cup V_2, E, L)$, we are interested in finding an index k such that the subsumption relationship also holds in the other direction.

Definition 6. *Let $\mathcal{G} = (V_1 \cup V_2, E, L)$ be a two-level graph and $u \in V_1$. We say that $\mathcal{G}$ is of bounded cycle depth w.r.t. u iff there is a $k \geq 0$ such that $\mathcal{G}_u^{(k)} \sqsubseteq \mathcal{G}_u^{(\ell)}$ holds for all $\ell > k$. In this case, the minimal such k is called the cycle depth of $\mathcal{G}$ w.r.t. u.*

The main decision problem considered in this paper is the following:

---

[2] Note that the two labeling functions agree on V_2, shared by $\mathcal{G}_u^{(k)}$ and $\mathcal{G}_u^{(\ell)}$.

Given: A two-level graph $\mathcal{G} = (V_1 \cup V_2, E, L)$ and a node $u \in V_1$.
Question: Is $\mathcal{G}$ of bounded cycle depth w.r.t. u?

Before stating the connection of this problem to the problem of deciding the existence of the lcs and the msc in $\mathcal{EL}$ w.r.t. descriptive semantics, let us consider three examples.

First, consider the two-level graph $\mathcal{G}_1$ on the left-hand side of Figure 3 (where $V_1 := \{u\}$ and $V_2 := \{v\}$). This graph is of bounded cycle depth w.r.t. u. In fact, already $k = 0$ satisfies Definition 6 since any infinite path starting with $u_0^{(\ell)}$ will eventually lead to v, and thus can be simulated by the path $u_0^{(0)} \xrightarrow{r} v \xrightarrow{r} v \xrightarrow{r} \cdots$.

Second, consider the two-level graph $\mathcal{G}_2$ on the right-hand side of Figure 3 (where $V_1 := \{u\}$ and $V_2 := \{v_1, v_2\}$). Though this graph looks quite similar to $\mathcal{G}_1$, it is not of bounded cycle depth. In fact, $\mathcal{G}_{2,u}^{(k)} \not\sqsubseteq \mathcal{G}_{2,u}^{(k+1)}$ for all $k \geq 0$. To see this, consider the path $p_1 : u_0^{(k+1)} \xrightarrow{r} \cdots \xrightarrow{r} u_k^{(k+1)} \xrightarrow{r} u_{k+1}^{(k+1)}$ of length $k + 1$ in $\mathcal{G}_{2,u}^{(k+1)}$. If this path is simulated by a path p_2 of length $k + 1$ in $\mathcal{G}_{2,u}^{(k)}$, then the last node of p_2 is either v_1 or v_2. Assume without loss of generality that it is v_1. If we continue the path p_1 by an infinite loop through v_2, then this infinite path p_1' can only be simulated in $\mathcal{G}_{2,u}^{(k)}$ by continuing to go through the node v_1. Thus, no synchronization occurs.

Third, the two-level graph $\mathcal{G}_3$ depicted in Figure 4 (where $V_1 = \{u_1, u_2\}$ and $V_2 = \{v\}$) is not of bounded cycle depth w.r.t. u_1, but shows a somewhat surprising phenomenon. Here we have $\mathcal{G}_{3,u_1}^{(k)} \sqsubseteq \mathcal{G}_{3,u_1}^{(k+1)}$ for all odd numbers k, but $\mathcal{G}_{3,u_1}^{(k)} \not\sqsubseteq \mathcal{G}_{3,u_1}^{(k+1)}$ if k is even. First, assume that k is odd. Then there are no infinite paths in $\mathcal{G}_{3,u_1}^{(k+1)}$ that use the node $u_{1,k+1}^{(k+1)}$ since this node does not have a successor node. As an easy consequence, every infinite path in $\mathcal{G}_{3,u_1}^{(k+1)}$ can be simulated by "the same" path in $\mathcal{G}_{3,u_1}^{(k)}$. In addition, the finite path to $u_{1,k+1}^{(k+1)}$ can

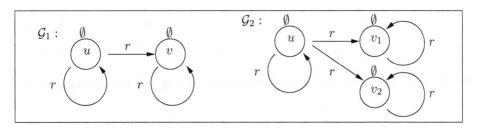

Fig. 3. Two two-level graphs, one of bounded and one of unbounded cycle depth.

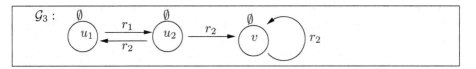

Fig. 4. Another two-level graph of unbounded cycle depth.

be simulated by a path in $\mathcal{G}_{3,u_1}^{(k)}$ that ends with v. Consequently, $\mathcal{G}_{3,u_1}^{(k)} \sqsubseteq \mathcal{G}_{3,u_1}^{(k+1)}$ for odd k. In contrast, if k is even, then $u_{1,k}^{(k+1)}$ has a successor node in $\mathcal{G}_{3,u_1}^{(k+1)}$ (namely $u_{2,k+1}^{(k+1)}$) reached by an edge with label r_1. Any node reachable from $u_{1,0}^{(k)}$ in $\mathcal{G}_{3,u_1}^{(k)}$ by a path of length k (i.e., $u_{1,k}^{(k)}$ or v) does not have a successor w.r.t. r_1. Thus, there is a path in $\mathcal{G}_{3,u_1}^{(k+1)}$ that cannot be simulated by a path in $\mathcal{G}_{3,u_1}^{(k)}$, which shows that $\mathcal{G}_{3,u_1}^{(k)} \not\sqsubseteq \mathcal{G}_{3,u_1}^{(k+1)}$ for even k.

The last example shows that, in order to find the number k required by Definition 6, one can*not* simply test subsumption between $\mathcal{G}_u^{(i+1)}$ and $\mathcal{G}_u^{(i)}$ for $i = 0, 1, 2, \ldots$ until $\mathcal{G}_u^{(i)} \sqsubseteq \mathcal{G}_u^{(i+1)}$, and then stop with output $k = i$.

The characterization of the lcs and the msc given in [1] and [2], respectively, can easily be reformulated in terms of the notions introduced above. As an easy consequence, the existence problem can be reduced to the main decision problem introduced in this paper (see [5] for detail).

Proposition 1. *The problems of deciding the existence of the lcs (msc) in $\mathcal{EL}$ with descriptive semantics can be reduced in polynomial time to the problem of deciding whether a two-level graph $\mathcal{G}$ is of bounded cycle depth. In addition, if the cycle depth of $\mathcal{G}$ is polynomial in the size of $\mathcal{G}$, then the lcs (msc) can be computed in polynomial time.*

3 Deciding if a Graph Is of Bounded Cycle Depth

Let $\mathcal{G} = (V_1 \cup V_2, E, L)$ be a two-level graph, and $u \in V_1$. We reduce the problem of deciding whether $\mathcal{G}$ is of bounded cycle depth w.r.t. u to the problem of deciding whether a certain formula $\phi_{\mathcal{G}}^u$ of monadic second-order logic (MSO) on infinite trees is satisfiable. As shown by Rabin [9], the satisfiability problem for MSO is decidable. In the following, we assume that the reader is familiar with MSO on infinite trees (see, e.g., [12] for an introduction). Before we define the formula $\phi_{\mathcal{G}}^u$, we describe the intuition underlying this reduction.

Encoding Synchronized Simulations by Infinite Trees. The main idea underlying our reduction is that all simulation chains starting with a given pair of nodes of a graph $\mathcal{G} = (V, E, L)$ and selected by some selection function (see Definition 2) can be represented by an infinite tree t. Basically, the nodes of this tree are labeled with pairs of nodes of $\mathcal{G}$. Assume that the node n of t has label (u, v). If $(u, r_1, u_1), \ldots, (u, r_p, u_p)$ are all the edges in $\mathcal{G}$ starting with u, then the node n has p successor nodes $n_1, \ldots, n_p$ that are respectively labeled with $(u_1, v_1), \ldots, (u_p, v_p)$, where v_i is the result of applying the selection function to the partial simulation chain determined by the path in t leading to the node n and the edge (u, r_i, u_i). Since in MSO one considers trees with a fixed branching factor, the node n may have some additional dummy successor nodes labeled with the dummy label $\sharp$. Note that the simulation relation Z itself is also encoded in the tree t: it consists of all tuples (u, v) such that $(u, v) \in V \times V$ is the label of a node n of t. Because of the definition of the successor nodes of the nodes in t, property (S2) in the definition of a simulation relation (Definition 1) is satisfied.

To ensure that Z also satisfies (S1), it is enough to require $L(u) \subseteq L(v)$ for all labels $(u, v) \in V \times V$ of nodes in t. Given two nodes u, v of $\mathcal{G}$, how can we ensure that the simulation relation Z encoded by such a tree t contains (u, v) and is (u, v)-synchronized? To ensure that $(u, v) \in Z$, we require that (u, v) is the label of the root of t. To ensure synchronization, we must require that on all infinite paths in the tree t, we encounter a label of the form (v', v') or $\sharp$. This can easily be expressed in MSO.

What we have said until now can be used to show that the following problem is decidable: given a graph $\mathcal{G}$ and nodes u, v in $\mathcal{G}$, is there a (u, v)-synchronized simulation Z such that $(u, v) \in Z$? However, decidability of this problem (in polynomial time) was already shown directly in [4] without the need for a reduction to the (complex) logic MSO.

What we actually want to decide here is whether a given two-level graph $\mathcal{G} = (V_1 \cup V_2, E, L)$ is of bounded cycle depth w.r.t. a node $u \in V_1$. For this, we must consider not $\mathcal{G}$ itself but rather unravelings $\mathcal{G}_u^{(k)}$ and $\mathcal{G}_u^{(\ell)}$ of $\mathcal{G}$. In addition, we need to express the quantification on the numbers k and ℓ ("there exists a k such that for all ℓ") by (second-order) quantifiers in MSO.

Encoding Unravelings $\mathcal{G}_u^{(k)}$ and $\mathcal{G}_u^{(\ell)}$ and the Quantification on k and ℓ. Assume that we have an infinite tree t encoding a (u, u)-synchronized simulation Z on the two-level graph $\mathcal{G}$, as described above. If (v_1, v_2) is the label of a node n on some level i of t, then there are paths of length i in $\mathcal{G}$ from u to v_1 and from u to v_2, respectively. The first (second) path corresponds to a path in $\mathcal{G}_u^{(\ell)}$ ($\mathcal{G}_u^{(k)}$) iff $i \leq \ell$ or $v_1 \in V_2$ ($i \leq k$ or $v_2 \in V_2$). Thus, the idea could be to introduce two second-order variables X and Y (with the appropriate quantifier prefix $\exists Y. \forall X.$), and then ensure that X contains exactly the nodes of t up to some level ℓ, and Y contains exactly the nodes of t up to some level k. In order to ensure that the paths in $\mathcal{G}$ encoded in the tree t really belong to $\mathcal{G}_u^{(\ell)}$ (when considering the first component of the node labels) and $\mathcal{G}_u^{(k)}$ (when considering the second component of the node labels), we must require that, for a node n labeled with (v_1, v_2), we have $X(n)$ or $v_1 \in V_2$, and $Y(n)$ or $v_2 \in V_2$. Unfortunately, sets containing exactly the nodes of an infinite tree up to some depth bound are not expressible in MSO[3]. However, for our purposes it turns out to be sufficient to ensure that X and Y are finite prefix-closed sets (i.e., if a node n that is not the root node belongs to one of them, then its predecessor also does). Both "prefix-closed" and "finite" can easily be expressed in MSO.

The Formal Definition. Let $\mathcal{G} = (V_1 \cup V_2, E, L)$ be a two-level graph, $u \in V_1$, and assume that b is the maximal number of successors of the nodes in $\mathcal{G}$. To define the formula $\phi_{\mathcal{G}}^u$, we consider the infinite tree with branching factor b (i.e., we have b successor functions $s_1, \ldots, s_b$ in the signature of MSO). As usual, we will denote second-order variables (standing for sets of nodes) by upper-case letters, and first-order variables (standing for nodes) by lower-case letters. The second-order variables used in the following are

---

[3] Since then one could also express that two nodes are on the same level, which is know to be inexpressible in MSO [12].

- the variables X and Y whose function was already explained above;
- variables $Q_{(u_1,u_2)}$ for $(u_1, u_2) \in (V_1 \cup V_2) \times (V_1 \cup V_2)$ and $Q_\sharp$. The values of these variables encode the selection function S by encoding all S-selected simulation chains. Intuitively, a node n of the tree belongs to $Q_{(u_1,u_2)}$ ($Q_\sharp$) iff it is labeled with (u_1, u_2) ($\sharp$);
- the variable P standing for an infinite path in the tree, which is used to express the synchronization property.

The formula $\phi_{\mathcal{G}}^u$ is defined as

$$\exists Y.(PrefixClosed(Y) \wedge Finite(Y) \wedge \forall X.(PrefixClosed(X) \wedge Finite(X) \Rightarrow \psi_{\mathcal{G}}^u)),$$

where $PrefixClosed(.)$ and $Finite(.)$ are the well-known MSO-formulae expressing that a set of nodes is prefix-closed and finite, respectively[4], and $\psi_{\mathcal{G}}^u$ consists of an existential quantifier prefix on the variables $Q_{(u_1,u_2)}$ for $(u_1, u_2) \in (V_1 \cup V_2) \times (V_1 \cup V_2)$ and $Q_\sharp$, followed by the conjunction $\vartheta_{\mathcal{G}}^u$ of the following formulae:

- *A formula expressing that any node has exactly one label.*

$$\forall x. \bigvee_{l_1 \in (V_1 \cup V_2) \times (V_1 \cup V_2) \cup \{\sharp\}} \left(Q_{l_1}(x) \wedge \bigwedge_{\substack{l_2 \in (V_1 \cup V_2) \times (V_1 \cup V_2) \cup \{\sharp\} \\ l_2 \neq l_1}} \neg Q_{l_2}(x) \right)$$

- *A formula expressing that the root has label (u, u).*

$$Q_{(u,u)}(root)$$

- *Formulae expressing the function of the sets X and Y. For all $(u', u'') \in V_1 \times (V_1 \cup V_2)$ the formula*

$$\forall x. Q_{(u',u'')}(x) \Rightarrow X(x)$$

and for all $(u', u'') \in (V_1 \cup V_2) \times V_1$ the formula

$$\forall x. Q_{(u',u'')}(x) \Rightarrow Y(x)$$

- *Formulae encoding the requirements on the selection function. Let $(u', u'') \in (V_1 \cup V_2) \times (V_1 \cup V_2)$, and let $(u', r_1, v_1'), \ldots, (u', r_p, v_p')$ be all the edges in E with source u'. First, for each $i, 1 \leq i \leq p$, we have one formula in the conjunction. If $v_i' \in V_2$, then we take the formula*

$$\forall x. Q_{(u',u'')}(x) \Rightarrow \left(\bigvee_{(u'',r_i,v'') \in E \wedge L(v_i') \subseteq L(v'')} Q_{(v_i',v'')}(s_i(x)) \right)$$

---

[4] Defining $PrefixClosed(.)$ is a simple exercise. A definition of $Finite(.)$ can be found in [12].

Otherwise (i.e., if $v'_i \in V_1$), then we take the formula

$$\forall x. \left(Q_{(u',u'')}(x) \wedge X(s_i(x))\right) \Rightarrow \left(\bigvee_{(u'',r_i,v'')\in E \wedge L(v'_i)\subseteq L(v'')} Q_{(v'_i,v'')}(s_i(x))\right)$$

Second, we need formulae that fill in the appropriate dummy nodes:

$$\forall x. Q_{(u',u'')}(x) \Rightarrow \left(\bigwedge_{j=p+1}^{j=b} Q_\sharp(s_j(x))\right)$$

and for all $i, 1 \leq i \leq p$, such that $v'_i \in V_1$

$$\forall x. \left(Q_{(u',u'')}(x) \wedge \neg X(s_i(x))\right) \Rightarrow Q_\sharp(s_i(x))$$

– *A formula expressing that dummy nodes have only dummy successors.*

$$\forall x. Q_\sharp(x) \Rightarrow \left(\bigwedge_{j=1}^{j=b} Q_\sharp(s_j(x))\right)$$

– *A formula expressing the synchronization property.*

$$\forall P. Path(P) \Rightarrow \exists x. P(x) \wedge \left(Q_\sharp(x) \vee \bigvee_{v\in V_2} Q_{(v,v)}(x)\right)$$

where $Path(.)$ is the well-known MSO-formula expressing that a set of nodes consists of the nodes on an infinite path starting with the root (see [12]).

Lemma 1. *Let $\mathcal{G} = (V_1 \cup V_2, E, L)$ be a two-level graph, and $u \in V_1$. Then $\mathcal{G}$ is of bounded cycle depth w.r.t. u iff the MSO-formula $\phi_{\mathcal{G}}^u$ is satisfiable.*

Since satisfiability in MSO on infinite trees is decidable, the lemma (whose proof can be found in [5]) implies decidability of bounded cycle depth.

Theorem 1. *The problem of deciding whether a two-level graph is of bounded cycle depth w.r.t. one of its nodes is decidable.*

Unfortunately, the reduction does not give us a polynomial (or even a singly exponential) complexity bound for this decision problem. This is due to the fact that the formula $\phi_{\mathcal{G}}^u$ contains several quantifier changes[5].

Together with Propositions 1, this theorem implies:

Corollary 1. *The existence of the lcs and the msc is decidable in $\mathcal{EL}$ with descriptive semantics.*

---

[5] In Rabin's decidability proof based on automata, every negation requires a worst-case exponential complementation operation, and expressing a universal quantifier by an existential one (as required by Rabin's decision procedure) introduces two negation signs.

4   A Polynomial Bound on the Cycle Depth

A given two-level graph need not be of bounded cycle depth, but if it is then we can show that its cycle depth is actually polynomial in the size of the graph.

Theorem 2. *Let $\mathcal{G} = (V_1 \cup V_2, E, L)$ be a two-level graph, $u \in V_1$, and let m be the cardinality of $V_1 \cup V_2$. Then $\mathcal{G}$ is of bounded cycle depth iff $\mathcal{G}$ has cycle depth d w.r.t. u for some $d \leq m^2$.*

The "if" direction of this theorem is trivial. To prove the "only-if" direction, assume that $k > m^2$ is such that $\mathcal{G}_u^{(k)} \sqsubseteq \mathcal{G}_u^{(\ell)}$ for all $\ell > k$. To show that the cycle depth of $\mathcal{G}$ w.r.t. u is at most m^2, it is sufficient to show that $\mathcal{G}_u^{(m^2)} \sqsubseteq \mathcal{G}_u^{(\ell)}$ holds for all $\ell > m^2$. To show this, it is in turn enough to show that $\mathcal{G}_u^{(m^2)} \sqsubseteq \mathcal{G}_u^{(k)}$. The fact that is enough is a consequence of the following two facts:

1. $\mathcal{G}_u^{(k)} \sqsubseteq \mathcal{G}_u^{(\ell)}$ is trivially true for all $\ell < k$ and it holds for all $\ell > k$ by our assumption on k.
2. The subsumption relation $\sqsubseteq$ is transitive (see [5]).

Thus, the above theorem is proved once we have shown the following lemma.

Lemma 2. *Let $\mathcal{G} = (V_1 \cup V_2, E, L)$ be a two-level graph containing the node $u \in V_1$, let m be the cardinality of $V_1 \cup V_2$, and let $k > m^2$ be such that $\mathcal{G}_u^{(k)} \sqsubseteq \mathcal{G}_u^{(\ell)}$ for all $\ell > k$. Then we have $\mathcal{G}_u^{(m^2)} \sqsubseteq \mathcal{G}_u^{(k)}$.*

Proof. By our assumption on k we know that $\mathcal{G}_u^{(k)} \sqsubseteq \mathcal{G}_u^{(2k)}$, i.e., there is a $(u_0^{(2k)}, u_0^{(k)})$-synchronized simulation Z such that $(u_0^{(2k)}, u_0^{(k)}) \in Z$. Let S be the corresponding selection function. As sketched in the previous section, the S-selected $(u_0^{(2k)}, u_0^{(k)})$-simulation chains can be encoded into an infinite tree.

To be more precise, let b be the maximal number of successors of a node in $\mathcal{G}$, and let $\mathcal{L}_{2k}$ ($\mathcal{L}_k$) be the set of all nodes up to level $2k$ (level k) of the infinite tree with branching factor b. Now, $\mathcal{G}_u^{(k)} \sqsubseteq \mathcal{G}_u^{(2k)}$ implies that the formula $\psi_{\mathcal{G}}^u$ is satisfiable with X replaced by $\mathcal{L}_{2k}$ and Y replaced by $\mathcal{L}_k$. We can use the sets assigned to the variables Q_l for $l \in (V_1 \cup V_2) \times (V_1 \cup V_2) \cup \{\sharp\}$ to label the nodes of the infinite tree with branching factor b by elements of $(V_1 \cup V_2) \times (V_1 \cup V_2) \cup \{\sharp\}$. Let t denote the labeled tree obtained this way. Our goal is to transform t into a new tree t' that encodes a $(u_0^{(k)}, u_0^{(m^2)})$-synchronized simulation containing $(u_0^{(k)}, u_0^{(m^2)})$. The main properties that this new tree must satisfy are:

1. If the node n of t' is labeled with an element of $(V_1 \cup V_2) \times V_1$, then n is of depth at most m^2.
2. If the node n of t' is labeled with $(u', v') \in V_1 \times (V_1 \cup V_2)$ and is of depth smaller than k, then its successor nodes must cover *all* the successors in $\mathcal{G}$ of u', i.e., not only the ones in V_2, but also the ones in V_1.
3. The synchronization property is satisfied, i.e., any infinite path in t' contains a node whose label is $\sharp$ or of the form (v', v') for some node $v' \in V_2$.

In order to satisfy the first property, we modify the tree t as follows. Assume that n is a node of t with label $(u', v') \in (V_1 \cup V_2) \times V_1$ that is on a level above m^2. By the definition of t, $v' \in V_1$ implies that n is at most at level k (since all such nodes must belong to $\mathcal{L}_k$). Now, consider the path in t from the root to n. Since this path is longer than m^2, there are two distinct nodes n_1, n_2 on this path such that their labels agree. Assume that n_1 comes before n_2 on this path. Then we replace the subtree at node n_1 by the subtree at node n_2.

We continue this replacement process until all nodes with a label in $(V_1 \cup V_2) \times V_1$ are on depth at most m^2. This process terminates since there were only finitely many such nodes in t (all of them have depth at most k), and the replacements do not increase the depth of a node, but strictly decrease the depth of at least one node with a label in $(V_1 \cup V_2) \times V_1$. In addition, since all nodes with a label in $(V_1 \cup V_2) \times V_1$ are of depth at most k in t, the depth of a given node can decrease by at most k over the whole replacement process.

Let t' denote the labeled tree obtained this way. Then we can show that t' satisfies the properties 1, 2, 3 mentioned above, and thus encodes a $(u_0^{(k)}, u_0^{(m^2)})$-synchronized simulation that contains $(u_0^{(k)}, u_0^{(m^2)})$ (see [5] for details).   □

One might think that this polynomial bound on the cycle depth of a two-level graph can be used to show that the problem of deciding whether a graph is of bounded cycle depth or not can also be decided in polynomial time. However, this does not appear to be the case. In fact, assume that $\mathcal{G} = (V_1 \cup V_2, E, L)$ is a two-level graph with m nodes, and let $u \in V_1$. Then we know that $\mathcal{G}$ is of bounded cycle depth iff $\mathcal{G}_u^{(m^2)} \sqsubseteq \mathcal{G}_u^{(\ell)}$ for all $\ell > m^2$. However, testing this directly is still not possible since we would need to check infinitely many subsumption relationships. We could, of course, also try to use Theorem 2 to modify the reduction given in Section 3. However, all we would gain by this is that we could avoid the existential quantification over Y; the (expensive) universal quantification over X would still remain.

Together with Propositions 1, Theorem 2 implies:

Corollary 2. *The lcs (msc) in $\mathcal{EL}$ with descriptive semantics can be computed in polynomial time, provided that it exists.*

5 Conclusion

We have introduced the notion "bounded cycle depth" of so-called two-level graphs, and have shown that the corresponding decision problem (i.e.: Given a two-level graph, is it of bounded cycle depth?) is decidable. In addition, we have shown that the cycle depth of a two-level graph of bounded cycle depth is polynomial in the size of the graph. These results solve the two main problems that were left open in the previous papers [1, 2] on the lcs and the msc in $\mathcal{EL}$ with descriptive semantics: the existence of the lcs (msc) is decidable, and if it exists, then it can be computed in polynomial time.

What remains open is the exact complexity of the decision problems. Though this may seem unsatisfactory from a theoretical point of view, it is probably not very relevant in practice. In fact, independent of whether the lcs of the concepts A, B defined in a terminology $\mathcal{T}$ exists or not, the results in [1] show how to compute common subsumers P_i ($i \geq 0$) of A, B in $\mathcal{T}$. The results of Section 4 imply that we can compute a number k that is polynomial in the size of $\mathcal{T}$ such that A, B in $\mathcal{T}$ have an lcs w.r.t. descriptive semantics iff P_k is the lcs. Thus, we may just dispense with deciding whether the lcs exists, and return P_k. If the lcs exits, then P_k is the lcs. Otherwise, P_k is a common subsumer, and we can take it as an approximation of the lcs. The same is true for the msc.

Another interesting question is whether two-level graphs and the problem of deciding whether they are of bounded cycle depth also has applications in other areas. Is the cycle depth of a two-level graph an artifact of the characterization of the lcs and the msc in $\mathcal{EL}$ with descriptive semantics given in [1, 2], or is it a natural notion that is of interest in its own right?

References

1. F. Baader. Computing the least common subsumer in the description logic $\mathcal{EL}$ w.r.t. terminological cycles with descriptive semantics. In *Proc. ICCS 2003*, Springer LNAI 2746, 2003.
2. F. Baader. The instance problem and the most specific concept in the description logic $\mathcal{EL}$ w.r.t. terminological cycles with descriptive semantics. In *Proc. KI 2003*, Springer LNAI 2821, 2003.
3. F. Baader. Least common subsumers and most specific concepts in a description logic with existential restrictions and terminological cycles. In *Proc. IJCAI 2003*, Morgan Kaufmann, 2003.
4. F. Baader. Terminological cycles in a description logic with existential restrictions. In *Proc. IJCAI 2003*, Morgan Kaufmann, 2003.
5. F. Baader. A graph-theoretic generalization of the least common subsumer and the most specific concept in the description logic $\mathcal{EL}$. LTCS-Report 04-02, TU Dresden, Germany, 2004. See http://lat.inf.tu-dresden.de/research/reports.html.
6. F. Baader, D. Calvanese, D. McGuinness, D. Nardi, and P.F. Patel-Schneider, editors. *The Description Logic Handbook: Theory, Implementation, and Applications*. Cambridge University Press, 2003.
7. F. Baader and R. Küsters. Computing the least common subsumer and the most specific concept in the presence of cyclic $\mathcal{ALN}$-concept descriptions. In *Proc. KI'98*, Springer LNAI 1504, 1998.
8. R. Küsters. *Non-standard Inferences in Description Logics*, Springer LNAI 2100, 2001.
9. M.O. Rabin. Decidability of second-order theories and automata on infinite trees. *Trans. of the Amer. Mathematical Society*, 141, 1969.
10. K.A. Spackman. Normal forms for description logic expressions of clinical concepts in SNOMED RT. *J. of the American Medical Informatics Association*, 2001. Symposium Supplement.
11. K.A. Spackman, K.E. Campbell, and R.A. Cote. SNOMED RT: A reference terminology for health care. *J. of the American Medical Informatics Association*, 1997. Fall Symposium Supplement.
12. W. Thomas. Automata on infinite objects. In *Handbook of Theoretical Computer Science*, Volume B. Elsevier Science Publishers, Amsterdam, 1990.

The Computational Complexity of the Minimum Weight Processor Assignment Problem

Hajo J. Broersma[1], Daniel Paulusma[2], Gerard J.M. Smit[2],
Frank Vlaardingerbroek[2], and Gerhard J. Woeginger[3]

[1] Department of Computer Science,
University of Durham, Science Labs, South Road, Durham DH1 3LE, England
hajo.broersma@durham.ac.uk
[2] Faculty of Electrical Engineering, Mathematics and Computer Science,
University of Twente, 7500 AE Enschede, The Netherlands
d.paulusma@math.utwente.nl, smit@cs.utwente.nl,
f.vlaardingerbroek@student.utwente.nl
[3] Department of Mathematics and Computer Science,
Eindhoven University of Technology, P.O. Box 513,
5600 MB Eindhoven, The Netherlands
gwoegi@win.tue.nl

Abstract. In portable multimedia systems a number of communicating tasks has to be performed on a set of heterogeneous processors in an energy-efficient way. We model this problem as a graph optimization problem, which we call the minimum weight processor assignment problem. We show that our setting generalizes several problems known in literature, including minimum multiway cut, graph k-colorability, and minimum (generalized) vertex covering. We show that the minimum weight processor assignment problem is NP-hard, even when restricted to instances where the (process) graph is a bipartite graph with maximum degree at most 3, or with only two processors, or with arbitrarily small weight differences, or with only two different edge weights. For graphs with maximum degree at most 2 (or in fact the larger class of degree-2-contractible graphs) we give a polynomial time algorithm. Finally we generalize this algorithm into an exact (but not efficient) algorithm for general graphs.

1 Introduction

In portable multimedia systems a number of communicating tasks has to be performed on a set of heterogeneous processors. The key issue in the design of portable multimedia systems is to find a good balance between flexibility and high-processing power on one side, and area and energy-efficiency of the implementation on the other side (cf. [7]).

In this paper we will model a particular problem from this area as a graph optimization problem, in which the vertices of the graph represent the tasks, and its edges represent the communication between two tasks. Each task must

J. Hromkovič, M. Nagl, and B. Westfechtel (Eds.): WG 2004, LNCS 3353, pp. 189–200, 2004.

be performed on exactly one processor. Each possible choice of processors involves a certain amount of costs reflecting the energy consumption involved in running a particular task on a particular processor, whereas the assignment of two communicating tasks to two processors involves a certain amount of costs, too, reflecting the energy involved in transferring data between the processors. The problem boils down to finding a "mapping" of the set of tasks to the set of processors that minimizes the total costs. We call this problem the MINIMUM WEIGHT PROCESSOR ASSIGNMENT problem (MWPA).

The paper is organized as follows. The next section gives a sketch of the background of the problem, and the motivation from current research in the area of Embedded Systems. The next section is followed by a section that provides the necessary definitions and gives a formal description of our model. In the fourth section we give a survey of NP-hard problems that all can be reduced to MWPA. We show that MWPA is already NP-hard if the set of processors is restricted to two, or if the (process) graphs are taken from the class of connected bipartite graphs with maximum degree at most 3. For the class of degree-2-contractible graphs we give a polynomial time algorithm. In the fifth section we generalize this algorithm into an exact (but not efficient) algorithm for any (process) graph.

2 Background and Motivation

2.1 General Background

Current research in the area of Embedded Systems is for a great deal driven by the explosive growth in the use of handheld mobile devices, such as cellular phones, personal digital assistants, digital cameras, global positioning systems, and so forth. Personal mobile computing (often referred to as ubiquitous computing) is likely to play a significant role in driving technology in the next decade. In this paradigm, the basic personal computing and communication device will be an integrated, battery-operated device, small enough to carry along all the time.

The technological developments needed to establish this paradigm of personal mobile computing lead to many challenging problems. In particular, these devices have limited battery resources, must handle diverse data types, and must operate in environments that are insecure, unplanned, and show different characteristics over time.

Traditionally, (embedded) systems that have demanding applications - e.g., driven by portability, performance, or cost - lead to the development of one or more custom processors or application-specific integrated circuits (ASICs) to meet the design objectives. However, the development of ASICs is expensive in time, manpower and money.

Another way to solve the problems has been to use general-purpose processors, i.e., trying to solve all kinds of applications running on a very high speed processor. A major drawback of using these general-purpose devices is that they are extremely inefficient in terms of utilising their resources.

In current developments the key issue to overcome the drawbacks of the aforementioned approaches is to use reconfigurable heterogeneous processors.

2.2   Current Research

To match the required computation with the architecture, in current research like the CHAMELEON project which is taking place at the University of Twente, an alternative approach is made in order to meet the requirements of future low-power hand-held systems. In the CHAMELEON project we apply reconfiguration at multiple levels of granularity. This means that the architecture contains different processing entities: e.g. a general-purpose processor (e.g. ARM core), a bit-level reconfigurable part (embedded FPGA) and several word-level reconfigurable parts (e.g. Montium tiles). The main philosophy used is that operations on data should be done at the place where it is most energy-efficient and where it minimizes the required communication. Partitioning is an important architectural decision, which dictates where applications can run, where data can be stored, and also the complexity of the mobile and the cost of communication services. Our approach is based on a dynamic (i.e. at run-time) matching of the architecture and the application.

2.3   Software Design Flow

The key issue in the design of portable multimedia systems is to find a good balance between flexibility and high-processing power on one side, and area and energy-efficiency of the implementation on the other side. The design of the above-mentioned architecture is useless without a proper tool chain supported by a solid design methodology. At various levels of abstraction, modern computing systems are defined in terms of processes and communication (or, at least, synchronisation) between processes. Many applications can be structured as a set of processes or threads that communicate via channels. These threads can be executed on various platforms (e.g. general purpose CPU, FPFA, FPGA, etc).

2.4   Process Graphs

We use a so-called Kahn based process graph model, which abstracts system functionality into a set of processes/tasks represented as vertices in a graph, and represents functional dependencies among processes (channels) with graph edges. The functionality of a process graph will be referred to as a set of (sub)tasks. This model emphasizes communication and concurrency between system processes. Edge labels are used to represent communication bandwidth requirements, while vertex labels may store a measure of process computational requirements. The costs associated with a process graph in the context of reconfiguration can be divided into communication costs between the processes, computational costs of the processes and initialization costs of the (sub)tasks. The costs can be expressed in terms of energy consumption, resource usage, and aspects of time (latency, jitter, etc). The mapping of applications (a set of communicating tasks) is done in two phases. In the first phase (macro-mapping) for each set of tasks the optimal (or near to optimal) processing entity is determined. This phase defines what is processed where and when. In the second phase (micro-mapping) for each task a detailed mapping is derived to the platform of choice. The problem of this paper is related to the first phase.

2.5   Macro-mapping

In a reconfigurable system, application instantiation consists first of all of finding a suitable partition of the system specification into parts that can be mapped onto the most appropriate resources of the system (processors, FPGAs, coarse-grain reconfigurable entities). Because of the dynamics of the mobile environment we would like to perform the macro-mapping at run-time. The search for the 'best' mapping is typically a very hard problem, due to the size of the search space. We will refrain from giving more technical details, but stick to the (sub)problem of mapping the tasks to the heterogeneous processors in such a way that the costs of running the tasks on the processors and transferring data between communicating tasks is minimized.

3   Preliminaries

For the modeling of the optimization problem described in the previous section we consider simple graphs, denoted by $G = (V_G, E_G)$, where V_G is a finite nonempty set of vertices and E_G is a set of unordered pairs of vertices, called edges.

For a vertex $u \in V_G$ we denote its neighborhood, i.e. the set of adjacent vertices, by $N(u) = \{v \,|\, (u,v) \in E_G\}$. The *degree* $\deg(u)$ of a vertex u is the number of edges incident with it. The symbol Δ_G denotes the maximum degree among all vertices of G. If all vertices in G have the same degree $k \in \mathbb{N}$, then G is called *k-regular*.

A graph G is called *connected* if for every pair of distinct vertices u and v, there exists a *path* connecting u and v, i.e., a sequence of distinct vertices starting with u and ending with v where each pair of consecutive vertices forms an edge of G.

A graph is called *bipartite* if it is simple and its vertices can be partitioned into two sets A and B such that each edge has one of its end vertices in the set A and the other in B.

If $e = (u,v) \in E_G$, the *contraction* G/e of G is the simple graph obtained from G by replacing u and v and the edges incident with u and v by one new vertex uv and edges joining uv with the vertices adjacent to u or v in G. For another graph H, G is said to be *contractible* to H if H can be obtained from G by successive contractions of edges. A graph G is called *degree-2-contractible* if G is contractible to the graph consisting of one vertex by successively contracting edges incident with vertices of degree 1 or 2. Note that the class of degree-2-contractible graphs is equal to the class of graphs that have tree-width at most two (cf. [2]).

Now let $G = (V_G, E_G)$ be a simple graph with vertex set V_G and edge set E_G. The vertices of G represent the tasks that have to be performed on a set P of processors. An edge $e = (u,v)$ exists if and only if there is communication between the processes of task u and task v.

Let the vertex weight $w_p^u \geq 0$ represent the costs of the process of task u, if u is performed on processor $p \in P$. This way we define a weight vector w^u of size $|P|$ for $u \in V_G$.

If in practice u cannot be performed on processor p, this can be expressed by setting $w_p^u = \infty$ (or a bounded, sufficiently large number M).

For $e = (u, v) \in E_G$ let the edge weight $w_{pq}^e \geq 0$ represent the communication costs between the processes of task u and v, if u is performed on processor $p \in P$ and v is performed on processor $q \in P$. This way we define a matrix W^e of size $|P| \times |P|$ for $e \in E_G$.

If in practice two tasks u and v cannot be performed on the same processor p simultaneously, we can model this by adding an edge $e = (u, v)$ and setting $w_{pp}^{(u,v)} = M$, where M is a sufficiently large number.

The graph G together with the weight vectors w^u, and weight matrices W^e is called a *weighted process graph*, and denoted by G_w.

We call a mapping $f : V_G \to P$ a *processor assignment* of G. Let $\mathcal{F}_G$ denote the set of all processor assignments of G. We define the *weight of a processor assignment* $f \in \mathcal{F}_G$ for a weighted process graph G_w as

$$w(f) = \sum_{v \in V_G} w_{f(v)}^v + \sum_{(u,v) \in E_G} w_{f(u)f(v)}^{(u,v)}.$$

A *minimum weight processor assignment* f^* is a processor assignment that has minimum weight, i.e., with $w(f^*) = \min\{w(f) \mid f \in \mathcal{F}_G\}$.

The MINIMUM WEIGHT PROCESSOR ASSIGNMENT problem (MWPA) is the problem of finding a minimum weight processor assignment for a given weighted process graph G_w and a set P of processors.

If any minimum weight processor assignment f will map task u on processor q, we say that u is *fixed* on q. If $w_p^u = 0$ for all processors $p \in P$, we will say that u is *free*.

4 Complexity Results

It is easy to prove that MWPA is an NP-hard problem. In fact several known NP-hard problems, such as the MINIMUM MULTIWAY CUT problem ([3]), can be used to show this. We give the reduction from the MINIMUM MULTIWAY CUT problem as an example. In later sections we show relations with other NP-hard problems which can be applied to show NP-hardness of MWPA for restricted graph classes, a limited number of processors, or arbitrarily close edge weights.

An instance of the MINIMUM MULTIWAY CUT problem is formed by an undirected graph G, a subset $S \subseteq V_G$ of terminals, and weights $c(e)$ for all $e \in E_G$. A *multiway cut* is a subset $F \subseteq E_G$ such that in the graph $G' = (V_G, E_G \backslash F)$ there is no path between any pair of terminals from S. The MINIMUM MULTIWAY CUT problem is the problem of finding a multiway cut F that minimizes the costs $c(F) = \sum_{e \in F} c(e)$ for a given weighted graph and a given set of terminals.

Observation 1 *The* MINIMUM MULTIWAY CUT *problem can be reduced to the* MINIMUM WEIGHT PROCESSOR ASSIGNMENT *problem.*

Proof. Given a graph G with a set $S = \{u_1, \ldots, u_s\}$ of terminals and weights $c(e)$ for $e \in E_G$ define a set of processors $P = \{p_1, \ldots, p_s\}$.

For $u \notin S$ let $w_p^u = 0$ for all $p \in P$. For $u_i \in S$ let $w_p^u = 0$ if $p = p_i$ and for $p \neq p_i$ let $w_p^u = M$, where M is a sufficiently large number. Then each minimum weight processor assignment fixes each terminal $u_i \in S$ to its 'own' processor p_i. For each $e \in E_G$ and $p, q \in P$ we define $w_{pq}^e = c(e)$ if $p \neq q$ and $w_{pq}^e = 0$ if $p = q$.

This way we have constructed an instance (G_w, P). Our claim is that finding a multiway cut with minimum costs comes down to solving MWPA on the instance (G_w, P).

Let f be a minimum weight processor assignment for (G_w, P). Then $F = \{e \in E_G \mid e = (u, v) \text{ with } f(u) \neq f(v)\}$ is a multiway cut with costs $c(F) = w(f)$. Suppose a multiway cut F' exists with $c(F') < c(F)$. We may assume that each vertex of $(V_G, E_G \backslash F')$ is connected to some u_i; otherwise there is a multiway cut $F'' \subset F'$ with $c(F'') \leq c(F')$. Now we define a processor assignment f' of (G_w, P) by $f'(u) = p_i$, if there exists a path from u to $u_i \in S$ in the graph $G' = (V_G, E_G \backslash F')$. Obviously, $w(f') = c(F') < w(f)$, a contradiction.

Hence solving MWPA is at least as hard as solving MINIMUM MULTIWAY CUT.   □

In [3] it is shown that the MINIMUM MULTIWAY CUT problem is already NP-hard for the class of graphs with three terminals. For two terminals it is polynomially solvable by standard network flow techniques. In the sequel we prove NP-hardness of MWPA for instances with two processors, and for instances with only free vertices instead of $|V_G| - 3$ free vertices and 3 fixed vertices.

4.1   Processor Graphs with Only Free Vertices

We show that MWPA is still NP-hard if the instances (G_w, P) are restricted to the class of weighted process graphs with only free vertices. For this purpose we make a reduction from the well-known problem GRAPH k-COLORABILITY, which is known to be NP-complete for each integer $k \geq 3$ (cf [5]).

GRAPH k-COLORABILITY (GkC)
Instance: A graph G and a set of colors $C = \{a_1 \ldots a_k\}$.
Question: Is G k-colorable, i.e., does there exist a mapping $g : V_G \rightarrow C$ such that $g(u) \neq g(v)$ for all $u, v \in V_G$ with $(u, v) \in E_G$?

Proposition 1. *Let $\mathcal{G}$ be a class of graphs for which GkC is NP-complete. Then MWPA is already an NP-hard problem for instances (G_w, P) in which G is a graph from $\mathcal{G}$ with $|V_G|$ free vertices, and sets P of $|P| = k \geq 3$ processors.*

Proof. Given a graph G with color set $\{a_1, \ldots, a_k\}$ we define a set of processors $P = \{p_1, \ldots, p_k\}$. For each edge $e \in E_G$ we define weights $w_{p_i p_i}^e = 1$ for all $p_i \in P$ and $w_{p_i p_j}^e = 0$ if $i \neq j$. All weights w_p^u are set to 0. Obviously, G is k-colorable if and only if (G_w, P) has a minimum weight processor assignment f with $w(f) = 0$.   □

GkC (with $k \geq 3$) is amongst others NP-complete for planar graphs and cographs (i.e., graphs which do not contain an induced path on 4 vertices). Hence, MWPA is NP-hard if instances (G_w, P) are restricted to one of these two classes (with $|P| \geq 3$).

Note that the above proof also shows that MWPA remains NP-hard when restricted to instances with only two different values for the edge weights (and no vertices with positive weights).

4.2 Sets of Exactly Two Processors

In Proposition 1 we have shown that MWPA is NP-hard for instances (G_w, P), where $|P| = k \geq 3$. Of course, the problem is trivially solvable in polynomial time if the set of processors only contains one processor. In this section we show that MWPA is NP-hard if the set of instances is restricted to instances with two processors. It turns out that the recently studied NP-hard problem MINIMUM GENERALIZED VERTEX COVER ([6]) is a special case in our model.

Instances of MINIMUM GENERALIZED VERTEX COVER are undirected graphs G with numbers $d_0(e) \geq d_1(e) \geq d_2(e) \geq 0$ for every edge $e \in E_G$, and costs $c(u)$ for every vertex $u \in V$. For $S \subseteq V_G$ and $e = (u, v) \in E_G$ define $c(e)$ to be the cost of e depending on the number of its end vertices that are included in S, i.e., $c(e) = d_0(e)$ if $u, v \in V_G \backslash S$, $c(e) = d_1(e)$ if $u \in S, v \in V_G \backslash S$ or $v \in S, u \in V_G \backslash S$, and $c(e) = d_2(e)$ if $u, v \in S$. A *generalized vertex cover* is a subset $S \subseteq V_G$. The MINIMUM GENERALIZED VERTEX COVER problem is the problem of finding a generalized vertex cover S that minimizes the costs $c(S) = \sum_{v \in S} c(u) + \sum_{e \in E_G} c(e)$ for a given graph G and cost function c.

By choosing $d_0(e) = 1$ and $d_1(e) = d_2(e) = 0$ for all $e \in E_G$, and $c(v) = \frac{1}{n}$ for all $v \in V_G$, it is clear that the MINIMUM GENERALIZED VERTEX COVER problem is a generalization of the well-known MINIMUM VERTEX COVER problem. In the proof of the proposition below we show that our setting includes both problems.

Proposition 2. MWPA *is* NP-*hard, even if the class of instances (G_w, P) is restricted to instances with two processors.*

Proof. Assume we are given a graph G with costs $c(v)$ for $v \in V_G$, and numbers $d_i(e)$ for $e \in E_G$ and $0 \leq i \leq 2$. Define a set of processors $P = \{p, q\}$. Let $w_p^u := c(u)$ and $w_q^u := 0$ for all $u \in V_G$. Let $w_{pp}^e := d_2(e), w_{pq}^e = w_{qp}^e := d_1(e)$, and $w_{qq}^e := d_0(e)$ for all $e \in E_G$.

This way we have obtained an instance (G_w, P) of MWPA. Each processor assignment f corresponds to a specific generalized vertex cover S with costs $c(S) = w(f)$, namely $S = \{u \in V_G \mid f(u) = p\}$. Vice versa, each generalized vertex cover S corresponds to a unique processor assignment f with costs $w(f) = c(S)$: Choose f given by $f(u) = p$ if $u \in S$, and $f(u) = q$ if $u \notin S$.

Hence the MINIMUM GENERALIZED VERTEX COVER problem is a special case of the MINIMUM WEIGHT PROCESSOR ASSIGNMENT problem.   □

4.3 Process Graphs with Arbitrarily Close Edge Weights

It is straightforward to see that MWPA is polynomially solvable for instances (G_w, P) with process graphs G_w with constant edge weights, or with edge weights $w_{pq}^e = w_p^u + w_q^v$ if $e = (u, v)$ and $p, q \in P$. In both cases a processor assignment f that maps each vertex u on a processor p for which w_p^u is minimal, is a minimum weight processor assignment.

However, from the proof of Proposition 1 it turns out that MWPA stays NP-hard if the class of instances is restricted to instances (G_w, P) with arbitrarily close weights: Choose in the proof of Proposition 1 weights $w^e_{p_i p_i} = \epsilon$, where $\epsilon > 0$ is an arbitrarily small value, and $w^e_{p_i p_j} = 0$ if $i \neq j$. As remarked before, this also implies NP-hardness of MWPA when restricted to instances with only two possible values for the edge weights.

4.4 Process Graphs That Are Degree-2-Contractible

In the theorem below we show that MWPA can be solved in polynomial time for instances (G_w, P) where G is a degree-2-contractible graph. So MWPA can be solved efficiently if for instance G_w is a tree or a unicyclic graph. However, already for the class of 3-regular bipartite graphs the problem turns out to be NP-hard.

For the NP-hardness construction we make a reduction from a particular variant of HYPERGRAPH 2-COLORABILITY. This is a well-known NP-complete problem (cf. [5]).

HYPERGRAPH 2-COLORABILITY (H2C)
Instance: A set $Q = \{q_1, \ldots, q_m\}$ and a set $\mathcal{S} = \{S_1, \ldots, S_n\}$ with $S_j \subseteq Q$ and $|S_j| = 3$ for $1 \leq j \leq n$.
Question: Is there a 2-coloring of $(Q, \mathcal{S})$, i.e., a partition of Q into $Q_1 \cup Q_2$ such that $Q_1 \cap S_j \neq \emptyset$ and $Q_2 \cap S_j \neq \emptyset$ for $1 \leq j \leq n$?

With such a hypergraph we associate its incidence graph I, which is a bipartite graph on $Q \cup \mathcal{S}$, where (q, S) forms an edge if and only if $q \in S$ (cf. Figure 1).

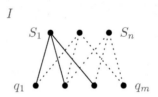

Fig. 1. The incidence graph of an instance of H2C.

Note that we may assume without loss of generality that I is connected.

Theorem 1. MWPA *can be solved in polynomial time for instances* (G_w, P) *in which* G_w *is a degree-2-contractible graph.* MWPA *is already* NP-*hard if the class of instances only contains instances* (G_w, P), *where* G_w *is a connected 3-regular bipartite graph.*

Proof. Let (G_w, P) form an instance of MWPA, where G is a degree-2-contractible graph. Suppose $v \in V_G$ and $\deg(v) = 1$. Let $u \in V_G$ be the only neighbor of v. Let $G' = G/(u, v) = (V_G \backslash v, E_G \backslash (u, v))$ be the graph G without the vertex v. For all $p \in P$ define

$$\tilde{w}^u_p = w^u_p + \min_{q \in P}\{w^v_q + w^{(u,v)}_{pq}\},$$

and do not change any other vertex or edge weights, i.e., $\tilde{w}_p^x = w_p^x$ for $x \in V_{G'} \setminus u, p \in P$ and $\tilde{w}_{pq}^e = w_{pq}^e$ for $e \in E_{G'}$ and $p, q \in P$. This way we have obtained an equivalent instance $(G'_{\tilde{w}}, P)$ of MWPA with one vertex less.

We continue with this procedure as long as we have a process graph with minimum degree equal to one. As long as we still have edges in our graph we proceed as follows.

Suppose $y \in V_G$ and $\deg(y) = 2$. Let $x, z \in V_G$ be the only two neighbors of v. We remove vertex y and edges (x, y) and (y, z), and add an edge (x, z) if it is not already in G. This way we have obtained a graph G^*. Assume $w_{pq}^{(x,z)} = 0$ for all $p, q \in P$ if (x, z) is not an edge in G. Then we define for all $p, q \in P$

$$\tilde{w}_{pq}^{(x,z)} = w_{pq}^{(x,z)} + \min_{r \in P} \{ w_r^y + w_{pr}^{(x,y)} + w_{rq}^{(y,z)} \},$$

and do not change any other vertex or edge weights, i.e., $\tilde{w}_p^u = w_p^u$ for $u \in V_{G^*} \setminus y, p \in P$ and $\tilde{w}_{pq}^e = w_{pq}^e$ for $e \in E_{G^*} \setminus (x, z)$ and $p, q \in P$. This way we have obtained an equivalent instance $(G^*_{\tilde{w}}, P)$ of MWPA with one vertex less.

As long as G^* has edges, we repeat the steps above.

Combining these two steps gives us a polynomial time algorithm for degree-2-contractible graphs. Note that this approach fails, if during the procedure we obtain a graph $\bar{G}$ that does not have a vertex of degree one or two. (See the next section for a more general algorithm.)

We now prove NP-hardness for the class of instances (G_w, P) where G is a connected 3-regular bipartite graph.

Let $(Q, \mathcal{S})$ be an instance of H2C. First, we construct its incidence graph I. Since we assume that each set S_j has exactly 3 elements, we obtain that I is 3-regular. We introduce a set of processors

$$P = \{ p_{qr} \mid (q, r) \in Q \times Q, q \neq r \} \ \bigcup \ \{ p_q \mid q \in Q \} \ \bigcup \ \{ p'_q \mid q \in Q \}.$$

Let M be a sufficiently large number. For each $S \in \mathcal{S}$ we define $w_p^S = 0$ if $p = p_{qr}$ and q, r are elements of S, and $w_p^S = M$ otherwise. For each $q \in Q$ we define $w_p^q = 0$ if $p = p_q$ or $p = p'_q$, and $w_p^q = M$ otherwise.

For each edge $e = (q, S) \in E_I$ we define the following edge weights. We make $w_{p_1 p_2}^e = 1$ if

- $p_1 = p_q$ and $p_2 = p_{rq}$ for some $r \in S$, or
- $p_1 = p'_q$ and $p_2 = p_{qr}$ for some $r \in S$.

We define $w_{p_1 p_2}^e = 0$ if

- $p_1 = p_q$ and $p_2 = p_{qr}$ for some $r \in S$, or
- $p_1 = p_q$ and $p_2 = p_{rt}$ for $r, t \in S \setminus q$, or
- $p_1 = p'_q$ and $p_2 = p_{rq}$ for some $r \in S$, or
- $p_1 = p'_q$ and $p_2 = p_{rt}$ for $r, t \in S \setminus q$.

Finally we define $w_{pq}^e = M$ for all other possibilities.

This way we have obtained a weighted process graph I_w. Our claim is that $(Q, \mathcal{S})$ is 2-colorable if and only if (I_w, P) has a minimum weight processor assignment f with $w(f) = 0$.

Suppose $Q_1 \cup Q_2$ is a 2-coloring of $(Q, \mathcal{S})$. If q is in Q_1 we define $f(q) = p_q$. Otherwise we let $f(q) = p'_q$. Since $Q_1 \cup Q_2$ is a 2-coloring, each $S \in \mathcal{S}$ contains a vertex q in Q_1 and a vertex r in Q_2. We define $f(S) = p_{qr}$. This way we have constructed a processor assignment f with $w(f) = 0$. Since all vertex and edge weights are positive, f is a minimum weight processor assignment.

To prove the reverse statement suppose a minimum weight processor assignment f exists with $w(f) = 0$. For all $q \in Q$ we do the following. If $f(q) = p_q$, we place q in Q_1. If $f(q) = p'_q$, then we place q in Q_2.

Now suppose $Q_1 \cup Q_2$ would not be a 2-coloring. Then a set $S = \{q, r, t\}$ in $\mathcal{S}$ exists that is fully contained in either Q_1 or Q_2. Suppose S is a subset of Q_1. By definition of Q_1, we obtain $f(q) = p_q$, $f(r) = p_r$, and $f(t) = p_t$ implying $w(f) > 0$. $\square$

5   An Exact Algorithm

Although MWPA is NP-hard for many graph classes, here we present a relatively simple algorithm that determines the weight of a minimum processor assignment for *any* weighted process graph G_w and set of processors P. We call this algorithm MINWEIGHT. Its running time is exponential. However, in practice it could compute solutions quite fast, as long as the input graphs have a small number of vertices with a high degree (greater than two) or a high number of fixed vertices.

Steps (3) and (4) in the algorithm have already been explained in the proof of Theorem 1. Step (5) handles vertices with degree greater than two. Each time this step is executed the number of computations increases exponentially.

Note that, in practice rules such as "delete forbidden processor/vertex combinations" can be added to increase the running time of MINWEIGHT in case there are many of such combinations. It is also easy to implement the algorithm in such a way that it gives as output a minimum weight processor assignment f with $w(f) = w^*$. Moreover, instead of using the same (Cartesian products of) processor sets for all vertices, in an implementation of the algorithm we could use different sets for different vertices in order to save on memory and computational steps.

6   Conclusion

In the previous sections we have introduced the MINIMUM WEIGHT PROCESSOR ASSIGNMENT problem (MWPA), motivated from current research in the area of Embedded Systems. We have analysed the complexity of MWPA, its relation to various other graph problems, and showed that it is NP-hard, even when restricted to very specific classes of instances. For a number of classes of instances we have shown that MWPA is polynomial. We presented a simple exact (exponential) algorithm, based on edge contractions, for solving MWPA for general

Algorithm MINWEIGHT

(**1**) FOR $(u,v) \notin E_G$ and $p,q \in P$ DO $w_{pq}^{(u,v)} := 0$.

(**2**) Choose a vertex $v \in V_G$.

(**3**) IF $\deg(v) = 1$,
THEN let $V_G := V_G \backslash v$, and $E_G := E_G \backslash (u,v)$ for $u \in N(v)$.
FOR $u \in N(v)$ and $p \in P$ DO

$$w_p^u := w_p^u + \min_{q \in P}\{w_q^v + w_{pq}^{(u,v)}\}.$$

(**4**) IF $\deg(v) = 2$,
THEN let $V_G := V_G \backslash v$ and $E_G := (E_G \cup \{(x,z)\}) \backslash \{(x,v),(v,z)\}$ for $x, z \in N(v)$.
FOR $x, z \in N(v)$ and $p, q \in P$ DO

$$w_{pq}^{(x,z)} := w_{pq}^{(x,z)} + \min_{r \in P}\{w_r^v + w_{pr}^{(x,v)} + w_{rq}^{(v,z)}\}.$$

(**5**) IF $\deg(v) \geq 3$,
THEN choose a vertex $u \in N(v)$. Let $G := G/(u,v)$.
Set $P := P \times P$.
FOR $(p,q) \in P$ DO $w_{(p,q)}^{uv} := w_p^u + w_{pq}^{(u,v)} + w_q^v$.
FOR $x \in V_G$ and $(p,q) \in P$ DO $w_{(p,q)}^x := w_p^x$.
FOR $e = (uv,x)$ with $x \in N(uv)$ and $(p,q),(r,s) \in P$ DO $w_{(p,q)(r,s)}^{(uv,x)} := w_{pr}^{(u,x)} + w_{qr}^{(v,x)}$.
FOR $e \in E_G$ not incident with uv and $(p,q),(r,s) \in P$ DO $w_{(p,q)(r,s)}^e := w_{pr}^e$.

(**6**) IF $|V_G| \geq 2$, THEN GOTO (2).

(**7**) Output $w^* := \min\{w_p^u \mid p \in P, u \in V_G\}$. STOP.

instances. We are currently investigating the running-time of the algorithm for real-life examples. After an extensive search initiated by a remark of Hans Bodlaender we became aware of a number of related papers. In some of them, the static assignment problem has been studied under different names, whereas this paper is motivated by the dynamic problem of allocating tasks in run-time. Part of the results of the present paper are covered. We refer the interested reader to [4] and the survey paper [1].

Acknowledgements

The authors thank Asaf Levin for explaining the relationship of the problem to the minimum multiway cut problem.

References

1. AARDAL, K. I., VAN HOESEL, S. P. M., KOSTER, A. M. C. A., MANNINO, C., SASSANO, A. Models and solution techniques for frequency assignment problems. *4OR 1*, 4 (2003), 261–317.

2. BODLAENDER A tourist guide through treewidth. *Acta Cybernetica 11* (1993), 1–21.
3. DAHLHAUS, E., JOHNSON, D. S., PAPADIMITRIOU, C. H., SEYMOUR, P. D., AND YANNAKAKIS, M. The complexity of multiterminal cuts. *SIAM Journal on Computing 23,* 4 (1994), 864–894.
4. FERNÁNDEZ-BACA, D. Allocating modules to processors in a distributed system. *IEEE Transactions on Software Engineering 15,* 11 (1989), 1427–1436.
5. GAREY, M. R., AND JOHNSON, D. S. *Computers and Intractability.* W. H. Freeman and Co., New York, 1979.
6. HASSIN, R., AND LEVIN, A. The minimum generalized vertex cover problem. In *Algorithms ESA 2003, 11th ESA '03, Budapest, Hungary* (2003), no. 2832 in Lecture Notes in Computer Science, Springer Verlag, 289–300.
7. SMIT, G.J.M., HAVINGA, P.J.M., SMIT, L.T., HEYSTERS, P.M., AND ROSIEN, M.A.J. Dynamic reconfiguration in mobile systems. In *Field-Programmable Logic and Applications. Reconfigurable Computing Is Going Mainstream, 12th FPL '02, Montpellier, France* (2002), no. 2438 in Lecture Notes in Computer Science, Springer Verlag, 171–181.

A Stochastic Location Problem
with Applications to Tele-diagnostic

Nicola Apollonio[1], Massimiliano Caramia[2], and Giuseppe F. Italiano[3]

[1] Dipartimento di Statistica, Probabilità, e Statistiche Applicate,
Università di Roma "La Sapienza", P.le A. Moro 5, 00185 Rome, Italy
nicola.apollonio@uniroma1.it
[2] Istituto per le Applicazioni del Calcolo "M. Picone",
Viale del Policlinico, 137 – 00161 Rome, Italy
m.caramia@iac.cnr.it
[3] Dipartimento di Informatica, Sistemi e Produzione,
University of Rome "Tor Vergata", Via del Politecnico, 1 – 00133 Rome, Italy
italiano@disp.uniroma2.it

Abstract. In this paper we study a stochastic location problem with applications to tele-diagnostic, locating the boundaries between polynomiality and NP-completeness, and providing efficient approximation algorithms.

1 Introduction

Many location problems have been widely studied in the literature (see e.g., [3, 4, 11–13]). In general, location problems are given in the following form: Given a graph G with edge set E, and a weight function $w : E \to R_+$, where $w(e)$ is the cost of traversing edge $e \in E$, place a number of facilities in the nodes of G and assign nodes of G to those facilities so as to minimize the sum of distances from each node to the associated facility. As a natural extension, many stochastic location problems have been proposed in the literature (see e.g., [5, 8–10]): they model situations where a facility is not able to satisfy all requests, but can provide a given service only with a certain probability. This seems to be important in many scenarios, and other types of stochastic location problems seem to arise from many new applications as well such as Web Caching (see, e.g., [1]).

In this paper, we consider another variety of stochastic location problem, motivated by applications arising from tele-diagnostics. We are given an undirected graph G, with vertex set $V(G) = V \cup \{\sigma\}$ ($|V(G)| = n + 1$) and edge set $E(G) = E$ ($|E(G)| = m$). Each node $v \in V$ in the network represents a site, and each site contains some physical devices. Such devices can be monitored and possibly repaired from a remote site in case they become faulty. We assume that there can be two kinds of faults in the system: *soft faults*, which can be repaired remotely from another site, and *severe faults* which cannot be repaired remotely and require further (possibly human) interventions. We assume that soft faults, i.e., faults that can be repaired remotely, happen with some fixed probability

J. Hromkovič, M. Nagl, and B. Westfechtel (Eds.): WG 2004, LNCS 3353, pp. 201–213, 2004.

λ, $0 < \lambda \leq 1$. Note that in the special case $\lambda = 1$ the problem is a standard P-median Problem (see, e.g., [6]).

An edge $e \in E$ of the form (x, y) represents a connection of cost $w(x, y) \geq 0$ from site x to site y in G. The graph has a special node $\sigma \in V(G)$, referred to as *supervisor*, whose main task is to supervise all the operations in the network. We would like to locate at most P *monitors* in the network, and assign nodes in the network to monitors so that each monitor will be responsible for the assigned nodes. In more details, if node v is assigned to monitor μ, then μ is able to monitor all devices in v via remote sensors. Whenever a device in v is faulty, then μ detects the fault immediately through those sensors. If the fault in v is soft, then μ can start a remote repair process on v. The cost of this process will be proportional to the weighted distance between v and μ, say $c_{v,\mu}$. Otherwise, if the fault is severe, then μ must transmit a detailed report about the fault to the supervisor σ: in this case, the cost of the process will be proportional to the distance between μ and σ, say $c_{\mu,\sigma}$. Note that in this model only monitors are allowed to communicate with the supervisor. Then the expected cost of the whole process is:

$$\lambda \cdot c_{v,\mu} + (1 - \lambda) \cdot c_{\mu,\sigma}.$$

We observe that this puts already in perspective the main differences between the problem considered in this paper and other (classical) location problems. First of all, there is a probability λ which plays an important role in the cost function. Second, in addition to the cost of assigning a node to a facility (monitor), there is an extra cost of $(1 - \lambda) \cdot c_{\mu,\sigma}$ (due to severe faults), which depends on the location of the facility, i.e., on the distance from the monitor μ to the supervisor σ.

We investigate the problem of locating monitors in the network so as to minimize the total communication costs. Throughout for shortness this problem will be referred to as TDP (Tele-Diagnostic Problem). We formalize TDP as a stochastic location problem, and study structural properties which allow us to locate the boundaries between polynomiality and NP-completeness for this problem. In particular, we show that if $(G, \sigma, c, \lambda, P)$ is a generic instance of our problem then,

(i) if $\deg_G(\sigma) > P$, TDP is NP-Complete, for all $\lambda \in (0, 1]$;
(ii) if $\deg_G(\sigma) \leq P$ then,
  ii.1 if $\lambda \leq 1/2$ TDP is solvable in polynomial time;
  ii.2 if $\lambda - 1/2 = O(n^{-h})$, for some fixed positive integer h TDP is NP-Complete.
(iii) TDP is 4-approximable that is there exists a deterministic polynomial-time algorithm that given $(G, \sigma, c, \lambda, P)$ returns a solution whose value is at most four times the value of the optimal solution of $(G, \sigma, c, \lambda, P)$.

The rest of the paper is organized as follows. In Section 2 we give notations and the problem formulation. Section 3 investigates structural properties of the problem, that lead to the complexity and approximability results in Section 4.

2 Definitions and Notations

We start with some preliminary notation and assumptions. Let $G = (V \cup \{\sigma\}, E)$ be a simple connected undirected graph with $n + 1$ vertices and m edges, where σ is a special node, called *supervisor*. Recall that a graph is simple if there are no parallel edges (different edges with the same endpoints) and it is connected if for every two distinct vertices there exists a path joining them. A path joining u and v in a graph $H = (V', E')$ is a sequence $v_0 e_0 v_1 e_1 v_2 e_2 \ldots v_{k-1} e_{k-1} v_k$, of distinct vertices of V' and distinct edges of E' such that $v_0 = u$, $v_k = v$ and $e_i = (v_i, v_{i+1})$, for $i = 0, \ldots, k$. If a path P joins u and v we say that u and v are the endpoints of P and that P is an uv path. For $v \in V \cup \{\sigma\}$ let $N(v)$ denote the set of nodes of G that are adjacent to v (the set of neighbors of v). A weight function $w : E \to R_+$ is defined, where $w(e)$ is the cost of traversing edge $e \in E$. For a path P in G the length of P is defined as the sum of the weights of the edges of P. For $(i, j) \in (V \cup \{\sigma\}) \times (V \cup \{\sigma\})$ let $c_{i,j}$ be the minimum cost of an ij path. An ij path achieving this minimum cost will be referred to as a shortest ij path. Given two sets A and B, we denote by B^A the sets of all functions from A to B. Recalling that λ is the probability of repairing a fault, for $(i, j) \in V \times V$ we let $b_{i,j}(\lambda) = \lambda \cdot c_{i,j} + (1 - \lambda) \cdot c_{j,\sigma}$. We write B_λ for the corresponding square matrix of order n. B_λ will be referred to as the *expected cost matrix*. Note that matrix B_λ is not in general symmetric but it easy to see that it still satisfies the triangular inequality as c does.

3 Structural Properties

Let Q be a possible candidate set of monitors. We now study some combinatorial properties of the solutions for TDP. For $Q \subseteq V$ let $\Omega(Q)$ be the set of all surjective mappings τ from V to Q fixing all points of Q, i.e.,

$$\Omega(Q) = \{\tau \in Q^V : \tau(V) = Q, \ \tau(i) = i, \forall i \in Q\}.$$

Since each $\tau \in \Omega(Q)$ is onto Q, τ defines a partition $\mathcal{D}_\tau := \{D_i, i \in Q\}$ of V, where $D_i = \{j \in V : \tau(j) = i\}$. Clearly, for each $i \in Q$, $i \in D_i$. If we define

$$z(Q, \tau; \lambda) = \sum_{i \in V} b_{i,\tau(i)}(\lambda)$$

TDP can be formulated as

$$\min_{Q \subseteq V} \{z(Q, \tau; \lambda) : |Q| \leq P, \tau \in \Omega(Q)\}. \tag{1}$$

Let $Q \subseteq V$ and for $i \in V \setminus Q$ let $\pi(i)$ be the node $j \in Q$ which minimizes the distance $b_{i,j}(\lambda)$ for a given i, i.e., $\pi(i)$ is such that $b_{i,\pi(i)}(\lambda) \leq b_{i,j}(\lambda)$ with $j \in Q$. Take $\pi(i) = i$, for $i \in Q$. By construction, $\pi \in \Omega(Q)$. Moreover, as $b_{i,\tau(i)}(\lambda) \geq b_{i,\pi(i)}(\lambda)$ for all $\tau \in \Omega(Q)$, one has

$$z(Q, \pi; \lambda) = \min_{\tau \in \Omega(Q)} z(Q, \tau; \lambda). \tag{2}$$

We call such a π a *good assignment* for Q and we write $z(Q;\lambda)$ for $z(Q,\pi;\lambda)$. After (2) we can write (1) as

$$\min_{Q\subseteq V,|Q|\leq P} z(Q;\lambda). \tag{3}$$

It is convenient to write $z(Q,\tau;\lambda)$ as

$$z(Q,\tau;\lambda) = b(Q,\tau;\lambda) + s(Q;\lambda) \tag{4}$$

where

$$b(Q,\tau;\lambda) = \sum_{i\in V\setminus Q} b_{i,\tau(i)}(\lambda)$$

and

$$s(Q;\lambda) = s(Q,\tau;\lambda) = \sum_{i\in Q} b_{i,i}(\lambda).$$

Note that for $\lambda = 1$ $s(Q;\lambda) = 0$. Clearly,

$$z(Q;\lambda) = b(Q;\lambda) + s(Q;\lambda) = \sum_{i\in V\setminus Q} b_{i,\pi(i)}(\lambda) + \sum_{i\in Q} b_{i,i}(\lambda) \tag{5}$$

where we have set $b(Q;\lambda) := b(Q,\pi;\lambda) = \min_{\tau\in\Omega(Q)} b(Q,\tau;\lambda)$.

Remark 1. Representation (5) (or more in general (4)) closely resembles the combinatorial formulation of a P-median Problem where the set of potential facilities and the set of customers both coincide with V (see for instance [6]). Indeed if the $b_{i,i}(\lambda)$'s were equal to zero for all $i \in V$, (5) would coincide with the objective function of some instance of the standard P-median Problem. As the $b_{i,i}(\lambda)$'s are in general nonzero, (4) and (5) represent $z(.,.;\lambda)$ as the sum of two terms $b(Q;\lambda)$ and $s(Q;\lambda)$. It is worth observing that while $b(Q;\lambda)$ does not increase with $|Q|$, $s(Q;\lambda)$ always increases with $|Q|$. When $b(Q;\lambda)$ decreases with $|Q|$ we may have a trade-off between $b(Q;\lambda)$ and $s(Q;\lambda)$. Furthermore, we observe that $z(.,.;\lambda)$ can be written as

$$z(Q;\lambda) = \lambda \sum_{i\in V\setminus Q} c_{i,\pi(i)} + (1-\lambda)\sum_{i\in Q} |D_i|c_{i,\sigma} \tag{6}$$

which shows that $z(.,.;\lambda)$ is not necessarily monotone in the cardinality of Q.

In the remainder of this section we will show that if $\lambda \leq 1/2$ then the candidate set Q of medians lies entirely in the neighborhood of σ, i.e., $N(\sigma)$. For $Q \subseteq V$ let π be one of its good assignment that is, $(Q,\pi;\lambda) = z(Q;\lambda)$. For $i \in Q$ let $\mu(i)$ be the node immediately preceding σ on a shortest path from i to σ. This correspondence defines a function $\mu : Q \to N(\sigma)$. Let $\overline{Q}$ be the image of Q under μ, i.e., $\overline{Q} = \{\mu(i) : i \in Q\}$. Define $\bar{\pi} \in \Omega(\overline{Q})$ as follows

$$\bar{\pi} = \begin{cases} i & \text{if } i \in \overline{Q} \\ \mu(\pi(i)) & \text{otherwise} \end{cases}$$

For a subset Q of V call the corresponding $\overline{Q}$ the *projection* of Q into $N(\sigma)$. Let $\mathcal{Q}$ be the set of those Q which are not subsets of $N(\sigma)$. We are interested in the following numbers. Let $\lambda(G, w)$ the largest number such that

$$z(Q; \lambda) - z(\overline{Q}; \lambda) \geq 0, \ \forall Q \in \mathcal{Q}, \ \forall \lambda \in [0, \lambda(G, w)]; \tag{7}$$

and let $\xi(G, w)$ be the largest number such that

$$\exists S \subseteq N(\sigma) : z(Q; \lambda) - z(S; \lambda) \geq 0, \ \forall Q \in \mathcal{Q}, \ \lambda \in [0, \xi(G, w)]. \tag{8}$$

Clearly,

$$\lambda(G, w) \leq \xi(G, w), \ \forall w \in R_+^{E(G)}.$$

Moreover, if we define $\lambda(G)$ and $\xi(G)$ respectively as the largest numbers for which (7) and (8) hold for all $w \in R_+^{E(G)}$, we have

$$\inf_{w \in R_+^{E(G)}} \lambda(G, w) = \lambda(G) \leq \xi(G) = \inf_{w \in R_+^{E(G)}} \xi(G, w).$$

Actually, it is worth observing that both λ and ξ depend on c through w and G. The following lemma is crucial to the main result of this section.

Lemma 1. *Let Q be a subset of V and let π, $\overline{Q}$ and $\bar{\pi}$ as above. Then,*

$$z(\overline{Q}, \bar{\pi}; \lambda) - z(Q; \lambda) = (\lambda - 1) \sum_{i \in V} c_{\bar{\pi}(i), \pi(i)} + \lambda \sum_{i \in V} (c_{i, \bar{\pi}(i)} - c_{i, \pi(i)}).$$

Proof. By definition

$$z(\overline{Q}, \bar{\pi}; \lambda) - z(Q, \pi; \lambda) = \sum_{i \in V} (b_{i, \bar{\pi}(i)}(\lambda) - b_{i, \pi(i)}(\lambda))(\lambda).$$

Since $c_{\pi(i), \sigma} - c_{\bar{\pi}(i), \sigma} = c_{\bar{\pi}(i), \pi(i)}$, as by construction $\bar{\pi}(i)$ lies on a shortest $\pi(i)\sigma$ path, one has

$$b_{i, \bar{\pi}(i)}(\lambda) - b_{i, \pi(i)}(\lambda) = (\lambda - 1) \cdot c_{\bar{\pi}(i), \pi(i)} + \lambda \cdot (c_{i, \bar{\pi}(i)} - c_{i, \pi(i)}),$$

hence, the thesis.

Theorem 1. *Let $G = (V \cup \{\sigma\}, E)$ be a graph with the special node σ distinguished. Then $\lambda(G) \geq 1/2$.*

Proof. By Lemma 1, for any given Q, if $\pi \in \Omega(Q)$, $\overline{Q}$ and $\bar{\pi} \in \Omega(\overline{Q})$ are determined by Q as in Lemma 1, the number

$$\lambda(Q, w) := \frac{\sum_{i \in V} c_{\bar{\pi}(i), \pi(i)}}{\sum_{i \in V} c_{\bar{\pi}(i), \pi(i)} + \sum_{i \in V} (c_{i, \bar{\pi}(i)} - c_{i, \pi(i)})} \tag{9}$$

is such that, if $\lambda \leq \lambda(Q, w)$ then, $(Q; \lambda) \geq z(\overline{Q}, \bar{\pi}; \lambda)$. Since by (2) $z(\overline{Q}, \bar{\pi}; \lambda) \geq z(\overline{Q}; \lambda)$, if we let

$$\varphi(G) := \inf_{w \in R_+^{E(G)}} \min_{Q \in \mathcal{Q}} \lambda(Q, w)$$

it follows that, $\varphi(G)$ is the largest number such that for $\lambda \leq \varphi(G)$ replacing (Q, π) by $(\bar{Q}, \bar{\pi})$ we get a not worse solution. Thus, $\varphi(G) \leq \lambda(G)$ and for all $Q \subseteq V$

$$\lambda \leq \varphi(G) \leq \lambda(G) \Rightarrow z(Q; \lambda) \geq z(\bar{Q}; \lambda).$$

Recall that $c_{i,j}$ equals the length of a ij shortest path. Hence, c satisfies the triangular inequality, namely $c_{h,j} \leq c_{h,i} + c_{i,j}$ holds for any three nodes $h, i, j \in V(G)$. Therefore, by splitting off $\sum_{i \in V}(c_{i,\bar{\pi}(i)} - c_{i,\pi(i)})$ we have

$$\sum_{i \in V}(c_{i,\bar{\pi}(i)} - c_{i,\pi(i)}) = \sum_{i \in \bar{Q}}(-c_{\bar{\pi}(i),\pi(i)}) + \sum_{i \in V \setminus \bar{Q}}(c_{i,\bar{\pi}(i)} - c_{i,\pi(i)})$$

as, for $i \in \bar{Q}$, $c_{i,\bar{\pi}(i)} - c_{i,\pi(i)} = -c_{\bar{\pi}(i),\pi(i)}$, since $c_{\bar{\pi}(i),\bar{\pi}(i)}$ vanishes over $\bar{Q}$. Moreover, by triangular inequality

$$\sum_{i \in V \setminus \bar{Q}}(c_{i,\bar{\pi}(i)} - c_{i,\pi(i)}) \leq \sum_{i \in V \setminus \bar{Q}} c_{\bar{\pi}(i),\pi(i)}. \tag{10}$$

Hence,

$$\sum_{i \in V}(c_{i,\bar{\pi}(i)} - c_{i,\pi(i)}) \leq \sum_{i \in \bar{Q}}(-c_{\bar{\pi}(i),\pi(i)}) + \sum_{i \in V \setminus \bar{Q}}(c_{\bar{\pi}(i),\pi(i)}). \tag{11}$$

By (9) and (11) it follows that

$$\lambda(Q, w) \geq \frac{\sum_{i \in V} c_{\bar{\pi}(i),\pi(i)}}{2 \sum_{i \in V \setminus \bar{Q}} c_{\bar{\pi}(i),\pi(i)}} = \frac{1}{2} + \frac{\sum_{i \in \bar{Q}} c_{\bar{\pi}(i),\pi(i)}}{2 \sum_{i \in V \setminus \bar{Q}} c_{\bar{\pi}(i),\pi(i)}}$$

and thus,

$$\lambda(G) \geq \varphi(G) \geq \frac{1}{2} + \inf_{w} \min_{Q} \frac{\sum_{i \in \bar{Q}} c_{\bar{\pi}(i),\pi(i)}}{2 \sum_{i \in V \setminus \bar{Q}} c_{\bar{\pi}(i),\pi(i)}}. \tag{12}$$

As an immediate consequence we have the following result.

Corollary 1. *Let $G = (V \cup \{\sigma\}, E)$ be a graph with the special node σ distinguished. Then for $\lambda \leq 1/2$ optimal solutions to any instance of TDP are entirely contained in $N_G(\sigma)$, for any given number P of available monitors.*

Remark 2. Referring to the proof of Theorem 1, we note that the weaker is the triangular inequality (10) the closer is $\lambda(G)$ to its lower bound $\varphi(G)$. For instance, take as $G(V \cup \{\sigma\}, E)$ a path on $n + 1$ nodes. Let σ be one of its end-nodes and let a be its unique neighbor. Let b be any other node of $V \setminus \{\sigma, a\}$. If we let $Q = \{b\}$, we have $\bar{Q} = \{a\}$. Moreover, for $i \in V \setminus \{\sigma\}$ if $\pi(i) = b$ then, $\bar{\pi}(i) = a$. Furthermore, if b is the other neighbor of a, (11) is satisfied with equality. It follows that, for all $w \in R_+^{E(G)}$, and for $Q = \{b\}$, where b is the other neighbor of a

$$\lambda(Q, w) \leq \frac{1}{2} + \frac{c_{a,b}}{2(n-1)c_{a,b}} \quad \text{hence,} \quad \varphi(G) \leq \frac{1}{2} + \frac{1}{2(n-1)}.$$

On the other hand, by (12), for all $i \in V \setminus \{\sigma, a\}$

$$\varphi \geq \frac{1}{2} + \frac{c_{i,a}}{2(n-1)c_{i,a}} = \frac{1}{2} + \frac{1}{2(n-1)} \quad \text{therefore,} \quad \lambda(G) = \frac{1}{2} + \frac{1}{2(n-1)},$$

since there exists exactly one $\bar{\pi} \in \Omega(\overline{Q})$ for each $Q \in \mathcal{Q}$.

4 Complexity Issues

Let $\{c_{i,j}\}_{i \in V(G), j \in V(G)}$ be the symmetric cost function matrix obtained by means of w, and, as before, throughout we let $b_{i,j}(\lambda) = \lambda c_{i,j} + (1-\lambda)c_{j,\sigma}$. We need the following two lemmas.

Lemma 2. *Let G and B_λ be defined as above and let u and v be nodes of $N(\sigma)$. Then, for all $\lambda \in (0,1)$*

$$b_{u,v}(\lambda) < b_{u,u}(\lambda) \Rightarrow b_{i,v}(\lambda) < b_{i,u}(\lambda) \text{ for all } i \in V$$

Proof. Suppose not. Hence, there exists an $i \in V$ such that $b_{i,u}(\lambda) \leq b_{i,v}(\lambda)$. It follows that $\lambda \cdot c_{i,u} + (1-\lambda) \cdot c_{u,\sigma} \leq \lambda \cdot c_{i,v} + (1-\lambda) \cdot c_{v,\sigma}$ which in turn implies

$$(1-\lambda) \cdot (c_{u,\sigma} - c_{v,\sigma}) \leq \lambda \cdot (c_{i,v} - c_{i,u}) \leq \lambda \cdot c_{u,v} \tag{13}$$

where the rightmost inequality holds due to $c_{i,u} + c_{u,v} \leq c_{i,v}$. On the other hand, since $b_{u,v}(\lambda) < b_{u,u}(\lambda)$, we have $\lambda \cdot c_{u,v} + (1-\lambda) \cdot c_{v,\sigma} < (1-\lambda) \cdot c_{u,\sigma}$ and, hence,

$$\lambda \cdot c_{u,v} < (1-\lambda) \cdot (c_{u,\sigma} - c_{v,\sigma})$$

which contradicts (13). $\qquad \square$

Lemma 3. *Let G be a connected graph of order n and let $w \in R_+^{E(G)}$ be a weight function. Then for any $\delta > 0$ there exists a graph G_δ with order q_δ depending on δ along with a weight function $\bar{w}$ with the following properties*

(a) *G_δ has a special node σ such that G is a subgraph induced by a subset of nodes in $V(G_\delta) \setminus (N(\sigma) \cup \{\sigma\})$;*
(b) *$\bar{w}$ extends w over $E(G_\delta)$;*
(c) *$\xi(G_\delta, \bar{w}) \leq 1/2 + \delta$.*

Proof. We may suppose, possibly after scaling, that $1 > \text{diam}(G) := \max_{ij} c_{ij}$. Let $\delta > 0$ be an arbitrarily small number. Throughout we write q for q_δ. Let $n_1 = \lceil \frac{n}{2\delta} \rceil$. Remark that if $\delta = O(n^{-h})$, for some positive integer h then, $n_1 = O(n^{h-1})$. Construct G_δ as follows: take a set V_1 disjoint from $V(G)$ with n_1 points. Let σ and σ_1 be any two points not in $V(G) \cup V_1$; join each node of V_1 to each node of $V(G)$, join σ_1 to each node of $V(G)$ and join σ to σ_1. Moreover, join two points of $V(G)$ if and only if they are joined in G. Now G_δ fulfils (a) with $q = n + n_1 + 2$. Let $\alpha > 1$ be a real number and denote by $\bar{w}(e)$ the length of an edge e. For $e \in E(G)$, let $\bar{w}(e) = w(e)$. For $e \in E(G_\delta) \setminus E(G)$, let $\bar{w}(e) = 1/2$ if e is of the form $e = (\sigma_1, v)$, with $v \in V(G)$. Take $\bar{w}(e) = \alpha$, otherwise. Then, clearly $\bar{w}$ extends w over $E(G_\delta)$. To show (c), we have to exhibit a solution $Q \subseteq V(G_\delta)$ not containing σ_1 such that if $\lambda \geq 1/2 + \delta$ then, $z(Q; \lambda) - z(\sigma_1; \lambda) < 0$. Indeed, since $\xi(G_\delta, \bar{w})$ is by definition the largest number such that there exists an optimal solution to $(G_\delta, P, \bar{w}, \lambda, \sigma)$ contained in $N(\sigma)$, if $\lambda \geq 1/2 + \delta$ implies $z(Q; \lambda) - z(\sigma_1; \lambda) < 0$ then, $\xi(G_\delta, \bar{w}) \leq 1/2 + \delta$. First of all observe that for each $\lambda \in [0, 1]$ no optimal solution can contain any node $v \in V_1$, otherwise replacing such a node v by any of its neighbors would lead to a strictly better solution. Now let us compare a solution $Q \subseteq V(G)$ with σ_1. It is not hard to see that

$$z(\sigma_1; \lambda) = (1 - \lambda) \cdot \alpha \cdot (q - 1) + n_1 \cdot \lambda \cdot \left(\alpha + \frac{1}{2}\right) + n \cdot \frac{\lambda}{2}$$

and, since in $V(G)$ nodes outside Q are at distance at most $\mathrm{diam}(G) < 1$, one has

$$z(Q; \lambda) \leq (1 - \lambda) \cdot \left(\alpha + \frac{1}{2}\right) \cdot (q - 1) + n_1 \cdot \lambda \cdot \alpha + \frac{\lambda}{2} + \lambda \cdot (n - |Q|) \leq$$

$$(1 - \lambda) \cdot \left(\alpha + \frac{1}{2}\right) \cdot (q - 1) + n_1 \cdot \lambda \cdot \alpha + \frac{\lambda}{2} + \lambda \cdot (n - 1).$$

It follows that

$$z(Q; \lambda) - z(\sigma_1; \lambda) \leq (1 - \lambda) \cdot \frac{1}{2} \cdot (q - 1) - n_1 \cdot \lambda \cdot \frac{1}{2} + (n - 1) \cdot \lambda \cdot \frac{1}{2}.$$

Therefore,

$$\lambda > \frac{q - 1}{q + n_1 - n} = \frac{q - 1}{2(q - n) + 2} \Rightarrow z(Q; \lambda) - z(\sigma_1; \lambda) < 0.$$

As for $n_1 = \lceil \frac{n}{2\delta} \rceil$ one has

$$\frac{q - 1}{2(q - n) + 2} \leq \frac{q - 1}{2(q - n)} \leq \frac{q}{2(q - n)} \leq \frac{1}{2} + \delta$$

and (c) follows.

Recall that an instance of the P-median Problem is a quintuplet (G, I, J, c, P) where:

– G is a connected graph;
– I, J are subsets of $V(G)$: I is the set of potential facilities while J is the set of clients;
– P is the maximum number of available facilities;
– c is a function from $V(G) \times V(G)$ into the set of nonnegative reals: $c_{i,j}$ represents the cost of connecting $j \in J$ to $i \in I$.

Given an instance of the P-median Problem one seeks a pair (Q, ψ) where $Q \subseteq I$ is a set of at most P facilities and $\psi : J \to Q$ is the assignment function such that $\sum_{j \in J} c_{j,\psi(j)}$ is minimized. It is well known that the P-median Problem is NP-Complete even when $I = J = V$ [7]. In this case we write (G, c, P) for the generic instance of such a P-median Problem.

Theorem 2. *Let $(G, \sigma, c, \lambda, P)$ be the generic instance of TDP then,*

   i) *if $\deg_G(\sigma) > P$, TDP is NP-Complete, for all $\lambda \in (0, 1]$;*
  ii) *if $\deg_G(\sigma) \leq P$ then,*
     ii.1) *if $\lambda \leq 1/2$ TDP is solvable in polynomial time;*
     ii.2) *if $\lambda \geq 1/2 + O(n^{-h})$, for some fixed positive integer h TDP is NP-Complete.*

Proof. i). If $\lambda = 0$ TDP is clearly polynomial solvable. To see this observe that (6), reduces to $\sum_{i \in Q} |D_i| c_{i,\sigma}$, which is minimized by the point $i \in V$ closest to σ. So we may assume $\lambda > 0$. Let (G, c, P) an instance of the P-median Problem. Add to $V(G)$ an extra node σ and join it with all the other nodes by edges having length $\mathrm{diam}(G) + 1$. By (6) z can be written as

$$z(Q; \lambda) = \lambda \sum_{i \in V \setminus Q} c_{i,\pi(i)} + (1 - \lambda) \sum_{i \in Q} |D_i| c_{i,\sigma} \tag{14}$$

for any $\lambda \in (0, 1]$. Since $c_{i,\sigma}$ is constant over $V(G)$, $z(., .; \lambda)$ attains its minimum when $f(Q; \lambda) = \lambda \sum_{i \in V \setminus Q} c_{i,\pi(i)}$ attains its minimum. Recall the definition of $\pi \in \Omega(Q)$. Moreover, since each path traversing σ has a too large length, no such a path is used to reach any facility. Hence, for all $\lambda \in (0, 1]$, $f(.; \lambda)$ is precisely the objective function of a standard P-median Problem on $(G, \lambda c, P)$ and $P < n =$ order of G.
ii) Suppose $\deg_G(\sigma) \leq P$.
ii.1) The problem reduces to minimum cost bipartite assignment problem.
ii.2) Let (G, c, P) an instance of the P-median Problem. Let $\delta = O(n^{-h})$ and take G_δ as in Lemma 3. Since $\delta = O(n^{-h})$, G_δ has size and order bounded by a polynomial in the size and the order of G. So it can be constructed in polynomial time from G. By Lemma 3 optimal solutions to TDP on $(G_\delta, \sigma, c, \lambda, P)$ are attained over $V(G)$. As in the proof of i), $z(., .; \lambda)$ attains the minimum if $f(.; \lambda)$ attains the minimum over $V(G)$. Again the minimization of $z(., .; \lambda)$ is equivalent to the minimization of $f(.; \lambda)$ on $(G, \lambda c, P)$.

Now we are going to investigate the approximability of this problem. To this end let us look more closely at the behaviour of the objective function $z(., .; \lambda)$. We know that $z(., .; \lambda)$ is not in general monotone. However, Lemma 2 allows us to write down some domination rules that help either in solving TDP or in approximating its solution.

Corollary 2. To Lemma 2. *Let $G = (V \cup \{\sigma\}, E)$, c be as in previous sections. For $\lambda \in [0, 1]$ let B^λ be the associated expected cost function. If $b_{i,i}(\lambda) > b_{i,j}(\lambda)$, for some i and j in V, then any optimal solution to TDP never contains i.*

Proof. Let Q be any solution to TDP and $\mathcal{D}_\pi = \{D_h, h \in Q\}$ be the associated partition where $\pi \in \Omega(Q)$ is a good assignment for Q. By Lemma 2 $b_{h,i}(\lambda) > b_{h,j}(\lambda)$ holds for all $h \in V \setminus \{i\}$, i.e., elements in column i are component-wise larger than those of column j. Suppose that $i \in Q$ and let k be such that $b_{i,k}(\lambda) \leq b_{i,h}(\lambda)$ for all $h \in V \setminus \{i\}$. By hypothesis $k \neq i$ and $b_{i,k}(\lambda) < b_{i,i}(\lambda)$.

Let $Q' := Q \cup \{k\} \setminus \{i\}$ and define τ as follows: if $\pi(h) \neq i$ then, $\tau(h) = \pi(h)$. If $\pi(h) = i$ let $\tau(h) = k$. Now, since if $h \notin D_i$ we have that $\pi(h) = \tau(h)$, and if $h \in D_i$ we have that $\pi(h) = i$ and $\tau(h) = k$, one has

$$z(Q; \lambda) - z(Q'; \lambda) \geq z(Q; \lambda) - z(Q', \tau; \lambda) = \sum_{h \in D_i} (b_{h,i}(\lambda) - b_{h,k}(\lambda)) > 0$$

Hence, Q cannot be optimal.

The above corollary allows us to slightly strengthen ii.1) of Theorem 2 via the next proposition. For $\lambda \in [0,1]$, call a node $i \in V$ *dominated* if there exists a node $j \in V \setminus \{i\}$ such that $b_{i,i}(\lambda) > b_{i,j}(\lambda)$. In other words, i is dominated if the minimum in i-th row of B_λ is attained by an off-diagonal element. Let U_λ be the set of non-dominated elements.

Remark 3. Let α, $\alpha:V \to Q$, be the *closest point mapping*, namely for all $i \in V$, $\alpha(i)$ is the node of Q such that $b_{i,\alpha(i)}(\lambda) \leq b_{i,j}(\lambda) \ \forall j \in Q$. In general, for $Q \subseteq V$, $\alpha \notin \Omega(Q)$. Indeed if Q contains some dominated point i, $\alpha(i) \neq i$. However, Corollary 2 states that if $Q \subseteq U_\lambda$, $\alpha \in \Omega(Q)$. Moreover, α is a good assignment π for all $Q \subseteq U_\lambda$. It follows that

$$z(Q; \lambda) = \sum_{i \in V} \min_{j \in Q} b_{i,j}(\lambda), \ \forall Q \subseteq U_\lambda. \tag{15}$$

Proposition 1. *For all $\lambda \in [0,1]$, U_λ is non-empty. In particular, if $\lambda \geq 1/2$, $U_\lambda = V$. Hence for all $\lambda \in [0,1]$, optimal solutions to TDP are contained in U_λ.*

Proof. Observe first that $b_{i,h}(\lambda) \geq b_{i,i}(\lambda)$ for all $h \in V$, i.e., the minimum on the i-th column is attained over a diagonal element. Indeed,

$$b_{h,i}(\lambda) = \lambda \cdot c_{h,i} + (1-\lambda) \cdot c_{i,\sigma} \geq (1-\lambda) \cdot c_{i,\sigma} = b_{i,i}(\lambda).$$

Moreover, for $i \neq j$ and $\lambda \geq \frac{1}{2}$,

$$
\begin{aligned}
b_{i,j}(\lambda) = \lambda \cdot c_{i,j} + (1-\lambda) \cdot c_{j,\sigma} &= (1-\lambda) \cdot (c_{i,j} + c_{j,\sigma}) + (2\lambda - 1) \cdot c_{i,j} \\
\text{by triangular inequality} &\geq (1-\lambda) \cdot c_{i,\sigma} + (2\lambda - 1) \cdot c_{i,j} \\
\text{since } \lambda \geq \tfrac{1}{2} \quad &\geq (1-\lambda) \cdot c_{i,\sigma} \\
&= b_{i,i}(\lambda).
\end{aligned}
$$

Hence, for $\lambda \geq \frac{1}{2}$, the minimum of each row of B_λ is attained over a diagonal element. So $U_\lambda = V$. It follows that U_λ is non-empty, for all $\lambda \in [0,1]$. By Corollary 2 optimal solutions to TDP must be contained in U_λ.

In view of Proposition 1 we can write (3) as

$$\min\{z(Q; \lambda) | Q \subseteq U_\lambda, |Q| \leq P\} \tag{16}$$

hence, we have,

Corollary 3. To ii.1 of Theorem 2. *Let $(G, \sigma, c, \lambda, P)$ be the generic instance of TDP. Let $k = |N_G(\sigma) \cap U_\lambda|$. Then if $P \geq k$ the problem is polynomial-time solvable.*

Proof. The proof follows by replacing $\deg_G(\sigma)$ with k in ii.1 of Theorem 2.

Let us consider the restriction v of z defined by (5) (or (2)) on the set U_λ. Denote this restriction by v. By definition $v: 2^{U_\lambda} \to R_+$, and $v(Q) = z(Q; \lambda)$ for all $Q \subseteq U_\lambda$, so

$$\min\{z(Q; \lambda) | Q \subseteq U_\lambda, |Q| \leq P\} = \min\{v(Q) | Q \subseteq U_\lambda, |Q| \leq P\}, \quad (17)$$

and, again by Proposition 1 we have

Corollary 4. *For $\lambda \in [0, 1]$, let U_λ and v be as above. Then v is a monotone non increasing set function on the set of all subsets of U_λ.*

Proof. Over 2^{U_λ}, $z(., .; \lambda)$ and v coincide. Hence v can be represented by (15), and this function is clearly non increasing.

Corollary 4, states via (16) and (17), that solving TDP is essentially the same as solving a deterministic P-median Problem (with set I of potential facilities identical to U_λ and set J of customers identical to V), up (a) and (b) below:

(a) B_λ is not in general symmetric;
(b) diagonal elements of B_λ are in general non-zero.

Given a minimization problem **P** with optimum value $OPT(I)$ and a non-negative real number $\alpha \geq 1$, an α-approximation algorithm for **P** is an algorithm that given any instance I of **P** returns a solution whose value is at most $\alpha OPT(I)$. The number $\alpha - 1$ is said to be the relative error of the approximation. In view of the above (a) and (b), we can not expect, in general, that an α-approximations algorithm for the P-median Problem directly translates into an α-approximation for TDP. Nevertheless the following result shows that a reduction preserving approximation can be given. An instance (G, I, J, c, P) of the P-median Problem is said to be metric if the connection costs satisfy the Triangular Inequality. Since there exists a 4-approximation algorithm (see [2, 6]) for the class of metric instances of the P-median Problem, from Theorem 3 below, it follows that an 4-approximation algorithm for TDP exists as well. As we have seen given an instance (G, I, J, c, P) of the P-median Problem one seeks a pair (Q, ψ) where $Q \subseteq I$ is a set of at most P facilities and $\psi : J \to Q$ is the assignment function such that $\sum_{i \in J} c_{i, \psi(i)}$ is minimized. Clearly, as for every $Q \subset I$ one has $\psi(i) \in Q$ for all $i \in J$ and

$$\sum_{i \in J} \min_{h \in Q} c_{i_h} \leq \sum_{i \in J} c_{i, \psi(i)},$$

an optimal set of facilities is a solution to the following problem

$$\min_{Q \subseteq I, |Q| \leq k} \sum_{i \in J} \min_{h \in Q} c_{i,h}. \quad (18)$$

Given an instance $(G, \sigma, P, w, \lambda)$ of TDP one seeks a pair (Q, τ) where Q is a subset of at most P vertices in V and $\tau : V \to Q$ is the assignment function such that

$$\sum_{i \in V(G) \setminus \{\sigma\}} \lambda c_{i,\tau(i)} + (1 - \lambda) c_{\tau(i),\sigma}$$

is minimized. We have already shown (see Lemma 1) that for every λ there exists a set $U_\lambda \subseteq V$ such that optimal solutions to TDP are entirely contained in U_λ. Moreover, by (15) and Corollary 4, TDP can be re-stated as:

$$\min_{Q \subseteq U_\lambda, |Q| \le k} \sum_{i \in V(G) \setminus \{\sigma\}} \min_{i \in Q} b_{i,j}(\lambda),$$

where we have set $b_{i,j}(\lambda) = \lambda c_{j,i} + (1 - \lambda) c_{i,\sigma}$.

Theorem 3. *The TDP Problem can be reduced to a Metric P-median and the reduction preserves approximations.*

Proof. Recall that an n-wheel is a graph on $n + 1$ nodes obtained from a simple cycle on n vertices by joining each vertex along the cycle to an extra vertex called the *center*. Let $(G, \sigma, P, w, \lambda)$ be an instance of TDP and c be the associated cost function. Let $V(G) = V \cup \{\sigma\}$ where $V = \{1, 2 \ldots, n\}$, and let U_λ be as above and $t = |U_\lambda|$. For $i \in U_\lambda$ let W_i be an n-wheel with center σ_i. Let the other vertices be labelled as $a_{i,j}, j = 1, 2, \ldots, n$ (see Figure 1). For $i = 1, \ldots, t$, W_i is weighted on the edges as follows: edges of the form $(\sigma_i, a_{i,j})$ have weight $(1 - \lambda) c_{i,\sigma}$, all the other edges have weight zero. Now for $j \in V$, join j to $a_{i,j}, i = 1, \ldots, t$, by an edge of weight $\lambda c_{i,j}$. Let $\tilde{G}$ be the weighted graph arising in this way. Clearly for $i \in U_\lambda$, $b_{i,j}(\lambda)$ equals the length of a shortest path in $\tilde{G}$ between j and σ_i. Set $I = \{\sigma_1, \ldots, \sigma_t\}$, $J = V$ and define $\tilde{c} : V(\tilde{G}) \times V(\tilde{G}) \to R^+$ as the shortest

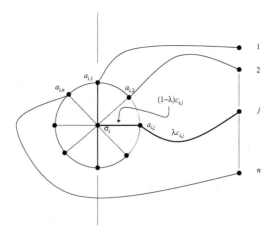

Fig. 1. The i-th wheel along with connections to the vertices in V.

path function in $\tilde{G}$ w.r.t to the weight given on the edges of $\tilde{G}$. Clearly $\tilde{c}$ satisfies the Triangular Inequality. Moreover,

$$\min_{Q \subseteq U_\lambda, |Q| \leq P} \sum_{i \in V} \min_{i \in Q} b_{i,j}(\lambda) = \min_{Q \subseteq I, |Q| \leq P} \sum_{j \in V} \min_{h \in Q} \tilde{c}_{h,j}.$$

It follows that optimal solutions to the P-median instance $(\tilde{G}, I, J, P, \tilde{c})$ defined above are optimal to TDP as well and conversely.

It follows by Theorem 3 and by the 4-approximation algorithm in [2] for the metric P-median Problem, that there exists a 4-approximation algorithm for TDP as well.

References

1. M. Abrams, C. R. Standridge, G. Abdulla, S. Williams, and E. A. Fox. Caching proxies: Limitations and potentials. In Proceedings of the Fourth International World Wide Web Conference, pages 119-133, Boston, MA, December 1995.
2. M. Charikar and S. Guha. Improved combinatorial algorithms for the facility location and k-median problems. In Proceedings of the 40th Annual Symposium on Foundations of Computer Science. IEEE Computer Society Press, Los Alamitos, Calif., 378-388, 1999.
3. H. W. Hamacher and S. Nickel. Classification of location models. Location Science, 6:229-242, 1998.
4. G. Y. Handler and P. B. Mirchandani, 1979, Location on Networks: Theory and Algorithms, M.I.T. Press, Cambridge, MA.
5. J. V. Jucker and R. C. Carlson. The simple plant-location problem under uncertainty. Operations Res, 24:1045-1055, 1977.
6. J.Kamal and V.V. Vazirani. Approximation algorithms for metric facility location and k-Median problems using the primal-dual schema and Lagrangian relaxation. J. ACM 48(2):274-296, 2001.
7. O. Kavir and S.L. Hakimi. The p-median problems. In: An Algorithmic Approach to Network Location Problems. SIAM Journal on Applied Mathematics, Philadelphia, 37, 539-560, 1979.
8. G. Laporte, F. V. Louveaux, and H. Mercure. Models and exact solutions for a class of stochastic location-routing problems. Eur. J. Oper. Res, 39(1):71-78, 1989.
9. F. V. Louveaux. Discrete stochastic location models. Annals of Operations Research, 6(4):23-34, 1986.
10. F. V. Louveaux and D. Peeters. A dual-based procedure for stochastic facility location. Oper. Res., 40(3):564-573, 1992.
11. R. F. Love, J. G. Morris and G. O. Wesolowsky, 1988, Facilities Location: Models and Methods, North Holland, New York.
12. P. B. Mirchandani, and R. L. Francis, 1990, Discrete Location Theory, John Wiley and Sons, Inc., New York.
13. S. Nickel. Discrete and Network Location Theory. Lecture Notes, Fachbereich Mathematik, Universität Kaiserslautern, 1999.

A Robust PTAS for Maximum Weight Independent Sets in Unit Disk Graphs

Tim Nieberg*, Johann Hurink, and Walter Kern

University of Twente
Faculty of Electrical Engineering, Mathematics & Computer Science
Postbus 217, NL-7500 AE Enschede
{T.Nieberg,J.L.Hurink,W.Kern}@utwente.nl

Abstract. A unit disk graph is the intersection graph of unit disks in the euclidean plane. We present a polynomial-time approximation scheme for the maximum weight independent set problem in unit disk graphs. In contrast to previously known approximation schemes, our approach does not require a geometric representation (specifying the coordinates of the disk centers).

The approximation algorithm presented is robust in the sense that it accepts any graph as input and either returns a $(1 + \varepsilon)$-approximate independent set or a certificate showing that the input graph is no unit disk graph. The algorithm can easily be extended to other families of intersection graphs of geometric objects.

1 Introduction

A unit disk graph (UDG) is the intersection graph of unit disks in the plane. In other words, $G = (V, E)$ is a UDG if there exists a map $f : V \to \mathbb{R}^2$ (a *geometric representation*) satisfying

$$(u, v) \in E \iff \|f(u) - f(v)\| \le 2, \tag{1}$$

where $\|.\|$ denotes the euclidean norm.

A subset of vertices in G is called independent if the vertices in this subset are pairwise not connected by an edge. The maximum independent set problem now consists of finding a such an independent subset of the vertices of maximum cardinality. For the maximum weight independent set problem, each vertex $v \in V$ is also assigned a weight $w_v > 0$, and the goal is to find an independent set of maximal total weight, i.e. of maximum sum of all weights of the vertices in the independent set.

In this document, we give a *polynomial-time approximation scheme* (PTAS) for the maximum (weight) independent set problem in unit disk graphs for the case that a geometric representation is not given, i.e. we seek for an algorithm which, given as input a UDG $G = (V, E)$ and a parameter $\varepsilon > 0$, computes an

* This work is partially supported by the European research project EYES (IST-2001-34734).

J. Hromkovič, M. Nagl, and B. Westfechtel (Eds.): WG 2004, LNCS 3353, pp. 214–221, 2004.

independent set $I \subset V$ of size (weight) at least $(1 + \varepsilon)^{-1}$ times the maximum size (weight) of an independent set in G. The running time of the algorithm is allowed to depend on ε, but should be polynomial in $n = |V|$ for fixed $\varepsilon > 0$.

Most of the work concerning approximation schemes in unit disk graphs has been done assuming a given geometric representation. The representation makes it possible to perform separation of the graph alongside a grid. This so called *shifting strategy* is presented in [1] and [7]. Combined with a dynamic programming approach, the shifting strategy is used by Erlebach et. al. [5] to give a PTAS for the maximum weight independent set in disk graphs of arbitrary diameter. Using quadtrees as separation, the runtime has been improved by Chan [3] to $n^{O(1/\varepsilon^{d-1})}$, where d gives the dimension of the euclidean space. The shifting strategy can also be applied to related problems like minimum vertex cover, and minimum dominating set [8]. Also, the minimum connected dominating set problem can be approximated by a PTAS with the help of separation using a grid structure [4].

The case where no geometric representation f for the UDG $G = (V, E)$ is available is significantly different: Computing a corresponding representation function f for a given UDG $G = (V, E)$ is NP-hard. Indeed, any polynomial time algorithm computing geometric representation functions for unit disk graphs could be used in a straightforward way to solve the UDG recognition problem (determine whether a given graph is a UDG), which is known to be NP-hard [2].

Without given geometric representation, a PTAS for the maximum (weight) independent set problem was not known previously. However, constant factor approximation algorithms have been given in the literature. A simple greedy strategy gives a 5-approximation for the maximum weight independent set, and a more sophisticated choice of a node to be greedily added to the partial set of independent nodes gives an approximation within a factor of 3 for the unweighted case [10]. Both algorithms work without given representation, however, the running time can be improved a lot when the representation is given.

Unit disk graphs form a subclass of the more general class of all undirected graphs. This raises the question of robustness (see [12]) for algorithms designed for the restricted domain of UDGs. Generally speaking, a robust algorithm $\mathcal{A}$ on a restricted class $\mathcal{U} \subset \mathcal{G}$ solves a problem for all instances in $\mathcal{U}$, but also accepts any instance in $\mathcal{G}$. For instances in $\mathcal{G} \setminus \mathcal{U}$, the algorithm $\mathcal{A}$ either solves this problem or provides a certificate showing that the input is not in $\mathcal{U}$. For the PTAS presented in this paper, the algorithm accepts any graph as input, e.g. given by an adjacency list or matrix, and either returns a $(1 + \varepsilon)$-approximate (weighted) independent set or a certificate to show that the graph does not belong to the class of unit disk graphs. In case the input graph is a unit disk graph, the algorithm always returns an independent set of desired quality.

The problem of finding a maximum (weight) independent set arises for example in the context of clustering in wireless ad-hoc networks [11]. The battery-operated nodes, each equipped with a radio transceiver of fixed transmission range, form a UDG representing the communication network. Nodes of a maximum independent set can be used as controlling instances, or clusterheads, of the other nodes within range of their radio. A maximum independent set also forms a dominating set. In the ad-hoc scenario, it is hard or costly to determine the exact position of each node and therefore for the resulting network only a representation by the nodes and communication links between them is known. The weights of those nodes may correspond to the residual energy in order to make nodes with more battery power clusterheads to prolong the operational lifetime of the network.

The remainder of this paper is organized as follows. The next section introduces the algorithm that gives the PTAS for the unweighted case of finding an independent set of maximum cardinality in UDGs without using a geometric representation. Section 3 then gives the modifications of the algorithm to efficiently approximate the maximum weight independent set. The algorithms are presented with the assumption (or "promise") that the input is a UDG. In Section 4, we show how to obtain a robust algorithm from the PTAS. In Section 5, we identify some other classes of geometric intersection graphs for which our approach also is efficient. The paper ends with a short conclusion and an outlook on future work.

2   The Approximation Algorithm

In this section, we introduce the approximation algorithm that forms the core of the robust PTAS for the maximum independent set problem on UDGs. The same algorithm is then adapted to the weighted version of the problem in Section 3.

The algorithm does not rely on a geometric representation, it thus accepts any graph as an input-instance. However, the statements concerning the running time depend on the assumption that the graph has a geometric representation.

Let $\varepsilon > 0$ and let $\rho := 1 + \varepsilon$ denote the desired approximation guarantee. Thus, given a unit disk graph $G = (V, E)$, we seek to construct an independent set $I \subset V$ of cardinality at least ρ^{-1} times $\alpha(G)$, the maximum size of an independent set in G.

The basic idea is simple. We start with an arbitrary node $v \in V$ and consider for $r = 0, 1, 2, \ldots$, the rth neighborhood

$$N^r = N^r(v) := \{w \in V \,|\, w \text{ has distance at most } r \text{ from } v\}.$$

Starting with N^0, we compute a maximum independent set $I_r \subset N^r$ for each $r = 0, 1, 2, \ldots$ as long as

$$|I_{r+1}| > \rho |I_r| \tag{2}$$

holds.

Let $\bar{r}$ denote the smallest $r \geq 0$ for which (2) is violated. Such an $\bar{r} \geq 0$ indeed exists and it is bounded by a constant (depending on ρ):

Lemma 1. *There exists a constant $c = c(\rho)$ such that $\bar{r} \leq c$.*

Proof. From (1), we conclude that any $w \in N^r$ satisfies

$$\|f(v) - f(w)\| \leq 2r.$$

So, the unit disks corresponding to nodes in I_r are pairwise disjoint and are all contained in a disk of radius $R = 2r + 1$ around $f(v)$. This implies

$$|I_r| \leq \pi R^2 / \pi = O(r^2). \tag{3}$$

On the other hand, by definition of $\bar{r}$, we have for $r < \bar{r}$

$$|I_r| > \rho|I_{r-1}| > \ldots > \rho^r|I_0| = \rho^r. \tag{4}$$

Comparing (3) and (4), the claim follows.   □

To achieve an independent set for the graph G, the above algorithm is iteratively applied to the graph $G' := G \setminus N^{\bar{r}+1}$, and the resulting independent set for G' is combined with $I_{\bar{r}}$ for an independent set in G. Note that, due to (3), we may compute I_r by complete enumeration in time $O(n^{C^2})$, where $C = O(r) = O(1/\varepsilon^2 \log 1/\varepsilon)$ for $r \leq \bar{r}$ (see Appendix). The algorithm evolving from the above description thus runs in polynomial time, as all other computations are dominated by this complexity. The correctness and approximation guarantee of the algorithm follows from the following theorem.

Theorem 1. *Suppose inductively that we can compute a ρ-approximate independent set $I' \subset V \setminus N^{\bar{r}+1}$ for G'. Then $I := I_{\bar{r}} \cup I'$ is a ρ-approximate independent set for G.*

Proof. Since each $v \in I' \subset V \setminus N^{\bar{r}+1}$ has no neighbor in $N^{\bar{r}}$, and thus not in $I_{\bar{r}} \subset N^{\bar{r}}$, I is an independent set.
Furthermore, by definition of $\bar{r}$, we have

$$|I_{\bar{r}+1}| \leq \rho|I_{\bar{r}}|.$$

In other words, the subgraph $G[N^{\bar{r}+1}]$ induced by $N^{\bar{r}+1}$ has a maximum independent set size bounded by

$$\alpha(G[N^{\bar{r}+1}]) \leq \rho|I_{\bar{r}}|.$$

Further, by assumption, I' is ρ-approximately optimal for $G' = G[V \setminus N^{\bar{r}+1}]$. Thus,

$$\alpha(G[V \setminus N^{\bar{r}+1}]) \leq \rho|I'|.$$

Adding the two inequalities, we obtain

$$\alpha(G) \leq \alpha(G[N^{\bar{r}+1}]) + \alpha(V \setminus G[N^{\bar{r}+1}]) \leq \rho|I|,$$

as claimed.   □

3   The Algorithm for the Weighted Problem

The approximation algorithm presented in the previous section for the maximum independent set problem on UDGs can easily be adapted for the case that each node $v \in V$ is also given a nonnegative weight w_v. In that case, we are seeking an independent set of maximum total weight in the unit disk graph G. In the following, we present the modified algorithm that returns an independent set of total weight at least $(1 + \varepsilon)^{-1}$ the maximum total weight of an independent set in the UDG given as input.

For a subset $I \subset V$ of vertices, let $W(I)$ denote the total weight of I, i.e. $W(I) = \sum_{i \in I} w_i$. Furthermore, let I^{OPT} be an optimal solution to the maximum weight independent set problem for the graph $G = (V, E)$.

The approximation algorithm again follows the idea of the algorithm in the previous section. This time, however, we start with a vertex of maximal weight $w_{\max} = \max\{w_i | i \in V\}$, and then compute the independent set $I_r \subset N^r$ of maximum weight as long as $W(I_{r+1}) > \rho W(I_r)$ holds. Let $\bar{r}$ denote the smallest $r \geq 0$ for which this criterion is violated.

Lemma 2. *There exists a constant $c = c(\rho)$ such that $\bar{r} \leq c$.*

Proof. Suppose $r < \bar{r}$. Adapting the proof of Lemma 1, we get

$$W(I_r) = \sum_{i \in I_r} w_i \leq \sum_{i \in I_r} w_{\max} = |I_r| w_{\max}, \tag{5}$$

and

$$W(I_r) > \rho W(I_{r-1}) > \ldots > \rho^r W(I_0) = \rho^r w_{\max} \tag{6}$$

respectively. Since $|I_r| = O(r^2)$, comparing (5) and (6) again yields the claim. □

The running time of this algorithm remains polynomial in the weighted case. Also, the approximation ratio can be guaranteed as follows.

Theorem 2. *The adapted algorithm yields an independent set of weight at least $\rho^{-1} = (1 + \varepsilon)^{-1}$ the weight of a maximum weight independent set.*

Proof. Let $V' := V \backslash N^{\bar{r}+1}$, and inductively assume $I' \subset V'$ to be a ρ-approximate independent weighted set in $G[V']$. Clearly, $I_{\bar{r}} \cup I'$ is an independent set in G. For the weighted independent set in the neighborhood $N^{\bar{r}+1}$, we have

$$W(I^{\mathrm{OPT}} \cap N^{\bar{r}+1}) \leq W(I_{\bar{r}+1}) \leq \rho W(I_{\bar{r}}).$$

For the weight of the set returned by the algorithm, $W(I_{\bar{r}} \cup I')$, it is

$$\begin{aligned}
W(I^{\mathrm{OPT}}) &= W((I^{\mathrm{OPT}} \cap N^{\bar{r}+1}) \cup (I^{\mathrm{OPT}} \cap V')) \\
&= W(I^{\mathrm{OPT}} \cap N^{\bar{r}+1}) + W(I^{\mathrm{OPT}} \cap V') \\
&\leq \rho W(I_{\bar{r}}) + \rho W(I') \\
&= \rho W(I_{\bar{r}} \cup I'). \qquad \square
\end{aligned}$$

4   Robustness

In this section, we show that the approximation algorithms of the previous two sections actually lead to a robust algorithm.

Definition 1. *Let $\mathcal{A}$ be an algorithm defined on $\mathcal{G}$, f be a function on $\mathcal{G}$, and $\mathcal{U} \subset \mathcal{G}$. Then $\mathcal{A}$ computes f robustly (on $\mathcal{U}$), if*

1. *for all instances $i \in \mathcal{U}$, the algorithm $\mathcal{A}$ returns $f(i)$, and*
2. *for all instances $i \in \mathcal{G} \backslash \mathcal{U}$, the algorithm $\mathcal{A}$ returns either $f(i)$, or a certificate showing that $i \notin \mathcal{U}$.*

Of course, the notion of a robust algorithm is especially interesting when $\mathcal{A}$ has polynomial running time with respect to the size of the input instance, and the decision whether an instance belongs to the subclass $\mathcal{U} \subset \mathcal{G}$ is not as easy to decide. In our situation, $\mathcal{G}$ is the set of all undirected graphs, f gives a $(1 + \varepsilon)$-approximation of the weight or cardinality of an independent set of maximum weight or size respectively, and $\mathcal{U}$ is the subclass of unit disk graphs.

In the previous sections, we have shown that the introduced approximation algorithms yield a PTAS for the case that the input instances represent a unit disk graph. We thus continue our discussion only for the case that the input instance is a graph for which there exists no geometric representation satisfying the characterization of a UDG.

Observe that in Theorems 1 and 2, we did not use any properties of a UDG. The algorithms thus always return a $(1 + \varepsilon)$-approximate independent set. However, the polynomial running time of the algorithms is a direct result from the fact that any independent set in the rth neighborhood is polynomially bounded in r, i.e. $|I_r| \leq (2r + 1)^2$. For a general graph, the running time may thus not be polynomial. So, during the execution of the algorithm, if an independent set of size $|I_r| > (2r + 1)^2$ can be found, this set is returned as certificate of non-membership in the class of unit disk graphs. This procedure, i.e. trying to find an independent set of size $(2r + 1)^2 + 1$, is also a task requiring polynomial running time.

5   Extensions

It is straightforward to verify that our arguments apply equally well to intersection graphs of some other geometrical objects related to unit disks. For example, the unit disks may be replaced by disks with fixed lower and upper bound on the radius (*bounded disk graphs*). Similarly, an extension to (fixed) dimension $d \geq 2$ is possible. Indeed, all that is needed in the proof is a polynomial bound on the maximum geometric diameter divided by the minimum volume of the objects under consideration. Quasi (Unit) Disk Graphs, as a more realistic model of a wireless communication graph [9] satisfy this characterization.

The algorithm can also be applied to λ-precision disk graphs though they have no bound on the radius of the disks and thus no bound on the minimum volume of the disks. A λ-precision disk graph is an intersection disk graph where two vertices are at least λ apart in a geometric representation that has been scaled so that the disks have a maximum radius of 1 [8]. In this case, the size of the rth neighborhood, $|N^r|$, is already polynomially bounded in r. Note that this is a different condition as the one given above: for example, in a UDG, two vertices may be arbitrarily close as the graph can contain arbitrarily large cliques.

The independent set created by the PTAS may not be maximal, i.e. there may exist vertices that can be added to the solution set returned by the PTAS without violating the independent set property. However, a simple greedy strategy on the nodes in $N^{r_1+1} \setminus N^{r_1}$ that are not connected by an edge to a node from the independent set can resolve this problem. The obtained independent set then also forms a dominating set in the graph.

6   Conclusion

In this paper, we present a new PTAS for the maximum independent set problem in UDGs that does not depend on a geometric representation of the vertices. The algorithm is extended to solve also the weighted version of the problem. Both algorithms accept any graph as input and either return a $(1 + \varepsilon)$-approximate independent set or a certificate showing that the input presented to the algorithm is not a UDG. This certificate is given by an independent set which is too large to be contained in a bounded area given by the neighborhood N^r. The running time of the algorithm is given by the time to compute a maximum independent set in the largest neighborhood to be considered during the execution, which can be done in $n^{O(1/\varepsilon^2 \log 1/\varepsilon)}$.

Some extensions to different, related families of geometric intersection graphs are presented as well, including bounded and quasi disk graphs.

References

1. B.S. Baker. Approximation algorithms for NP-complete problems on planar graphs. *Journal of the ACM*, 41(1):153–180, 1994.
2. H. Breu and D.G. Kirkpatrick. Unit disk graph recognition is NP-hard. *Computational Geometry. Theory and Applications*, 9(1-2):3–24, 1998.
3. T.M. Chan. Polynomial-time approximation schemes for packing and piercing fat objects. *Journal of Algorithms*, 46:178–189, 2003.
4. X. Cheng, X. Huang, D. Li, W. Wu, and D.-Z. Du. A polynomial-time approximation scheme for the minimum-connected dominating set in ad hoc wireless networks. *Networks*, 42:202–208, 2003.
5. T. Erlebach, K. Jansen, and E. Seidel. Polynomial-time approximation schemes for geometric graphs. In *Proceedings of the 12th ACM-SIAM symposium on discrete algorithms (SODA'01)*, pages 671–679, Washington, DC, 2001.
6. R.L. Graham, D.E. Knuth, and O. Potashnik. *Concrete Mathematics*. Addison-Wesley, 2nd edition, 1998.
7. D.S. Hochbaum and W. Maass. Approximation schemes for covering and packing problems. *Journal of the ACM*, 32(1):130–136, 1985.
8. H.B. Hunt III, M.V. Marathe, V. Radhakrishnan, S. S. Ravi, D.J. Rosenkrantz, and R.E. Stearns. NC-approximation schemes for NP- and PSPACE-hard problems for geometric graphs.
9. F. Kuhn, R. Wattenhofer, and A. Zollinger. Ad-hoc networks beyond unit disk graphs. In *1st ACM DIALM-POMC Joint Workshop on Foundations of Distributed Computing*, San Diego, USA, 2003.
10. M.V. Marathe, H. Breu, H.B. Hunt III, S. S. Ravi, and D.J. Rosenkrantz. Simple heuristics for unit disk graphs. *Networks*, 25:59–68, 1995.

11. T. Nieberg, S. Dulman, P. Havinga, L.v. Hoessel, and J. Wu. Collaborative algorithms for communication in wireless sensor networks. In T. Basten, M. Geilen, and H. De Groot, editors, *Ambient Intelligence: Impact on Embedded System Design.* Kluwer Academic Publishers, 2003.
12. V. Raghavan and J. Spinrad. Robust algorithms for restricted domains. In *Proceedings of the twelfth annual ACM-SIAM symposium on Discrete algorithms*, pages 460–467. Society for Industrial and Applied Mathematics, 2001.

Appendix

Lemma 3. *For $0 < \varepsilon < \frac{1}{10}$, the value $c = \frac{1}{\varepsilon^2} \ln \frac{1}{\varepsilon}$ satisfies*

$$(2c + 1)^2 < (3c)^2 < (1 + \varepsilon)^c.$$

Proof. The first inequality is obviously true for the given choice of ε and c. The second inequality is equivalent to

$$2 \ln 3c < c \ln(1 + \varepsilon),$$

(taking the log). This, again, is equivalent with

$$\frac{2}{c}(\ln 3 + \ln c) + \varepsilon^2 < \ln(1 + \varepsilon) + \varepsilon^2.$$

In [6], it is shown that

$$\varepsilon \le \ln(1 + \varepsilon) + \varepsilon^2$$

holds for all $0 \le \varepsilon < \frac{1}{2}$. Thus, it remains to show that $\frac{2}{c}(\ln 3 + \ln c) + \varepsilon^2 < \varepsilon$. Note that, for $\varepsilon < \frac{1}{10}$, it is $\ln \frac{1}{\varepsilon} > 1$, and $\ln x < x$ for $x > 1$. Substituting $c = \frac{1}{\varepsilon^2} \ln \frac{1}{\varepsilon}$, we get

$$\frac{2}{c}(\ln 3 + \ln c) + \varepsilon^2 = \frac{2\varepsilon^2}{\ln \frac{1}{\varepsilon}}(\ln 3 + \ln(\frac{1}{\varepsilon^2} \ln \frac{1}{\varepsilon})) + \varepsilon^2$$

$$= 2\varepsilon^2(\frac{\ln 3}{\ln \frac{1}{\varepsilon}} + 2\frac{\ln \frac{1}{\varepsilon}}{\ln \frac{1}{\varepsilon}} + \frac{\ln \ln \frac{1}{\varepsilon}}{\ln \frac{1}{\varepsilon}} + \frac{1}{2})$$

$$< 2\varepsilon^2(\ln 3 + 2 + \frac{\ln \frac{1}{\varepsilon}}{\ln \frac{1}{\varepsilon}} + \frac{1}{2})$$

$$< 2\varepsilon^2(\ln 3 + 3.5).$$

Since $\varepsilon < \frac{1}{10}$, the inequality

$$2\varepsilon^2(\ln 3 + 3.5) < 10\varepsilon^2 < \varepsilon$$

holds and the claim follows. □

Tolerance Based Algorithms for the ATSP

Boris Goldengorin[1], Gerard Sierksma[1], and Marcel Turkensteen[1]

Faculty of Economic Sciences, University of Groningen,
P.O. Box 800, 9700 AV Groningen, The Netherlands
{b.goldengorin,g.sierksma,m.turkensteen}@eco.rug.nl

Abstract. In this paper we use arc tolerances, instead of arc costs, to improve Branch-and-Bound type algorithms for the Asymmetric Traveling Salesman Problem (ATSP). We derive new tighter lower bounds based on exact and approximate bottleneck upper tolerance values of the Assignment Problem (AP). It is shown that branching by tolerances provides a more rational branching process than branching by costs. Among others, we show that branching on an arc with the bottleneck upper tolerance value is the best choice, while such an arc appears quite often in a shortest cycle of the current AP relaxation. This fact shows why branching on shortest cycles was always found as a best choice. Computational experiments confirm our theoretical results.

1   Introduction

The Traveling Salesman Problem (TSP) is the problem of finding a shortest tour through a given number of cities such that every city is visited exactly once. The travel costs $c(i, j)$ are *symmetric* if traveling from city i to city j costs just as much as traveling from city j to city i, and *asymmetric* if there is an arc (i, j) such that $c(i, j) \neq c(j, i)$. The TSP is a typical NP-hard optimization problem and solving instances with a large number of cities is very difficult if not impossible. Recent developments in polyhedral theory and heuristics have significantly increased the size of instances which can be solved to optimality. The best known exact algorithms are based on the either branch-and-bound method for the Asymmetric TSP (ATSP) (see Fischetti et al. [3]) or branch-and-cut method for the Symmetric TSP (STSP) using the double index formulation of the problem (see Naddef [11]). The state-of-the-art of heuristics for the STSP and the ATSP is presented in Johnson and McGeoch [5] and Johnson et al. [4], respectively. Current algorithms, except Helsgaun's version of the Lin-Kernighan heuristic (see Helsgaun [8]), do not use criteria based on the tolerance values of the corresponding TSP relaxations to find a common arc (edge) of the optimal solutions to the relaxed TSP and the TSP itself.

In this paper we generalize Helsgaun's idea of using the tolerances to solve combinatorial optimization problems (COPs) with monotone objective functions and non-embedded sets of feasible solutions, i.e. sets of feasible solutions such that for any feasible solution S all its proper subsets $A \subset S$ are not feasible. Many problems have as underlying model this COP, among them being the traveling

J. Hromkovič, M. Nagl, and B. Westfechtel (Eds.): WG 2004, LNCS 3353, pp. 222–234, 2004.

salesman, quadratic assignment, linear ordering, assignment, 1-tree, max-flow, shortest path, and matching problems.

The paper is organized as follows. Section 2 shows that the tolerance values of an optimal solution to COP indicate the multiplicity (uniqueness) of the whole set of optimal solutions. Section 3 presents the relationships between extremal values of upper and lower tolerances. Section 4 empirically justifies the usefulness of upper tolerance values for constructing exact and heuristic algorithms for the ATSP. Section 5 derives new tighter lower bounds and branching rules for the ATSP which allow us to reduce the solution tree size for the ATSP substantially. Conclusions and future research directions appear in Section 7.

2   Tolerances for Combinatorial Optimization Problems

In this section we introduce the notion of upper and lower tolerances and indicate some useful properties of their extremal values. All necessary proofs can be found in Goldengorin and Sierksma [6].

A *Combinatorial Optimization Problem* COP $(\mathcal{E}, C, \mathcal{D}, f_C)$ is the problem of finding

$$S^* \in \arg opt\{f_C(S) \mid S \in \mathcal{D}\},$$

where $C : \mathcal{E} \to \Re$ is the given *instance* of the problem with a *ground set* $\mathcal{E}$ satisfying $|\mathcal{E}| = m$ $(m \geq 1)$, $\mathcal{D} \subseteq 2^{\mathcal{E}}$ is the *set of feasible solutions*, and $f_C : 2^{\mathcal{E}} \to \Re$ is the *objective function* of the problem. By $\mathcal{D}^* = \arg opt\{f_C(S) \mid S \in \mathcal{D}\}$ the set of optimal solutions is denoted. It is assumed that $\mathcal{D}^* \neq \emptyset$, and that $\cup \mathcal{D} \neq \emptyset$. In the remaining part of this paper we take $opt = \min$.

Let $g \in \mathcal{E}$, and $\alpha \geq 0$. By $C_{\alpha,g} : \mathcal{E} \to \Re$ we denote the instance defined as $C_{\alpha,g}(e) = C(e)$ for each $e \in \mathcal{E} \setminus \{g\}$, and $C_{\alpha,g}(g) = C(g) + \alpha$. Take any $S^* \in \mathcal{D}^*$. The *upper tolerance*, $u_{S^*}(e)$, of e with respect to S^* is defined as

$$u_{S^*}(e) = \max\{\alpha \geq 0 : S^* \in \arg\min\{f_{C_{\alpha,e}}(S) : S \in \mathcal{D}\}\},$$

and the *lower tolerance*, $l_{S^*}(e)$, with respect to S^* as

$$l_{S^*}(e) = \max\{\alpha \geq 0 : S^* \in \arg\min\{f_{C_{-\alpha,e}}(S) : S \in \mathcal{D}\}\}.$$

I.e. $u_{S^*}(e)$ is the maximal increase of $C(e)$ under which S^* stays optimal, and $l_{S^*}(e)$ is the maximal decrease of $C(e)$ under which S^* stays optimal.

We also assume that f_C is *monotone*, meaning that for each $S \in 2^{\mathcal{E}}$ and each $\alpha > 0$, it holds that

$$f_{C_{\alpha,e}}(S) > f_{C_{0,e}}(S) \quad \text{and} \quad f_{C_{-\alpha,e}}(S) < f_{C_{0,e}}(S).$$

Sum functions with $f_C(S) = \sum_{e \in S} C(e)$, *bottleneck functions* with $f_C(S) = \max_{e \in S} C(e)$, and *product functions* with $f_C(S) = \prod_{e \in S} C(e)$ and $C(e) \geq 1$ for each $e \in \mathcal{E}$ are all monotone functions.

We call a set $\mathcal{D}$ of feasible solutions *non-embedded* if for each $S_1, S_2 \in \mathcal{D}$ with $S_1 \neq S_2$, it holds that neither $S_1 \subset S_2$ nor $S_2 \subset S_1$.

The following theorem can be seen as a generalization of Libura's theorem on tolerances (see, Libura [10]) derived for the TSP. We will use the following extra notations. Let $e \in \mathcal{E}$. Then $\mathcal{D}_+(e) = \{S \in \mathcal{D} : e \in S\}$, and $\mathcal{D}_-(e) = \{S \in \mathcal{D} : e \notin S\}$. Clearly, $\mathcal{D} = \mathcal{D}_-(e) \cup \mathcal{D}_+(e)$ and $\mathcal{D}_-(e) \cap \mathcal{D}_+(e) = \emptyset$ for all $e \in \mathcal{E}$. $\mathcal{D}_+^*(e)$ and $\mathcal{D}_-^*(e)$ are the sets of optimal solutions containing e and not containing e, respectively.

Theorem 1. *Consider a COP $(\mathcal{E}, C, \mathcal{D}, f_C)$ with monotone f_C. For each $S^* \in \mathcal{D}^* \neq \emptyset$ the following holds:*

1. $e \in \cap \mathcal{D}^*$ *iff* $u_{S^*}(e) = f_C(S) - f_C(S^*) > 0$ *for each* $S \in \mathcal{D}_-^*(e)$, $l_{S^*}(e) = \infty$;
2. $e \in \mathcal{E} \setminus \cup \mathcal{D}^*$ *iff* $u_{S^*}(e) = \infty$, $l_{S^*}(e) = f_C(S) - f_C(S^*) > 0$ *for each* $S \in \mathcal{D}_+^*(e)$;
3. $e \in S^* \setminus \cap \mathcal{D}^*$ *iff* $u_{S^*}(e) = 0$, $l_{S^*}(e) = \infty$;
4. $e \in \cup \mathcal{D}^* \setminus S^*$ *iff* $u_{S^*}(e) = \infty$, $l_{S^*}(e) = 0$;

Proof. Case 1. Sufficiency. For sake of simplicity we prove only the case when f_C is a sum function. In general the values of $u_{S^*}(e)$ should be adjusted according to a specific presentation of f_C. Consider an element $g \in \cap \mathcal{D}^*$ and the partition of $\mathcal{D} = \mathcal{D}_-(g) \cup \mathcal{D}_+(g)$ such that $\mathcal{D}_-(g) \cap \mathcal{D}_+(g) = \emptyset$. If the value of $C_{\alpha,g}(g) = C(g) + \alpha$ increases by increasing the value of $\alpha > 0$ and the values of $C_{\alpha,g}(e) = C(e)$ for each $e \in \mathcal{E} \setminus \{g\}$ remain unchanged, then the values of all feasible solutions belonging to $\mathcal{D}_+(g)$ are also increased in the same way since f_C is monotone. But the values of $f_C(S)$ for each $S \in \mathcal{D}_-(g)$ are still the same. Hence S^* remains optimal as long as the increase of $C_{\alpha,g}(g)$ is not greater than the difference $f_C(S) - f_C(S^*) > 0$ for each $S \in \mathcal{D}_-^*(e)$.

Necessity. By contradiction. Assume that $u_{S^*}(g) > 0$ but $g \in = \mathcal{E} \setminus \cap \mathcal{D}^*$. If $g \in S^* \setminus \cap \mathcal{D}^*$ and the value of $C_{\alpha,g}(g)$ increases then the value of $f_C(S^*)$ increasing also and S^* became non-optimal under assumption that $\cup \mathcal{D}^* \setminus S^* \neq \emptyset$. Hence, $u_{S^*}(g) = 0$. If $g \in \cup \mathcal{D}^* \setminus S^*$ then by increasing the value of $C_{\alpha,g}(g)$ we loose the optimality of an optimal solution $S_1^* \neq S^*$. Again, $u_{S^*}(g) = 0$. If $g \in \mathcal{E} \setminus \cup \mathcal{D}^*$ then we have the case 3 of this theorem.

It is obvious that any decrement of $C_{-\alpha,g}(g)$ by increasing the value of $\alpha > 0$ does not change the optimality of S^*. Hence, $l_{S^*}(g) = \infty$ for each $g \in S^*$, see case 3. Proofs of cases 2, 3 and 4 can be obtained in a similar manner.   ∎

If $|\mathcal{D}^*| = 1$, then this theorem boils down to Libura's theorem on tolerances. If $\mathcal{D}_-(e) = \emptyset$ for some $e \in \mathcal{E}$, then $u_{S^*}(e) = \min\{f_C(T) : T \in \mathcal{D}_-(e)\} - f_C(S^*) = \min\{\emptyset\} = \infty$ (by definition). Similarly, for $\mathcal{D}_+(e) = \emptyset$ we take $l_{S^*}(e) = \infty$.

Note that the finite values of the upper and the lower tolerances are nonnegative and independent on the chosen $S^* \in \mathcal{D}^*$. Hence, we may write $u(e)$ and $l(e)$ instead of $u_{S^*}(e)$ and $l_{S^*}(e)$, respectively, when they are finite and positive.

One of the major problems, when solving an NP-hard COP by means of Branch-and-Bound (BnB) approach, is the choice of the branching element which keeps the search tree as small as possible. Using tolerances we are able to ease this choice. Namely, if there is an element from the optimal solution of the current relaxation with a finite positive upper tolerance, then this element is in all optimal solutions of the current relaxed problem (see Theorem 1(1)). Hence, branching on this element means that we enter a common part in all search trees emanating from this current stage. Therefore, branching on an element with a positive upper tolerance is not only necessary for finding a feasible solution to the original NP-hard instance but also is the best choice. Of course, if this situation is not at hand we still have the above mentioned major problem.

In the following sections we study extremal values of tolerances. The purpose is to reduce the search tree sizes of BnB type algorithms for the ATSP. For an extensive account on properties of upper and lower tolerances in the context of sensitivity analysis, see Greenberg [7] and references therein.

3   Extremal Values of Upper and Lower Tolerances

In this section it will be shown that under very natural conditions, it holds that $\min\{u_{S^*}(e) : e \in S^*\} = \min\{l_{S^*}(e) : e \in \mathcal{E} \setminus S^*\}$.

Theorem 2. *Consider a $COP(\mathcal{E}, C, \mathcal{D}, f_C)$ with monotone f_C and non-embedded $\mathcal{D}$. Then*

$$u_{\min} = \min\{u_{S^*}(e) : e \in S^*\} = l_{\min} = \min\{l_{S^*}(e) : e \in \mathcal{E} \setminus S^*\},$$

for each $S^* \in \mathcal{D}^*$.

Proof. If $E_m = \{e \in S^* : u_{S^*}(e) = u_{\min}\}$ and $G_m = \{e \in \mathcal{E} \setminus S^* : l_{S^*}(e) = l_{\min}\}$, then by Theorem 1(1,2) $u_{\min} = f_C[S^*_-(e_m)] - f_C(S^*)$ and $l_{\min} = f_C[S^*_+(g_m)] - f_C(S^*)$. Hence, the equality $u_{\min} = l_{\min}$ is equivalent to the equality $f^u_{\min} = f_C[S^*_-(e_m)] = f^l_{\min} = f_C[S^*_+(g_m)]$. In other terms, $f^u_{\min} = \min\{f_C[S^*_-(e)] : e \in S^*\}$ and $f^l_{\min} = \min\{f_C[S^*_+(e)] : e \in \mathcal{E} \setminus S^*\}$ as well as $E_m = \{e \in S^* : f_C[S^*_-(e)] = f^u_{\min}\}$ and $G_m = \{e \in \mathcal{E} \setminus S^* : f_C[S^*_+(e)] = f^l_{\min}\}$. Denote by $e_m \in E_m$ and by $g_m \in G_m$. The proof of this theorem is based on the following two inequalities $f^l_{\min} \leq f^u_{\min}$ and $f^l_{\min} \geq f^u_{\min}$. Combining both inequalities we obtain a proof.

Let us prove only the first inequality because the proof of the second inequality can be done in a similar way. Note that $S^*_-(e_m)$ differs from S^* at least by one element, say $t^* \in S^*_-(e_m)$ such that $t^* \notin S^*$, since $S^*_-(e_m)$ and S^* are non-embedded. Hence, $S^*_-(e_m) \in \mathcal{D}_+(t^*)$. Note that $f_C[S^*_+(t^*)] = \min\{f_C[S] : S \in \mathcal{D}_+(t^*)\} \leq f_C[S^*_-(e_m)]$ implies that $f^l_{\min} = f_C[S^*_+(g_m)] \leq f_C[S^*_+(t^*)] \leq f_C[S^*_-(e_m)] = f^u_{\min}$.   ∎

Theorem 2 allows us to compute the minimal upper and lower tolerances by solving at most $\min\{O(|S^*|), O(|\mathcal{E} \setminus S^*|)\}$ (sub)COPs. Note that the minimum is not always attained on $|S^*|$. For example consider a sparse TSP instance of 100 cities with 120 arcs. Then $|S^*| = 100$, and $|\mathcal{E} \setminus S^*| = 20$. This can be a considerable improvement in comparison to the situation where only Theorem 1 (including Libura's theorem) is available. In the latter case, $O(|\mathcal{E}|)$ (sub)COPs needed to be solved. Also in terms of (sub)COPs the complexity of computing the minimal values of upper and lower tolerances is $O(|\mathcal{E}|)$. Note that the conditions from Theorem 2 hold for "regular" COPs, such that for any feasible solution S all its proper subsets $A \subset S$ are not feasible. The sets of feasible solutions of many COPs with a monotone objective function have that property. The irregular situation of Theorem 2 occurs in problems with embedded sets of feasible solutions (for example location problems; see Goldengorin and Sierksma [6]).

Define the finite values of $u_{\max} = \max\{u_{S^*}(e) : e \in S^*\}$ and $l_{\max} = \max\{l_{S^*}(e) : e \in \mathcal{E} \setminus S^*\}$ for $S^* \in \mathcal{D}^*$. If an optimal AP solution a^* is used to obtain a shortest ATSP tour h^*, which type of tolerances should be used, upper or lower? The arcs which should be considered for inclusion are the arcs in $e \in \mathcal{E} \setminus a^*$ that are also in an shortest tour h^*. Lower tolerances are computed for all arcs $e \in \mathcal{E} \setminus a^*$, but if upper tolerances are computed, only a subset $A_u \subseteq \mathcal{E} \setminus a^*$ is considered for inclusion. Here, $A_u = \{e \in \mathcal{E} \setminus a^* : l_{a^*}(e) \leq u_{max}\}$. Table 1 shows that in general, the arcs entering h^* are contained in A_u. Since in general, the complexity of computing all lower tolerances is higher, we use only the upper tolerances in case of the ATSP.

Table 1. Fraction of arcs e in ATSP solution such that $l_{a^*}(e) > u_{max}$.

Instance type	Fraction of arcs
ATSPLIB	2.335%
Degree of symmetry 0.33	6.006%
Degree of symmetry 0.66	6.878%
Full symmetry	19.149%
Usual random	0.000%
Degree of sparsity 50%	5.576%

4   Statistical Analysis of Arc Tolerances and Costs

In this section, we explore whether tolerance based Branch and Bound (BnB) algorithms are more effective for the ATSP than cost based BnB algorithms. We present a statistical analysis of optimal AP and ATSP to explore which arcs in an optimal AP solution a^* that also appear in an optimal solution h^* of the ATSP instance with the same cost matrix. Table 2 shows that the average percentage of common arcs in AP and ATSP solutions varies between 40 and 80%. Similar research shows that the Minimum 1-Trees and optimal STSP tours have between 70% and 80% of the edges in common [8].

BnB methods make a sequence of steps in which parts of the AP solution at hand are included and excluded, until an optimal solution of the ATSP is found. If a BnB algorithm predicts correctly which element to delete or to insert, then its search tree will be small. Most algorithms base the prediction whether an AP arc is in an optimal ATSP solution on its cost value. The question is: do predictions improve if they are based on the upper tolerance values of the AP?

We explore whether there are relationships between the cost values and the upper tolerance values of arcs and their appearance in a fixed shortest complete

Table 2. Fraction of common arcs in optimal AP and ATSP solutions.

Instance type	Fraction AP and ATSP
ATSPLIB	53.52%
Degree of symmetry 0.33	69.29%
Degree of symmetry 0.66	51.10%
Full symmetry	43.44%
Asymmetric random	80.49%
Degree of sparsity 50%	86.27%
Degree of sparsity 75%	84.23%
Degree of sparsity 90%	83.46%

tour h^*. These relationships are measured with *correlations*, which require two continuous variables, and with the *adjusted Rand index*, which measures the relationship between two partitions [9]. The costs and the upper tolerances are continuous variables, and they are compared with a partition of a^*.

In order to calculate the correlations, we define the following measure for all arcs $e \in a^*$: $IN(e) = 1$ if $e \in h^*$ and $IN(e) = 0$ if $e \notin h^*$. We compare the absolute correlations between costs and the measure, and between upper tolerances and the measure. If the absolute correlation is high, say for costs, then we may expect that high cost arcs in an AP solution are in none of the optimal ATSP solutions, and therefore, if an algorithm excludes high cost arcs from an AP solution, it will arrive at an optimal ATSP solution quickly. Table 3 shows that the correlations of tolerances are larger in absolute terms than the correlations of costs. Hence, one may expect that including or excluding an arc from the AP solution into the ATSP solution is done more accurately when tolerances are used instead of costs.

In the adjusted Rand index analysis, we create partitions based on upper tolerances and costs. Firstly, the arcs in a fixed optimal AP solution a^* are partitioned into two subsets: the subset of arcs IN_1 which are also in a fixed shortest Hamiltonian tour h^*, and the subset of arcs IN_0 which are not in h^*. Call this partition $IN = \{IN_0, IN_1\}$. We try to replicate IN with partitions C and U based on the cost and tolerance values of the arcs, respectively. Define

Table 3. Absolute correlations and adjusted Rand indices.

Instance type	Correlations		Adjusted Rand indices	
	Tolerance	Cost	Tolerance	Cost
ATSPLIB	0.358	0.078	0.113	-0.003
Degree of symmetry 0.33	0.368	0.057	0.152	0.007
Degree of symmetry 0.66	0.340	0.034	0.188	0.028
Full symmetry	0.086	0.068	0.158	0.013
Asymmetric random	0.345	0.101	0.287	0.039
Sparsity 50%	0.274	0.091	0.361	0.017
Sparsity 75%	0.274	0.091	0.252	0.033
Sparsity 90%	0.274	0.091	0.219	0.032

$C := \{C_0, C_1\}$, where $C_0 = \{e \in a^* \text{ s.t. } c(e) \geq c^*\}$, $C_1 := \{e \in a^* \text{ s.t. } c(e) < c^*\}$, and choose c^* in such a way that $|C_0| = |IN_0|$. Arcs are partitioned into a set of low cost arcs C_1 and a class of high cost arcs C_0. If it is true that all high cost arcs are not in the shortest tour, then the sets IN_0 and C_0 and the sets IN_1 and C_1 coincide and cost values lead to a perfect prediction. Similarly, define $U = \{U_0, U_1\}$, where $U_0 := \{e \in a^* \text{ s.t. } u_{a^*}(e) < u^*\}$, $U_1 := \{e \in a^* \text{ s.t. } u_{a^*}(e) \geq u^*\}$, and choose u^* in such a way that $|U_0| = |IN_0|$. The *adjusted Rand index* [9] measures how similar the partitions U and C are to the partition IN.

The more similar two partitions are, the higher the adjusted Rand index between both partitions is. An adjusted Rand index of 1 indicates that for each nonempty class A_i of partition A, there exists a class B_j of partition B such that $A_i = B_j$. The expected adjusted Rand index is 0, if both partitions assign objects to classes randomly having the original number of objects in each class.

The adjusted Rand indices between IN and C and between IN and U are shown in Table 3. The adjusted Rand indices are larger for the tolerance based partitions U than for the cost based partitions C, which confirms that predictions are better if they are based on upper tolerance values.

5 New Lower Bound for the ATSP

In Theorem 3 we derive a new tighter lower bound for the ATSP based on the tolerance values for the corresponding AP.

Theorem 3. *Let a^* and h^* be optimal solutions to the AP and ATSP with optimal values $f_C(a^*)$ and $f_C(h^*)$, respectively, whereas both problems have the same the cost matrix C. Assume that a^* consists of $k > 1$ cycles, $a^* = \{C_1, \ldots, C_k\}$ and $C_i = \{e_1^i, \ldots, e_{t(i)}^i\}$ is the set of arcs in i-th cycle C_i with upper tolerance values $u(e_1^i), \ldots, u(e_{t(i)}^i)$ $(i = 1, \ldots, k)$. Define $u(i) = \min\{u(e_p^i) : p = 1, \ldots, t(i)\}$ for $i = 1, \ldots, k$, and by $u_e = \max\{u(i) : i = 1, \ldots, k\}$. The following inequalities hold:*

$$f_C(a^*) \leq lb_e = f_C(a^*) + u_e \leq f_C(h^*).$$

Proof. Since $u_e \geq 0$ we have that $f_C(a^*) \leq f_C(a^*) + u_e$. Note that to find an optimal solution h^* to the ATSP we have to exclude at least one arc from each cycle C_i. If we delete an arc e_i such that $u(e_i) = u(i)$ then by Theorem 1(1) we have that $u(e_i) = f_C[a_-^*(e_i)] - f_C(a^*)$, and $f_C[a_-^*(e_i)]$ is the smallest value defined on the set $\mathcal{A}_-(e_i) = \{a \in \mathcal{A} : e_i \notin a\}$, with $\mathcal{A}$ the collection of all assignments. Hence $f_C[a_-^*(e_i)] = f_C(a^*) + u(e_i) \leq f_C(h^*)$ for each $i = 1, \ldots, k$. Therefore, $\max\{f_C[a_-^*(e_i)] : i = 1, \ldots, k\} \leq f_C(h^*)$ implies that $f_C(a^*) \leq f_C(a^*) + u_e \leq f_C(h^*)$. Note that in case of multiple solutions to the AP, i.e. $|\mathcal{A}^*| > 1$, the conditions of $u_e > 0$ and $a^* = \{C_1, \ldots, C_k\}$ with $k(> 1)$ cycles, imply that $h^* \notin \mathcal{A}^*$.   ∎

It is well known (see e.g., Balas and Toth [1], pages 370-371) that for finding an optimal solution $a_-^*(e_i)$ based on the given AP solution a^*, it is enough to use only one labelling procedure in the Hungarian method, which can be done in time $O(n2)$. Hence, the time complexity of our new lower bound lb_e is $2O(n3)$, where $O(n3)$ is the time complexity of Hungarian method for solving the AP.

Further we refer to lb_e and u_e as to the Exact Bottleneck Bound (EB) and Exact Bottleneck Tolerance (ET), respectively. We also use the Approximate Bottleneck Bound (AB), and denote it by $lb_a = f_C(a^*) + u_a$. The corresponding Approximate Bottleneck Tolerance (AT) is denoted by $u_a = u(i_0)$ with $i_0 \in$ $\arg\min\{|C_i| : i = 1, \ldots, k\}$. It is clear that $u_a \leq u_e$. For different types of ATSP instances the relative reductions $r_e = \frac{u_e}{f_C(h^*) - f_C(a^*)} \times 100\%$ and $r_a = \frac{u_a}{f_C(h^*) - f_C(a^*)} \times 100\%$ of the gap $f_C(h^*) - f_C(a^*)$ are shown in Table 4. The results show that on average around 50% of the gap is bridged for random instances. As a consequence, the exclusion of an ET arc from a randomly generated instance brings the algorithm in large steps, on average 50% of the gap, towards an optimal ATSP solution h^*. Also, the reductions achieved by the AT are almost as large as the AT reductions, and therefore, the AT is a good approximation.

For Example 2 from Balas and Toth [1] (see page 381) we obtain that $f_C(a^*) = 17$, $lb_a = 21$, $lb_e = 25$, and $f_C(h^*) = 26$. Here, $r_e = \frac{8}{26-17} \times 100\% = 88.8\%$, $r_e = \frac{4}{26-17} \times 100\% = 44.4\%$, and $lb_e = 25$ is the best value among all bounds discussed in Balas and Toth [1].

Similar new bounds can be obtained for the STSP by taking into account the upper tolerance values for edges incident with vertices of degree at least three and the lower tolerance values for edges incident with vertices of degree one in the corresponding 1-tree solution.

Table 4. Relative reductions of the "AP - ATSP" gap.

Instance	r_e	r_a
ATSPLIB	19.97%	6.39%
Degree of symmetry 0.33	34.62%	17.07%
Degree of symmetry 0.66	26.66%	12.27%
Full symmetry	21.64%	14.61%
Asymmetric random	50.47%	43.31%
Degree of sparsity 50%	56.50%	48.99%
Degree of sparsity 75%	45.78%	40.45%
Degree of sparsity 90%	49.86%	35.45%

6   Computational Experiments with ATSP Instances

The tested set of BnB algorithms for solving various types of TSP instances is shown in Table 5. Here "Normal" branching rule means branching by a shortest cycle from the AP solution in a non-increasing order of arc costs. All other branching rules mean branching by a cycle on which either ET or AT is attained but in a non-increasing order of arc upper tolerances.

We have selected instances from the ATSPLIB (see Reinelt [12]) that are solvable within reasonable time limits. The random instances have degree of

Table 5. Variants of BnB algorithms.

Algorithm	Lower bound	Branching rule	Comments
0	$f_C(a^*)$	Normal	basic algorithm
1	$f_C(a^*)$	ET	efficiency of exact branching
2	$f_C(a^*)$	AT	efficiency of appr. branching
3	lb_e	Normal	efficiency of exact bound
4	lb_e	ET	efficiency of exact BnB
5	lb_e	AT	not considered
6	lb_a	Normal	efficiency of appr. bound
7	lb_a	ET	not considered
8	lb_a	AT	efficiency of appr. BnB

symmetry 0, 0.33, 0.66, and 1, where the *degree of symmetry* is defined as the fraction of off-diagonal entries of the cost matrix $\{c_{ij}\}$ that satisfy $c_{ij} = c_{ji}$. The sparse random instances have degrees of sparsity of 50%, 75%, and 90%, where the *degree of sparsity* is the percentage of arcs that is missing in an instance. The sizes of the randomly generated instances are reported in Table 6. For each problem set and for all instance sizes, 10 instances have been generated. The experiments are conducted on a Pentium 4 computer with 256 MB RAM memory and 2 GHz speed. In all tables $s(i)$ and $t(i)$ with $i = 0, 1, \ldots, 8$ are the sizes of the search trees, and solution times, respectively.

Table 6. Size n of the instances used in the experiments.

Tables	Sparse	Usual random	Degree of symmetry 0.33 and 0.66	Full symmetry
1,2,3,4	$n = 60, 70, 80$	$n = 60, 70, 80$	$n = 60, 70, 80$	$n = 60, 70$
8		$n = 60,\ldots,200$		
10		$n = 60,\ldots,1000$		
11	$n = 100, 200, 400$		$n = 60, 70, 80$	$n = 60, 70, 80$

There are two effects that reduce the size of the search tree of tolerance based BnB algorithms. The first effect arises from the improved choice of the branching variable shown in Section 4. The second effect is caused by the improvement in the lower bounds obtained using AB and EB; see Section 5. Tables 7 and 8 show that the increase in lower bounds are the main cause of the large search tree reductions. Since the EB lower bound is tighter, the reductions of the ET based branching rules are larger than the reductions of AT based branching rules. The joint use of tolerance based lower bounds and branching rules often creates larger reductions compared to the cases where only the bound or the branching rule is used. Therefore, we concentrate on Algorithms 0, 4, and 8 in the experiments below.

The question is whether the search tree reductions of Algorithms 4 and 8 are sufficient to compensate for the time invested in the tolerance calculations.

Table 7. Search tree sizes for ATSPLIB instances.

Instance	n	$s(0)$	ET				AT			
			$s(1)$	$s(3)$	$s(4)$	$\frac{s(0)}{s(4)}$	$s(2)$	$s(6)$	$s(8)$	$\frac{s(0)}{s(8)}$
ft53	53	20111	*	7039	*	*	89511	17703	20955	0.96
ft70	70	25831	3058035	5619	5178853	0.14	22843	6717	5097	5.07
ftv33	34	7065	8926	1983	1696	4.17	6007	3137	1843	3.83
ftv35	36	6945	16432	2553	2824	2.46	12047	3219	2477	2.80
ftv38	39	6195	10175	2235	1494	4.14	14663	2821	2723	2.28
ftv44	45	619	610	187	130	4.76	937	249	247	2.51
ftv47	48	29025	42581	8017	5206	5.58	48345	9703	9595	3.03
ftv55	56	92447	88698	12413	8554	10.81	114641	26483	31717	2.91
ftv64	65	43441	100265	9449	8364	5.19	162639	11007	22863	1.90
ftv70	71	253873	532743	25939	56271	4.51	136296	52289	23275	10.91

*Memory exhausted

Table 8. Search tree sizes for random instances.

n	$s(0)$	ET				AT			
		$s(1)$	$s(3)$	$s(4)$	$\frac{s(0)}{s(4)}$	$s(2)$	$s(6)$	$s(8)$	$\frac{s(0)}{s(8)}$
60	3808	2695	978	323	11.79	1032	2832	221	17.23
70	4528	3511	1138	312	14.51	1286	1781	247	18.33
80	9014	3784	2414	217	41.53	2494	2746	256	35.21
100	9002	2298	1978	174	51.74	2188	5306	135	66.68
200	36390	20002	7114	858	42.41	17612	7612	796	45.72

Tables 9, 10, and 11 show that Algorithm 8 obtains the fastest solution times for asymmetric instances, sparse instances, and instances with degree of symmetry 0.33 and 0.66, but the solution times of Algorithm 0 are better for ATSPLIB (except the instances ftv33, ftv44, and ftv70) and fully symmetric instances. Algorithm 4, which uses the EB lower bound and the ET branching rule, generally requires too much tolerance calculation time to be competitive, in spite of its small search trees.

7   Summary and Future Research Directions

In this paper we present the relationships between the extremal values of upper and lower tolerances which constitute a background for improvements of exact and heuristic algorithms for solving the ATSP. In case of the ATSP Theorem 2 shows that for efficient implementation of the upper and lower tolerance values in the BnB type algorithms it is enough to use only the upper tolerance values. For example, in Helsgaun's implementation of the Lin-Kernighan heuristic for the STSP the main improvement is based on the first five smallest lower tolerance values (see Helsgaun [8]). Theorem 2 shows that it is enough to compute for Helsgaun's implementation only $O(n)$ upper tolerance values instead of $O(n2)$ lower tolerance values.

Table 9. AT and ET versus cost based branching for ATSPLIB instances.

Instance	n	$s(0)$	$t(0)$	$s(4)$	$t(4)$	$s(8)$	$t(8)$
ft53	53	20111	2.31	*	*	20955	6.65
ft70	70	25831	3.85	5178853	775.05	5097	2.03
ftv33	34	7065	0.22	1696	0.99	1843	0.16
ftv35	36	6945	0.22	2824	1.81	2477	0.33
ftv38	39	6195	0.22	1494	1.26	2723	0.49
ftv44	45	205	0.06	130	0.16	247	0.06
ftv47	48	29025	1.32	5206	7.64	9595	2.03
ftv55	56	92447	4.51	8554	18.35	31717	13.63
ftv64	65	43441	3.13	8364	27.47	22863	15.99
ftv70	71	253873	23.08	56271	250.27	23275	13.30

* Memory exhausted

Table 10. AT and ET versus cost based branching for asymmetric random instances.

n	$s(0)$	$t(0)$	$s(4)$	$t(4)$	$s(8)$	$t(8)$
60	3808	0.33	323	1.21	221	0.27
70	4528	0.38	312	1.76	247	0.27
80	9014	1.26	217	2.36	256	0.38
100	9002	1.92	174	2.64	135	0.22
200	36390	33.	858	73.	796	11.
300	178498	481.	936	287.	1506	66.
400	284994	1410.	742	541.	1216	120.
500	434576	3687.	1878	2684.	2253	439.
1000	922890	39516.	1421	15569.	3739	5360.

Table 11. AT and ET versus cost based branching for symmetric and sparse instances.

Instance	$s(0)$	$t(0)$	$s(4)$	$t(4)$	$s(8)$	$t(8)$
Degree of sparsity 50%	368736	1341.	2687	1173.	1785	70.
Degree of sparsity 75%	386468	1467.	3259	1247.	2432	117.
Degree of sparsity 90%	423284	1669.	3466	1909.	2521	141.
Degree of symmetry 0.33	58878	8.08	2919	20.00	4173	2.91
Degree of symmetry 0.66	202894	32.42	7990	54.51	32914	18.57
Full symmetry	13390054	1759.	454961	3036.	11382356	3631.

We also present an experimental analysis of the tolerance based BnB type algorithms for the ATSP. These algorithms reduce the search tree sizes substantially, and the computation times are reduced for random instances including instances with symmetry 0.33, 0.66 and sparse instances. Our experiments show that the only improved values of lower bounds (see Theorem 3) based on the value of the Assignment Problem (AP) and bottleneck upper tolerance values crucially reduce the search tree sizes for all tested instances. The exact bottleneck upper tolerance (ET) value provides the largest possible increase of the current AP value. A good approximation of the ET is attained on a shortest cycle in the AP solution. Not only is branching on the shortest subcycle efficient in the sense that a small number of subproblems are generated at each branching step, there is also an arc with the ET value or a good approximation of it in the shortest subcycle. Section 6 shows that branching on the ET arc decreases the sizes of the search trees. Even the normal cost based algorithm benefits from this effect.

An interesting direction of research is to develop book-keeping techniques that accelerate tolerances computations, and lead to reducing the solution times of the ATSP instances. Other directions of research are to incorporate a concept of bottleneck tolerances based on both upper and lower tolerance values as well as to different types of heuristics. We plan to experiment with these algorithms in a followup of this work.

Acknowledgments

The authors are very thankful to the Program Committee and three anonymous referees for their helpful comments and suggestions. The first author would like to thank Prof. dr. Jop Sibeyn for providing him with a pleasant atmosphere in the Computer Science Institute of Halle University, Germany and useful comments on an early draft of this paper.

References

1. E. Balas, P. Toth. Branch and bound methods. Chapter 10 in: The Traveling Salesman Problem. E.L. Lawler, J.K. Lenstra, A.H.G. Rinnooy Kan, D.B. Shmoys (Eds.). John Wiley & Sons, Chichester, 1985.
2. W.J.Cook, W.H. Cunningham, W.R. Pulleyblank, A. Schrijver. Combinatorial Optimization. John Wiley & Sons, Chichester, 361–402, 1998.
3. M. Fischetti, A. Lodi, P. Toth. Exact methods for the asymmetric traveling salesman problem. Chapter 2 in: The Traveling Salesman Problem and Its Variations. G. Gutin, A.P. Punnen (Eds.). Kluwer, Dordrecht, 169–194, 2002.
4. D.S. Johnson, L.A. McGeoch. Experimental analysis of heuristics for the STSP. Chapter 9 in: The Traveling Salesman Problem and Its Variations. G. Gutin, A.P. Punnen (Eds.). Kluwer, Dordrecht, 369–444, 2002.
5. D.S. Johnson, G. Gutin, L.A. McGeoch, A. Yeo, W. Zhang, A. Zverovich. Experimental analysis of heuristics for the ATSP. Chapter 10 in: The Traveling Salesman Problem and Its Variations. G. Gutin, A.P. Punnen (Eds.). Kluwer, Dordrecht, 445–489, 2002.
6. B. Goldengorin, G. Sierksma. Combinatorial optimization tolerances calculated in linear time. SOM Research Report 03A30, University of Groningen, Groningen, The Netherlands, 2003 (http://www.ub.rug.nl/eldoc/som/a/03A30/03a30.pdf).
7. H. Greenberg. An annotated bibliography for post-solution analysis in mixed integer and combinatorial optimization. In: D. L. Woodruff (Ed.). Advances in computational and stochastic optimization, logic programming, and heuristic search. Kluwer Academic Publishers, Dordrecht, 97–148, 1998.
8. K. Helsgaun. An effective implementation of the Lin-Kernigan traveling salesman heuristic. European Journal of Operational Research 126 106-130, 2000.

9. L.J. Hubert, P. Arabie, *Comparing Partitions*, Journal of Classification **2** 193–218, 1985.
10. M. Libura. Sensitivity analysis for minimum hamiltonian path and traveling salesman problems. Discrete Applied Mathematics **30** 197–211, 1991.
11. D. Naddef. Polyhedral theory and branch-and-cut algorithms for the symmetric TSP. Chapter 2 in: The Traveling Salesman Problem and Its Variations. G. Gutin, A.P. Punnen (Eds.). Kluwer, Dordrecht, 29–116, 2002.
12. G. Reinelt. TSPLIB – a Traveling Salesman Problem Library. ORSA Journal on Computing **3** 376–384, 1991.

Finding k Disjoint Triangles in an Arbitrary Graph[*]

Mike Fellows[1], Pinar Heggernes[2], Frances Rosamond[1],
Christian Sloper[2], and Jan Arne Telle[2]

[1] School of Electrical Engineering and Computer Science
University of Newcastle, Australia
{mfellows,fran}@cs.newcastle.edu.au
[2] Department of Informatics
University of Bergen, Norway
{pinar,sloper,telle}@ii.uib.no

Abstract. We consider the *NP*-complete problem of deciding whether an input graph on n vertices has k vertex-disjoint copies of a fixed graph H. For $H = K_3$ (the triangle) we give an $O(2^{2k \log k + 1.869k} n^2)$ algorithm, and for general H an $O(2^{k|H| \log k + 2k|H| \log |H|} n^{|H|})$ algorithm. We introduce a preprocessing (kernelization) technique based on crown decompositions of an auxiliary graph. For $H = K_3$ this leads to a preprocessing algorithm that reduces an arbitrary input graph of the problem to a graph on $O(k^3)$ vertices in polynomial time.

1 Introduction

For a fixed graph H and an input graph G, the H-packing problem asks for the maximum number of vertex-disjoint copies of H in G. The K_2-packing (edge packing) problem, which is equivalent to maximum matching, played a central role in the history of classical computational complexity. The first step towards the dichotomy of "good" (polynomial-time) versus "presumably-not-good" (*NP*-hard) was made in a paper on maximum matching from 1965 [10], which gave a polynomial time algorithm for that problem. On the other hand, the K_3-packing (triangle packing) problem, which is our main concern in this paper, is *NP*-hard [12].

Recently, there has been a growing interest in the area of exact exponential-time algorithms for *NP*-hard problems. When measuring time in the classical way, simply by the size of the input instance, the area of exact algorithms for *NP*-hard problems lacks the classical dichotomy of good (P) versus presumably-not-good (*NP*-hard) [16]. However, if in the area of exact algorithms for *NP*-hard problems we instead measure time in the parameterized way, then we retain the classical dichotomy of good (*FPT* - Fixed Parameter Tractable) versus presumably-not-good ($W[1]$-hard) [8]. It therefore seems that the parameterized

---

[*] This work was initiated while the first and third authors were visiting the University of Bergen.

J. Hromkovič, M. Nagl, and B. Westfechtel (Eds.): WG 2004, LNCS 3353, pp. 235–244, 2004.
© Springer-Verlag Berlin Heidelberg 2004

viewpoint gives a richer complexity framework. In fact, a formal argument for this follows from the realization that the non-parameterized viewpoint, measuring time by input size, is simply a special case of the parameterized viewpoint with the parameter chosen to be the input size. Parameterized thusly, any problem is trivially *FPT* and the race for the best *FPT* algorithm is precisely the same as the race for the best non-parameterized exact algorithm. Note that for any optimization or decision problem, there are many interesting possibilities for choice of parameter, that can be guided by both practical and theoretical considerations, see for example [11] for a discussion of five different parameterizations of a single problem. In our opinion, the relevant discussion for the field of exact algorithms for *NP*-hard problems is therefore not "parameterized or non-parameterized?" but rather "which parameter?".

In this paper our focus is on parameterized algorithms for deciding whether a graph G has k disjoint copies of K_3, with the integer k being our parameter. On input (G, k), where G is a graph on n vertices, an *FPT* algorithm for this problem is one with runtime $O(n^\alpha f(k))$, for a constant α and an unrestricted function $f(k)$. We want, of course, both α and the growth rate of $f(k)$ to be as small as possible.

A practical spinoff from the field of parameterized algorithms for *NP*-hard problems has been a theoretical focus on the algorithmic technique of preprocessing, well-known from the heuristic algorithms community. In fact, the parameterized problems having *FPT* algorithms are *precisely* the parameterized problems where preprocessing can in polynomial time reduce a problem instance (G, k) to a kernel, *i.e.*, a decision-equivalent problem instance (G', k') where the size of G' is bounded by a function of k (only), and where also $k' \leq k$ [9]. One direction of this fact is trivial, since any subsequent brute-force algorithm on (G', k') would give an overall *FPT* algorithm. In the other direction, assume we have an *FPT* algorithm with runtime $O(n^\alpha f(k))$ and consider an input (G, k) on n vertices. If $n \geq f(k)$ then the runtime of the *FPT* algorithm on this instance is in fact polynomial and can be seen as a reduction to the trivial case. On the other hand, if $n \leq f(k)$ then the instance (G, k) already satisfies the kernel requirements. Note that in this case the kernel size $f(k)$ is exponential in k, and a smaller kernel is usually achievable. For this reason, in the field of parameterized algorithms for *NP*-hard problems, it can be argued that there are two distinct races [11]:

- Find the fastest *FPT* algorithm for the problem.
- Find the smallest polynomial-time computable kernelization for the problem.

In this paper, we enter the parameterized K_3-packing problem into both these races, giving on the one hand an $O(2^{2k \log k + 1.869k} n^2)$ *FPT* algorithm, and on the other hand an $O(k^3)$ kernelization. Our *FPT* algorithm is derived by an application of a fairly new technique known as greedy localization [14], and our kernelization algorithm by a non-standard application of the very recently introduced notion of Crown Reduction Rules [4, 5, 11]. We end the paper by asking how well these two results on K_3-packing generalize to H-packing. It turns out that the *FPT* algorithm generalizes quite easily, giving *FPT* algorithms for deciding whether an input graph G has k disjoint copies of an arbitrary connected H. However, we presently do not see how to generalize the kernelization algorithm.

Just in time for the final version of this paper we realized that Theorem 6.3 in [2] can be used to give a $2^{O(k)}$ algorithm for graph packing using color coding. However, we still believe our result to be of practical interest as the constants in color coding can be impractical.

The next section gives some basic graph terminology. We then proceed in Sections 3, 4 and 5 with the kernelization results, before continuing with the *FPT* algorithm in Section 6 for K_3 and in Section 7 for general H.

2   Preliminaries

We assume simple, undirected, connected graphs $G = (V, E)$, where $|V| = n$. The neighbors of a vertex v are denoted by $N(v)$. For a set of vertices $A \subseteq V$, $N(A) = \{v \notin A \mid uv \in E \text{ and } u \in A\}$, and the subgraph of G induced by A is denoted by $G(A)$. For ease of notation, we will use informal expressions like $G \setminus u$ to denote $G(V \setminus \{u\}, E)$, $G \setminus U$ to denote $G(V \setminus U, E)$, and $G \setminus e$ to denote $(V, E \setminus \{e\})$, where u is a vertex, U is a vertex set, and e is an edge in G. A subset S of V is a *separator* if $G \setminus S$ is disconnected.

An *H-packing* W of G is a collection of disjoint copies of graph H in G. We will use $V(W)$ to denote the vertices of G that appear in W, and $E(W)$ to denote the edges. A *matching* is a K_2-packing.

We will in the following two sections describe a set of reduction rules. If any of these rules can be applied to G, we say that G is *reducible*, otherwise *irreducible*.

3   Reduction Rules for K_3-Packing

Let us start with a formal definition of the problem that we are solving:

k-K_3-PACKING (TRIANGLE PACKING)
INSTANCE: Graph $G = (V, E)$
PARAMETER: k
QUESTION: Does G have k disjoint copies of K_3?

We say that a graph G has a k-K_3-packing if the answer to the above question is "yes." In this section, we identify vertices and edges of the input graph that can be removed without affecting the solution of the k-K_3-PACKING problem.

Definition 1. *If vertices a, b, and c induce a K_3, we say that vertex a sponsors edge bc. Likewise, edge bc sponsors vertex a.*

We start with two simple observations that also give preprocessing rules useful to delete vertices and edges that cannot participate in any triangle.

Reduction Rule 1. *If $e \in E$ has no sponsor then G has a k-K_3-packing $\Leftrightarrow$ $G \setminus e$ has a k-K_3-packing.*

Reduction Rule 2. *If $u \in V$ has no sponsor then G has a k-K_3-packing $\Leftrightarrow$ $G \setminus u$ has a k-K_3-packing.*

Both observations are trivially true, and let us remove vertices and edges from the graph so that we are left with a graph containing only vertices and edges that could potentially form a K_3.

Reduction Rule 3. *If $u \in V$ sponsors at least $3k - 2$ disjoint edges then G has a k-K_3-packing $\Leftrightarrow G \setminus u$ has a $(k - 1)$-K_3-packing.*

Proof. ($\Rightarrow$:) This direction is clear as removing one vertex can decrease the number of K_3s by at most one.
($\Leftarrow$:) If $G \setminus u$ has a $(k - 1)$-K_3-packing S, then S can use vertices from at most $3(k - 1) = 3k - 3$ of the disjoint edges sponsored by u. This leaves at least one edge that can form a K_3 with u, thus raising the number of K_3s to k.   □

4   Reducing Independent Sets – Crown Reduction

In this section we will first give a trivial reduction rule that removes a specific type of independent sets. This reduction rule is then generalized and replaced by a more powerful rule that allows us to reduce any 'large' independent set in the graph.

Reduction Rule 4. *If $\exists u, v \in V$ such that $N(u) = N(v) = \{a, b\}$ and $ab \in E$ then G has a k-K_3-packing $\Leftrightarrow G \setminus u$ has a k-K_3-packing.*

Proof. This is trivial as it is impossible to use both u and v in any K_3-packing.
□

This reduction rule identifies a redundant vertex and removes it. The vertex is redundant because it has a stand-in that can form a K_3 in its place and there is no use for both vertices. Generalizing, we try to find a set of vertices such that there is always a distinct stand-in for each vertex in the set.

Definition 2. *A* crown decomposition *(H, C, R) in a graph $G = (V, E)$ is a partitioning of the vertices of the graph into three sets H, C, and R that have the following properties:*

1. *H (the head) is a separator in G such that there are no edges in G between vertices belonging to C and vertices belonging to R.*
2. *$C = C_u \cup C_m$ (the crown) is an independent set in G.*
3. *$|C_m| = |H|$, and there is a perfect matching between C_m and H.*

Crown-decomposition is a recently introduced idea that supports nontrivial and powerful preprocessing (reduction) rules for a wide variety of problems, and that performs very well in practical implementations [4, 11, 3]. It has recently been shown that if a graph admits a crown decomposition, then a crown decomposition can be computed in polynomial time [1]. The following theorem can be deduced from [4, page 7], and [11, page 8].

Theorem 1. *Any graph G with an independent set I, where $|I| \geq \frac{n}{2}$, has a crown decomposition (H, C, R), where $H \subseteq N(I)$, that can be found in linear time, given I.*

For most problems, including k-K_3-PACKING, it is not clear how a crown decomposition can directly provide useful information. We introduce here the idea of creating an auxiliary graph model where a crown decomposition in the auxiliary graph is used to identify preprocessing reductions for the original graph.

For k-K_3-PACKING we will show that an auxiliary graph model can be created to reduce large independent sets in the problem instance. Consider an independent set I in a graph G. Let E_I be the set of edges that are sponsored by the vertices of I.

The auxiliary model that we consider is a bipartite graph G_I where we have one vertex u_i for every vertex v_i in I and one vertex f_j for every edge e_j in E_I. For simplicity, we let both sets $\{e_j \mid e_j \in E_I\}$ and $\{f_j \mid e_j \in E_I\}$ be denoted by E_I. The edges of G_I are defined as follows: let $u_i f_j$ be an edge in G_I if and only if u_i sponsors f_j.

We now prove the following generalization of Reduction Rule 4. This rule now replaces rule 4.

Reduction Rule 5. *If G_I has a crown decomposition $(H, C_m \cup C_u, R)$ where $H \subseteq E_I$ then G has a k-K_3-packing $\Leftrightarrow G \setminus C_u$ has a k-K_3-packing.*

Proof. Assume on the contrary that G_I has a crown decomposition $(H, C_m \cup C_u, R)$, where $H \subseteq E_I$ and G has a k-K_3-packing W^* but $G \setminus C_u$ has no k-K_3-packing. This implies that some of the vertices of C_u were used in the k-K_3-packing W^* of G.

Let H^* be the set of vertices in H whose corresponding edges in G use vertices from $C = C_m \cup C_u$ to form K_3s in the k-K_3-packing W^* of G. Note that vertices in C_u can only form K_3s with edges of G that correspond to vertices in H. Observe that each edge corresponding to a vertex in H^* uses exactly one vertex from C. Further, $|H^*| \leq |H|$. By these two observations it is clear that every edge whose corresponding vertex is in H^* can be assigned a vertex from C_m to form a K_3. Thus C_u is superfluous, contradicting the assumption. $\square$

Observation 1. *If a bipartite graph $G = (V \cup V', E)$ has two crown decompositions (H, C, R) and (H', C', R') where $H \subseteq V$ and $H' \subseteq V$, then G has a crown decomposition $(H'' = H \cup H', C'' = C \cup C', R'' = R \cap R')$.*

It is easy to check that all properties of a crown decomposition hold for (H'', C'', R'').

Lemma 1. *If G has an independent set I such that $|I| > 2|E_I|$ then we can in polynomial time find a crown decomposition $(H, C_m \cup C_u, R)$ where $H \subseteq E_I$, and $C_u \neq \emptyset$.*

Proof. Assume on the contrary that G has an independent set I such that $|I| > 2|E_I|$ but G has no crown decomposition with the properties stated in the lemma.

By Theorem 1 the bipartite model G_I as described above has a crown decomposition $(H, C = C_m \cup C_u, R)$ where $H \subseteq N(I)$ and consequently $C \subseteq I$. If $|I \setminus C| > |E_I|$ then $G_I \setminus C$ has a crown decomposition (H', C', R'), where $H' \subset N(I)$. By Observation 1 (H, C, R) and (H', C', R') could be combined to form a bigger crown. Let $(H'', C'' = C_m'' \cup C_u'', R'')$ be the largest crown decomposition that can be obtained by repeatedly finding a new crown in $I \setminus C$ and combining it with the existing crown decomposition to form a new head and crown.

By our assumption $C_u'' = \emptyset$. Since $|C_m''| = |H''| \leq E_I$ and it follows from Theorem 1 that $|I \setminus C_m''| \leq |E_I|$ (otherwise a new crown could be formed), we have that $|I| = |C_m''| + |I \setminus C_m''| \leq |E_I| + |E_I| \leq 2|E_I|$ contradicting the assumption that $|I| > 2|E_I|$. $\square$

5   Computing a Cubic Kernel

We now introduce a polynomial time algorithm that either produces a k-K_3-packing or finds a valid reduction of any input graph $G = (V, E)$ of at least a certain size. We show that this algorithm gives an $O(k^3)$ kernel for k-K_3-PACKING.

The algorithm has the following steps:

1. Reduce by Rule 1 and 2 until neither apply.
2. Greedily, find a maximal K_3-packing W in G. If $|V(W)| \geq 3k$ then ACCEPT.
3. Find a maximal matching Q in $G \setminus V(W)$. If a vertex $v \in V(W)$ sponsors more than $3k - 3$ matched edges, then v can be reduced by Reduction Rule 3.
4. If possible, reduce the independent set $I = V \setminus (V(W) \cup V(Q))$ with Reduction Rule 5.

We now give the following lemma to prove our result:

Lemma 2. *If $|V| > 108k^3 - 72k^2 - 18k$ then the preprocessing algorithm will either find a k-K_3-packing or it will reduce $G = (V, E)$.*

Proof. Assume on the contrary to the stated lemma that $|V| > 108k^3 - 72k^2 - 18k$, but that the algorithm produced neither a k-K_3-packing nor a reduction of G.

By the assumption the maximal packing W is of size $|V(W)| < 3k$.

Let Q be the maximal matching obtained by step 2 of the algorithm.

Claim. $|V(Q)| \leq 18k^2 - 18k$

> *Proof.* Assume on the contrary that $|V(Q)| > 18k^2 - 18k$. Observe that no edge in $G \setminus V(W)$ can sponsor a vertex in $G \setminus V(W)$ as this would contradict that W is maximal, therefore all edges in the the maximal matching Q are sponsored by at least one vertex in $V(W)$. If $|V(Q)| > 18k^2 - 18k$, Q contains more than $9k^2 - 9k$ edges. Thus at least one vertex $v \in V(W)$ sponsors more than $\frac{9k^2 - 9k}{3k} = 3k - 3$ edges. Consequently v should have been removed by Reduction Rule 3, contradicting the assumption that no reduction of G took place. We have reached a contradiction, thus the assumption that $|V(Q)| > 18k^2 - 18k$ must be wrong. $\square$

Let $I = V \setminus (V(W) \cup V(Q))$. Note that I is an independent set.

Claim. $|I| \leq 108k^3 - 90k^2$

Proof. Assume on the contrary that $|I| > 108k^3 - 90k^2$. Observe that each edge that is sponsored by a vertex of I is either in the subgraph of G induced by $V(W)$, or is an edge between $V(W)$ and $V(Q)$. The are at most $|E_I| = |V(Q)| \cdot |V(W)| + |V(W)|^2 \leq (18k^2 - 18^k) \cdot 3k + (3k)^2 \leq 54k^3 - 45k^2$ such edges.

By Lemma 1 there are no more than $2|E_I| = 108k^3 - 90k^2$ vertices in I, which contradicts the assumption that $|I| > 108k^3 - 90k^2$.   □

Thus the total size $|V| = |V(W)| + |V(Q)| + |I| \leq 3k + 18k^2 - 18k + 108k^3 - 90k^2 = 108k^3 - 72k^2 - 18k$. This contradicts the assumption that $|V| > 108k^3 - 72k^2 - 18k$.   □

Corollary 1. *Any instance* (G, k) *of* k-K_3-PACKING *can be reduced to a problem kernel of size* $O(k^3)$.

Proof. This follows from Lemma 2, as we can repeatedly run the algorithm until it fails to reduce the graph further. By Lemma 2 the resulting graph is then of size $O(k^3)$.   □

Note that a $O(k^3)$ kernel gives us a trivial *FPT*-algorithm by testing all $O(\binom{k^3}{3k})$ subsets in a brute force manner. This leads to an $O(2^{9k \log k} + poly(n, k))$ algorithm. However, we will show in the next section that another *FPT* technique yields a faster algorithm.

6   Winning the *FPT* Runtime Race

In this section we give a faster *FPT*-algorithm using the technique of "greedy localization" and a bounded search tree.

We begin with the following crucial observation.

Observation 2. *Let W be a maximal K_3-packing, and let W^* be a k-K_3-packing. Then for each K_3 T of W^* we have that $V(T) \cap V(W) \neq \emptyset$.*

Proof. Assume on the contrary that there exists a K_3 T in W^* such that $V(T) \cap V(W) = \emptyset$. This implies that $V(T) \cup V(W)$ is a K_3-packing contradicting that W is a maximal packing.   □

Theorem 2. *It is possible to determine whether a graph $G = (V, E)$ has a k-K_3-packing in time $O(2^{2k \log k + 1.869k} n^2)$.*

Proof. Let W be a maximal K_3-packing. If $|V(W)| \geq 3k$ we have a K_3-packing. Otherwise, create a search tree T. At each node we will maintain a collection $S^i = S_1^i, S_2^i, \ldots, S_k^i$ of vertex subsets. These subsets represent the k triangles of the solution, and at the root node all subsets are empty.

From the root node, create a child i for every possible subset W_i of $V(W)$ of size k. Let the collection at each node i contain k singleton sets, each containing a vertex of W_i.

We say that a collection $S^i = S_1^i, S_2^i, \ldots, S_k^i$ is a *partial solution* of a k-K_3-packing W^* with k disjoint triangles $W_1^*, W_2^*, \ldots, W_k^*$ if and only if $S_j^i \subseteq V(W_j^*)$ for $1 \leq j \leq k$.

For a child i, consider its collection $S_i = S_1^i, S_2^i, \ldots, S_k^i$. Add vertices to S_1^i such that S_1^i induces a K_3 in G, continue in a greedy fashion to add vertices to S_2^i, S_3^i and so on. If we can complete all k subsets we have a k-K_3 packing. Otherwise, let S_j^i be the set first set which is not possible to complete, and let V' be the vertices we have added to S^i so far. We can now make the following claim.

Claim. If $S^i = S_1^i, S_2^i, \ldots, S_k^i$ is a partial solution then there exists a vertex $v \in V'$ such that $S^i = S_1^i, \ldots, (S_j^i \cup \{v\}), \ldots, S_k^i$ is a partial solution.

> *Proof.* Assume on the contrary that $S^i = S_1^i, S_2^i, \ldots, S_k^i$ is a partial solution but that there exists no vertex $v \in V'$ such that $S^i = S_1^i, (S_j^i \cup \{v\}), \ldots, S_k^i$ is a partial solution. This implies that $V(W_j^*) \cap V' = \emptyset$, but then we could add $V(W_j^*) \setminus S_j^i$ to S_j^i to form a new K_3, thus contradicting that it was not possible to complete S_j^i. □

We now create one child u of node i for every vertex in $u \in V'$. The collection at child u is $S^i = S_1^i, (S_j^i \cup \{u\}), \ldots, S_k^i$. This is repeated at each node l, until we are unable to complete any set in node l's collection, i.e. $V' = \emptyset$.

By Observation 2 we know that if there is k-K_3-packing then one of the branchings from the root node will have a partial solution. Claim 1 guarantees that this solution is propagated down the tree until finally completed at level $2k$.

At each level the collections S at the nodes grow in size, thus we can have at most $2k$ levels in the search tree. Observe that at height h in the search tree $|V'| < 2k - h$, thus fan-out at height h is limited to $2k - h$. The total size of the tree is then at most $\binom{3k}{k} 2k \cdot (2k - 1) \cdot \ldots = \binom{3k}{k} \cdot 2k! = \frac{(3k)!}{k!}$. Using Stirling's approximation and suppressing some constant factors we have $\frac{(3k)!}{k!} \approx 3.654^k \cdot k^{2k} = 2^{2k \log k + 1.869k}$. At each node we need $O(n^2)$ time to maximize the sets. Hence, the total running time is $O(2^{2k \log k + 1.869k} n^2)$ □

Note that it is, of course, possible to run the search tree algorithm from this section on the kernel obtained in the previous section. The total running time is then $O(2^{2k \log k + 1.869k} k^6 + p(n, k))$. This could be useful if n is much larger than k as the additive exponential (rather than multiplicative) factor becomes significant.

7   Packing Arbitrary Graphs

In their paper from 1978 Hell and Kirkpatrick [12] prove that k-H-packing for any connected graph H of 3 or more vertices is NP-complete. We will in this section show that our search tree technique for k-K_3-packing easily generalizes to arbitrary graphs H, thus proving that packing any subgraph is in *FPT*.

k-H-PACKING
INSTANCE: Graph $G = (V, E)$
PARAMETER: k
QUESTION: Does G have at least k disjoint copies of H?

Theorem 3. *It is possible to determine whether a graph $G = (V, E)$ has a k-H-packing in time $O(2^{k|H|\log k + 2k|H|\log|H|} n^{|H|})$.*

Proof. The proof is analogous to the proof of Theorem 2. However, as we no longer can depend upon perfect symmetry in H (since H is not necessarily complete), we must maintain a collection of ordered sequences at each tree-node. Each sequence represents a partial H-subgraph.

The possible size of V' increases to $k|H| - k$. Then when we want to determine which v of V' to add to the sequence, we must try every v in every position in H. Thus the fan-out at each node increases to $k|H|^2 - k|H|$. The height of the tree likewise increases to at most $k|H| - k$. Thus the new tree size is $\binom{k|H|}{k}(k|H|^2 - k|H|)^{k|H|-k}$, which is strictly smaller than $k^{k|H|}|H|^{2k|H|}$ or $2^{k|H|\log k + 2k|H|\log|H|}$. $\square$

8   Summary and Open Problems

Our main results in the two *FPT* races are:

(1) We have shown an $O(k^3)$ problem kernel for the problem of packing k triangles.

(2) We have shown that for any fixed graph H, the problem of packing k Hs is in *FPT* with a parameter function of the form $O(2^{O(k\log k)})$ and more practical constants than [2].

In addition to "upper bound" improvements to these initial results, which would be the natural course for further research – now that the races are on – it would also be interesting to investigate lower bounds, if possible.

It would be interesting to investigate the "optimality" of the form of our *FPT* results in the sense of [6, 7]. Can it be shown that there is no $O(2^{o(k)})$ *FPT* algorithm for k-H-PACKING unless *FPT= M*[1]?

Many parameterized problems admit linear problem kernels. In fact, it appears that most naturally parameterized problems in *APX* are in *FPT* and have linear problem kernels. However, it seems unlikely that *all FPT* problems admit linear kernels. We feel that k-K_t-PACKING is a natural candidate for an *FPT* problem where it may not be possible to improve on $O(k^t)$ kernelization. Techniques for the investigation of lower bounds on kernelization are currently lacking, but packing problems may be a good place to start looking for them.

References

1. F. AbuKhzam and H. Suters, Computer Science Department, University of Tennessee, Knoxville, private communications, Dec. 2003.
2. N. Alon, R. Yuster, U. Zwick. Color-Coding. J. ACM, pp. 844-856, 1995

3. F. Abu-Khzam, R. Collins, M. Fellows and M. Langston. Kernelization Algorithms for the Vertex Cover Problem: Theory and Experiments. *Proceedings ALENEX 2004*, Springer-Verlag, *Lecture Notes in Computer Science* (2004), to appear.
4. B. Chor, M. Fellows, and D. Juedes. Saving k Colors in Time $O(n^{5/2})$. Manuscript, 2003.
5. B. Chor, M. Fellows, and D. Juedes. Linear Kernels in Linear Time, or How to Save k Colors in $O(n^2)$ steps. Proceedings of WG2004, Springer-Verlag, *Lecture Notes in Computer Science* (2004)
6. L. Cai and D. Juedes. On the existence of subexponential parameterized algorithms. *Journal of Computer and System Sciences* 67 (2003).
7. R. Downey, V. Estivill-Castro, M. Fellows, E. Prieto-Rodriguez and F. Rosamond. Cutting Up is Hard to Do: the Parameterized Complexity of k-Cut and Related Problems. Electronic Notes in Theoretical Computer Science 78 (2003), 205–218.
8. R. Downey and M. Fellows. *Parameterized Complexity* Springer-Verlag (1999).
9. R. Downey, M. Fellows and U. Stege, Parameterized Complexity: A Framework for Systematically Confronting Computational Intractability, in: *Contemporary Trends in Discrete Mathematics*, (R. Graham, J. Kratochvil, J. Nesetril and F. Roberts, eds.), AMS-DIMACS Series in Discrete Mathematics and Theoretical Computer Science 49, pages 49-99, 1999.
10. J.Edmonds. Paths, trees and flowers, *Can.J.Math.*, 17, 3, pages 449-467, 1965.
11. M. Fellows. Blow-ups, Win/Wins and Crown Rules: Some New Directions in *FPT*. *Proceedings WG 2003*, Springer Verlag LNCS 2880, pages 1-12, 2003.
12. P. Hell and D. Kirkpatrick. On the complexity of a generalized matching problem. *Proceedings of 10th ACM Symposium on theory of computing*, pages 309-318, 1978.
13. C. A. J. Hurkens, and A. Schrijver. On the size of systems of sets every t of which have an SDR, with an application to the worst-case ratio of heuristics for packing problems, *SIAM J. Disc. Math. 2*, pages 68-72, 1989.
14. W. Jia, C. Zhang and J. Chen. An efficient parameterized algorithm for m-set packing, *Journal of Algorithms*, 50(1):106–117, 2004.r.
15. V. Kann. Maximum bounded 3-dimensional matching is MAX SNP-complete, *Inform. Process. Lett.* 37, pages 27-35, 1991.
16. G. Woeginger. Exact algorithms for NP-hard problems: A survey, *Combinatorial Optimization - Eureka! You shrink!*, M. Juenger, G. Reinelt and G. Rinaldi (eds.). LNCS 2570, Springer, pages 185-207, 2003.

Exact (Exponential) Algorithms for the Dominating Set Problem

Fedor V. Fomin[1,*], Dieter Kratsch[2], and Gerhard J. Woeginger[3]

[1] Department of Informatics, University of Bergen, N-5020 Bergen, Norway
fomin@ii.uib.no
[2] LITA, Université de Metz, 57045 Metz Cedex 01, France
kratsch@sciences.univ-metz.fr
[3] Department of Mathematics and Computer Science, TU Eindhoven,
P.O. Box 513, 5600 MB Eindhoven, The Netherlands
g.j.woeginger@tue.nl

Abstract. We design fast exact algorithms for the problem of computing a minimum dominating set in undirected graphs. Since this problem is NP-hard, it comes with no big surprise that all our time complexities are exponential in the number n of vertices. The contribution of this paper are 'nice' exponential time complexities that are bounded by functions of the form c^n with reasonably small constants $c < 2$: For arbitrary graphs we get a time complexity of 1.93782^n. And for the special cases of split graphs, bipartite graphs, and graphs of maximum degree three, we reach time complexities of 1.41422^n, 1.73206^n, and 1.51433^n, respectively.

1 Introduction

Nowadays, it is common believe that NP-hard problems can not be solved in polynomial time. For a number of NP-hard problems, we even have strong evidence that they cannot be solved in sub-exponential time. For these problems the only remaining hope is to design exact algorithms with good exponential running times. How good can these exponential running times be? Can we reach 2^{n^2} for instances of size n? Can we reach 10^n? Or even 2^n? Or can we reach c^n for some constant c that is very close to 1? The last years have seen an emerging interest in attacking these questions for concrete combinatorial problems: There is an $O^*(1.2108^n)$ time algorithm for independent set (Robson [13]); an $O^*(2.4150^n)$ time algorithm for graph coloring (Eppstein [4]); an $O^*(1.4802^n)$ time algorithm for 3-Satisfiability (Dantsin & al. [2]). We refer to the survey paper [14] by Woeginger for an up-to-date overview of this field. In this paper, we study the *dominating set* problem from this exact (exponential) algorithms point of view.

Basic Definitions. Let $G = (V, E)$ be an undirected, simple graph without loops. We denote by n the number of vertices of G. The open *neighborhood* of a vertex v is denoted by $N(v) = \{u \in V : \{u, v\} \in E\}$, and the closed

---

* F. Fomin is supported by Norges forskningsråd project 160778/V30.

J. Hromkovič, M. Nagl, and B. Westfechtel (Eds.): WG 2004, LNCS 3353, pp. 245–256, 2004.

neighborhood of v is denoted by $N[v] = N(V) \cup \{v\}$. The degree of a vertex v is $|N(v)|$. For a vertex set $S \subseteq V$, we define $N[S] = \bigcup_{v \in S} N[v]$ and $N(S) = N[S] - S$. The subgraph of G induced by S is denoted by $G[S]$. We will write $G - S$ short for $G[V - S]$. A set $S \subseteq V$ of vertices is a *clique*, if any two of its elements are adjacent; S is *independent*, if no two of its elements are adjacent; S is a *vertex cover*, if $V - S$ is an independent set.

Throughout this paper we use the so-called *big-Oh-star* notation, a modification of the big-Oh notation that suppresses polynomially bounded terms: We will write $f = O^*(g)$ for two functions f and g, if $f(n) = O(g(n)\text{poly}(n))$ holds with some polynomial $\text{poly}(n)$. We say that a problem is solvable in *sub-exponential* time in n, if there is an effectively computable monotone increasing function $g(n)$ with $\lim_{n\to\infty} g(n) = \infty$ such that the problem is solvable in time $O(2^{n/g(n)})$.

The Dominating Set Problem. Let $G = (V, E)$ be a graph. A set $D \subseteq V$ with $N[D] = V$ is called a *dominating set* for G; in other words, every vertex in G must either be contained in D or adjacent to some vertex in D. A set $A \subseteq V$ *dominates* a set $B \subseteq V$ if $B \subseteq N[A]$. The *domination number* $\gamma(G)$ of a graph G is the cardinality of a smallest dominating set of G. The *dominating set problem* asks to determine $\gamma(G)$ and to find a dominating set of minimum cardinality. The dominating set problem is one of the fundamental and well-studied classical NP-hard graph problems (Garey & Johnson [6]). For a large and comprehensive survey on domination theory, we refer the reader to the books [8, 9] by Haynes, Hedetniemi & Slater. The dominating set problem is also one of the basic problems in parameterized complexity (Downey & Fellows [3]); it is contained in the parameterized complexity class W[2]. Further recent investigations of the dominating set problem can be found in Albers & al. [1] and in Fomin & Thilikos [5].

Results and Organization of This Paper. What are the best time complexities for dominating set in n-vertex graphs that we can possibly hope for? Well, of course there is the trivial $O^*(2^n)$ algorithm that simply searches through all the 2^n subsets of V. But can we hope for a sub-exponential time algorithm, maybe with a time complexity of $O^*(2^{\sqrt{n}})$? Section 2 provides the answer to this question: No, probably not, unless some very unexpected things happen in computational complexity theory ... Hence, we should only hope for time complexities of the form $O^*(c^n)$, with some small value $c < 2$. And indeed, Section 3 presents such an algorithm with a time complexity of $O^*(1.93782^n)$. This algorithm combines a recursive approach with a deep result from extremal graph theory. The deep result is due to Reed [12], and it provides an upper bound on the domination number of graphs of minimum degree three.

Furthermore, we study exact exponential algorithms for the dominating set problem on some special graph classes: In Section 4, we design an $O^*(1.41422^n)$ time algorithm for split graphs, and an $O^*(1.73206^n)$ time algorithm for bipartite graphs. In Section 5, we derive an $O^*(1.51433^n)$ time algorithm for graphs of maximum degree three. Note that for these three graph classes, the dominating set problem remains NP-hard (Garey & Johnson [6], Haynes, Hedetniemi & Slater [9]).

2   A Negative Observation

We will show that the existence of a sub-exponential time algorithm for the dominating set problem would be highly unlikely. Our (straightforward) argument exploits the structural similarities between the dominating set problem and the vertex cover problem: "Given a graph, find a vertex cover of minimum cardinality".

Proposition 1. *Let* $G = (V, E)$ *be a graph. Let* G^+ *be the graph that results from* G *by adding for every edge* $e = \{u, v\} \in E$ *a new vertex* $x(e)$ *together with the two new edges* $\{x(e), u\}$ *and* $\{x(e), v\}$.

Then the graph G *has a vertex cover of size at most* k, *if and only if the graph* G^+ *has a dominating set of size at most* k.

Proposition 2. *(Johnson & Szegedy [11])*
If the vertex cover problem on graphs of maximum degree three can be solved in sub-exponential time, then also the vertex cover problem on arbitrary graphs can be solved in sub-exponential time.

Proposition 3. *(Impagliazzo, Paturi & Zane [10])*
If the vertex cover problem (on arbitrary graphs) can be solved in sub-exponential time, then the complexity classes SNP and SUBEXP satisfy SNP $\subseteq$ SUBEXP (and this is considered a highly unlikely event in computational complexity theory).

Now suppose that the dominating set problem is solvable in sub-exponential time. Take an instance $G = (V, E)$ of the vertex cover problem with maximum degree at most three, and construct the corresponding graph G^+. Note that G^+ has at most $|V| + |E| \le 5|V|/2$ vertices; hence, its size is linear in the size of G. Solve the dominating set problem for G^+ in sub-exponential time. Proposition 1 yields a sub-exponential time algorithm for vertex cover in graphs with maximum degree at most three. Propositions 2 and 3 yield that SNP $\subseteq$ SUBEXP.

3   An Exact Algorithm for Arbitrary Graphs

In this section we present the main result of our paper. It is the first exact algorithm for the dominating set problem breaking the natural $\Omega(2^n)$ barrier for the running time: We present an $O^*(1.93782^n)$ time algorithm to compute a minimum dominating set on any graph. Our algorithm heavily relies on the following result of Reed to restrict the search space.

Proposition 4. *(Reed [12])*
Every graph on n *vertices with minimum degree at least three has a dominating set of size at most* $3n/8$.

In fact, we will tackle the following generalization of the dominating set problem: An input for this generalization consists of a graph $G = (V, E)$ and a subset $X \subseteq V$. We say that a set $D \subseteq V$ dominates X, if $X \subseteq N[D]$. The goal is to find a dominating set D for X of minimum cardinality. (Obviously, setting $X := V$ yields the classical dominating set problem). We will derive an exact $O^*(1.93782^n)$ time algorithm for this generalization.

The algorithm is based on the so-called pruning the search tree technique. The idea is to branch into subcases and to remove all vertices of degree one and two, until we terminate with a graph with all vertices of degree zero or at least three. Denote by V' the set of all vertices of degree at least three in this final graph. Let $t = |V'|$ and let $G' = G[V']$. Then Proposition 4 yields that there exists some vertex set in G' with at most $3t/8$ vertices that dominates all vertices of G'; consequently, there exists also a dominating set for $X' = X \cap V'$ of size at most $3t/8$ in G'. We simply test all possible subsets with up to $3t/8$ vertices to find a minimum dominating set D' for X' in G'. By using Stirling's approximation $x! \approx x^x e^{-x} \sqrt{2\pi x}$ for factorials, and by suppressing some polynomial factors, we see that the number of tested subsets is at most

$$\binom{t}{3t/8} = \frac{(t)!}{(3t/8)!\,(5t/8)!} = O^*(8^t \cdot 3^{-3t/8} \cdot 5^{-5t/8}) = O^*(1.93782^t),$$

where $8/(3^{3/8} \cdot 5^{5/8})$ is approximately 1.9378192. This test can be done in time $O^*(\sum_{i=1}^{3t/8} \binom{t}{i}) = O^*(1.93782^t)$. Finally, we add all degree zero vertices of X to the set D' to obtain a minimum dominating set of G.

Now let us discuss the branching into subcases. While there is a vertex of degree one or two, we pick such a vertex, say v, and we recurse distinguishing four cases depending on the degree of v and whether $v \in X$ or not.

Case A: The Vertex v Is of Degree One and $v \in V - X$. In this case there is no need to dominate the vertex v and there always exists a minimum dominating set for X that does not contain v. Then a minimum dominating set for $X - \{v\}$ in $G - \{v\}$ is also a minimum dominating set for X in G, and thus we recurse on $G - \{v\}$ and $X - \{v\}$.

Case B: The Vertex v Is of Degree One and $v \in X$. Let w be the unique neighbor of v. Then there always exists a minimum dominating set for X that contains w, but does not contain v. If D' is a minimum dominating set for $X - N[w]$ in $G - \{v, w\}$ then $D' \cup \{w\}$ is a minimum dominating set for X in G, and thus we recurse on $G - \{v, w\}$ and $X - N[w]$.

We need the following auxiliary result.

Lemma 1. *Let v be a vertex of degree 2 in G, and let u_1 and u_2 be its two neighbors. Then for any subset $X \subseteq V$ there is a minimum dominating set D for X such that one of the following holds.*

(i) $u_1 \in D$ and $v \notin D$;
(ii) $v \in D$ and $u_1, u_2 \notin D$;
(iii) $u_1 \notin D$ and $v \notin D$.

Proof. If there exists a minimum dominating set D for X that contains u_1 then there exists a minimum dominating set D' for X that contains u_1 but not v. In fact, if $v \in D$, then $D' = (D - \{v\}) \cup \{u_2\}$ is a dominating set for X and

$|D'| \leq |D|$. Similarly, if there exists a minimum dominating set for X that contains u_2 then there exists a minimum dominating set for X that contains u_2 but not v.

Thus we are left with five possibilities how v, u_1, u_2 might show up in a minimum dominating set D for X: (a) $u_1, u_2, v \notin D$; (b) $v \in D$ and $u_1, u_2 \notin D$; (c) $u_1 \in D$ and $v, u_2 \notin D$; (d) $u_2 \in D$ and $v, u_1 \notin D$; (e) $u_1, u_2 \in D$ and $v \notin D$. Now (i) is equivalent to (c) or (e), (ii) is equivalent to (b), and (iii) is equivalent to (a) or (d). This concludes the proof.   □

Now consider a vertex v of degree two. Depending on whether $v \in X$ or not we branch in different ways. Additionally, the search is restricted to those minimum dominating sets D satisfying the conditions of Lemma 1.

Case C: The Vertex v of Degree 2 and $v \in V - X$. Let u_1 and u_2 be the two neighbors of v in G. By Lemma 1, we can branch into three subcases for a minimum dominating set D:

(C.1): $u_1 \in D$ and $v \notin D$. In this case if D' is a minimum dominating set for $X - N[u_1]$ in $G - \{u_1, v\}$ then $D' \cup \{u_1\}$ is a minimum dominating set for X in G, and thus we recurse on $G - \{u_1, v\}$ and $X - N[u_1]$.

(C.2): $v \in D$ and $u_1, u_2 \notin D$. In this case if D' is a minimum dominating set for $X - \{u_1, u_2\}$ in $G - \{u_1, v, u_2\}$ then $D' \cup \{v\}$ is a minimum dominating set for X in G, and thus we recurse on $G - \{u_1, v, u_2\}$ and $X - \{u_1, u_2\}$.

(C.3): $u_1 \notin D$ and $v \notin D$. In this case a minimum dominating set for X in $G - \{v\}$ is also a minimum dominating set for X in G, and thus we recurse on $G - \{v\}$ and X.

Case D: The Vertex v Is of Degree 2 and $v \in X$. Let u_1 and u_2 denote the two neighbors of v in G. Again according to Lemma 1, we branch into three subcases for a minimum dominating set D:

(D.1): $u_1 \in D$ and $v \notin D$. In this case if D' is a minimum dominating set for $X - N[u_1]$ in $G - \{u_1, v\}$ then $D' \cup \{u_1\}$ is a minimum dominating set for X in G. Thus we recurse on $G - \{u_1, v\}$ and $X - N[u_1]$.

(D.2): $v \in D$ and $u_1, u_2 \notin D$. In this case if D' is a minimum dominating set for $X - \{u_1, v, u_2\}$ in $G - \{u_1, v, u_2\}$ then $D' \cup \{v\}$ is a minimum dominating set for X in G. Thus we recurse on $G - \{u_1, v, u_2\}$ and $X - \{u_1, v, u_2\}$.

(D.3): $u_1 \notin D$ and $v \notin D$. Then $v \in X$ implies $u_2 \in D$. Now we use that if D' is a minimum dominating set for $X - N[u_2]$ in $G - \{v, u_2\}$ then $D' \cup \{u_2\}$ is a minimum dominating set for X in G. Thus we recurse on $G - \{v, u_2\}$ and $X - N[u_2]$.

To analyse the running time of our algorithm we denote by $T(n)$ the worst case number of recursive calls performed by the algorithm for a graph on n vertices. Each recursive call can easily be implemented in time polynomial in the size of the graph passed to the recursive call. In cases A and B we have $T(n) \leq T(n-1)$, in case C we have $T(n) \leq T(n-1) + T(n-2) + T(n-3)$ and in case D we have $T(n) \leq 2 \cdot T(n-2) + T(n-3)$. Standard calculations yield that the worst behavior of $T(n)$ is within a constant factor of α^n, where α is the largest root of $\alpha^3 = \alpha^2 + \alpha + 1$, which is approximately 1.8393. Thus $T(n) = O^*(1.8393^n)$. Therefore, the most time consuming part of the algorithm is the procedure of checking all subsets of size at most $3t/8$ where $t \le n$. As already discussed, this can be performed in $O^*(1.93782^n)$ steps by a brute force algorithm.

Summarizing, we have proved the following theorem.

Theorem 1. *A minimum dominating set of a graph on n vertices can be computed in time $O^*(1.93782^n)$ time. (The base of the exponential function in the running time is $8/(3^{3/8} \cdot 5^{5/8}) \approx 1.9378192$.)*

4   Split Graphs and Bipartite Graphs

In this section we present an exponential algorithm for the minimum set cover problem obtained by dynamic programming. This algorithm will then be used as a subroutine in exponential algorithms for the NP-hard minimum dominating set problems on split graphs and on bipartite graphs.

Let X be a ground set of cardinality m, and let $T = \{T_1, T_2, \ldots, T_k\}$ be a collection of subsets of X. We say that a subset $T' \subseteq T$ *covers* a subset $S \subseteq X$, if every element in S belongs to at least one member of T'. A minimum set cover of (X, T) is a subset T' of T that covers the whole set X. The minimum set cover problem asks to find a minimum set cover for given (X, T). Note that a minimum set cover of X can trivially be found in time $O^*(2^k)$ by checking all possible subsets of T.

Lemma 2. *There is an $O(mk\, 2^m)$ time algorithm to compute a minimum set cover for an instance (X, T) with $|X| = m$ and $|T| = k$.*

Proof. Let (X, T) with $T = \{T_1, T_2, \ldots, T_k\}$ be an instance of the minimum set cover problem over a ground set X with $|X| = m$. We present an exponential algorithm solving the problem by dynamic programming.

For every nonempty subset $S \subseteq X$, and for every $j = 1, 2, \ldots, k$ we define $F[S; j]$ as the minimum cardinality of a subset of $\{T_1, \ldots, T_j\}$ that covers S. If $\{T_1, \ldots, T_j\}$ does not cover S then we set $F[S; j] := \infty$.

Now all values $F[S; j]$ can be computed as follows. In the first step, for every subset $S \subseteq X$, we set $F[S; 1] = 1$ if $S \subseteq T_1$, and $F[S; 1] = \infty$ otherwise. Then in step $j + 1$, $j = 1, 2, \ldots, k - 1$, $F[S; j + 1]$ is computed for all $S \subseteq X$ in $O(m)$ time as follows:

$$F[S; j + 1] = \min\{F[S; j],\ F[S - T_{j+1}; j] + 1\}.$$

This yields an algorithm to compute $F[S; j]$ for all $S \subseteq X$ and all $j = 1, 2, \ldots, k$ of overall running time $O(mk\, 2^m)$. In the end, $F[X; k]$ is the cardinality of a minimum set cover for (X, T).   □

Now we shall use Lemma 2 to establish an exact exponential algorithm to solve the NP-hard minimum dominating set problem for split graphs. Let us recall that a graph $G = (V, E)$ is a split graph if its vertex set can be partitioned into a clique C and an independent set I.

Theorem 2. *There is an $O(n^2\, 2^{n/2}) = O^*(1.41422^n)$ time algorithm to compute a minimum dominating set for split graphs.*

Proof. If G is a complete graph or an empty graph, then the dominating set problem on G is trivial. If $G = (V, E)$ is not connected, then all of its components are isolated vertices except possibly one, say $G' = (V', E)$. If D' is a minimum dominating set of the connected split graph G' then $D' \cup (V - V')$ is a minimum dominating set of G.

Thus we may assume that the input graph $G = (V, E)$ is a connected split graph with a partition of its vertex set into a clique C and an independent set I where $|I| \geq 1$ and $|C| \geq 1$. Such a partition can be found in linear time (Golumbic [7]). A connected split graph has a minimum dominating set D such that $D \subseteq C$: consider a minimum dominating set D' of G with $|D' \cap I|$ as small as possible; then a vertex $x \in D' \cap I$ can be replaced by a neighbor $y \in C$. $N[x] \subseteq N[y]$ implies that $D'' := (D' - \{x\}) \cup \{y\}$ is a dominating set, and either $|D''| < |D'|$ (if $y \in D'$), or $|D''| = |D'|$ and $|D'' \cap I| < |D' \cap I|$–both contradicting the choice of D'.

Let $C = \{v_1, v_2, \ldots, v_k\}$. For every $j \in \{1, 2, \ldots, k\}$ we define $T_j = N(v_j) \cap I$. Clearly, $D \subseteq C$ is a dominating set in G if and only if $\{T_i \colon v_i \in D\}$ covers I. Hence the minimum dominating set problem for G can be reduced to the minimum set cover problem for (I, T) with $|I| = n - k$ and $|T| = k$. For $k \leq n/2$ this problem can be solved by trying all possible subsets in time $O(n\, 2^k) = O(n\, 2^{n/2})$. For $k > n/2$, by Lemma 2, the problem can be solved in time $O((n - k)k\, 2^{n-k}) = O(n^2\, 2^{n/2})$.

Thus a minimum dominating set of G can be computed in time $O(n^2\, 2^{n/2})$. $\qquad \square$

A modification of the technique used to prove Theorem 2, can be used to obtain faster algorithms for graphs with large independent set.

Theorem 3. *There is an $O(nz \cdot 3^{n-z})$ time algorithm to compute a minimum dominating set for graphs with an independent set of size z. In particular, there is an $O(n^2 \cdot 3^{n/2}) = O^*(1.73206^n)$ time algorithm to compute a minimum dominating set for bipartite graphs.*

Proof. Let $G = (V, E)$ be a graph with an independent set of size z. Note that such an independent set can be identified in $O^*(1.2108^n)$ time by the algorithm of Robson [13].

Let $R = V - I$ denote the set of vertices outside the independent set. In an initial phase, we fix for every subset $X \subseteq R$ some corresponding vertex set $I_X \subseteq I$ via the following three steps.

1. Determine $Y = I - N[X]$.
2. Compute a vertex set $Z \subseteq N[X] \cap I$ of minimum cardinality subject to $R - N[X] - N[Y] \subseteq N(Z)$.
3. Set $I_X = Y \cup Z$.

First, we observe that $Y \subseteq I$ and $Z \subseteq I$ yield $I_X \subseteq I$. Secondly, $I \subseteq Y \cup N[X]$ implies that I is dominated by $X \cup I_X$, and $R - N[X] - N[Y] \subseteq N(Z)$ implies that

R is dominated by $X \cup I_X$. Consequently, the set $X \cup I_X$ forms a dominating set for the graph G. Thirdly, we claim that among all dominating sets D for G with $D \cap R = X$, the dominating set $X \cup I_X$ has the smallest possible cardinality: Indeed, $D \cap R = X$ means that the vertices in $Y = I - N[X]$ can only be dominated, if they are contained in D; hence $Y \subseteq D$. Furthermore, the vertices in $R - N[X] - N[Y]$ must all be dominated through some vertices in $N[X] \cap I$; in the second step, we determine the smallest possible subset $Z \subseteq N[X] \cap I$ with this property. Summarizing, for finding a minimum dominating set for G, it is sufficient to look through all the 2^{n-z} sets $X \cup I_X$.

What is the time complexity of this approach? The only (exponentially) expensive step for determining the sets I_X is the computation of the sets Z. And this expensive step boils down to solving a set covering problem that consists of a ground set $R - N[X] - N[Y]$ with at most $|R - X| \leq n - z - |X|$ elements, and that consists of a collection of $|N[X] \cap I| \leq z$ subsets. By Lemma 2, such a set covering problem can be solved in $O(nz \cdot 2^{n-z-|X|})$ time. The overall time for solving all set covering problems for all subsets $X \subseteq R$ is proportional to $\sum_{k=1}^{n-z} \binom{n-z}{k} nz \cdot 2^{n-z-k}$. This yields an overall time complexity of $O(nz \cdot 3^{n-z})$.

$\square$

Note that for graphs with an independent set of size $z \geq 0.39782 \cdot n$, the running time of the algorithm in Theorem 3 is better than the running time of the algorithm for general graphs from Section 3.

5   Graphs of Maximum Degree Three

Computer experiments suggest that exact exponential algorithms like the trivial $O^*(2^n)$ time algorithm, or like our $O^*(1.93782^n)$ algorithm from Section 3 have the slowest running times for fixed values of n, if the input graphs have large domination numbers. One possible explanation is that the algorithm has to spend a lot of time on checking that no vertex subset of size $\gamma(G) - 1$ is dominating (even in case a true minimum dominating set is detected at an early stage). Since graphs of maximum degree three have high domination numbers, the algorithms for general graphs do not behave well on these graphs.

In this section, we design a better exact algorithm for graphs of maximum degree three, by using the pruning a search tree technique and a structural property of minimum dominating sets in graphs of maximum degree three provided in the following lemma.

Lemma 3. *Let $G = (V, E)$ be a graph of maximum degree three. Then there is a minimum dominating set D of G with the following two properties:*

(i) *every connected component of $G[D]$ is either an isolated vertex, or an iso-lated edge, and*

(ii) *if two vertices $x, y \in D$ form an isolated edge in $G[D]$, then x and y have degree three in G, and $N(x) \cap N(y) = \emptyset$.*

Proof. Let D be a minimum dominating set of G with the maximum number of isolated vertices in $G[D]$. If $G[D]$ has a vertex x of degree three, then $D - \{x\}$

is a smaller dominating set of G, which is a contradiction. Thus the maximum degree of $G[D]$ is two.

Assume $G[D]$ has a vertex y of degree two. If the degree of y in G is two, then $D - \{y\}$ is a smaller dominating set of G, a contradiction. Otherwise let z be the unique neighbor of y in G that is not in D. If $z \in N[D - \{y\}]$ then $D - \{y\}$ is a smaller dominating set of G, another contradiction. Finally, if $z \notin N[D - \{y\}]$ then $D_1 := (D \cup \{z\}) - \{y\}$ is another minimum dominating set in G with a larger number of isolated vertices in $G[D_1]$ than in $G[D]$. This contradiction concludes the proof of property (i).

To prove property (ii), let us first show that any two adjacent vertices $x, y \in D$ have degree three in G. For the sake of contradiction, assume that y has degree less than three in G. Clearly y cannot have degree one, otherwise $D - \{y\}$ is a dominating set, a contradiction. Suppose y has degree two, and let $z \neq x$ be the second neighbor of y. If $z \in N[D - \{y\}]$ then $D - \{y\}$ is a dominating set of smaller size than D, a contradiction. If $z \notin N[D - \{y\}]$, then $D_2 := (D - \{y\}) \cup \{z\}$ is a minimum dominating set in G with a larger number of isolated vertices in $G[D_2]$ than in $G[D]$, another contradiction.

Finally, we prove that $N(x) \cap N(y) = \emptyset$ in G. For the sake of contradiction, assume that $N(x) \cap N(y) \neq \emptyset$. If $N[x] \subseteq N[y]$ then $D - \{x\}$ is a dominating set, and if $N[y] \subseteq N[x]$ then $D - \{y\}$ is a dominating set. In both cases this contradicts our choice of D. Hence $N(x) = \{y, w, u\}$ with $N(x) - N(y) = \{w\}$ and $N(x) \cap N(y) = \{u\}$. If $w \in N[D - \{x\}]$ then $D - \{x\}$ is a dominating set, another contradiction. If $w \notin N[D - \{x\}]$ then $D_3 := (D - \{x\}) \cup \{w\}$ is a minimum dominating set in G with a larger number of isolated vertices in $G[D_3]$ than in $G[D]$, the final contradiction. □

Now we construct a search tree algorithm using the restriction of the search space guaranteed by Lemma 3, i.e. for a graph $G = (V, E)$ of maximum degree three only vertex sets $D \subseteq V$ satisfying the properties of of Lemma 3 have to be inspected. W.l.o.g. we assume that the input graph is connected.

Theorem 4. *There is a $O^*(1.51433^n)$ time algorithm to compute a minimum dominating set on graphs of maximum degree three. (The base of the exponential function in the running time is the largest real root $\alpha \approx 1.51433$ of $\alpha^6 = \alpha^3 + 2\alpha^2 + 4$.)*

Proof. The algorithm is based on the pruning a search tree technique. The idea is to branch into subcases until we obtain a graph of maximum degree two, and for such a graph a minimum dominating set can be computed in linear time since each of its connected components is either an induced path P_k ($k \geq 1$) or an induced cycle C_k ($k \geq 3$). In this way we obtain all minimum dominating sets of G satisfying the properties of Lemma 3

More precisely, the input graph $G = (V, E)$ and $D = \emptyset$ correspond to the root of the search tree. To each node of the search tree corresponds an induced subgraph $G[V']$ of G and a partial dominating set $D \subseteq V - V'$ of G already chosen to be part of the dominating set obtained in any branching from this node. To each leaf of the tree corresponds a subgraph $G[V']$ of maximum degree two. For each node of the search tree to which a subgraph $G[V']$ of maximum degree three corresponds the algorithm proceeds as follows: It chooses a neighbour (called x below) of a vertex of degree three such that x has smallest possible degree; then it inspects x and branches in various subcases. Suppose $(G[V'], D)$ corresponds to a node of the search tree and that $G[V']$ has maximum degree two. Then a linear time algorithm will be invoked to find a minimum dominating set D' of $G[V']$, and thus $D \cup D'$ is a dominating set of G. Finally the algorithm chooses a smallest set among all dominating sets of G obtained in this way and outputs it as a minimum dominating set of G.

To show that this algorithm has running time $O^*(1.51433^n)$ we have to study its branching into subcases. We denote by $T(n)$ the worst case number of recursive calls performed by the algorithm for a graph on n vertices.

The algorithm will pick a vertex x of degree three at most once, and this can only happen at the very beginning and only if all vertices of the input graph have degree three. Thus this branching is of no interest for the analysis of the overall running time of our algorithm.

We shall distinguish two cases: x has degree one or x has degree two. For each case the algorithm chooses one or two vertices to be added to the partial dominating set D and recurses on some smaller induced subgraphs. Based on Lemma 3 each connected component of $G[D]$ can be supposed to be a K_1 or a K_2. (Note that our analysis deals with the subgraph $G[V']$ that corresponds to the current node of the search tree.)

Case 1: x Is a Vertex of Degree One in $G[V']$. Let y be a degree three neighbour of x. Let z_1 and z_2 be the other neighbours of y. Clearly there is a minimum dominating set of $G[V']$ not containing x, and thus we may choose $x \notin D$ and $y \in D$. This leaves two possible subcases for the choice of the vertices to be added to D.

Subcase 1.A: $y \in D$ Isolated Vertex in $G[D]$. We add y to the dominating set D and recurse on $G - N[y]$. Since y has degree three the number of recursive calls on this subcase is $T(n-4)$.

Subcase 1.B: $y, z_i \in D$, $i \in \{1, 2\}$, isolated edge in $G[D]$. Then we may obtain 2 subcases as follows: Add y, z_i, $i \in \{1, 2\}$, to D and recurse on $G - (N[y] \cup N[z_i])$. By property (ii) of Lemma 3, this requires that z_i has degree three, hence we remove 6 vertices and the number of recursive calls on this subcase is at most $2\,T(n-6)$.

In total, in Case 1 we obtain the recurrence $T(n) \le T(n-4) + 2 \cdot T(n-6)$. Standard calculations yield that the worst behavior of $T(n)$ is within a constant factor of α^n. This α is the largest real root of $\alpha^5 = \alpha + 2$, which is approximately 1.26717. Thus $T(n) = O^*(1.26717^n)$.

Case 2: x Is a Vertex of Degree Two in $G[V']$. Let y_1 and y_2 be the neighbours of x. W.l.o.g. let y_1 be a degree three vertex.

Case 2.1: y_1 and y_2 Are Adjacent in $G[V']$. Then there is a minimum dominating set of $G[V']$ not containing x, and thus either y_1 or y_2 must be added to D.

Case 2.1.1: y_2 Has Degree Two. Hence w.l.o.g. $y_1 \in D$ and $y_2 \notin D$. Thus either y_1 is an isolated vertex in $G[D]$, or $y, z \in D$ where z is the third neighbour of y_1. Thus we obtain the recurrence $T(n) \le T(n-4) + \cdot T(n-6)$. Thus $T(n) = O^*(1.15097^n)$, where $\alpha \approx 1.15097$ is the largest real root of $\alpha^5 = \alpha + 2$.

Case 2.1.2: y_2 Has Degree Three. For $i = 1, 2$, let z_i be the third neighbour of y_i. Then either $y_1 \in D$ or $y_2 \in D$ is an isolated verted in $G[V']$, or $y_i, z_i \in D$ is an isolated edge in $G[V']$. Then we recurse on $G - N[y_i]$ and remove 4 vertices, or we recurse on $G - (N[y_i] \cup N[z_i])$ and remove 6 vertices. Consequently we obtain the recurrence $T(n) \le 2 \cdot T(n-4) + 2 \cdot T(n-6)$. Thus $T(n) = O^*(1.33015^n)$, where $\alpha \approx 1.33015$ is the largest real root of $\alpha^6 = 2\alpha^2 + 2$.

Case 2.2: y_1 and y_2 Are Not Adjacent in $G[V']$. Since x has degree two either $x \in D$ is an isolated vertex in $G[D]$ or $x \notin D$.

Case 2.2.1: y_2 Has Degree Two. Let z_{11} and z_{12} the other neighbours of y_1, and let z_2 be the other neighbour of y_2.

Subcase 2.2.1.A: $x \in D$ Isolated Vertex in $G[D]$. We add x to the dominating set D and recurse on $G - N[x]$. Since x has degree two the number of recursive calls on this subcase is $T(n-3)$.

Subcase 2.2.1.B: $y_i \in D$ Isolated Vertex in $G[D]$. For $i = 1, 2$, we add y_i to the dominating set D and recurse on $G - N[y_i]$. Since y_1 has degree three and y_2 has degree two, the number of recursive calls on this subcase is at most $T(n-3) + T(n-4)$.

Subcase 2.2.1.C: $y_1, z_{1j} \in D$, $j \in \{1, 2\}$, Isolated Edge in $G[D]$. Then we may obtain 2 subcases as follows: Add y_1, z_{1j}, $j \in \{1, 2\}$, to D and recurse on $G - (N[y_1] \cup N[z_{1j}])$. By property (ii) of Lemma 3, this requires that z_{1j} has degree three, hence we remove 6 vertices and the number of recursive calls on this subcase is at most $2T(n-6)$.

In total, in Case 2.2.1 we obtain the recurrence $T(n) \le 2 \cdot T(n-3) + T(n-4) + 2 \cdot T(n-6)$. As we have seen before, the worst behavior of $T(n)$ is within a constant factor of α^n. This α is the largest real root of $\alpha^6 = 2\alpha^3 + \alpha^2 + 2$, which is approximately 1.48613. Thus $T(n) = O^*(1.48613^n)$.

Case 2.2.2: y_2 Has Degree Three. Let z_{11} and z_{12} be the other neighbours of y_1, and let z_{21} and z_{22} be the other neighbours of y_2.

Subcase 2.2.2.A: $x \in D$ Isolated Vertex in $G[D]$. We add x to the dominating set D and recurse on $G - N[x]$. Since x has degree two the number of recursive calls on this subcase is $T(n-3)$.

Subcase 2.2.2.B: $y_i \in D$ Isolated Vertex in $G[D]$. For $i = 1, 2$, we add y_i to the dominating set D and recurse on $G - N[y_i]$. y_1 and y_2 have degree three, thus the number of recursive calls on this subcase is at most $2T(n-4)$.

Subcase 2.2.2.C: $y_i, z_{ij} \in D$, $i, j \in \{1, 2\}$, Isolated Edge in $G[D]$. Then we may obtain 4 subcases as follows: Add y_i, z_{ij}, $i, j \in \{1, 2\}$, to D and recurse on $G - (N[y_{ij}] \cup N[z_{ij}])$. This requires that z_{ij} has degree three, hence we remove 6 vertices and the number of recursive calls on this subcase is at most $4T(n-6)$.

In total, in Case 2.2.2 we obtain the recurrence $T(n) \leq T(n-3) + 2 \cdot T(n-4) + 4 \cdot T(n-6)$. The worst behavior of $T(n)$ is within a constant factor of α^n, where α is the largest real root of $\alpha^6 = \alpha^3 + 2\alpha^2 + 4$, which is approximately 1.5143218. Thus $T(n) = O^*(1.51433^n)$. □

References

1. J. ALBER, H. L. BODLAENDER, H. FERNAU, T. KLOKS, AND R. NIEDERMEIER. *Fixed parameter algorithms for dominating set and related problems on planar graphs.* Algorithmica 33, 2002, pp. 461–493.

2. E. DANTSIN, A. GOERDT, E. A. HIRSCH, R. KANNAN, J. KLEINBERG, C. PAPADIMITRIOU, P. RAGHAVAN, AND U. SCHÖNING. *A deterministic $(2 - 2/(k+1))^n$ algorithm for k-SAT based on local search.* Theoretical Computer Science 289, 2002, pp. 69–83.

3. R. G. DOWNEY AND M. R. FELLOWS. *Parameterized complexity.* Monographs in Computer Science, Springer-Verlag, New York, 1999.

4. D. EPPSTEIN. *Small maximal independent sets and faster exact graph coloring.* Proceedings of the 7th Workshop on Algorithms and Data Structures (WADS'2001), LNCS 2125, Springer, 2001, pp. 462–470.

5. F. V. FOMIN AND D. M. THILIKOS. *Dominating sets in planar graphs: Branchwidth and exponential speed-up.* Proceedings of the 14th ACM-SIAM Symposium on Discrete Algorithms (SODA'2003), 2003, pp. 168–177.

6. M. R. GAREY AND D. S. JOHNSON. *Computers and intractability. A guide to the theory of NP-completeness.* W.H. Freeman and Co., San Francisco, 1979.

7. M. C. GOLUMBIC. *Algorithmic graph theory and perfect graphs.* Academic Press, New York, 1980.

8. T. W. HAYNES, S. T. HEDETNIEMI, AND P. J. SLATER. *Fundamentals of domination in graphs.* Marcel Dekker Inc., New York, 1998.

9. T. W. HAYNES, S. T. HEDETNIEMI, AND P. J. SLATER. *Domination in graphs: Advanced Topics.* Marcel Dekker Inc., New York, 1998.

10. R. IMPAGLIAZZO, R. PATURI, AND F. ZANE. *Which problems have strongly exponential complexity?* Journal of Computer and System Sciences 63, 2001, pp. 512–530.

11. D. S. JOHNSON AND M. SZEGEDY. *What are the least tractable instances of max independent set?* Proceedings of the 10th ACM-SIAM Symposium on Discrete Algorithms (SODA'1999), 1999, pp. 927–928.

12. B. REED. *Paths, stars and the number three.* Combinatorics, Probability and Computing 5, 1996, pp. 277–295.

13. J.M. ROBSON. *Algorithms for maximum independent sets.* Journal of Algorithms 7, 1986, pp. 425–440.

14. G. J. WOEGINGER. *Exact algorithms for NP-hard problems: A survey.* Combinatorial Optimization: "Eureka, you shrink", LNCS 2570, Springer, 2003, pp. 185–207.

Linear Kernels in Linear Time, or How to Save k Colors in $O(n^2)$ Steps

Benny Chor[1], Mike Fellows[2], and David Juedes[3]

[1] School of Computer Science, Tel Aviv University, Tel Aviv, 69978, Israel
benny@cs.tau.ac.il
[2] School of EE & CS, University of Newcastle, Callaghan NSW 2308, Australia
mfellows@cs.newcastle.edu.au
[3] School of EE & CS, Ohio University, Athens, Ohio 45701, USA
juedes@ohio.edu

Abstract. This paper examines a parameterized problem that we refer to as $n - k$ GRAPH COLORING, i.e., the problem of determining whether a graph G with n vertices can be colored using $n - k$ colors. As the main result of this paper, we show that there exists a $O(kn^2 + k^2 + 2^{3.8161k}) = O(n^2)$ algorithm for $n - k$ GRAPH COLORING for each fixed k. The core technique behind this new parameterized algorithm is kernalization via maximum (and certain maximal) matchings.

The core technical content of this paper is a near linear-time kernelization algorithm for $n - k$ CLIQUE COVERING. The near linear-time kernelization algorithm that we present for $n - k$ CLIQUE COVERING produces a linear size $(3k - 3)$ kernel in $O(k(n + m))$ steps on graphs with n vertices and m edges. The algorithm takes an instance $\langle G, k \rangle$ of CLIQUE COVERING that asks whether a graph G can be covered using $|V| - k$ cliques and reduces it to the problem of determining whether a graph $G' = (V', E')$ of size $\leq 3k - 3$ can be covered using $|V'| - k'$ cliques. We also present a similar near linear-time algorithm that produces a $3k$ kernel for VERTEX COVER. This second kernelization algorithm is the *crown reduction rule*.

1 Introduction

Graph coloring is one of the hardest NP-complete problems under a variety of measures. It is well-known [1] that determining whether a graph can be colored using $k = 3$ colors is NP-complete, and the best-known polynomial-time algorithm for graph coloring, due to Halldórsson [2], produces a coloring that is only guaranteed to be within $O(|V|(\log\log|V|)^2/\log^3|V|)$ of optimal. It is known that approximating $\chi(G)$, the *chromatic number* of a graph G, to the ratio $|V|^\delta$ is NP-hard [3, 4]. Furthermore, it is known from the work of Feige and Kilian [5] that no polynomial-time algorithm can approximate $\chi(G)$ to within $|V|^{1-\epsilon}$ for any $\epsilon > 0$ unless NP $\subseteq$ ZPP. Hence, it does not appear likely that we will be able to find a much better approximation algorithm for graph coloring.

This extended abstract explores graph coloring from another, more tractable, perspective. Since it is easy to see that any graph $G = (V, E)$ can be colored

J. Hromkovič, M. Nagl, and B. Westfechtel (Eds.): WG 2004, LNCS 3353, pp. 257–269, 2004.
© Springer-Verlag Berlin Heidelberg 2004

via a trivial coloring that uses n colors and colors each vertex using a separate color, the alternative perspective is to ask whether G can be colored using $n - k$ colors. We refer to this problem as $n - k$ GRAPH COLORING. The parameter k corresponds to the how many colors can be "saved" when coloring a graph $G = (V, E)$ over the trivial coloring of G. It is easy to see that this problem is NP-complete. However, in contrast to the usual optimization measure for graph coloring, this version is much easier to approximate. A sequence of papers by Demange, Grisoni and Paschos [6], Hassin and Lahav [7], and Halldórsson [8, 9], and Duh and Fürer [10] examined approximation algorithms for graph coloring under this non-standard measure. The best-known polynomial-time approximation algorithm, due to Duh and Fürer [10], approximates $n - k$ GRAPH COLORING to the ratio $\frac{360}{280} \approx 1.246$.

This paper explores exact algorithms for $n - k$ GRAPH COLORING. In particular, we show that it is possible to determine whether a graph G can be colored using $n - k$ colors in $O(n^2)$ steps for any fixed k. Hence, $n - k$ GRAPH COLORING is *fixed parameter tractable* [11]. Our algorithm exploits the following, well-known result concerning the relationship between graph coloring and clique covering, $\chi(G) = \bar{\chi}(\bar{G})$, i.e., that the minimum number of colors needed to color G is equal to the minimum number of cliques needed to cover the complement of G. The main technical contribution of this work is a linear-time algorithm that kernelizes instances of $n - k$ CLIQUE COVERING, i.e., the problem of determining whether a graph G with n vertices can be covered by $n - k$ cliques. The general kernelization approach proceeds as follows.

1. Compute a maximal matching M with no M augmenting path of length 3 or shorter on the input graph G.
2. Identify an independent set I of vertices in the reduced input graph by examining the maximal matching M.
3. Compute a maximum matching M' in the bipartite graph formed by I its edges to the rest of the graph.
4. Eliminate all vertices not covered by either the first or second matching.

This general kernelization approach can be used to achieve a kernel of size $\leq 3k - 3$ for $n - k$ CLIQUE COVERING. We use a similar approach to give a *crown reduction rule* for VERTEX COVER that achieves a $\leq 3k$ kernel.

Once a graph G is kernelized, determining whether G has a clique covering of size $n - k$ can be solved in $f(k) = O(k^2 + 2^{3.8161k})$ steps by first converting the kernelized instance $\langle G', k \rangle$ into its complement $\langle \bar{G}', k \rangle$ and then applying the best-known exact algorithm for graph coloring due to Eppstein [12]. This approach leads to parameterized algorithms running in time $O(k(n + m) + k^2 + 2^{3.8161k})$ for $n - k$ CLIQUE COVERING and time $O(k \cdot n^2 + k^2 + 2^{3.8161k})$ for $n - k$ GRAPH COLORING. The algorithm for $n - k$ GRAPH COLORING is depicted in Figure 1.

We note that the use of maximum matchings in fixed parameter tractable algorithms is not new. For instance, Papadimitriou and Yannakakis [13] used a maximum matching based approach to show that VERTEX COVER can be solved in polynomial-time for all $k \leq \log n$. Their work implicitly gives a $O(3^k n)$

parameterized algorithm for VERTEX COVER. Similarly, the best-known parameterized algorithm for VERTEX COVER [14] constructs a $2k$ kernel for VERTEX COVER using a maximum matching based technique of Nemhauser and Trotter [15]. However, our kernelization technique appears to be more general since it is easily applied to both VERTEX COVER and $n - k$ CLIQUE COVERING. Moreover, this approach achieves linear kernels in near linear-time.

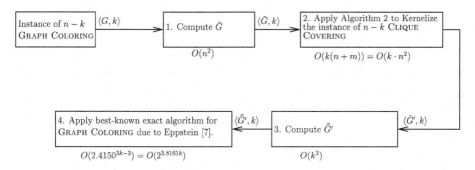

Fig. 1. A $O(kn^2 + k^2 + 2^{3.8161k})$ algorithm for $n - k$ GRAPH COLORING that uses the identity $\chi(G) = \bar{\chi}(\bar{G})$ twice.

The extended abstract is organized as follows. In section 2, we review necessary notation and prior work on maximum and maximal matchings. In section 3, we examine the combinatorial properties of maximal matchings without short augmenting paths. In section 4, we give the kernelization algorithm for $n - k$ CLIQUE COVERING and prove that the algorithm converts an instance of $\langle G, k \rangle$ to an equivalent instance $\langle G', k \rangle$ of size $\leq 3k - 3$. In section 5, we use the same technique to give a *crown reduction rule* for VERTEX COVER that produces a $\leq 3k$ kernel. In section 6, we use the kernelization algorithm and the best-known exact algorithm for graph coloring by Eppstein [12] to give fast parameterized algorithms for $n - k$ CLIQUE COVERING and $n - k$ GRAPH COLORING that run in the times mentioned above. Finally, in section 7, we provide some concluding remarks.

2   Preliminaries

In this section, we review the notation and the relevant prior work on maximum matchings that we use in this paper. Interested readers are directed to the survey paper by Galil [16] or the comprehensive book by Lovász and Plummer [17] for further details.

Given a graph $G = (V, E)$, a *matching* is a set of edges $M \subseteq E$ such that no two edges in M share an endpoint. A vertex $v \in V$ is said to be *covered* if it is an endpoint of an edge in M. If v is not covered, then we say that v is *exposed*.

Most algorithms that find maximum matchings do so by explicitly constructing *augmenting paths* [16]. Given a matching M, an *M-alternating path* is a simple path $P = \langle v_1, v_2, \ldots, v_n \rangle$ in G such that P consists of edges that alternate between edges in M and edges outside of M. An M-*augmenting path* is an M-alternating path that begins and ends with an exposed vertex. It is well-known that the existence of an M-augmenting path P implies that M is not a maximum matching since it is possible to create a larger matching by swapping the unmatched edges in P for the matched edges in P to create a larger matching. This observation leads to the following theorem, due to Berge [18].

Theorem 1. *M is a maximum matching in G if and only if there is no M-augmenting path in G.*

Fast algorithms to compute maximum matchings were first discovered for bipartite graphs. The best-known algorithm for maximum matching in bipartite graphs is due to Hopcroft and Karp [19] and takes time $O(m\sqrt{n})$ on graphs with m edges and n vertices. The best-known algorithm for maximum matching in general graphs comes from the work of Micali and Vazirani [20] and also takes time $O(m\sqrt{n})$.

To achieve linear-time kernelization here, we will employ Hopcroft and Karp's algorithm on bipartite graphs and a somewhat straightforward algorithm for finding maximal matchings without short augmenting paths in general graphs. To see that Hopcroft and Karp's algorithm runs in linear-time in our case, we need to briefly describe the operation of this algorithm. The algorithm by Hopcroft and Karp employs an $O(m)$ algorithm to find a maximal set of vertex disjoint augmenting paths of minimum length. The algorithm proceeds in multiple passes where at each pass the algorithm augments the matching M via the maximal set of vertex disjoint M-augmenting paths of minimum length. This process continues until no M-augmenting path is found, and hence by Theorem 1 M is a maximum matching in G. The algorithm executes at most $O(\sqrt{n})$ passes because, as Hopcroft and Karp [19] prove in Theorem 2 and Corollaries 3 and 4, once a maximal set of vertex disjoint M augmenting paths of length l is found and M is augmented to M', the shortest M'-augmenting path in the graph has length $l + 1$ or longer.

In the bipartite graphs that we use here, one of the two partitions will be of size $2k$ or smaller. Since no M augmenting path in such a graph can be of length longer than $4k$ edges, the algorithm of Hopcroft and Karp performs at most $4k$ augmenting passes. Therefore, Hopcroft and Karp's algorithm will complete in $O(km)$ steps.

The ideas behind Hopcroft and Karp's algorithm can be used to find maximal matchings in general graphs with no augmenting path of length 3 in $O(n + m)$ steps. To see this, consider the following approach. First, compute a maximal matching M in G using the standard $O(n+m)$ greedy algorithm. Next, compute a maximal set S of vertex disjoint M-augmenting paths of length 3 in G by examining each edge $e \in M$. For each edge $e = (u, v) \in M$, compute c_u, the number of exposed vertices connected to u and c_v the number of exposed vertices connected to v. If either c_u or c_v equals zero, no augmenting path of length 3 that uses e exists. If $c_u = c_v = 1$, then a path exists if and only if these vertices are different. Finally, if $c_u \geq 2$ and $c_v \geq 1$ (or vice-versa), then an augmenting path through e must exist and can be found easily. Mark the endpoints of any such found augmenting path as "covered." It is easy to see that this approach finds a maximal set of vertex disjoint augmenting paths of length 3 in time $O(n + m)$ since each edge in G is visited at most $O(1)$ times. Finally, the results of Hopcroft and Karp tell us that augmenting the matching M with S produces a matching M' with no augmenting path of length 3. Such matchings have properties that we exploit in the next section.

3 Maximal Matchings Without Short Augmenting Paths and Their Combinatorial Properties

Given a maximal matching M in a graph $G = (V, E)$, it is natural to partition V into the *covered* vertices and the *exposed* vertices. In this fashion, we partition V into $I_M = \{v \in V | \{u, v\} \in M\}$ and $O_M = V - I_M$. The set O_M forms an independent set because if there were an edge between any pair of vertices in O_M, then M would not be a maximal matching.

Given a matching M with no augmenting path of length 3, we partition the matched edges into classes to exploit certain combinatorial properties. In this respect, we partition M into three classes C_1, C_2, and C_3 based on how the endpoints of the edges in M connect to the exposed vertices. Figures 2–4 illustrate the three classes of matched edges.

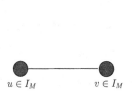

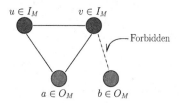

Fig. 2. The class 1 "boring" edges.   **Fig. 3.** The class 2 edges.

We refer to the first class of matched edges C_1 as the "boring edges" since they are not connected to any exposed vertex. In the second class of matched edges C_2, both endpoints are connected to vertices in O_M. As shown in Figure 3, both endpoints must be connected to the same vertex $a \in O_M$, or there will exist an M-augmenting path in G of length 3. Finally, the third class of edges in M, C_3, contains edges where only one of the two endpoints are connected to exposed vertices. This case is shown in Figure 4.

The main combinatorial property of maximal matchings M with no augmenting paths of length 3 that we use in this paper is the following lemma that relates the size of M to the size of the maximum matching M' in the bipartite graph formed by letting $V_1 = O_M$, $V_2 = N(O_M)$, and only including edges from O_M to $N(O_M)$ in G. We call this graph $G[O_M, N(O_M)]$.

Lemma 1. Let M be a maximal matching in G with no length 3 M-augmenting path. Let M' be a maximum matching in $G[O_M, N(O_M)]$. Then $|M'| \leq |M|$.

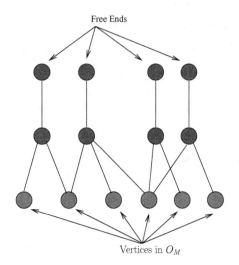

Fig. 4. The class 3 edges.

Proof. The bipartite graph $G' = G[O_M, N(O_M)]$ contains edges from exposed vertices to the end points of matched edges in M. It is clear by inspection that no more than one endpoint of matched edges in M can appear as an endpoint of a matched edge in G' since (i) for each edge in C_3 only one endpoint is connected to any vertex in O_M, and (ii) for each edge in C_2, both endpoints are connected to a single vertex a in O_M and hence at most one of these endpoints can be matched to a. It follows that $|M'| \leq |M|$.

4   A Kernelization Algorithm for Clique Covering

In this section, we develop a kernelization algorithm for CLIQUE COVERING that exploits the combinatorial properties of maximal matchings developed in the previous section. The correctness of our approach depends on the following generic results concerning independent sets, maximum matchings, and clique coverings.

In the following lemma, we call a clique "trivial" if it contains only a single vertex.

Lemma 2. *Let $G = (V, E)$ be a graph, and let I be an independent set of vertices in G. If M' is a maximum matching in $G[I, V - I]$, then in any clique covering C of G at most $|M'|$ vertices in I appear in non-trivial cliques in C.*

Proof. Assume that there are k vertices in I that appear in non-trivial cliques in C. Since I is an independent set, no two of these vertices can appear in the same clique. So, let $C_1, \ldots, C_k$ be the k cliques that contain these vertices. For the vertex $v_i \in I$ that appears in C_i, pick another vertex $u_i \in (V - I) \cap C_i$. Such a vertex must exist because C_i is non-trivial and I is an independent set. Furthermore, the edges $\{u_1, v_1\}, \{u_2, v_2\}, \ldots, \{u_k, v_k\}$ form a matching in $G[I, V - I]$ because the C_i's are disjoint. Since M' is a maximum matching, it follows that $k \leq |M'|$.

Lemma 3. *Let $G = (V, E)$ be a graph, let I be an independent set in G, and let M' be a maximum matching in $G[I, V - I]$. For every clique covering C of G, there exists a clique covering C' of G such that $|C'| \leq |C|$ and each vertex in $M'[I]$ (those vertices in I that are covered by M') appears in a non-trivial clique in C'.*

Proof. Let $C = \{C_1, \ldots, C_m\}$ be a clique covering of G. Examine each vertex $v_i \in M'[I]$. Let u_i be the other endpoint of the matched edge containing v_i. If v_i is contained in a trivial clique, delete u_i from its clique (it is still a clique, or we have deleted a clique), and add it to v_i's clique. This does not increase the total number of cliques. Continue this process until all of the vertices $v_i \in M'[I]$ are contained in non-trivial cliques. Notice that this process may create trivial cliques of vertices in $M'[I]$. However, this process must eventually halt, since we increase the total number of vertices in $M'[I]$ that appear in non-trivial cliques with their paired vertex during each iteration. Moreover, these new cliques cannot be "destroyed." This approach is given in Algorithm 1. To prove the correctness of this algorithm, it suffices to show that $c = |\{v_i \in M'[I] \mid \{u_i, v_i\} \in M' \text{ and } u_i, v_i \in C_j \text{ for some } C_j \in C\}|$ is a loop invariant for the algorithm. The algorithm will halt since c increases by one during each iteration of the loop. When the loop halts, the value of c must be $|M'[I]|$, and hence every vertex in $M'[I]$ appears in a non-trivial clique in C'. The algorithm creates no new cliques (though it may destroy some trivial cliques), hence the total number of cliques in C' is less than or equal to $|C|$.

Combining Lemma 2 and 3, we observe that if I is an independent set and M' is a maximum matching in the bipartite graph $G[I, V - I]$, then there is an optimal clique covering where all of the vertices in I that are not covered by M' appear in trivial cliques. Hence, these vertices can be deleted without affecting the relative size of the optimal clique covering. So, if we let G' be the graph

---

Algorithm 1 Creating Non-trivial Cliques

Input: A Clique Covering C of G
Output: A Clique Covering C' of G such that $|C'| \leq |C|$ and each vertex in $M'[I]$ is in a non-trivial clique.
set $c = |\{v_i \in M'[I] \mid |\{u_i, v_i\} \in M' \text{ and } u_i, v_i \in C_j \text{ for some } C_j \in C\}|$;
while (there exists a $v_i \in M'[I]$ in a trivial clique C_l) **do**
    Find $\{u_i, v_i\} \in M'$.
    Find the C_k such that $u_i \in C_k$;
    Remove u_i from C_k;
    add u_i to C_l;
    $c = c + 1$;
end while
return C;

---

formed by removing all of the vertices in I that are not covered by M', then G has a clique covering of size $n - k$ if and only if G' has a clique covering of size $n' - k$. This approach forms the basis for the main technical contribution of this paper, the kernelization algorithm for $n - k$ CLIQUE COVERING given in the following theorem.

Theorem 2. *There exists an algorithm running in $O(k(m+n))$ steps that, when given an instance $\langle G, k \rangle$ of $n - k$ CLIQUE COVERING with n vertices and m edges either determines that G can be covered using $n - k$ cliques or produces an equivalent instance $\langle G', k \rangle$, where G' has $\leq 3k - 3$ vertices.*

Proof. We present Algorithm 2 below. To see that the algorithm runs in $O(k(m + n))$ steps, consider the running time of each step. Step 1 completes in $O(m + n)$ steps, as described at the end of section 2. Steps 2–6 can also be computed in $O(m + n)$ steps or faster. As explained at the end of section 2, step 7 takes $O(k(m + n))$ steps because $N(O_M)$ has at most $2k$ vertices, and hence the maximum bipartite matching algorithm of Hopcroft and Karp performs at most $4k$ passes. The final two steps can also be implemented in $O(m + n)$ steps.

Algorithm 2 Kernelization for $n - k$ CLIQUE COVERING.

Input: An instant $\langle G, k \rangle$ of $n - k$ CLIQUE COVERING
Output: Either "yes" if G has an $|V| - k$ clique covering or an equivalent instance $\langle G', k \rangle$ of $n - k$ CLIQUE COVERING
 1: Compute a maximal matching M in G with no M augmenting path of length 3 as described in section 2.
 2: **if** $|M| \geq k$ **then**
 3:    return "yes" since G has a $n - k$ clique covering consisting of k cliques of size 2 and $n - 2k$ trivial cliques.
 4: **end if**
 5: Partition V into into covered (I_M) and exposed (O_M) vertices.
 6: Compute the bipartite graph $G^* = G[O_M, N(O_M)]$.
 7: Apply the algorithm of Hopcroft and Karp [19] to compute a maximum matching M' of G^*.
 8: Delete from G all vertices in O_M that are not covered by M'. Call this graph G'.
 9: Return $\langle G', k \rangle$.

To see that this algorithm produces the correct result, consider the following brief argument. Since O_M is an independent set and Algorithm 2 deletes from G all vertices in O_M that are not covered by M', it is clear from Lemma 2 and 3 that G' has a clique covering of size $n' - k$ if and only if the original instance G has a clique covering of size $n - k$. To see that G' has the correct size, we employ results from the previous section.

Since $|O_M| = n - 2|M|$, we delete exactly $n - 2|M| - |M'|$ vertices from G to form G' in step 8 of the algorithm. Since $|M'| \leq |M|$ by Lemma 1, it follows that we delete at least $n - 3|M|$ vertices from G. Since G has n vertices to begin with, this leaves at most $3|M|$ vertices in G'. Since $|M| < k$, it follows that G' has a most $3k - 3$ vertices. This completes the proof.

5   The Crown Reduction Rule for Vertex Cover

To show the generality of the kernelization approach taken in the previous section, we show how to apply this technique to give a linear-time kernelization algorithm for instances of VERTEX COVER. The key to our kernelization for VERTEX COVER is the following observation. Let I be an independent set in G, and assume that there is a matching of size $|N(I)|$ in the bipartite graph $G[I, N(I)]$ formed by I, the neighborhood of I, and the edges between them in G. Under this assumption, at least $|N(I)|$ vertices from $I \cup N(I)$ must appear in any vertex cover of G. Hence, it suffices to include all of $N(I)$ in the vertex cover and delete all of $I \cup N(I)$ from G. This is the *crown reduction rule*. Here, we show how to efficiently find such an I via a maximum and a maximal matching.

Let I be an independent set of vertices from a graph $G = (V, E)$. Then, the *bipartite subgraph of G formed by the sets I and $V - I$*, denoted by $G[I, V - I]$, is the graph $G' = (V, E')$, where $E' = \{\{u, v\} \mid u \in I, v \in V - I, \text{ and } \{u, v\} \in E\}$. Let M be a maximum matching in $G[I, V - I]$. Then, we write $M[I]$ for the set of vertices in I that are covered by edges in M and $M[V - I]$ for the set of vertices in $V - I$ that are covered by edges in M. If S is a set of endpoints of edges in M, then we write $M_-[S]$ for the set of vertices matching the vertices in S, i.e.,

$$M_-[S] = \{u \mid \{u, v\} \in M \text{ and } v \in S\}.$$

As in section 4, we show that it is possible to eliminate vertices in I that do not appear in $M[I]$. As mentioned above, we can eliminate all of the vertices in I when M is an *upper perfect matching*, i.e., when $|M| = |N(I)|$, or, in other words, all of the vertices on one side of the bipartite graph are matched. While a maximum matching in $G[I, V - I]$ for an arbitrary independent set I may not be an upper perfect matching, we show below that there must exist an upper perfect matching for a set I', where $I - M[I] \subseteq I' \subseteq I$. This requires some explanation.

Let I be an independent set in G, and let M be a maximum matching in the bipartite subgraph $G[I, V - I]$. Then, it is possible to select a subset of $M[V - I]$ that has an upper perfect matching with a subset of the vertices in I. To begin, we note the following easy result.

Lemma 4. $N(I - M[I]) \subseteq M[V - I]$, *i.e., vertices in I that are not covered by the matching must connect to only endpoints in the matching.*

Proof. Assume that a vertex $v \in I - M[I]$ connects to a vertex $u \in N(I) - M[V - I]$. Then, we can add the edge $\{u, v\}$ to the matching. Hence, the matching is not maximal (and hence not a maximum matching). This is a contradiction.

We next define a collection of subsets of $M[V - I]$.

$$C^0(M) = N(I - M[I])$$
$$C^i(M) = N(M_-[C^{i-1}(M)])$$
$$C(M) = \bigcup_{i=0}^{\infty} C^i(M).$$

The following results hold.

Lemma 5.   *1. For each i, $C^i(M) \subseteq M[V - I]$.*
2. For each $i > 0$, $C^{i-1}(M) \subseteq C^i(M)$.
3. For each i and for each vertex $v \in C^i(M)$, there exists an M-alternating path from a vertex $u \in I - M[I]$ to v.

Proof. (By induction on i) When $i = 0$, Lemma 4 tells us that $C^0(M) \subseteq M[V - I]$. Furthermore, each vertex $v \in C^0(M)$ is attached to a vertex $u \in I - M[I]$. This single edge is an M-alternating path from u to v.

Now, assume that conditions 1–3 are true for some $i \geq 0$. To see that condition (2) is true for $C^{i+1}(M)$, note that every vertex in $C^i(M)$ is an endpoint in the matching. Hence, $C^i(M) \subseteq N(M_-[C^i(M)]) = C^{i+1}(M)$. To see that condition (3) is true, assume that $v \in C^{i+1}(M) - C^i(M)$. Hence, there exists a vertex $u \in C^i(M)$ such that $\{u, v_1\} \in M$, and $v \in N(v_1)$. Since $v \in C^{i+1}(M) - C^i(M)$, it is clear that $\{v, v_1\} \notin M$. Moreover, since there exists an M alternating path from a vertex $v_2 \in I - M[I]$ to u (by the inductive hypothesis), adding the edges $\{u, v_1\}$, $\{v_1, v\}$ forms an M alternating path from v_2 to v. Finally, to see that condition (1) is true, notice that if $v \notin M[V - I]$, then the M alternating path from v_2 to v is an M augmenting path. Hence, the original matching M is not a maximum matching.

As we show above, the $C^i(M)$'s are ordered by the subset relation. As we show next, there must be a fixed point in these sets.

Lemma 6. *If $C^i(M) = C^{i+1}(M)$, then $C^i(M) = C^{i+k}(M)$ for all $k \geq 1$.*

Proof. (By induction on k.) When $k = 1$, the lemma holds trivially. Now, assume that the lemma holds when $k \geq 1$. Then,

$$C^{i+k+1}(M) = N(M_-[C^{i+k}(M)]) = N(M_-[C^i(M)]) = C^{i+1}(M) = C^i(M).$$

Lemma 5 and 6 imply a straightforward method for computing $C(M)$; compute $C^i(M)$ until $C^{i+1}(M) = C^i(M)$. The end result of this computation is $C(M)$. This computation must complete after $|M|$ passes, since $|C(M)| \leq |M|$. Moreover, notice that computing $C^{i+1}(M)$ from $C^i(M)$ can be done in $O(n+m)$ steps since it simply involves computing neighborhoods. Hence, the entire computation of $C(M)$ takes $O(|M|(n + m))$ steps, where $n = |V|$ and $m = |E|$.

Now, we can give a subset of $G[I, V - I]$ that has an upper perfect matching. Define

$$I' = M_-[C(M)] \cup (I - M[I]).$$

Lemma 7. $N(I') = C(M)$.

Proof. $N(I') = N(I - M[I]) \cup N(M_-[C(M)]) = C^0(M) \cup C(M) = C(M)$.

Lemma 8. *The bipartite graph $G[I', V - I']$ has an upper perfect matching.*

Proof. Lemma 7 tells us that $N(I') = C(M)$. Since I' contains all of the vertices in $M_-[C(M)]$, the edges from $C(M)$ to $M_-[C(M)]$ form an upper perfect matching of $G[I', V - I']$.

We now show that we can eliminate all of the vertices in I from G (and perhaps some others). The single reduction rule that we use is the following.

Crown Reduction Rule: Given an independent set I in G, if the bipartite graph $G[I, V - I]$ has an upper perfect matching (i.e., a matching of size $|N(I)|$), then delete I and $N(I)$ from G to form G' and set $k' = k - |N(I)|$.

The crown reduction rule is a generalization of the reduction rule that can be used to eliminate degree 1 vertices for VERTEX COVER.

Lemma 9. *G' has a vertex cover of size k' iff G has a vertex cover of size k.*

Proof. Assume that G' has a vertex cover V^* of size k'. Then, $V^* \cup N(I)$ is a vertex cover of G of size $k' + |N(I)| = k$, To see this, notice that the vertices in $N(I)$ covers the edges in the bipartite subgraph formed by I and $N(I)$. Moreover, the vertices in $N(I)$ also cover all edges from $N(I)$ to the rest of G.

Similarly, assume that G has a vertex cover V^* of size k and define $\hat{V}^* = V^* - (N(I) \cup I)$. This is a vertex cover of G'. Each matched edge connecting the vertices in $N(I)$ to the vertices in I must be covered by either a unique vertex in $N(I)$ or a unique vertex in I. Since I is an independent set, these sets are disjoint. Therefore, $|\hat{V}^*| \leq |V^*| - |N(I)| = k - |N(I)| = k'$.

The crown reduction rule leads to the following kernelization algorithm for VERTEX COVER.

Theorem 3. *There exists an algorithm running in time $O(k(n+m))$ steps that takes an instance $\langle G, k \rangle$ of VERTEX COVER with n vertices and m edges and either returns "no" (G has no vertex cover of size k) or produces an equivalent instance $\langle G', k' \rangle$, where $k'' \leq k$ and $|V'| \leq 3k$.*

Proof. The algorithm proceeds as follows.

Algorithm 3 Linear-Time Kernelization Algorithm for VERTEX COVER.

1: Compute a maximal matching M in G with no M-augmenting path of length 3 using the algorithm described in section 2.
2: **if** $|M| > k$ **then**
3:     return "no," because G needs at least $k + 1$ vertices in any vertex cover.
4: **end if**
5: Otherwise, partition V into I_M and O_M.
6: Compute the bipartite graph $G^* = G[O_M, N(O_M)]$.
7: Compute a maximum matching M' in the bipartite graph G^*.
8: Compute the crown $C(M')$, and let $I' = M'_-[C(M')] \cup (O_M - M'[O_M])$.
9: Apply the crown reduction rule. This rule deletes $C(M')$ and I' from G.
10: return $\langle G', k' \rangle$.

Notice that the correctness of this algorithm follows almost immediately from Lemma 9. To see that G' has $\leq 3k$ vertices, notice that (i) after step 3, we know that $|M| \leq k$, (ii) Lemma 1 tells us that $|M'| \leq |M|$, and (iii) the crown reduction rule removes at least $|O_M| - |M'| \geq |O_M| - |M|$ vertices from G. Since $n = |O_M| + 2|M|$ and $|M| \leq k$, $|V'| \leq n - |O_M| - |M| = 3|M| \leq 3k$.

To see that Algorithm 3 runs in time $O(k(m + n))$, notice that (i) step 1 takes $O(m + n)$ steps, (ii) steps 2–6 can be performed in $O(n + m)$ steps, (iii) step 7 takes $O(k(n + m))$ steps since $|N(O_M)| \leq 2k$ because $|M| \leq k$, (iv) step 8 takes $O(k(n + m))$ steps by the argument prior to Lemma 7, and (v) steps 9 and 10 can be performed in $O(n + m)$ steps. This completes the proof.

6   Parameterized Algorithms for Graph Coloring and Clique Covering

In this section, we combine the results of sections 2, 3, and 4 to prove our main result.

Theorem 4. $n - k$ CLIQUE COVERING *can be solved in time* $O(k(n+m)+k^2 + 2^{3.8161k})$ *steps on instances* $\langle G, k \rangle$, *where* G *has* m *edges and* n *vertices.*

Proof. Given an instance $\langle G, k \rangle$ of $n - k$ CLIQUE COVERING, apply Algorithm 2 from section 4. The algorithm either determines that there is a clique covering of size $n-k$ of G, or it kernelizes $\langle G, k \rangle$ to an equivalent instance $\langle G', k \rangle$ such that G' has a clique covering of size $n'-k$ if and only if G has a clique covering of size $n-k$. Moreover, we have that $n' \leq 3k - 3$. This takes $O(k(m + n))$ steps. Next, apply the relationship between graph coloring and clique covering, and convert $\langle G', k \rangle$ into an equivalent instance $\langle \bar{G}', k \rangle$ of $n - k$ GRAPH COLORING. This takes $O(k^2)$ steps. Finally, apply the best-known exact algorithm from graph coloring, due to Eppstein [12]. This algorithm runs in $O(2.4150^n) = O(2.4150^{3k-3}) = O(2^{3.8161k})$ steps. This completes the algorithm and the proof.

Theorem 5. $n - k$ GRAPH COLORING *can be solved in time* $O(kn^2 + k^2 + 2^{3.8161k})$ *on instances* $\langle G, k \rangle$, *where* G *has* m *edges and* n *vertices.*

Proof. It suffices to employ the identity $\chi(G) = \bar{\chi}(\bar{G})$. Hence, G can be colored using $n - k$ colors if and only if $\bar{G}$ can be covered using $n - k$ cliques. Therefore, it suffices to first compute $\bar{G}$ and to then employ the algorithm given in Theorem 4. The result follows.

7   Concluding Remarks

This paper provides an alternate approach to exact algorithms for graph coloring. The approach provided here can be used to design faster exact algorithms for graphs with high chromatic numbers $(\chi(G) \geq 2/3n)$ because the parameterized algorithm given here runs faster than the $O(2.4150^n)$ exact algorithm when $k \leq 1/3n$. Perhaps this new parameterized algorithm can be used in some inventive way to improve the running-time of the exact algorithm for graph coloring for all graphs.

Acknowledgments

The authors would like to thank Christian Sloper, Dan Xiao, and David Fleeman for their helpful comments on earlier drafts of this paper.

References

1. Garey, M.R., Johnson, D.S.: Computers and Intractability: a guide to the theory of NP-completeness. W.H. Freeman (1979)
2. Halldórsson, M.M.: A still better performance guarantee for approximate graph coloring. Information Processing Letters **45** (1993) 19–23
3. Lund, C., Yannakakis, M.: On the hardness of approximating minimization problems. Journal of the ACM **41** (1994) 960–981
4. Bellare, M., Goldreich, O., Sudan, M.: Free bits, PCPs and non-approximability – towards tight results. SIAM Journal on Computing **27** (1998) 804–915
5. Feige, U., Kilian, J.: Zero knowledge and the chromatic number. In: Proceedings of the Eleventh Annual IEEE Conference on Computational Complexity, IEEE Computer Society Press (1996) 278–289
6. Demange, M., Grisoni, P., Paschos, V.T.: Approximation results for the minimum graph coloring problem. Information Processing Letters **50** (1994) 19–23
7. Hassin, R., Lahav, S.: Maximizing the number of unused colors in the vertex coloring problem. Information Processing Letters **52** (1994) 87–90
8. Halldórsson, M.M.: Approximating discrete collections via local improvement. In: Proceedings of the Sixth ACM-SIAM Symposium on Discrete Algorithms, ACM Press (1995) 160–169
9. Halldórsson, M.M.: Approximating k-set cover and complementary graph coloring. In: Proceedings of the 5th IPCO Conference on Integer Programming and Combinatorial Optimization. Volume 1084 of LNCS., Springer-Verlag (1996) 118–131
10. c. Duh, R., Fürer, M.: Approximation of k-set cover by semi-local optimization. In: Proceedings of the Twenty-Ninth Annual ACM Symposium on Theory of Computing, ACM Press (1997) 256–264
11. Downey, R., Fellows, M.R.: Parameterized Complexity. Springer-Verlag (1999)
12. Eppstein, D.: Small maximal independent sets and faster exact graph coloring. In: Algorithms and Data Structures: 7th International Workshop, WADS 2001. Volume 2125 of Lecture Notes in Computer Science., Springer-Verlag (2001) 462–470
13. Papadimitriou, C.H., Yannakakis, M.: On limited nondeterminism and the complexity of the V-C dimension. Journal of Computer and System Sciences **53** (1996)
14. Chen, J., Kanj, I., Jia, W.: Vertex cover:further observations and further improvements. Journal of Algorithms **41** (2001) 280–301
15. Nemhauser, G.L., Trotter, L.E.: Vertex packings: Structural properties and algorithms. Mathematical Programming **8** (1975) 232–248
16. Galil, Z.: Efficient algorithms for finding maximum matching in graphs. ACM Computing Surveys **18** (1986) 23–38
17. Lovász, L., Plummer, M.D.: Matching Theory. Volume 29 of Annals of Discrete Mathematics. North Holland, Amsterdam (1986)
18. Berge, C.: Two theorems in graph theory. Proceedings of the National Academy of Sciences U.S.A. **43** (1957) 842–844
19. Hopcroft, J.E., Karp, R.M.: An $n^{5/2}$ algorithm for maximum matchings in bipartite graphs. SIAM Journal on Computing **2** (1973) 225–231
20. Micali, S., Vazirani, V.V.: An $O(\sqrt{|v|} \cdot |E|)$ algorithm for finding maximum matching in general graphs. In: 21st Annual Symposium on Foundations of Computer Science, Syracuse, New York, IEEE (1980) 17–27

Planar Graphs, via Well-Orderly Maps and Trees

Nicolas Bonichon[1], Cyril Gavoille[1], Nicolas Hanusse[1],
Dominique Poulalhon[2], and Gilles Schaeffer[3]

[1] Laboratoire Bordelais de Recherche en Informatique (LaBRI)
Université Bordeaux 1, 33405 Talence Cedex, France
{bonichon,gavoille,hanusse}@labri.fr
[2] Laboratoire d'Informatique Algorithmique, Fondements et Applications (LIAFA)
case 7014, 2, place Jussieu, 75251 Paris Cedex 05, France
dominique.poulalhon@liafa.jussieu.fr
[3] Laboratoire d'Informatique de l'École Polytechnique (LIX)
École polytechnique, 91128 Palaiseau Cedex, France
gilles.schaeffer@lix.polytechnique.fr

Abstract. The family of well-orderly maps is a family of planar maps with the property that every connected planar graph has at least one plane embedding which is a well-orderly map. We show that the number of well-orderly maps with n nodes is at most $2^{\alpha n + O(\log n)}$, where $\alpha \approx 4.91$. A direct consequence of this is a new upper bound on the number $p(n)$ of unlabeled planar graphs with n nodes, $\log_2 p(n) \leqslant 4.91n$.

The result is then used to show that asymptotically almost all (labeled or unlabeled), (connected or not) planar graphs with n nodes have between $1.85n$ and $2.44n$ edges.

Finally we obtain as an outcome of our combinatorial analysis an explicit linear time encoding algorithm for unlabeled planar graphs using, in the worst-case, a rate of 4.91 bits per node and of 2.82 bits per edge.

1 Introduction

Counting the number of (non-isomorphic) planar graphs with n nodes is a well-known long-standing unsolved graph-enumeration problem (cf. [1]). There is no known closed formula, neither asymptotic for unlabeled planar graphs.

There are only upper and lower bounds on the *growth rate* of the sequence of numbers $p(n)$ of unlabeled planar graphs. This growth rate, defined as $\mu = \lim_{n \to \infty} p(n)^{1/n}$, currently ranges between 27.2268 and 32.1556 (a superadditivity argument shows that such a limit exists [2, 3]).

The lower bound on μ comes from an asymptotic on the number of labeled planar graphs. This asymptotic is on the form $n! \lambda^{n+o(n)}$ [2, 3], and in [4], a precise estimation of λ is given: $27.2268 < \lambda < 27.2269$. The upper bound on μ, due to [5], comes from succinct encoding of plane planar graphs. More precisely, after a suitable embedding and triangulation of the planar graph, it is shown that such embeddings can be represented by a binary string of length at most $5.007n$ bits. Such representation implies that $p(n) \leqslant 2^{5.007n} \approx (32.1556)^n$.

J. Hromkovič, M. Nagl, and B. Westfechtel (Eds.): WG 2004, LNCS 3353, pp. 270–284, 2004.

Technically, enumerating unlabeled graphs is more difficult than counting the labeled version. And, as pointed out in [6], almost all labeled 2- and 1-connected planar graphs have exponentially large automorphism groups. In other words, Wright's Theorem [7] does not hold for random planar graphs, the asymptotic number of labeled and unlabeled planar graphs differ in more than the $n!$ term, i.e., $\lambda < \mu$. So, an asymptotic on the number of labeled planar graphs would not give a sharp lower bound on the growth rate of $p(n)$. The situation from the upper bound side is not better. There are many ways to embed a planar graph, and to recover the graph from a suitable triangulation requires deep understanding of plane triangulations, in particular their enumeration given several parameters depending on the input graph.

Besides the pure combinatorial aspect, the "encoding" approach is also relevant in Computer Science where a lot of attention is given to efficiently represent discrete objects. At least two field of applications of high interests are concerned with succinct planar graph representation: Computer Graphics [8–10] and Networking [11–14].

1.1 Related Works

Obviously, without sharp asymptotic formula, properties and behavior of large random objects cannot be described precisely. For lack of an adequate model, very little is known on random planar graphs. However, random generation of planar graphs has been investigated in the last decade.

Using a simple Markov chain, Denis et al. [2] showed, that, experimentally, random labeled planar graphs have $2n$ edges. In fact, Bodirsky et al. [15] have designed the first polynomial time (uniform) random generator of labeled planar graphs. Although limited in their experiments (mainly by the time complexity of this algorithm), they showed that actually the number of edges in a random labeled planar graph is more than $2n$. The best proved bounds on the number of edges in a random labeled planar graph are $1.85n$ [16] and $2.54n$ [5], for the unlabeled case $1.70n$ and $2.54n$, by [5].

Succinct representation of n-node m-edge planar graphs has a long history. Turán [17] pioneered a $4m$ bit encoding, that has been improved later by Keeler and Westbrook [18] to $3.58m$. Munro and Raman [19] then proposed a $2m + 8n$ bit encoding based on the 4-page embedding of planar graphs (see [20]). In a series of articles, Lu et al. [21, 22] refined the coding to $4m/3 + 5n$ thanks to orderly spanning trees, a generalization of Schnyder's trees [23].

1.2 Our Results

Any planar embedding of a n-node planar graph with n nodes can be seen as a subgraph of a n-node triangulation of the plane. Once given a triangulation and a set of edges to keep (or to remove), a planar map and the corresponding graph can be constructed. The converse is false in general. There is no known method to uniquely associate a triangulation to a planar graph.

However, in [5], a linear time algorithm is given to construct a triangulation of the plane in a canonical way for any planar graph, once given a planar embedding. The reader should keep in mind that there is a-priori no unique embedding for a planar graph. Some of planar embeddings have interesting graph properties based on the Schnyder's partition [23] of triangulations into trees. A new class of planar embeddings are proposed in [5]: the *well-orderly maps*, a more restrictive version of the *orderly maps* of Chuang et al. [21]. The two main properties of well-orderly maps that can be exploited for graph coding are: 1) every planar graph admits such an embedding, and 2) given a well-orderly map, we can uniquely associate a triangulation.

The main result of this paper is to give a good approximation of the number of well-orderly maps. As a byproduct, it gives a new upper bound on the number of planar graphs. More interestingly, the combinatorial analysis allow to us to give an explicit coding of such maps (and thus of planar graphs), as a function of n and m, the number of edges: 4.91 bits per node, and 2.82 bit per edge (clearly, $2.82m$ bits is always smaller than $4m/3 + 5n$ bits). It follows also a new bound on the number of edges of a random planar graph (labeled or not).

The paper is organized as follows. We describe in Section 2 the relationships between well-orderly maps, super-triangulations and realizers. The new coding is presented in Section 3, and in Section 4 are given the applications to the number of unlabeled planar graphs and to the number of edges in random planar graphs. Another application of our results is an upper bound on the minimal grid area of random triangulation of the plane. We show that in average, plane triangulations can be drawn on grids of area at most $\frac{7}{8}n \times \frac{7}{8}n$ and $\frac{11}{16}n \times \frac{5}{6}n$.

2   Encoding Planar Graphs with Minimal Realizers

In this section we collect some results from [5] about planar graphs, well-orderly maps, super-triangulations and realizers. In the last paragraph, these results are used to prove a new representation theorem.

2.1   Planar Graphs and Well-Orderly Maps

A *planar map* (or *plane graph*) is an embedding of a connected planar graph on the plane so that edges meet only at their endpoints. When cutting the plane along the edges, the remaining components are called the faces. Apart from the unbounded component, all these faces are homeomorphic to discs. A planar map is *rooted* if one of its edges is distinguished and oriented. This determines a *root edge*, a *root node* (its origin) and a *root face* (to its left), also called *external face* or *outerface*. A *triangulation* of the plane (or a maximal plane graph) is a planar map without loops or multiple edges such that all faces are triangles.

A plane tree is, as usual, a rooted tree such that the siblings of a node are linearly ordered. Equivalently it is a planar map with one face. Among the nodes of a tree, we distinguish the root, the inner nodes and the leaves. A spanning tree of a planar map is a subset of edges that forms a tree connecting all nodes.

Let T be a rooted spanning tree of a planar map H, and let $v_1, \ldots, v_n$ be the clockwise preordering of the nodes in T. Two nodes are *unrelated* if neither of them is an ancestor of the other in T. An edge of H is unrelated if its endpoints are unrelated.

A node v_i is *orderly* in H with respect to T if the edges incident to v_i in H form the following four (possibly empty) blocks in clockwise order around v_i (see Fig. 2(b)):

- $B_P(v_i)$: the edge incident to the parent of v_i in T;
- $B_<(v_i)$: edges that are unrelated in T and incident to nodes v_j with $j < i$;
- $B_C(v_i)$: edges that are incident to the children of v_i in T; and
- $B_>(v_i)$: edges that are unrelated in T and incident to nodes v_j with $j > i$.

A node v_i is *well-orderly* if it is orderly and if the clockwise first edge $(v_i, v_j) \in B_>(v_i)$, if it exists, verifies that the parent of v_j is an ancestor of v_i.

A rooted spanning tree T of H is a *well-orderly tree* of H if all the nodes of T are well-orderly in H with respect to T. A planar map H is a *well-orderly map of root v* if it contains a well-orderly tree of root v.

Theorem 1 ([5]). *Let G be a connected planar graph, and let v be any node of G. Then G has a well-orderly map of root v, which can be computed in linear time. Moreover, a well-orderly map of root v has a unique well-orderly tree of root v, which can also be computed in linear time.*

In Fig. 1 two orderly trees $\overline{T}_0$ span the same triangulation but only one is the well-orderly tree.

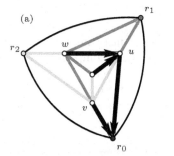 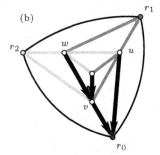

Fig. 1. Two realizers for a triangulation. The tree $\overline{T}_0$ rooted in r_0 (the tree with bold edges augmented with the edges (r_0, r_1) and (r_0, r_2)) is well-orderly in (b), but only orderly in (a) since node v is not well-orderly.

Observe that by definition of well-orderly nodes, an edge of H which is related with respect to a well-orderly tree T (i.e. one endpoint is a descendant of the other one in T) must belong to the tree T: indeed all edges are either unrelated or connect a node to its father. In particular all the edges incident in H to the root of T are in T.

2.2   Minimal Realizers and Super-triangulations

A *realizer* of a triangulation is a partition of its interior edges (the edges that do not lie on the external face) into three sets T_0, T_1, T_2 of directed edges such that the following conditions hold for each interior node v (see Fig. 2(a)):

- the clockwise order of the edges incident with v is: leaving in T_0, entering in T_1, leaving in T_2, entering in T_0, leaving in T_1 and entering in T_2;
- there is exactly one leaving edge incident with v in each of the sets T_0, T_1, and T_2.

Hereafter, when $R = (T_0, T_1, T_2)$ is a realizer, R also denotes the underlying triangulation.

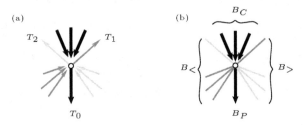

Fig. 2. Relationship between realizer and orderly tree: (a) edge-orientation rule around a node for a realizer, and (b) blocks ordering around an orderly node.

Observe that if (T_0, T_1, T_2) is a realizer, then (T_1, T_2, T_0) and (T_2, T_0, T_1) are also realizers. This cyclic permutation of the three sets of edges does not in general provide all the distinct realizers of a given triangulation. Fig. 1 depicts two realizers for a same triangulation.

Schnyder showed in [23] that if (T_1, T_2, T_3) is a realizer then each set T_i induces a tree rooted in one node of the external face and spanning all interior nodes. Moreover, for each T_i, we denote by $\overline{T}_i$ the tree composed of T_i augmented with the two edges of the external face incident to the root of T_i. For every non-root node $u \in T_i$, we denote by $p_i(u)$ the parent of u in T_i.

A realizer $S = (T_0, T_1, T_2)$ is a *super-triangulation* of a graph G if:

1. $V(S) = V(G)$ and $E(G) \subseteq E(S)$;
2. $E(T_0) \subseteq E(G)$;
3. $\overline{T}_0$ is a well-orderly tree of S; and
4. for every inner node v of T_2, $(v, p_1(v)) \in E(G)$.

Lemma 1 ([5]). *Let H be a well-orderly map, and T its unique well-orderly tree of root r_0. Assume that T has at least two leaves. Let r_2 and r_1 be the clockwise first and last leaves of T respectively. Then, there is a unique super-triangulation (T_0, T_1, T_2) of the underlying graph of H, preserving the embedding H, and such that each T_i has root r_i. Moreover, $T_0 = T \setminus \{r_1, r_2\}$ and the super-triangulation is computable in linear time.*

There is an alternative characterization of super-triangulation in terms of minimal realizers. A *cw-triangle* (or clockwise triangle), is a triple of nodes

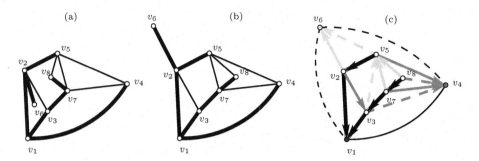

Fig. 3. (a) A planar graph G with an embedding which is not well-orderly. An easy way to see that it is not a well-orderly, is to remark that the edges $(v_1, v_2), (v_1, v_3), (v_1, v_4), (v_2, v_6)$ must be in any spanning tree of G rooted at v_1 such that G has only parent edges and unrelated edges. In such trees, v_2 is clearly not an orderly node. (b) A well-orderly map of G. (c) A super-triangulation of G (dotted edges are not in G).

(u, v, w) (not necessarily corresponding to a face) of a realizer such that $p_2(u) = v$, $p_1(v) = w$, and $p_0(w) = u$. A *minimal realizer* is a realizer that does not contain any clockwise triangle. In the realizer depicted in Fig. 1(a), (u, v, w) forms a cw-triangle, whereas the realizer of Fig. 1(b) has no cw-triangle.

Lemma 2 ([5]). *Let $S = (T_0, T_1, T_2)$ be any realizer. The following statements are equivalent:*

1. *S is a super-triangulation for some graph G.*
2. *S is a minimal realizer.*
3. *The tree $\overline{T}_i$ is well-orderly in S, for every $i \in \{0, 1, 2\}$.*

2.3   Results of the Paper

Theorem 2 (Coding version). *The following encoding sequence hold:*

- *Any connected planar graph can be embedded as a well-orderly map.*
- *Any well-orderly map can be represented as a minimal realizer (T_1, T_2, T_3) with a subset of marked edges included in the sets of edges of T_2 and of edges (u, v) of T_1 such that u is a leaf of T_1.*

Our first new result in this paper is that in fact the second encoding is almost tight.

Theorem 3 (Counting version). *Let H_n (resp. $H_{n,m}$) denote the set of well-orderly maps with n nodes (resp. with n nodes and m edges), and $R_{n,\ell}$ denote the set of minimal realizers (T_0, T_1, T_2) with n nodes and l leaves in T_2. Then*

$$\frac{1}{8} \sum_{\ell=1}^{n-3} |R_{n,\ell}| 2^{n+\ell} \leqslant |H_n| \leqslant \sum_{\ell=1}^{n-3} |R_{n,\ell}| 2^{n+\ell}.$$

$$\frac{1}{8} \sum_{\ell=\max\{1,2n-m-6\}}^{n-3} |R_{n,\ell}| \binom{n+\ell}{m-2n+6+\ell} \leqslant |H_{n,m}|.$$

$$|H_{n,m}| \leqslant \sum_{\ell=\max\{1,2n-m-6\}}^{n-3} |R_{n,\ell}| \binom{n+\ell}{m-2n+6+\ell}.$$

3   Counting and Coding Trees

In this section we briefly recall a result from [24] about minimal realizers and plane trees. An encoding of well-orderly maps follows.

3.1   Minimal Realizers and Plane Trees

A tree is planted if it is rooted on a leaf. Let $\mathcal{B}_n$ be the set of planted plane trees with n nodes and $2n$ leaves such that each node is adjacent to 2 leaves. Given a planted tree T of $\mathcal{B}_n$, its canonical orientation shall be toward the root for all inner edges, and toward the leaf for all dangling edges.

Fig. 4. On the left, a planted tree of $\mathcal{B}_n$ (the root is indicated by a square). Then from left to right, the partial closure of the tree.

A triple (e_1, e_2, e_3) of edges of a map M is an *admissible triple* if $e_1 = (v_0, v_1)$, $e_2 = (v_1, v_2)$ and $e_3 = (v_2, v_3)$ appear consecutively in the clockwise direction around the infinite face and if v_3 is a leaf. The *local closure* of M at the admissible triple (e_1, e_2, e_3) is obtained by fusing the leaf v_3 on node v_0 so as to create triangular face. Observe that by construction the orientation of the dangling edge prevents the formation of cw-triangles.

The *local closure* of a tree T of $\mathcal{B}_n$ is the map obtained by performing iteratively the local closure of any available admissible triple in a greedy way. As shown in [24], the local closure is well defined independently of the order of local closures. Moreover all bounded faces of the resulting map are triangular and the outer face has the structure shown on Fig. 5 (left hand side). In particular there are exactly two *canonical* dangling edges in the infinite face that are immediately followed by dangling edges in the clockwise direction around the infinite face. A tree T is *balanced* if its root is one of the two canonical leaves. Finally,

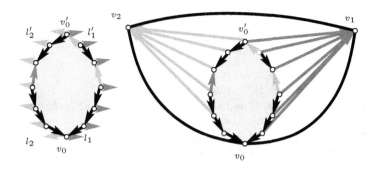

Fig. 5. The structure after a partial closure, and the complete closure.

the *complete closure* of a balanced tree T is the map obtained from the partial closure of T by fusing each remaining non canonical leaf with following canonical leaf in clockwise direction and adding a root edge, as illustrated by Fig. 5 (right hand side).

Theorem 4 ([24]). *Complete closure is one-to-one correspondence between balanced trees with $n-2$ and triangulations with n nodes. Moreover, the orientation of inner edges of the triangulation that is induced by the tree corresponds, via the coloration rule of Fig. 2(a) to a minimal realizer of the triangulation.*

Observe that the color of edges can be deduced from their orientation directly on the balanced tree from the application of the rule of Fig. 2(a).

The following new lemma will serve to predict entering edges created by complete closure at a node.

Lemma 3. *Let v be an inner node of a balanced tree B. Let $e_1 = (v, u)$ and $e_2 = (v, w)$ be two consecutive edges around v in the clockwise order. During the closure algorithm, no edges will be inserted between e_1 and e_2 if and only if:*

(a) w is a leaf of B, or
(b) w is an inner node of B and the node t such that the edge $e_3 = (w, t)$ is the next edge around w after e_2 in the clockwise order is a leaf of B.

Proof. Let v an inner node of a balanced tree B. Let us consider two consecutive edges (v, u), (v, w) around v in the clockwise order. If w is a leaf, then during the closure it will merge with a node w' and close a triangular face enclosing the corner between (v, u), (v, w). No other edge can thus arrive at this corner. Assume now that w is an inner node of B. Let (w, t) be the next edge around w in the clockwise order. If t is a leaf of B then it will merge with u to form a triangular face and again no edge can arrive in the corner between (v, u), (v, w). In the other cases, (v, w) is an inner edge followed by another inner edge (w, t). Since an edge that forming a triangular face that encloses the corner between (v, u), (v, w) must from w, the corner is not enclosed. But at the end of the partial closure, there are no more pairs of consecutive inner edges: some edge must be arrived in the corner. □

Lemma 4. *Let $R = (T_0, T_1, T_2)$ be the minimal realizer encoded by a balanced tree B. A node v of B is a leaf of T_2 if and only if v has no incoming edge colored 2 in B and,*

1. *the parent edge of v in B is colored 2, or*
2. *the parent edge of v in B is colored 1, or*
3. *the parent edge of v in B is colored 0 and v is the last child with an edge colored 0 in clockwise order around $P_B(v)$ and*
   (a) *the parent edge of $P_B(v)$ is colored 0, or*
   (b) *the parent edge of $P_B(v)$ is colored 2.*

The number of vertices of B satisfying these conditions is denoted $\ell(B)$.

From Lemma 4 and Theorem 2, we obtain:

Theorem 5. *Any well-orderly map with n nodes can be coded by a pair (B, W) where B is balanced tree of $\mathcal{B}_{n-2}$ and W a bit string of length $n + \ell(B)$. Encoding and decoding takes linear time.*

3.2   A Context-Free Grammar for Colored Trees

We shall now give a recursive decomposition of trees in which the parameter ℓ of Lemma 4 can be followed.

To do this we consider the three sets $\mathcal{F}_i$, for $i = 0, 1, 2$ of trees with a root edge of color i. To a tree T of $\mathcal{F}_i$, $i = 1, 2$, we associate the parameter $k(T) = \ell(T)$. To a tree T of $\mathcal{F}_0$ we associate the parameter $k(T)$ defined as $\ell(T)$ except for the root node which contributes to $k(T)$ as soon as it has no incoming edge of color 2, and a second parameter $k'(T)$ defined as $\ell(T)$ except for the root node which never contributes.

The decomposition is obtained, classically, at the root node: a tree with root edge of color 0 is made of a root node that carries, in clockwise order, a sequence of subtrees of root color 1, an outgoing edge of color 2, a sequence of subtrees of root color 0, an outgoing edge of color 1, and a sequence of subtrees of root color 2. The parameter ℓ is almost additive on subtrees . However, due to Rule 3 in Lemma 4, the root of a subtree with root edge of color 0 may or may not be susceptible to contribute depending how it is attached. In other terms, depending of how it is attached, a subtree T' with root color 0 contributes $k(T')$ or $k'(T')$.

On Fig. 6 the decomposition is pictured schematically: an incoming edge represents a tree, a triangle represents a possibly empty sequences of subtrees, and color correspond to root colors. For color 0, plain and dashed lines respectively indicate positions where the contribution is given by parameters k or k'. Finally root nodes that contribute to the parameters are pictured in a box.

3.3   Generating Functions of Trees
and the Asymptotic Number of Well-Orderly Maps

The reader can refer to [25] for a general presentation of enumeration of decomposable structures using grammars and generating series.

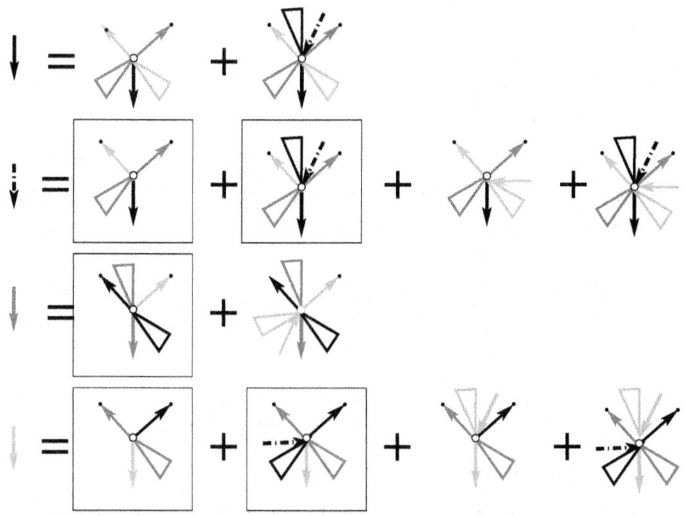

Fig. 6. A decomposition of colored trees allowing to track the contributions to ℓ.

Let us consider the generating function $F_i(z, u)$ of trees with root color i, $i = 0, 1, 2$ with respect to the number of edges and the parameter k, and $F'_0(z, u)$ of trees with root color 0 with respect to the number of edges and the parameter k':

$$F_i \equiv F_i(z, u) = \sum_{T \in \mathcal{F}_i} z^{|T|} u^{k(T)} \quad \text{and} \quad F'_0 \equiv F'_0(z, u) = \sum_{T \in \mathcal{F}_0} z^{|T|} u^{k'(T)}.$$

Recall that with respect to additive parameters, the generating function of a possibly empty sequence of elements of a set S is the quasi-inverse $1/(1 - f)$ of the generating function f of S. Therefore the previous decomposition translates into the following system of equations:

$$\begin{cases} F_0 = \dfrac{z\left(1 + \frac{F'_0}{1-F_0}\right)}{(1 - F_1)(1 - F_2)}, \\[2ex] F'_0 = \dfrac{z\left(u + \frac{F_2}{1-F_2}\right)\left(1 + \frac{F'_0}{1-F_0}\right)}{1 - F_1}, \\[2ex] F_1 = \dfrac{z\left(u + \frac{F_2}{1-F_2}\right)}{(1 - F_1)(1 - F_0)}, \\[2ex] F_2 = \dfrac{z\left(u + \frac{F_2}{1-F_2}\right)\left(1 + \frac{F'_0}{1-F_0}\right)}{1 - F_1}, \end{cases} \quad \text{or} \quad \begin{cases} F_0 = \dfrac{z\left(1 + \frac{F_2}{1-F_0}\right)}{(1 - F_1)(1 - F_2)}, \\[2ex] F_1 = \dfrac{z\left(u + \frac{F_2}{1-F_2}\right)}{(1 - F_1)(1 - F_0)}, \\[2ex] F_2 = \dfrac{z\left(u + \frac{F_2}{1-F_2}\right)\left(1 + \frac{F_2}{1-F_0}\right)}{1 - F_1}, \end{cases}$$

where the observation that $F'_0(z, u) = F_2(z, u)$ in the left hand side system yields the right hand side one. This system of equations completely defines the generating series $F_0(z, u)$. By elimination an algebraic equation for $F_0(z, u)$ is immediately obtained, which is of degree 4.

We are particularly interested in the case $u = 2$, since the coefficient f_n of z^n in

$$F(z) = F_0(z, 2) = \sum_{T \in \mathcal{F}_0} z^{|T|} 2^{\ell(T)},$$

counts n-node trees weighted by $2^{\ell(u)}$, and thus overcount n-nodes balanced trees with the same weight. According to Theorem 3, upon multiplying by 2^n, this yields an upper bound on the number of well-orderly maps with n nodes.

From elementary complex analysis, we have that $\log f_n \sim \log(\rho^{-n})$ where ρ is the radius of convergence of the series $F(z) = \sum_n f_n z^n$. Applying the implicit function theorem to the system defining $F(z)$, we can compute its radius of convergence and obtain:

$$\rho = (\sqrt{189 + 114\sqrt{3}} - 6\sqrt{3} - 9)/4.$$

From Theorem 5 we obtain:

Theorem 6. *The number of well-orderly maps with n nodes satisfies*

$$\frac{1}{n} \log_2 |H_n| \leqslant 1 + \log_2 1/\rho + o(1) \approx 4.9098 + o(1).$$

3.4   A Code for Colored Trees

Let S be a binary string. We denote by $\#S$ the number of binary strings having the same length and the same number of ones than S. More precisely, if S is of length x and has y ones, then we set $\#S := \binom{x}{y}$.

Lemma 5. [5] *Any binary string S of length n can be coded into a binary string of length $\log_2(\#S) + o(n)$. Moreover, knowing n, coding and decoding S can be done in linear time, assuming a RAM model of computation on $\Omega(\log n)$ bit words.*

Lemma 6. *Let B be a balanced tree such that the corresponding realizer $R = (T_0, T_1, T_2)$ has i_2 inner nodes in the tree T_2. The balanced tree B can be encoded with 5 binary strings S_1, S_2, S_3, S_4 and S_5 and 4 integers $a_0, a_0', a_1, i_2 \leqslant n$ such that:*

$$\#S_1 = \binom{n-a_0}{i_2-a_0}, \#S_2 = \binom{n-a_1}{a_0'}, \#S_3 = \binom{n+a_1}{a_1}, \#S_4 = \binom{a_1+a_0+a_0'}{a_0} \text{ and } \#S_5 = \binom{n-a_1-a_0'}{n-a_1-a_0'-i_2}.$$

Lemma 7. *Let H be an m-edge well-orderly map. H can be encoded with 6 binary strings (5 for the minimal realizer and a last one to store the missing edges) and 4 integers $a_0, a_1, a_0', i_2 \in [0, n]$ such that:* $\#S_1 = \binom{n-a_0}{i_2-a_0}$, $\#S_2 = \binom{n-a_1}{a_0'}$, $\#S_3 = \binom{n+a_1}{a_1}$, $\#S_4 = \binom{a_1+a_0+a_0'}{a_0}$, $\#S_5 = \binom{n-a_1-a_0'}{n-a_1-a_0'-i_2}$, $\#S_6 = \binom{2n-i2}{m-n-i_2}$.

Proof. With $S_1 - S_5$ a minimal realizer is encoded (Lemma 6). The last string indicates the edges to delete in order to rebuild the well-orderly map: for each v, one is used to indicate if the edge $(v, p_2(v))$ has to be removed and for each leaf v of T_2, one bit is used to indicate if the edge $(v, p_1(v))$ has to be removed.   □

4   Applications

In view of Theorems 2 and 6, the number of connected planar graphs is at most $2^{4.9098n}$. As shown in [5], the numbers of connected and general planar graphs differ by at most a polynomial term in n.

Theorem 7. *The number $p(n)$ of unlabeled planar graphs on n nodes satisfies, for every n large enough:*

$$\log_2 p(n) \leqslant \alpha n + O(\log n) \quad with \quad \alpha \approx 4.9098.$$

This result is completed by the (known) lower bound $\log_2 p(n) \geqslant \beta n + O(\log n)$, with $\beta \approx 4.767$.

The length of the coding of well-orderly map depends of the number of the edges of the well-orderly map.

The following two results are obtained from the analysis of the length of the code of Lemma 7. The length of this code depends on the number of edges of the well-orderly map (see Fig. 7).

Theorem 8. *Almost all unlabeled and almost all labeled planar graphs on n nodes have at least $1.85n$ edges and at most $2.44n$ edges. Moreover, the result holds also for unlabeled connected and labeled connected planar graphs.*

Theorem 9. *Every connected m-edge planar graph can be encoded in linear time with at most $2.82m + o(m)$ bits.*

5   The Average Size of Planar Drawings

Theorem 10. *The average number of leaves in a tree of a minimal realizer is $5n/8 + o(n)$ and the average number of 3-colored faces in a minimal realizer is $n/8 + o(n)$.*

Proof. Using classical techniques on generating function, we obtain that the average number of leaves of the tree T_0 of a minimal realizer is $5n/8 + o(n)$. By symmetry, this result is clearly true for the two other trees of the realizer. Since for any realizer, $\ell_0 + \ell_1 + \ell_2 + \Delta = 2n - 5$, where ℓ_i is the number of leaves in T_i and Δ is the number of 3-colored faces of the realizer [26], the second result comes directely.   □

In [27] a straight-line drawing algorithm base on minimal realizers is presented. This algorithm first computes the minimal realizer of a triangulation of the graph. Then the graph is drawn on a grid $(n - 1 - \Delta) \times (n - 1 - \Delta)$, where Δ is the number of 3-colored faces of the so obtained minimal realizer. Our analysis gives an average complexity of such drawings:

Corollary 1. *The average grid size required (i.e., the average width and the average height) to draw a triangulation is at most $(\frac{7n}{8} + o(n)) \times (\frac{7n}{8} + o(n))$.*

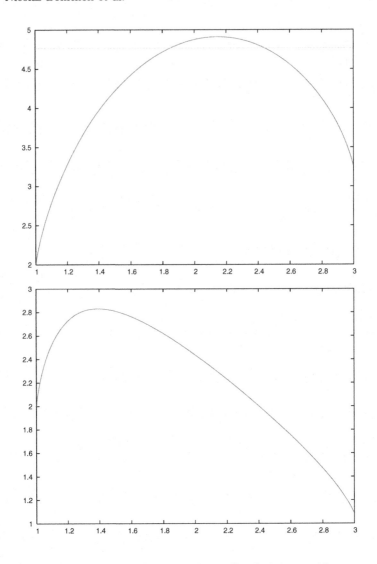

Fig. 7. (a) Number of bits necessary to encode a well-orderly map with $m = \alpha n$ edges, where $1 \leqslant \alpha \leqslant 3$. (b) Coding analyses: Number of bits per edges of a well-orderly map with $m = \alpha n$ edges, where $1 \leqslant \alpha \leqslant 3$.

In [28] a polyline drawing algorithm also based on minimal realizers is proposed. The graph is then drawn on a grid $(n - \lfloor \frac{\ell}{2} \rfloor - 1) \times \ell$, where ℓ is the number of leaves of the tree T_0 of the obtained minimal realizer $R = (T_0, T_1, T_2)$. Our analysis gives an average complexity of such drawings:

Corollary 2. *The average grid size required to draw a triangulation is at most* $\left(\frac{11n}{16} + o(n) \right) \times \left(\frac{5n}{8} + o(n) \right)$.

References

1. Liskovets, V.A., Walsh, T.R.: Ten steps to counting planar graphs. Congressus Numerantium **60** (1987) 269–277
2. Denise, A., Vasconcellos, M., Welsh, D.J.: The random planar graph. Congressus Numerantium **113** (1996) 61–79
3. McDiarmid, C.J., Steger, A., Welsh, D.J.: Random planar graphs (2001) Preprint.
4. Giménez, O., Noy, M.: Estimating the growth constant of labelled planar graphs. In: 3rd Colloquium on Mathematics and Computer Science: Algorithms, Trees, Combinatorics and Probabilities, Birkhäuser (2004)
5. Bonichon, N., Gavoille, C., Hanusse, N.: An information-theoretic upper bound of planar graphs using triangulation. In: 20th Annual Symposium on Theoretical Aspects of Computer Science (STACS). Volume 2607 of Lecture Notes in Computer Science., Springer (2003) 499–510
6. Bender, E.A., Gao, Z., Wormald, N.C.: The number of labeled 2-connected planar graphs. The Electronic Journal of Combinatorics **9** (2002) R43
7. Wright, E.M.: Graphs on unlabelled nodes with a given number of edges. Acta Math. **126** (1971) 1–9
8. Khodakovsky, A., Alliez, P., Desbrun, M., Schröder, P.: Near-optimal connectivity encoding of 2-manifold polygon meshes. Graphical Models (2002) To appear in a special issue.
9. King, D., Rossignac, J.: Guaranteed 3.67V bit encoding of planar triangle graphs. In: 11th Canadian Conference on Computational Geometry. (1999) 146–149
10. Rossignac, J.: Edgebreaker: Connectivity compression for triangle meshes. IEEE Transactions on Visualization and Computer Graphics **5** (1999) 47–61
11. Frederickson, G.N., Janardan, R.: Efficient message routing in planar networks. SIAM Journal on Computing **18** (1989) 843–857
12. Gavoille, C., Hanusse, N.: Compact routing tables for graphs of bounded genus. In: 26th International Colloquium on Automata, Languages and Programming (ICALP). Volume 1644 of LNCS., Springer (1999) 351–360
13. Lu, H.I.: Improved compact routing tables for planar networks via orderly spanning trees. In: 8th Annual International Computing & Combinatorics Conference (COCOON). Volume 2387 of LNCS., Springer (2002) 57–66
14. Thorup, M.: Compact oracles for reachability and approximate distances in planar digraphs. In: 42th Annual IEEE Symposium on Foundations of Computer Science (FOCS), IEEE Computer Society Press (2001)
15. Bodirsky, M., Gröpl, C., Kang, M.: Generating labeled planar graphs uniformly at random. In: 30th International Colloquium on Automata, Languages and Programming (ICALP). Volume 2719 of LNCS. (2003) 1095–1107
16. Gerke, S., McDiarmid, C.J.: On the number of edges in random planar graphs. Combinatorics, Probability & Computing (2002) To appear.
17. Turán, G.: Succinct representations of graphs. Discrete Applied Mathematics **8** (1984) 289–294
18. Keeler, K., Westbrook, J.: Short encodings of planar graphs and maps. Discrete Applied Mathematics **58** (1995) 239–252
19. Munro, J.I., Raman, V.: Succinct representation of balanced parentheses, static trees and planar graphs. In: 38th Annual IEEE Symposium on Foundations of Computer Science (FOCS), IEEE Computer Society Press (1997) 118–126
20. Yannakakis, M.: Embedding planar graphs in four pages. Journal of Computer and System Sciences **38** (1989) 36–67

21. Chiang, Y.T., Lin, C.C., Lu, H.I.: Orderly spanning trees with applications to graph encoding and graph drawing. In: 12th Symposium on Discrete Algorithms (SODA), ACM-SIAM (2001) 506–515

22. Chuang, R.C.N., Garg, A., He, X., Kao, M.Y., Lu, H.I.: Compact encodings of planar graphs via canonical orderings and multiple parentheses. In: 25th International Colloquium on Automata, Languages and Programming (ICALP). Volume 1443 of LNCS., Springer (1998) 118–129

23. Schnyder, W.: Embedding planar graphs on the grid. In: 1st Symposium on Discrete Algorithms (SODA), ACM-SIAM (1990) 138–148

24. Poulalhon, D., Schaeffer, G.: Optimal coding and sampling of triangulations. In: 30th International Colloquium on Automata, Languages and Programming (ICALP). Volume 2719 of LNCS., Springer (2003) 1080–1094

25. Goulden, I., Jackson, D.: Combinatorial Enumeration. John Wiley & Sons (1983)

26. Bonichon, N., Le Saëc, B., Mosbah, M.: Wagner's theorem on realizers. In: 29th International Colloquium on Automata, Languages and Programming (ICALP). Volume 2380 of LNCS., Springer (2002) 1043–1053

27. Zhang, H., He, X.: Compact visibility representation and straight-line grid embedding of plane graphs. In: Workshop on Algorithms and Data Structures (WADS). Volume 2748 of LNCS., Springer (2003) 493–504

28. Bonichon, N., Le Saëc, B., Mosbah, M.: Optimal area algorithm for planar polyline drawings. In: 28th International Workshop, Graph - Theoretic Concepts in Computer Science (WG). Volume 2573 of LNCS., Springer (2002) 35–46

Efficient Computation of the Lovász Theta Function for a Class of Circulant Graphs

Valentin E. Brimkov[1], Reneta P. Barneva[2],
Reinhard Klette[3], and Joseph Straight[2]

[1] Fairmont State University, Fairmont, West Virginia 26554-2470, USA
vbrimkov@fairmontstate.edu
[2] State University of New York, Fredonia, NY 14063, USA
{barneva,straight}@cs.fredonia.edu
[3] CITR Tamaki, University of Auckland, Building 731, Auckland, New Zealand
r.klette@auckland.ac.nz

Abstract. We consider the problem of estimating the Shannon capacity of a circulant graph $C_{n,J}$ of degree four with n vertices and chord length J, $2 \leq J \leq n$, by computing its Lovász theta function $\theta(C_{n,J})$. We present an algorithm that takes $O(J)$ operations if J is an odd number, and $O(n/J)$ operations if J is even. On the considered class of graphs our algorithm strongly outperforms the known algorithms for theta function computation.

1 Introduction

In the present paper we consider the problem of estimating the Shannon capacity of a circulant graph of degree four by computing its Lovász theta function. In a famous paper of 1956 [1] Shannon first studied the amount that an information channel can communicate without error. He introduced the notion of zero-error capacity of a graph, known thereafter as the Shannon capacity. It was understood quite early that the exact determination of the Shannon capacity is a very difficult problem, even for small and simple graphs (see [2, 3]). In 1979 Lovász [4] introduced a function $\theta(G)$ with the aim of estimating the Shannon capacity. Despite a lot of work in the field, very little is known about classes of graphs for whose theta function either a formula or a very efficient (e.g., linear) algorithm is available. An example for such a result is Lovász's formula $\theta(C_n) = \frac{n\cos\frac{\pi}{n}}{1+\cos\frac{\pi}{n}}$ for an odd cycle C_n with n nodes [4]. Recently Brimkov et al. [5] generalized this last result by obtaining formulas for $\theta(G)$ for the special cases of circulant graphs of degree four with chord length two and three.

Various applications of circulant graphs are known in counting and combinatorics [6], as well as in telecommunication networks, VLSI design, and distributed computing [7–10]. Low-degree circulants provided a basis for some classical parallel and distributed systems [11, 12], as well as for certain data alignment networks for complex memory systems [13]. Specifically, circulant graphs of degree four have been used in the design of local networks and interconnection subsystems

J. Hromkovič, M. Nagl, and B. Westfechtel (Eds.): WG 2004, LNCS 3353, pp. 285–295, 2004.

[14, 7]. Recent work [15] presents a class of such graphs with minimal topological distances. These graphs (called Midimew networks) have been used as a basis for constructing an optimal interconnection network for parallel computers with a very high degree of fault-tolerance [15], as well as for designing networks for massively parallel computers [16] or optimal VLSI [17].

Our interest in Shannon capacity and Lovász theta function of circulant graphs is also driven by possible applications to error-free communication of data describing the structure of a digital line, the latter being the most fundamental primitive in computer graphics and image analysis. Computer representation of digital lines has been an active research topic for nearly half a century (see the recent survey [18] and the bibliography therein). In [19] Dorst and Duin have developed the theory of spirographs in order to establish links between digital straight lines and number theory. Spirographs are diagrams that model the distribution of the integer points constituting a digital line and, as a matter of fact, appear to be circulant graphs of degree two or four (see [20]).

In the present paper we use a geometric approach to construct a very efficient algorithm for computing the theta function of *arbitrary* circulant graphs of degree four. For a circulant graph $C_{n,J}$ with n vertices and chord length J, $2 \leq J \leq n$, the algorithm performs $O(J)$ operations when J is odd and $O(n/J)$ operations when J is even, and appears to be strongly superior to the known algorithm from [21] for computing the theta function for general graphs, whose time complexity is of the order $O(n^4)$. If n is even and J odd, then $C_{n,J}$ appears to be perfect and $\Theta(C_{n,J}) = n/2$. We also obtain as a corollary that for the above-mentioned class of optimal chordal rings (Midimew networks), the algorithm computes the theta-function with $O(\sqrt{n})$ operations.

The paper is organized as follows. In the next section we recall some graph-theoretic notions and results to be used in the sequel. In Section 3 we introduce some geometrical constructions used in designing our algorithm. The basic results are presented in Section 4. We conclude with some remarks in the final Section 5.

2   Some Graph-Theoretic Notions and Facts

Here we recall some well-known definitions from graph theory. (See [22] for details.) Let $G(V, E)$ be a simple graph. The *complement graph* of G is the graph $\bar{G}(V, \bar{E})$, where $\bar{E}$ is the complement of E to the set of edges of the complete graph on V. An *automorphism* of the graph G is a permutation p of its vertices such that two vertices $u, v \in V$ are adjacent iff $p(u)$ and $p(v)$ are adjacent. G is *vertex symmetric* if its automorphism group is vertex transitive, i.e., for given $u, v \in V$ there is an automorphism p such that $p(u) = v$. By $\omega(G)$ and $\chi(G)$ we denote the clique and the chromatic numbers of G, respectively. For any graph G we have $\omega(G) \leq \chi(G)$. We also have the following classical result.

Theorem 1. *[22] A connected simple graph G with maximal degree d is d-colorable, unless $d \neq 2$ and G is a $(d + 1)$-clique, or $d = 2$ and G is an odd cycle.*

An *independent set* of G is a set of vertices no two of which are adjacent. The cardinality of a maximal independent set is called the *independence number* of G and denoted $\alpha(G)$. A graph $G'(V', E')$ is an *induced subgraph* of $G(V, E)$, if E' contains all edges from E that join vertices from $V' \subseteq V$. G is *perfect* if $\omega(G_A) = \chi(G_A)$, $\forall A \subseteq V$, where G_A is the induced subgraph of G on A.

An $n \times n$ matrix $A = (a_{i,j})_{i,j=0}^{n-1}$ is called *circulant* if its entries satisfy $a_{i,j} = a_{0,j-i}$, where the subscripts belong to the set $\{0, 1, \ldots, n-1\}$ and are calculated modulo n. In other words, any row of a circulant matrix can be obtained from the first one by a number of consecutive cyclic shifts, and thus the matrix is fully determined by its first row. A *circulant graph* is a graph with a circulant adjacency matrix. By $C_{n,J}$ we will denote a circulant graph of degree four, with vertex set $\{0, 1, \ldots, n-1\}$ and edge set $\{(k, k+1 \bmod n), (k, k+J \bmod n), k = 0, 1, \ldots, n-1\}$, where $1 < J \leq \frac{n-1}{2}$ is the *chord length*. See for illustration Fig. 2a presenting the circulant graph $C_{13,2}$. The Midimew networks mentioned in the Introduction are special circulant graphs of the form $C_{N,2h+1}$, where $N = h^2 + (h+1)^2$ and h is the graph diameter.

Now consider a graph G whose vertices are letters from a given alphabet and where adjacency indicates that two letters can be confused. In this setting, the maximal number of one-letter messages that can be communicated without danger of confusion equals the independence number $\alpha(G)$. Then the maximal number of k-letter messages that can be safely communicated is $\alpha(G^k)$, where G^k is the k-th power of G. It follows that $\alpha(G^k) \geq \alpha(G)^k$, as equality does not hold, in general [4]. The Shannon capacity of G is then defined as the limit $\Theta(G) = \lim_{k \to \infty} \sqrt[k]{\alpha(G^k)}$. It satisfies $\Theta(G) \geq \alpha(G)$, where equality does not need to occur. Shannon proved that if G is perfect, then $\Theta(G) = \alpha(G)$. As already mentioned, in order to estimate $\Theta(G)$, Lovász devised a function $\theta(G)$, known thereafter also as the Lovász number. Several equivalent definitions of the Lovász number are available [23]. We present here the one which requires only little technical machinery.

Definition 1. *Given a graph G, let $\mathbf{A}$ be the family of symmetric matrices A such that $a_{ij} = 0$ if v_i and v_j are adjacent in G. Let $\lambda_1(A) \geq \lambda_2(A) \geq \ldots \geq \lambda_n(A)$ be the eigenvalues of A. Then $\theta(A) = \max_{A \in \mathbf{A}} \{1 - \frac{\lambda_1(A)}{\lambda_n(A)}\}$.*

Combining the fact that $\Theta(G) \leq \theta(G)$ with the easy lower bound $\Theta(C_5) \geq \sqrt{5}$, Lovász was able to determine the capacity of the pentagon C_5, which turns out to be $\sqrt{5}$. We know very little about the Shannon capacity of other non-perfect graphs. For instance, $\Theta(C_7)$ is still unknown. $\theta(G)$, however, is computable in polynomial time with arbitrary precision, although being "sandwiched" between the clique number $\omega(G)$ and the chromatic number $\chi(G)$, whose computation is NP-hard for general graphs. More precisely, we have $\omega(G) \leq \theta(\bar{G}) \leq \chi(G)$. Because of this remarkable property and due to the relations to communication issues, the Lovász number is a subject of active study. For various results and applications see the surveys by Knuth [23] and Alizadeh [24] and the bibliography therein. See also [25–29] for a sample of the diversity of results and applications of $\Theta(G)$ and $\theta(G)$. Here we list a simple proposition for future reference.

Proposition 1 (see [23]). *For every graph G with n vertices, $\theta(G) \cdot \theta(\bar{G}) \geq n$. If G is vertex symmetric, then $\theta(G) \cdot \theta(\bar{G}) = n$.*

The algorithm for theta function computation described in the following sections applies to circulant graphs of degree four. We notice that all circulants of order ≤ 5 except the pentagon are perfect and their Shannon capacity is trivially determined. We also have $\Theta(C_{5,1}) = \theta(C_{5,1}) = \sqrt{5}$. Thus we can consider circulant graphs of order larger than 5. It is easy to see that for circulants $C_{n,J}$ with $n \geq 6$ it holds $\omega(C_{n,J}) \geq 2$ and $\chi(G) \leq 4$, hence $2 \leq \theta(\bar{C}_{n,J}) \leq 4$. Since the circulant graphs are vertex symmetric, by Proposition 1 we obtain the bounds $n/2 \geq \theta(C_{n,J}) \geq n/4$. In the subsequent sections we design an efficient algorithm for the exact computation of $\theta(C_{n,J})$.

3   LP Formulation of $\theta(C_{n,j})$ and Certain Subsidiary Geometrical Constructions

Taking advantage of the particular properties of circulant matrices whose eigenvalues can be expressed in closed form, one can easily generalize the approach of [4]. Then the validity of the following minmax formulation of the θ-function of circulant graphs of degree 4 can be derived.

Lemma 1 (see [5]). *Let $f_0(x, y) = n + 2x + 2y$ and for some fixed value of J*

$$f_k(x, y) = 2x \cos \frac{2\pi k}{n} + 2y \cos \frac{2\pi kJ}{n}, \quad k = 1, 2, \ldots, n - 1.$$

Then

$$\theta(C_{n,J}) = \min_{x,y} \max_k \left\{ f_k(x, y), k = 0, 1, \ldots, \left\lfloor \frac{n}{2} \right\rfloor \right\}. \tag{1}$$

This is in turn equivalent to the following Linear Programming (LP) problem:

$$\theta(C_{n,J}) = \min\{z \ : \ f_k(x, y) - z \leq 0, \ k = 0, 1, \ldots, \lfloor \tfrac{n}{2} \rfloor, z \geq 0\}. \tag{2}$$

We now observe that the equalities

$$f_1(x, y) - z = 0, f_2(x, y) - z = 0, \ldots, f_{n-1}(x, y) - z = 0$$

define planes through the origin. Having in mind the specific coefficients of these planes in the different ortants, as well as the relations between the coefficients of two consecutive planes, one can see that the set

$$\max_k \{ f_k(x, y), k = 1, 2, \ldots, n - 1 \}$$

is a polyhedral surface, namely a polyhedral cone C with its apex at the origin. The cone belongs to the positive halfspace $z \geq 0$ and the Oz axis is contained inside the cone. The faces of the cone are portions of planes with equations

$$z = f_k(x, y), \quad k = 1, 2, \ldots, n - 1.$$

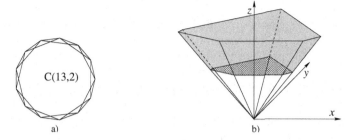

Fig. 1. a) The graph $C_{13,2}$. b) The polyhedral cone related to $C_{13,2}$, cut at $z = 2$.

Thus the rays of C are intersections of planes obtained for pairs of indices i, j, $1 \leq i, j \leq n-1$. Other intersections are not of interest, since they all fall "below" the conic surface $\max_k\{f_k\}$ and thus are not part of it.

A more careful analysis can show that only one of the planes forming the cone has two positive coefficients, and only one of them has a positive coefficient for y and a negative coefficient for x. In the other cases we may have arbitrary many planes. For example, this is the case in the ortant $x \leq 0, y \leq 0, z \geq 0$ (i.e., if the two coefficients are negative).

We now consider the plane $f_0(x, y) = n + 2x + 2y$. Its intersection with the cone C produces a new polyhedral surface. Roughly speaking, this is the upper part of the cone C, i.e., the one above the plane f_0 (see Fig. 1b). As it will turn out later, a part of the cone C will be "cut out" and thus some of the planes (forming the faces of C) will be eliminated.

Clearly, the intersection points of the plane f_0 with C are the possible candidates for solution of the problem. The theta function is the intersection point with minimal z. Consider the intersection of C and f_0. This intersection is the boundary of some 2D convex polyhedron P (possibly unbounded). As mentioned above, the solution is at some of the vertices of this intersection. Let this be the point $A = (x_0, y_0, z_0)$ (and thus $\vartheta = z_0$). Let us now assume that we have intersected C by the plane $z = z_0$ (parallel to the xy-plane). The intersection is a (bounded) convex polygon Q_{z_0}. By construction, it follows that the polyhedron P and the polygon Q_{z_0} intersect at a single point, i.e., the point $A = (x_0, y_0, z_0)$. We will determine A using the sides of Q_{z_0}, rather than the sides of P. Since the coefficients of x and y of the plane $z = n + 2x + 2y$ are equal (indeed they are both equal to 2), it is not difficult to see that A will be the vertex of Q_{z_0}, obtained as the intersection of the two sides of Q_{z_0} which "sandwich" the straight line in $z = z_0$ passing through A and with a slope of 45 degrees. These lines have equations $2x \cos \alpha + 2y \cos(\alpha J) = z_0$ and $2x \cos \beta + 2y \cos(\beta J) = z_0$, where $\alpha = \frac{2\pi k_1}{n}$ and $\beta = \frac{2\pi k_2}{n}$, for some indices k_1 and k_2. Once k_1 and k_2 are known, z_0 can be computed by solving the linear system

$$\begin{cases} z = 2x \cos \alpha + 2y \cos(2\alpha) \\ z = 2x \cos \beta + 2y \cos(2\beta) \\ z = n + 2x + 2y. \end{cases}$$

Note that one can use any horizontal intersection of the cone, since all such intersections are homothetic to each other.

Through a more detailed analysis of the structure of the admissible region defined by the linear constraints, in the next section we propose an efficient computation of $\theta(C_{n,J})$. The general idea is to reduce significantly the set of constraints in the LP problem (2) and then apply existing efficient algorithms for LP in 3D (e.g., the Megiddo's algorithm [30] which performs in time linear with respect to the number of constraints of the problem). We measure the complexity of a computation by counting the number of arithmetic operations in the set $S = \{+, -, *, /, \lfloor . \rfloor, \cos(\cdot)\}$ as a function of the number of constraints in the three-dimensional LP problem (2).

4   Computation of $\theta(G)$

Relying on (2), we will focus on the geometric unintuitive regularities of the polygon defined by the lines l_k of equation

$$x \cos(\alpha_k) + y \cos(J\alpha_k) = 1, \tag{3}$$

with $\alpha_k = \frac{2\pi}{n} k$. We will refer to angle α_k as to the angle of line l_k. We distinguish two cases: J even and J odd.

4.1   Odd Chord Lengths

As a first result let us prove the following lemma for arbitrary circulant graphs. It provides an immediate solution to the case of $C_{n,J}$ with n even and J odd.

Denote by $C(n; J_1, J_2, \ldots, J_m)$ a circulant graph with vertex set $\{0, 1, \ldots, n-1\}$ and edge set $\{(k, k+1 \bmod n), (k, k+J_j \bmod n), k = 0, 1, \ldots, n-1, j = 1, 2, \ldots, m\}$, where $J_1, J_2, \ldots, J_m$ are the *chord lengths* satisfying $2 < J_1 < \ldots < J_m$ and $1 < J_j \leq \frac{n-1}{2}$ for $j = 1, 2, \ldots, m$.

Lemma 2. *Consider a circulant graph $C(n; J_1, J_2, \ldots, J_m)$. Assume that n is even and all chord lengths J_j, $j = 1, 2, \ldots, m$, are odd. Then $C(n; J_1, J_2, \ldots, J_m)$ is perfect and $\theta(C(n; J_1, J_2, \ldots, J_m)) = n/2$.*

Proof. Since every circulant graph is vertex symmetric, we have

$$\theta(C(n; J_1, \ldots, J_m)) \cdot \theta(\bar{C}(n; J_1, \ldots, J_m)) = n.$$

Thus it is enough to show that $\theta(\bar{C}(n; J_1, \ldots, J_m)) = 2$. Bearing in mind the inequality $\omega(G) \leq \theta(\bar{G}) \leq \chi(G)$ which applies to any graph G, we obtain that it is enough to show that

$$\omega(C(n; J_1, \ldots, J_m)) = \chi(C(n; J_1, \ldots, J_m)) = 2.$$

In fact, the clique number is 2 since for $n \geq 6$ and $J_j \geq 3$, the minimal cycle in $C(n; J_1, \ldots, J_m)$ has length at least 4 (which bound is reached for $J_j = 3$). It is also not hard to see that the vertices of $C(n; J_1, \ldots, J_m)$ can be alternatively colored with two colors only, if n is even and the J_j's are odd, which completes the proof. $\qquad\square$

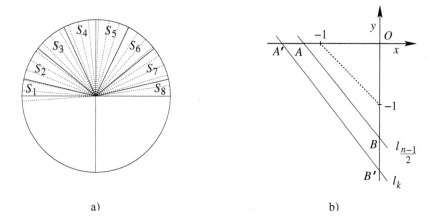

a)   b)

Fig. 2. a) Pictorial description of the S-intervals for $j = 7$. b) Position of line $l_{(n-1)/2}$ with respect to any l_k with $\alpha_{l_k} \in S_1$.

Now we obtain the following main result.

Theorem 2. *Let n, J be two odd numbers with $J \leq \frac{n-1}{2}$. Then $\theta(C_{n,J})$ can be computed with $O(J)$ operations by solving a 3D LP problem having $O(J)$ constraints.*

Proof. We prove that we can identify in constant time a set of at most $\lfloor \frac{J}{2} \rfloor + 1$ lines that define the polygon Q_{z_0}.

Let $S = \{S_1, S_2, \ldots, S_{J+1}\}$ be a set of adjacent intervals covering $[0, \pi]$, defined as

$$
\begin{aligned}
S_1 &= [\pi - \tfrac{\pi}{2J}, \pi], \\
S_{J+1} &= [0, \tfrac{\pi}{2J}], \\
S_{k+1} &= [\pi - (2k+1)\tfrac{\pi}{2J}, \pi - (2k-1)\tfrac{\pi}{2J}], \text{ for } k = 1, 2, \ldots, J-1.
\end{aligned}
$$

So, S_1, S_{J+1} are intervals of width $\frac{\pi}{2J}$, whereas $S_2, S_3, \ldots, S_J$ are intervals of width $\frac{\pi}{J}$ (see Fig 2a). A quick analysis of the function $\cos(J\alpha)$ reveals that

a) it is periodic of period $2\pi/J$;

b) it nullifies on $S_k \cap S_{k+1}$, for $k = 1, 2, \ldots, J$, and

c) it is negative on the odd numbered intervals attaining -1 on their middle points.

Let us define

$$a(k) = 1/\cos(\alpha_k), b(k) = 1/\cos(J\alpha_k),$$

where $\alpha_k = \frac{2\pi}{n}k$, be the x and y axes cuts, respectively, with a line l_k defined by (3).

Consider line $l_{\lfloor \frac{n}{2} \rfloor}$ in the interval S_1. This verifies

$$
\begin{cases}
a(\lfloor \frac{n}{2} \rfloor) = \frac{1}{\cos(\pi - \pi/n)} = \max\{a(k) \mid a(k) < 0 \wedge k \geq 0\} < -1 \\[2mm]
b(\lfloor \frac{n}{2} \rfloor) = \frac{1}{\cos(J(\pi - \frac{\pi}{n}))} = \frac{1}{\cos(\pi - J\frac{\pi}{n})}
\end{cases}
$$

This line defines a face of Q_{z_0} because it is the one that intersects the Ox axis in the closest point to $(-1, 0)$ (see Fig. 2b). Furthermore, its inclination with respect to the Oy axis is less than 45 degrees and all other lines in S_1 have a lower x-cut and y-cut therefore falling outside Q_{z_0}.

Consider the even numbered intervals. Lines whose angle τ falls in these intervals all have positive y coordinate because $\cos(J\tau) > 0$. Their x-coordinate can be either positive, and in that case they would not even cross the third quadrant, or negative, in which case it must be

$$\frac{1}{\cos(\tau)} < \frac{1}{\cos(\alpha_{(n-1)/2})} = a(\lfloor \frac{n}{2} \rfloor).$$

So, we can conclude that those lines can not affect the solution.

Consider the odd numbered intervals S_{2k-1}, for $k = 1, 2, \ldots, (J+1)/2$ and let $\beta^{(k)} = \pi - (2k - 2)\pi/J$. Angle $\beta^{(1)}$ is π, whereas, for $k > 1$, angles $\beta^{(k)}$ correspond to the centers of intervals S_{2k-1} and all verify $\cos(J\beta^{(k)}) = -1$. We can observe that the only lines l_k that intersect $l_{\lfloor \frac{n}{2} \rfloor}$ in the third quadrant are those for which

$$b(\lfloor \frac{n}{2} \rfloor) < b(k) < -1. \tag{4}$$

Now, focus for a moment on the function $f(x) = 1/\cos(kx)$. It is periodic and assumes the same values within the odd numbered intervals

$$S_{2k-1} = [\beta^{(k)} - \pi/2J, \beta^{(k)} + \pi/2J].$$

Furthermore it is increasing over $[\beta^{(k)} - \pi/2J, \beta^{(k)}]$, decreasing over $[\beta^{(k)}, \beta^{(k)} + \pi/2J]$ and verifies:

$$f(\beta^{(k)}) = -1, \quad \lim_{x \to (\beta^{(k)} - \pi/2J)+} = \lim_{x \to (\beta^{(k)} + \pi/2J)-} = -\infty.$$

Observe that $b(k) = f(2k\pi/n)$, i.e., condition (4) can be rephrased as

$$f\left(\pi - \frac{\pi}{n}\right) < f\left(\frac{2\pi}{n} \cdot k\right) < -1 .$$

The behavior of $f(x)$ on the interval $[\pi - \pi/2J, \pi + \pi/2J]$ is the same as for all the other odd numbered intervals $[\beta^{(k)} - \pi/2J, \beta^{(k)} + \pi/2J]$. Hence, the condition

$$f\left(\pi - \frac{\pi}{n}\right) < f(x) < -1$$

will be verified only for

$$|x - \beta^{(k)}| < \frac{\pi}{n}, \tag{5}$$

where $\{\beta^{(k)} \mid k = 1, 2, \ldots, (J+1)/2\}$ is the set of the solutions to equation $f(x) = 1/\cos(Jx) = -1$ on $[0, \pi]$.

Given that the angle does not vary with continuity but assumes only a discrete set of values $\alpha_k = 2\pi k/n$ for $0 \le k \le (n-1)/2$, we can see that, if for some u, α_u satisfies condition (5), then $\alpha_{u+1} = \alpha_u + 2\pi/n$ does not.

Thus we can deduce that for each odd numbered interval S_{2k-1} there can be at most one line verifying condition (4). Since we have $\lfloor \frac{J}{2} \rfloor$ odd numbered intervals to consider, there will be at most as many lines to select.

Now the obtained linear program can be solved in $O(J)$ time by the Megiddo linear programming algorithm which has linear complexity when the number of variables is fixed.   □

Corollary 1. *The theta-function of a Midimew network $C_{N,J}$ can be computed with $O(\sqrt{N})$ operations.*

Proof. Follows from the fact that Midimew networks are circulant graphs of the form $C_{N,J}$ with $J = 2h+1$ and $N = h^2 + (h+1)^2$. This implies $J < \sqrt{2N}$. Then $\theta(C_{N,J})$ can be computed in $O(J) = O(\sqrt{N})$ time.   □

4.2   Even Chord Lengths

Consider now the case when the chord length J is an even number. We have the following theorem.

Theorem 3. *Let n be a positive integer and J an even number with $J \leq \frac{n-1}{2}$. Then $\theta(C_{n,J})$ can be computed with $O(n/J)$ operations by solving a 3D LP problem having $O(n/J)$ constraints.*

Proof. Consider once more the family of intervals $S = \{S_1, S_2, \ldots, S_{J+1}\}$. Now the $J/2$ even numbered ones, S_{2k}, for $k = 1, 2, \ldots, J/2$, are those in which $\cos(J\alpha)$ is negative. Let $\beta^{(k)}$ denote the angle corresponding to the center of S_{2k}. Thus $\cos(J\beta^{(k)}) = -1$ for all k. First, notice that each interval contains no more than $\lceil \frac{n}{J} \rceil$ lines.

Let us focus on $S_1 \cup S_2$ and define an enumeration of consecutive (with respect to the corresponding increasing angle) lines $l_1, l_2, \ldots, l_s$, where l_1 is the line whose angle $\alpha^{(1)}$ is the closest from below to the center of S_2, and l_s is the line whose angle is the largest within S_1, i.e, $l_{\lfloor n/2 \rfloor}$. It is not hard to see that those lines define a set C_1 of segments that, together with the x and y negative axes, bound a convex polygon Q.

We needed to consider also the lines in S_1, because those have x-coordinate that happen to be very close to point $(-1, 0)$ and, as proven above, they contribute to shaping polygon Q. Furthermore, all the other lines in S_2 whose angle is smaller than l_1's angle, must have lower x-cut and y-cut and therefore do not intersect this polygon.

We can apply the same idea to the other even numbered intervals S_{2k}, $k = 2, 3, \ldots, J/2$, and define the corresponding finite sequences of lines $l_1^{(k)}, l_2^{(k)}, \ldots,$ $\ldots, l_{s_k}^{(k)}$, now ending with lines whose angle is within S_{2k}. (Note the asymmetry in the definition of the lines $\{l_i^{(k)}\}$; in that $l_{s_k}^{(k)}$ has the largest angle in S_{2k}, whereas the sequence l_i is not limited to S_2 but goes on until the exhaustion of the interval S_1 adjacent to S_2.)

Now, as in the case for J odd, it turns out that for all k only $l_1^{(k)}$ might intersect Q. Furthermore, this would occur only when the angle $\alpha^{(k)}$ of $l_1^{(k)}$ satisfies

$$| \alpha^{(k)} - \beta^{(k)} | < | \alpha^{(1)} - \beta^{(1)} | \leq \pi/n . \tag{6}$$

As a consequence, the search for the solution can be restricted to the vertices of the polygon formed by the two axes, the lines in $S_1 \cup S_2$, plus, possibly, the lines whose angle verifies property (6). Thus the total number of lines to be considered is $O(n/J)$. Then the obtained LP problem can be solved in $O(n/J)$ time by the Megiddo linear programming algorithm. $\square$

5   Concluding Remarks

We have presented efficient ways to compute the theta function of circulant graphs of degree four. In particular, the problem can be reduced to a 3-variable LP problem having at most $O(J)$ constraints when J is odd, whereas for J even the bound on the number of significant constraints was shown to be $O(n/J)$. Consequently, an application of the Megiddo algorithm allows to compute $\theta(C_{n,J})$ with $O(J)$ or $O(n/J)$ operations depending on the evenness of J. Megiddo algorithm solves any LP problem in linear time with respect to the number of constraints, provided that the number of variables is fixed. It is indeed known that its complexity would include an implicit factor of the order of $O(2^{s^2})$, where s is the number of variables, which however is a small number for the considered dimension. Work in progress aims at providing efficient computation of the theta-function of circulant graphs of higher degree or of other interesting classes of graphs.

Acknowledgements

We thank Bruno Codenotti, Valentino Crespi and Mauro Leoncini for a number of discussions that motivated this research. We are grateful to the Referees for their useful remarks and suggestions. The first author thanks Boris Goldengorin for several interesting and helpful discussions during WG 2004.

References

1. Shannon, C.E., The zero-error capacity of a noisy channel, *IRE Trans. Inform. Theory* **IT-2** (1956) 8-19
2. Haemers, W., An upper bound for the Shannon capacity of a graph, Colloq. Math. Soc. János Bolyai **25** (1978) 267-272
3. Rosenfeld, M., On a problem of Shannon, *Proc. Amer. Mat. Soc.* **18** (1967) 315-319
4. Lovász, L., On the Shannon capacity of a graph, *IEEE Trans. on Inf. Theory*, **25** (1979) 1-7
5. Brimkov, V.E., B. Codenotti, V. Crespi, M. Leoncini, On the Lovász number of certain circulant graphs, In: *Algorithms and Complexity*, Bongiovanni, G., G. Gambosi, R. Petreschi (Eds.), LNCS No 1767 (2000) 291-305
6. Minc, H., Permanental compounds and permanents of (0,1) circulants, *Linear Algebra and its Applications* **86** (1987) 11-46
7. Bermond, J.-C., F. Comellas, D.F. Hsu, Distributed loop computer networks: A survey, *J. of Parallel and Distributed Computing* **24** (1995) 2-10

8. Leighton, F.T., *Introduction to parallel algorithms and architecture: Arrays, trees, hypercubes*, M. Kaufmann (1996)

9. Liton, B., B. Mans, On isomorphic chordal rings, Proc. of the Seventh Australian Workshop on Combinatorial Algorithms (AWOCA'96), Univ. of Sydney, BDCS-TR-508 (1996) 108-111.

10. Mans, B., Optimal distributed algorithms in unlabel tori and chordal rings, *J. of Parallel and Distributed Computing* **46**(1) (1997) 80-90

11. Bouknight, W.J., S.A. Denenberg, D.E. McIntyre, J.M. Randall, A.H. Samel, D.L. Slotnick, The Illiac IV System, Proc. IEEE **60**(4) (1972) 369-378

12. Wilkov, R.S., Analysis and design of reliable computer networks, *IEEE Trans. on Communications* **20** (1972) 660-678

13. Wong, C.K., D. Coppersmith, A combinatorial problem related to multimodule memory organization, *Journal of the ACM* **21**(3) (1974) 392-402

14. Adám, A., Research problem 2-10, *J. Combinatorial Theory* **393** (1991) 1109-1124

15. Beivide, R., E. Herrada, J.L. Balcázar, A. Arruabarrena, Optimal distance networks of low degree for parallel computers, *IEEE Trans. on Computers* **C-30**(10) (1991) 1109-1124

16. Yang, Y., A. Funashashi, A. Jouraku, H. Nishi, H. Amano, T. Sueyoshi, Recursive diagonal torus: an interconnection network for massively parallel computers, *IEEE Trans. on Parallel and Distributed Systems* **12**(7) (2001) 701-715

17. Huber, K., Codes over tori, *IEEE Trans. on Information Theory* **43**(2) (1997) 740-744

18. Rosenfeld, A., R. Klette, Digital straightness, *Electronic Notes in Theoretical Computer Science* **46** (2001) http://www.elsevier.nl/locate/entcs,volume46.html

19. Dorst, L., R.P.W. Duin, Spirograph theory: a framework for calculations on digitized straight lines, *IEEE Trans. Pattern Analysis and Machine Intelligence*, **6** (1984) 632-639

20. Brimkov, V.E., R. Barneva, R. Klette, J. Straight, Lovász theta-function of a class of graphs representing digital lines, Preproceedings of *WG 2004 - 30th International Workshop on Graph-Theoretic Concepts in Computer Science*, Bad Honnef, Germany, June 21-23, 2004

21. Alizadeh, F. et al., SDPPACK user's guide, http://www.cs.nyu.edu/faculty/overton/sdppack,sdppack.html

22. Berge, C., Graphs, North-Holland Mathematical Library, 1985

23. Knuth, D.E., The sandwich theorem, *Electronic J. Combinatorics*, **1** (1994) 1-48

24. Alizadeh, F., Interior point methods in semidefinite programming with applications to combinatorial optimization, *SIAM J. Optimization* **5**(1) (1995) 13-51

25. Alon, N., On the capacity of digraphs, *European J. Combinatorics* **19** (1998) 1-5

26. Alon, N., A. Orlitsky, Repeated communication and Ramsey graphs, *IEEE Trans. on Inf. Theory* **33** (1995) 1276-1289

27. Ashley, J.J., P.H. Siegel, A note on the Shannon Capacity of run-length-limited codes, *IEEE Trans. on Inf. Theory* **IT-33** (1987) 601-605

28. Farber, M., An analogue of the Shannon capacity of a graph, *SIAM J. on Alg. and Disc. Methods* **7** (1986) 67-72

29. Feige U., Randomized graph products, chromatic numbers, and the Lovász θ-function, *Proc of the 27th STOC* (1995) 635-640

30. Megiddo, N., Linear programming in linear time when the dimension is fixed, *J. of ACM*, **31** 1 (1984) 114-127

Unhooking Circulant Graphs:
A Combinatorial Method for Counting
Spanning Trees and Other Parameters[*]

Mordecai J. Golin and Yiu Cho Leung

Dept. of Computer Science, Hong Kong University of Science and Technology,
Clear Water Bay, Kowloon, Hong Kong
{golin,cscho}@cs.ust.hk

Abstract. It has long been known that the number of spanning trees in circulant graphs with fixed jumps and n nodes satisfies a recurrence relation in n. The proof of this fact was algebraic (relating the products of eigenvalues of the graphs' adjacency matrices) and not combinatorial. In this paper we derive a straightforward combinatorial proof of this fact. Instead of trying to decompose a large circulant graph into smaller ones, our technique is to instead decompose a large circulant graph into different *step graph* cases and then construct a recurrence relation on the step graphs. We then generalize this technique to show that the numbers of Hamiltonian Cycles, Eulerian Cycles and Eulerian Orientations in circulant graphs also satisfy recurrence relations.

1 Introduction

The purpose of this paper is to develop a *combinatorial* derivation of the recurrence relations on the number of spanning trees on circulant graphs. We then extend the technique developed in order to derive recurrence relations on other parameters of circulant graphs.

We start with some definitions and background. The n node *undirected circulant graph* with jumps $s_1, s_2, \ldots s_k$, is denoted by $C_n^{s_1, s_2, \cdots, s_k}$. This is the $2k$ regular graph[1] with n vertices labeled $\{0, 1, 2, \cdots, n-1\}$, such that each vertex i $(0 \leq i \leq n-1)$ is adjacent to $2k$ vertices $i \pm s_1$, $i \pm s_2$, $\cdots$, $i \pm s_k$ mod n. The simplest circulant graph is the n vertex cycle C_n^1. The next simplest is the *square of the cycle* $C_n^{1,2}$ in which every vertex is connected to its two neighbors and neighbor's neighbors. Figure 1 illustrates three circulant graphs.

For connected graph G, $T(G)$ denotes the number of spanning trees in G. Counting $T(G)$ is a well studied problem, both for its own sake and because it

---

[*] Partially supported by HK CERG grants HKUST6162/00E, HKUST6082/01E and HKUST6206/02E. A full version of this paper is available at [6].

[1] If $\gcd(n, s_1, s_2, \cdots, s_k) > 1$ then the graph is disconnected and contains no spanning trees. Therefore, for the purposes of this extended abstract, we assume that $\gcd(s_1, s_2, \cdots, s_k) = 1$, forcing the graph to be connected. Also note that if $n \leq 2s_k$ it is possible that the graph is a *multigraph* with some repeated edges.

J. Hromkovič, M. Nagl, and B. Westfechtel (Eds.): WG 2004, LNCS 3353, pp. 296–307, 2004.
© Springer-Verlag Berlin Heidelberg 2004

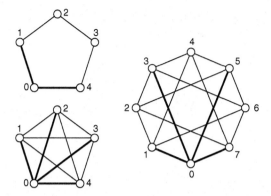

Fig. 1. Three examples of circulant graphs: C_5^1, $C_5^{1,2}$, $C_8^{1,3}$.

has practical implications for network reliability, e.g., [5]. For any *fixed* graph G, Kirchhoff's *Matrix-Tree Theorem* [8] efficiently permits calculating $T(G)$ by evaluating a co-factor of the *Kirchhoff matrix* of G (this essentially calculates the determinant of matrix related to the adjacency matrix of G.)

The interesting problem is in calculating the number of spanning trees in graphs chosen from defined *classes* as a function of a parameter. When G is a circulant graph the behavior of $T(G)$ as a function of n has been well studied. The canonical result is that $T(C_n^{1,2}) = nF_n^2$, F_n the *Fibonacci* numbers, i.e., $F_n = F_{n-1} + F_{n-2}$ with $F_1 = F_2 = 1$. This was originally conjectured by Bedrosian [2] and subsequently proven by Kleitman and Golden [9]. The same formula was also conjectured by Boesch and Wang [3] (without the knowledge of [9]). Different proofs can been found in [1, 4, 11]. Formulas for $T(C_n^{1,3})$ and $T(C_n^{1,4})$ are provided in [10]. These were later generalized in [12] to prove the following general theorem: *For any fixed* $1 \leq s_1 < s_2 < \cdots < s_k$,

$$T(C_n^{s_1, s_2, \cdots, s_k}) = na_n^2,$$

where a_n satisfies a recurrence relation of order $2^{s_k - 1}$ with constant coefficients. Knowing the *existence* and *order* of the recurrence relation permits explicitly constructing it by using Kirchhoff's theorem to evaluate $T(C_n^{s_1, s_2, \cdots, s_k})$ for $n = 1, 2, \ldots, 2^{s_k - 1}$ and solving for the coefficients of the recurrence relation.

With the exception of that in [9] all of the proofs above work as follows

- Let $s_1, s_2, \ldots s_k$ be fixed.
- Find the eigenvalues of the adjacency matrix of $C_n^{s_1, s_2, \ldots s_k}$. This can be done because the adjacency matrix is a *circulant matrix* and eigenvalues of circulant matrices are well understood.
- Express $T(C_n^{s_1, s_2, \cdots, s_k})$ as a product function of these eigenvalues.
- Simplify this product to show that $\sqrt{T(C_n^{s_1, s_2, \cdots, s_k})}/n$, as a function of n, satisfies a recurrence relation of the given order.

The major difficulty with this technique is that, even though it proves the *existence* of the proper order recurrence relation, it does not provide any combinatorial interpretation, e.g., some type of inclusion-exclusion counting argument, as to why this relation is correct.

As mentioned above, Kleitman and Golden's derivation of $T(C_n^{1,2}) = nF_n^2$, in [9] is an exception to this general technique; their proof is a very clever, fully combinatorial one. Unfortunately, it is also very specific to the special case $C_n^{1,2}$ and can not be extended to cover any other circulant graphs. The major impediment to deriving a general combinatorial proof is that, at first glance, it is difficult to see how to decompose $T(C_n^{s_1,s_2,\cdots,s_k})$ in terms of $T(C_m^{s_1,s_2,\cdots,s_k})$ where $m < n$; larger circulant graphs just do not seem to be able to be decomposed into smaller ones.

The main motivation of this paper was to develop a *combinatorial* derivation of the fact that $T(C_n^{s_1,s_2,\cdots,s_k})$, as a function of n, satisfies a recurrence relation. Our general technique is *unhooking*, i.e., removing all edges

$$\{(i,j) : n - s_k \le i < n \text{ and } 0 \le j < s_k\}$$

from the graph, creating a new *step graph* $L_n^{s_1,s_2,\cdots,s_k}$. We then define a *fixed* number of classes of forests of $L_n^{s_1,s_2,\cdots,s_k}$ and *combinatorially* derive a system of recurrences counting the number of forests in each class. We then relate this to the original problem by writing $T(C_n^{s_1,s_2,\cdots,s_k})$ as a linear combination of the number of forests in each class. Technically, we define a $(m \times 1)$-vector $(m,$ the number of forest classes, will be defined later) $\boldsymbol{T}(L_n^{s_1,s_2,\cdots,s_k})$ denoting the number of forests in each class; a $m \times m$ matrix A denoting the system of recurrence relations; and a $(1 \times m)$ row vector $\boldsymbol{\beta}$ such that

$$T(C_n^{s_1,s_2,\cdots,s_k}) = \boldsymbol{\beta} \cdot \boldsymbol{T}(L_n^{s_1,s_2,\cdots,s_k}), \quad \text{and} \quad \boldsymbol{T}(L_n^{s_1,s_2,\cdots,s_k}) = A \cdot \boldsymbol{T}(L_{n-1}^{s_1,s_2,\cdots,s_k}).$$

Given these matrix equations, standard techniques, e.g., solving for the generating functions, permit us to derive an order m constant coefficient recurrence relation for $T(C_n^{s_1,s_2,\cdots,s_k})$.

This technique of unhooking circulant graphs, i.e., developing a system of recurrences on the resultant step graphs and then writing the final result as a function of the step-graph values, is actually quite general and can be used to enumerate many other parameters of circulant graphs. In this extended abstract, we further describe how it can be used to derive recurrence relations for the number of Hamiltonian cycles. In the full version of this paper we also describe how to derive recurrence relations for Eulerian cycles and Eulerian Orientations as well. To the best of our knowledge, this is the first time that techniques for deriving recurrence relations for these other functions of circulant graphs have been developed.

The remainder of the paper is structured as follows. In the first part of section 2 we use our unhooking technique to re-derive the formula $T(C_n^{1,2}) = nF_n^2$. This introduces all of the basic ideas and techniques which are then generalized into a technique for deriving recurrence relations for all $T(C_n^{s_1,s_2,\cdots,s_k})$ as a function of n. In section 3 we discuss Hamiltonian cycles. Finally, in Section 4, we conclude some comments and open questions.

2   Counting Spanning Trees

2.1   Analyzing $T(C_n^{1,2})$

Let $C_n^{s_1,s_2,\ldots,s_k} = (V, E_C)$ be a circulant graph; $V = \{0, 1, \ldots, n-1\}$ and $E_C = \{(i,j) : i - j \bmod n \in \{s_1, s_2, \ldots, s_k\}\}$.

The associated *Step Graph* $L_n^{s_1,s_2,\ldots,s_k}$ is defined by $L_n^{s_1,s_2,\ldots,s_k} = (V, E_L)$ where $E_L = \{(i,j) : i - j \in \{s_1, s_2, \ldots, s_k\}\}$. For example, the difference between $C_5^{1,2}$ and $L_5^{1,2}$ is $E_C - E_L = \{\{0,4\}, \{0,3\}, \{1,4\}\}$ (See Figure 2).

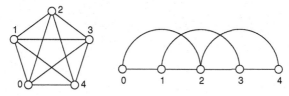

Fig. 2. $C_5^{1,2}$ and $L_5^{1,2}$.

The step graph can be thought of as being obtained from the circulant graph by unhooking the edges that cross over the interval $(n-1, 0)$ in the circulant graph.

For the rest of this subsection we restrict ourselves to the graphs $C_n^{1,2}$ and $L_n^{1,2}$. In the next subsection we will sketch how to generalize the approach to any circulant graph.

The difference between $C_n^{1,2}$ and $L_n^{1,2}$ is the set of edges $E_C - E_L = \{\{0, n-1\}, \{0, n-2\}, \{1, n-1\}\}$ Any spanning tree T of $C_n^{1,2}$ is a collection of $n-1$ edges of E_C; it may or may not contain some edges from $E_C - E_L$.

The main idea behind the counting method is to remove all edges in $E_C - E_L$ from T. Depending upon which edges were in the spanning tree, T can either remain the same or become a disconnected forest of $C_n^{1,2}$. In any case, since we have removed all edges in $E_C - E_L$ what remains is a *forest of* $L_n^{1,2}$ (See Figure 3).

Note that the spanning trees of $C_n^{1,2}$ can be partitioned into eight separate classes, depending upon which, if any of the 3 edges in $E_C - E_L = \{\{0, n-1\}, \{0, n-2\}, \{1, n-1\}\}$ the tree contains. For example, one set of the partition contains all the spanning trees which contain the edge $\{0, n-1\}$ but not $\{0, n-2\}$ and $\{1, n-1\}$. Thus, the number of spanning trees of $C_n^{1,2}$ will be the sum of the numbers of the spanning trees in these eight partitions.

More formally, for $S \subseteq E_C - E_L$ let

$$C_S(n) = \{T : T \text{ a spanning tree of } C_n^{1,2} \text{ s.t. } T \cap (E_C - E_L) = S\}$$

be the set of spanning trees containing only S. Then $T\left(C_n^{1,2}\right) = \sum_S |C_S(n)|$.

We now examine each set in the partition separately. We take the set previously mentioned again as an example, i.e. $C_{\{\{0,n-1\}\}}$, in which all trees in the set contain only $\{0, n-1\}$ but not $\{0, n-2\}$ and $\{1, n-1\}$.

After removing $\{0, n-1\}$ each tree in this set becomes a forest in $L_n^{1,2}$ containing exactly two components, one component containing node 0 and the other

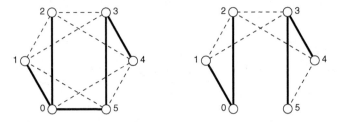

Fig. 3. Removing edges in $E_C - E_L$ from the spanning tree of $C_6^{1,2}$ leaves a disconnected forest of $L_6^{1,2}$. Solid edges are the ones in the tree; dashed ones are existing edges not in the tree. The spanning tree illustrated on the left is in the set $C_{\{\{0,n-1\}\}}$. The forest on the right is a member of $F_{\{0,1\}\{n-1,n-2\}}(n)$.

containing node $n-1$. These can be further divided into the following four classes of forests with two components in $L_n^{1,2}$:

1. one component contains node 0, the other contains $1, n-2, n-1$
2. one component contains node 0, 1 the other contains $n-2, n-1$
3. one component contains node 0, $n-2$, the other contains $1, n-1$
4. one component contains node 0, 1, $n-2$, the other contains $n-1$

This partition is reversible; that is, by adding edge $\{0, n-1\}$ to any of these forests we create the corresponding spanning tree of $C_n^{1,2}$. Thus, summing up the number of forests in the four classes gives us exactly the number of spanning trees of $C_n^{1,2}$ that contain $\{0, n-1\}$ but not $\{0, n-2\}$ and $\{1, n-1\}$.

Extending the above example note that removing all edges in $E_C - E_L = \{\{0, n-1\}, \{0, n-2\}, \{1, n-1\}\}$ from a spanning tree of $C_n^{1,2}$ will result in a forest of $L_n^{1,2}$ that contains 1, 2, 3 or 4 components such that each component (tree) in the forest contains at least one of the four vertices $n-2, n-1, 0, 1$. For later use we will call such forests *legal* and classify the legal forests of $L_n^{1,2}$ by considering how the four vertices are partitioned among the connected components of the forest (we do not consider non-legal forests of $L_n^{1,2}$).

More formally, let $\mathcal{P}$ be the set of partitions of $\{n-2, n-1, 0, 1\}$. For $X \in \mathcal{P}$ define $|X|$ to be the number of sets in X.

Now let $F_X(n)$ be the set containing all forests in $L_n^{1,2}$ with $|X|$ components such that $u, v \in \{n-2, n-1, 0, 1\}$ are in the same component of the forest if and only if they are in the same set of X.

For example, $F_{\{0\}\{1,n-1\}\{n-2\}}(n)$ is the set of spanning forests of $L_n^{1,2}$ with three components s.t. one component contains node 0, another component contains nodes 1 and $n-1$, and the last component contains node $n-2$.

Finally, set $T_X(n) = |F_X(n)|$ to be the number of such forests. Using this notation we can rewrite the discussion above as

$$|C_{\{0,n-1\}}| = T_{\{0\},\{1,n-2,n-1\}}(n) + T_{\{0,1\},\{n-2,n-1\}}(n)$$
$$+ T_{\{0,n-2\},\{1,n-1\}}(n) + T_{\{0,1,n-2\},\{n-1\}}(n).$$

The important observation here is that if we fix $X \in \mathcal{P}$ and $S \subseteq E_C - E_L$ then adding the set of edges S into a forest in class $F_X(n)$ results in exactly one of the following three consequences and we can determine which of the consequence occurs simply by checking X and S (independent of n)

1. The resulting forest is disconnected.
2. The resulting set of edges contains at least one cycle.
3. The forest becomes a spanning tree of $C_n^{1,2}$ in set $\mathcal{C}_S$.

For example suppose $S = \{\{0, n-1\}, \{0, n-2\}\}$ and

$$X_1 = \{\{0\}, \{1\}, \{n-1\}, \{n-2\}\}, \quad X_2 = \{\{0,1\}, \{n-1, n-2\}\},$$

$$X_3 = \{\{0\}, \{1, n-2\}, \{n-1\}\}.$$

Adding S to a forest in $F_{X_1}(n)$ will leave the forest disconnected; adding S to a forest in $F_{X_2}(n)$ will create a cycle; adding S to a forest in $F_{X_3}(n)$ will create a spanning tree.

We can therefore define

$$\alpha_{S,X} = \begin{cases} 1 \text{ if adding } S \text{ to forest in } F_X(n) \text{ yields a spanning tree} \\ 0 \text{ otherwise} \end{cases} \quad (1)$$

and find that $|\mathcal{C}_S(n)| = \sum_{X \in \mathcal{P}} \alpha_{S,X} T_X(n)$ so

$$T\left(C_n^{1,2}\right) = \sum_S |\mathcal{C}_S(n)| = \sum_{X \in \mathcal{P}} \left(\sum_S \alpha_{S,X} \right) T_X(n). \quad (2)$$

Now define $\boldsymbol{T}\left(L_n^{1,2}\right)$ to be the column vector of all of the $T_X(n)$ ordered as follows:

$$\boldsymbol{T}\left(L_n^{1,2}\right) = \begin{pmatrix} T_{\{0,1,n-2,n-1\}}(n) \\ T_{\{0\}\{1,n-2,n-1\}}(n) \\ T_{\{1\}\{0,n-2,n-1\}}(n) \\ T_{\{n-2\}\{0,1,n-1\}}(n) \\ T_{\{n-1\}\{0,1,n-2\}}(n) \\ T_{\{0,1\}\{n-2,n-1\}}(n) \\ T_{\{0,n-2\}\{1,n-1\}}(n) \\ T_{\{0,n-1\}\{1,n-2\}}(n) \\ T_{\{0\}\{1\}\{n-2,n-1\}}(n) \\ T_{\{0\}\{n-2\}\{1,n-1\}}(n) \\ T_{\{0\}\{n-1\}\{1,n-2\}}(n) \\ T_{\{1\}\{n-1\}\{0,n-2\}}(n) \\ T_{\{1\}\{n-2\}\{0,n-1\}}(n) \\ T_{\{n-2\}\{n-1\}\{0,1\}}(n) \\ T_{\{0\}\{1\}\{n-2\}\{n-1\}}(n) \end{pmatrix}$$

Each entry of $\boldsymbol{T}\left(L_n^{1,2}\right)$ is the number of forests of $L_n^{1,2}$ in the corresponding class. Now, for $X \in \mathcal{P}$ set $\beta_X = \sum_S \alpha_{S,X}$ and $\boldsymbol{\beta} = (\beta_X)_{X \in \mathcal{P}}$. In this notation, (2) simply states that $T(C_n^{1,2}) = \boldsymbol{\beta} \cdot \boldsymbol{T}\left(L_n^{1,2}\right)$. Mechanically working out the values of the β_X from (1) gives

$$T(C_n^{1,2}) = \begin{pmatrix} 1 & 2 & 1 & 1 & 2 & 3 & 1 & 2 & 2 & 1 & 3 & 1 & 1 & 2 & 1 \end{pmatrix} \cdot \boldsymbol{T}\left(L_n^{1,2}\right). \quad (3)$$

Until now we have only seen that $T(C_n^{1,2})$ can be written in terms of vector $\boldsymbol{T}\left(L_n^{1,2}\right)$ but this still doesn't say anything about a formula for $T(C_n^{1,2})$. The important observation at this point is that, unlike for circulant graphs, it is quite easy to write a matrix recurrence relation for $\boldsymbol{T}\left(L_n^{1,2}\right)$. In fact, we will be able to write a one-step recurrence of the form $\boldsymbol{T}\left(L_n^{1,2}\right) = A\boldsymbol{T}\left(L_{n-1}^{1,2}\right)$ where A is some fixed integer matrix.

To see this, suppose that we remove node n along with its incident edges from a legal forest in $L_{n+1}^{1,2}$. What remains is a legal forest in $L_n^{1,2}$. We can therefore build all the legal forests of $L_{n+1}^{1,2}$ by knowing the legal forests of $L_n^{1,2}$.

Constructing from the other direction note that the only edges connecting to n in $L_n^{1,2}$ are $\{n, n-1\}$ and $\{n, n-2\}$. Suppose that we add node n and a set of edges $U \subseteq \{\{n, n-1\}, \{n, n-2\}\}$ to a forest of $L_n^{1,2}$ in class $F_X(n)$. The resulting graph will either have a cycle or be a forest in a particular class $F_{X'}(n+1)$ where X' is only determined by X and U (See Figure 4).

Let us now define

$$a_{X',X} = |\{U \subseteq \{\{n, n-1\}, \{n, n-2\}\} : \text{adding } U \text{ to } F_X(n) \text{ yields } F_{X'}(n+1)\}| \quad (4)$$

to be the number of different sets U that can be added to a forest in $F_X(n)$ to yield a forest in $F_{X'}(n+1)$. These $a_{X,X'}$ (which are independent of n) can be mechanically calculated by checking all cases.

Then $T_{X'}(n+1) = \sum_X a_{X',X}T_X(n)$. So, letting $A = (a_{X',X})_{X',X\in\mathcal{P}}$, we find that, for $n \geq 4$, $\boldsymbol{T}\left(L_{n+1}^{1,2}\right) = A\boldsymbol{T}\left(L_n^{1,2}\right)$ and we have derived a system of recurrence relations on the $T_X(n)$.

For our particular case we have worked through the calculations to find A.[2] Combining A with (3) yields a recurrence relation for $T(C_n^{1,2})$. This is a very standard technique so we only sketch the idea here. For all $X \in \mathcal{P}$ create the generating functions $T_X(z) = \sum_n T_X(n)z^n$. $\boldsymbol{T}(L_n^{1,2}) = A\boldsymbol{T}(L_{n-1}^{1,2})$ then corresponds to a system of simultaneous equations *on the generating functions*, and we can use a procedure akin to Gaussian elimination to solve for closed forms of all of the generating functions. Because of the way in which they are derived, all of the generating functions will be rational functions in z, i.e., in the form $P_X(z)/Q_X(z)$ where $P_X(z)$ and $Q_X(z)$ are polynomials in z. Now set $T(z) = \sum_n T(C_n^{1,2})z^n = \sum_X \beta_X T_X(z)$. As the (weighted) sum of rational functions, $T(z)$ will also be a rational function in z. The fact that $T(z)$ is rational then permits us to recover a recurrence relation on $T(C_n^{1,2})$. Performing the above steps yield

$$T(C_n^{1,2}) = 4T(C_{n-1}^{1,2}) - 10T(C_{n-3}^{1,2}) + 4T(C_{n-5}^{1,2}) - T(C_{n-6}^{1,2})$$

with initial values $36, 125, 384, 1183, 3528, 10404$ for $n = 4, 5, 6, 7, 8, 9$ respectively for which it can be verified that the solution is $T(C_n^{1,2}) = nF_n^2$. We have therefore just given another combinatorial proof of the result due to Kleitman and Golden [9].

---

[2] The full matrix A is given in [6].

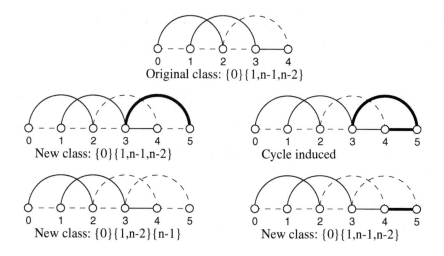

Fig. 4. Different ways to add node 5 to a forest of $L_5^{1,2}$ to generate different classes of forests of $L_6^{1,2}$. Bold edges are the ones added with node 5.

2.2   The General Case

In the previous subsection we developed machinery for counting the number of spanning trees in $C_n^{1,2}$. It is not difficult to see how to generalize this to count the number of spanning trees in $C_n^{s_1,s_2,\ldots,s_k}$. Since this is very similar to the previous section we only sketch the steps.

We start by defining, for all $S \subseteq E_C - E_L$,

$$\mathcal{C}_S(n) = \{T : T \text{ a spanning tree of } C_n^{s_1,s_2,\ldots,s_k} \text{ s.t. } T \cap (E_C - E_L) = S\}$$

as the set of spanning trees containing only S. Then $T\left(C_n^{s_1,s_2,\ldots,s_k}\right) = \sum_S |\mathcal{C}_S(n)|$.

Let $W_{s_k} = \{0, 1, \ldots, s_k - 1\} \cup \{n - s_k, n - s_k + 1, \ldots, n - 1\}$. Define $\mathcal{P}_{s_k}$ to be the set of all partitions of W_{s_k}. A *legal forest* of $L_n^{s_1,s_2,\ldots,s_k}$ is one in which every component in the forest contains at least one element in W_{s_k}. For $X \in \mathcal{P}_{s_k}$ define $F_X(n)$ to be the set of all legal forests in $L_n^{s_1,s_2,\ldots,s_k}$ with $|X|$ components such that $u, v \in W_{s_k}$ are in the same component of the forest if and only if they are in the same set of X. Set $T_X(n) = |F_X(n)|$.

We generalize (1) to

$$\alpha_{S,X} = \begin{cases} 1 \text{ if adding } S \text{ to forest in } F_X(n) \text{ yields a spanning tree of } C_n^{s_1,s_2,\ldots,s_k} \\ 0 \text{ otherwise} \end{cases}$$

$$(5)$$

and find that, as before, $|\mathcal{C}_S(n)| = \sum_{X \in \mathcal{P}_{s_k}} \alpha_{S,X} T_X(n)$ so

$$T\left(C_n^{s_1,s_2,\ldots,s_k}\right) = \sum_S |\mathcal{C}_S(n)| = \sum_{X \in \mathcal{P}_{s_k}} \left(\sum_S \alpha_{S,X}\right) T_X(n). \qquad (6)$$

Let $T\left(L_n^{s_1,s_2,\ldots,s_k}\right)$ be the column vector $(T_X(n))_{X\in\mathcal{P}_{s_k}}$, set $\beta_X = \sum_S \alpha_{S,X}$ and define $\boldsymbol{\beta} = (\beta_X)_{X\in\mathcal{P}_{s_k}}$. Then

$$T(C_n^{s_1,s_2,\ldots,s_k}) = \boldsymbol{\beta} \cdot T(L_n^{s_1,s_2,\ldots,s_k}). \tag{7}$$

Exactly as before we can set

$$a_{X',X} = |\{U \subseteq \cup_{i=1}^k \{\{n,n-s_i\}\} : \text{adding } U \text{ to } F_X(n) \text{ yields } F_{X'}(n+1)\}| \tag{8}$$

and mechanically calculate the $a_{X',X}$ values. Then, letting $A = (a_{X',X})_{X',X\in\mathcal{P}_{s_k}}$, we have for $n \geq 2s_k$,

$$T\left(L_{n+1}^{s_1,s_2,\ldots,s_k}\right) = A T\left(L_n^{s_1,s_2,\ldots,s_k}\right). \tag{9}$$

Combining (7) and (9) proves what we want; that $T(C_n^{s_1,s_2,\ldots,s_k})$ can be expressed in terms of a recurrence relation.

3   Hamiltonian Cycles of $C_n^{1,2}$

The unhooking technique developed in the previous section is quite general and can be used to count various other parameters of circulant graphs. In this section we sketch how use it to derive a recurrence relation on the number of Hamiltonian cycles $H(C_n^{1,2})$, in $C_n^{1,2}$. The generalization to deriving a recurrence relation on the number of Hamiltonian cycles $H(C_n^{s_1,s_2,\ldots,s_k})$ in any $C_n^{s_1,s_2,\ldots,s_k}$ will be straightforward.

First note that, as in the spanning tree case, we can partition the Hamiltonian cycles of $C_n^{1,2}$ into eight different classes, depending upon which, if any of the 3 edges in $E_C - E_L = \{\{0,n-1\},\{0,n-2\},\{1,n-1\}\}$ the cycle contains.

For $S \subseteq E_C - E_L$ let

$$\mathcal{H}_S(n) = \left\{H : H \text{ is a Hamiltonian cycle of } C_n^{1,2} \text{ s.t. } H \cap (E_C - E_L) = S\right\}.$$

Then $H\left(C_n^{1,2}\right) = \sum_S |\mathcal{H}_S(n)|$.

Now suppose that we are given some Hamiltonian cycle $H \in \mathcal{H}_S(n)$. After removing the edges in S from H we observe that one of the following three cases must occur:

1. $H - S$ is still a Hamiltonian cycle (of $L_n^{1,2}$).
2. $H - S$ is a Hamiltonian path of $L_n^{1,2}$ with endpoints in $\{0,1,n-2,n-1\}$
3. $H - S$ is the union of disjoint simple paths in $L_n^{1,2}$ with endpoints in $\{0,1,n-2,n-1\}$. (See Figure 5).

In the third case, we are considering that if a node is left isolated without any incident edges in $H - S$ then it is in its own path (note that this can only happen to nodes 0 and $n - 1$). Also, note that in the second and third case, just by knowing the edges in S it is possible to know what the endpoints of the disjoint paths are (and what, if any, isolated vertices exist).

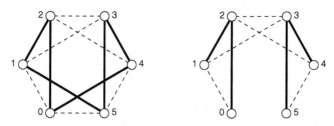

Fig. 5. Decomposition of Hamiltonian cycle $C_6^{1,2}$ to disjoint simple paths in $L_6^{1,2}$.

This observation leads us to define a *legal path decomposition* in $L_n^{1,2}$ to be a disjoint set of paths containing all vertices in V such that all endpoints of the paths are in $\{0, 1, n-2, n-1\}$ and only 0 and $n-1$ are allowed to be isolated vertices. We can classify the legal path decompositions by their endpoints. Define $H_{\{u_1,v_1\},\{u_2,v_2\},...,\{u_w,v_w\}}(n)$ to be the number of subgraphs of $L_n^{1,2}$ with w connected components such that all w components are simple paths with end-points $\{u_1, v_1\}, \{u_2, v_2\}, \ldots, \{u_w, v_w\}$ respectively, e.g., $H_{\{1,n-1\}\{0,0\}}(n)$ is the number of all subgraphs of $L_n^{1,2}$ with two components; one component being a path with end-points 1 and $n-1$ and the second component being the single vertex 0. Define one more special case, $H_\emptyset(n)$, to be the number of Hamiltonian cycle of $L_n^{1,2}$. We then define $\boldsymbol{H}(L_n^{1,2})$ to be the column vector:

$$\boldsymbol{H}(L_n^{1,2}) = \begin{pmatrix} H_{\{0,1\}}(n) \\ H_{\{0,n-2\}}(n) \\ H_{\{0,n-1\}}(n) \\ H_{\{1,n-2\}}(n) \\ H_{\{1,n-1\}}(n) \\ H_{\{n-2,n-1\}}(n) \\ H_{\{0,1\}\{n-1,n-1\}}(n) \\ H_{\{0,n-2\}\{n-1,n-1\}}(n) \\ H_{\{1,n-2\}\{0,0\}}(n) \\ H_{\{1,n-2\}\{n-1,n-1\}}(n) \\ H_{\{1,n-2\}\{0,0\}\{n-1,n-1\}}(n) \\ H_{\{1,n-1\}\{0,0\}}(n) \\ H_{\{n-2,n-1\}\{0,0\}}(n) \\ H_{\{0,1\}\{n-2,n-1\}}(n) \\ H_{\{0,n-2\}\{1,n-1\}}(n) \\ H_{\{0,n-1\}\{1,n-2\}}(n) \\ H_\emptyset(n) \end{pmatrix}$$

Let $\mathcal{P}$ be the indices of these items. For $X = \{u_1, v_1\}, \{u_2, v_2\}, \ldots, \{u_w, v_w\} \in \mathcal{P}$ we say that a legal path decomposition is of type X if it is decomposed into simple paths with end-points $\{u_1, v_1\}, \{u_2, v_2\}, \ldots, \{u_w, v_w\}$. For any $S \subseteq E_C - E_L$ and $X \in \mathcal{P}$ define

$$\alpha_{S,X} = \begin{cases} 1 \text{ if adding } S \text{ to path decomposition of type } X \text{ yields a HC} \\ 0 \text{ otherwise} \end{cases} \tag{10}$$

so

$$H(C_n^{1,2}) = \sum_S |\mathcal{H}_S(n)| = \sum_{X \in \mathcal{P}} \left(\sum_S \alpha_{S,X} \right) H_X(n). \tag{11}$$

Now, for $X \in \mathcal{P}$ set $\beta_X = \sum_S \alpha_{S,X}$ and define $\boldsymbol{\beta} = (\beta_X)_{X \in \mathcal{P}}$. From (11) $H(C_n^{1,2}) = \boldsymbol{\beta} \cdot \boldsymbol{H}\left(L_n^{1,2}\right)$. Evaluating $\boldsymbol{\beta}$ yields

$$H(C_n^{1,2}) = (\,0\ 1\ 1\ 0\ 1\ 0\ 1\ 0\ 0\ 0\ 1\ 0\ 1\ 1\ 0\ 1\ 1\,) \cdot \boldsymbol{H}(L_n^{1,2}) \tag{12}$$

Note that adding node n and edge set $U \subseteq \{\{n, n-1\}, \{n, n-2\}\}$ to a legal path decomposition of type X on $L_n^{1,2}$ either does not yield a legal path decomposition or yields a decomposition of type X' on $L_{n+1}^{1,2}$ where X' is fully determined by X and U. Following the ideas in the previous section we therefore define

$$a_{X',X} = |\{U \subseteq \{\{n, n-1\}, \{n, n-2\}\} : \text{adding } U \text{ to decomposition}$$
$$\text{of type } X \text{ yields } X'\}| \tag{13}$$

where $a_{X,X'}$ can be mechanically calculated by checking all cases. Then $H_{X'}(n+1) = \sum_X a_{X',X} H_X(n)$. So, letting $A = (a_{X',X})_{X',X \in \mathcal{P}}$, we find that for $n \geq 4$, $\boldsymbol{H}\left(L_{n+1}^{1,2}\right) = A\,\boldsymbol{H}\left(L_n^{1,2}\right)$. Calculating this A (it appears in [6]), combining with (12) and simplifying as before yields the recurrence

$$H(C_n^{1,2}) = 2H(C_{n-1}^{1,2}) - H(C_{n-3}^{1,2}) - H(C_{n-5}^{1,2}) + H(C_{n-6}^{1,2})$$

with initial values $9, 12, 16, 23, 29, 41$ for $n = 4, 5, 6, 7, 8, 9$ respectively.

Although we only derived a recurrence for $H(C_n^{1,2})$ the technique developed can easily be generalized to derive a recurrence on $H(C_n^{s_1, s_2, \ldots, s_k})$ in much the same way that the technique for calculating $T(C_n^{1,2})$ in Section 2.1 was generalized to calculate $T(C_n^{s_1, s_2, \ldots, s_k})$ in section 2.2. The important changes are (i) to extend the definition of a *legal path decomposition* to $L_n^{s_1, s_2, \ldots, s_k}$ to be a disjoint set of paths containing all vertices in V such that all endpoints of the paths are in $\{0, 1, \ldots, s_k\} \cup \{n - s_k, \ldots, n - 2, n - 1\}$ and (ii) to set

$$a_{X',X} = |\{U \subseteq \cup_{i=1}^k \{\{n, n - s_i\}\} : \text{adding } U \text{ to decomposition}$$
$$\text{of type } X \text{ yields } X'\}|. \tag{14}$$

Everything else is the same as in the derivation for $H(C_n^{1,2})$ and will yield $H(C_n^{s_1, s_2, \ldots, s_k}) = \boldsymbol{\beta} \cdot \boldsymbol{H}\left(L_n^{s_1, s_2, \ldots, s_k}\right)$ and $\boldsymbol{H}\left(L_{n+1}^{s_1, s_2, \ldots, s_k}\right) = A\,\boldsymbol{H}\left(L_n^{s_1, s_2, \ldots, s_k}\right)$.

4   Conclusion

In this paper we developed the first general *combinatorial* technique for showing that the number of spanning trees in circulant graphs satisfies a recurrence relation. This contrasts to the only previously known general method which used algebraic (spectral) methods.

Our basic approach, unhooking, permits decomposing a problem on *circulant* graphs into many problems on *step* graphs. We then used the fact that step graphs are much more amenable to recursive decomposition to yield our results.

A nice consequence of our technique is that it can be easily modified to work for many other parameters of circulant graphs, e.g., to show that the number of Hamiltonian cycles, Eulerian tours and Eulerian orientations in these graphs also obey a recurrence relation. To the best of our knowledge this is the first time these parameters have been analyzed. We also point out that, even though our technique was described only for *undirected* circulant graphs, it is quite easy to extend it to *directed* circulant graphs as well.

We conclude with an open question. Our analysis implicitly assumed that $s_1, s_2, \ldots, s_k$, the jumps in the circulant graph, are *fixed*. Recent work [7] has shown that in many cases when the s_i are functions of n, then the number of spanning trees also satisfies a recurrence relation. For example, $T(C_{2n}^{1,n}) = \frac{n}{2}[(\sqrt{2} + 1)^n + (\sqrt{2} - 1)^n]^2$. The proofs of such results are, again, algebraic, involving evaluating products of the eigenvalues of the graph's adjacency matrix. Unfortunately, due to the structure of these graphs, the unhooking technique is not applicable. It is still open as to whether there is any combinatorial derivation of the number of spanning trees of such *non-fixed-jump* circulant graphs.

References

1. G. Baron, H. Prodinger, R. F. Tichy, F. T. Boesch and J. F. Wang. "The Number of Spanning Trees in the Square of a Cycle," *Fibonacci Quarterly*, **23.3** (1985), 258-264.

2. S. Bedrosian. "The Fibonacci Numbers via Trigonometric Expressions," *J. Franklin Inst.* **295** (1973), 175-177.

3. F. T. Boesch, J. F. Wang. "A Conjecture on the Number of Spanning Trees in the Square of a Cycle," In: *Notes from New York Graph Theory Day V*, New York: New York Academy Sciences, 1982. p. 16.

4. F. T. Boesch, H. Prodinger. "Spanning Tree Formulas and Chebyshev Polynomials," *Graphs and Combinatorics*, **2**, (1986), 191-200.

5. C. J. Colbourn. *The combinatorics of network reliability*, Oxford University Press, New York, (1987).

6. M. J. Golin and Yiu Cho Leung. *Unhooking Circulant Graphs: A Combinatorial Method for Counting Spanning Trees and Other Parameters* Technical Report HKUST-TCSC-2004-??. Available at *http://www.cs.ust.hk/tcsc/RR/*,

7. M. J. Golin, Y.P. Zhang. "Further applications of Chebyshev polynomials in the derivation of spanning tree formulas for circulant graphs," in *Mathematics and Computer Science II: Algorithms, Trees, Combinatorics and Probabilities*, 541-552. Birkhauser-Verlag. Basel. (2002)

8. G. Kirchhoff. "Über die Auflösung der Gleichungen, auf welche man bei der Untersuchung der linearen Verteilung galvanischer Ströme geführt wird," *Ann. Phys. Chem.* **72** (1847) 497-508.

9. D. J. Kleitman, B. Golden. "Counting Trees in a Certain Class of Graphs," *Amer. Math. Monthly,* **82** (1975), 40-44.

10. X. Yong, Talip, Acenjian. "The Numbers of Spanning Trees of the Cubic Cycle C_N^3 and the Quadruple Cycle C_N^4," *Discrete Math.*, **169** (1997), 293-298.

11. X. Yong, F. J. Zhang. "A simple proof for the complexity of square cycle C_p^2," *J. Xinjiang Univ.*, **11** (1994), 12-16.

12. Y. P. Zhang, X. Yong, M. J. Golin. "The number of spanning trees in circulant graphs," *Discrete Math.*, 223 (2000) 337-350.

Computing Bounded-Degree Phylogenetic Roots of Disconnected Graphs*

Zhi-Zhong Chen[1],[**] and Tatsuie Tsukiji[2]

[1] Dept. of Math. Sci., Tokyo Denki Univ., Hatoyama, Saitama 350-0394, Japan
chen@r.dendai.ac.jp
[2] Dept. of Info. Sci., Tokyo Denki Univ., Hatoyama, Saitama 350-0394, Japan
tsukiji@j.dendai.ac.jp

Abstract. The PHYLOGENETIC kTH ROOT PROBLEM (PRk) is the problem of finding a (phylogenetic) tree T from a given graph $G = (V, E)$ such that (1) T has no degree-2 internal nodes, (2) the external nodes (*i.e.* leaves) of T are exactly the elements of V, and (3) $(u, v) \in E$ if and only if the distance between u and v in tree T is at most k, where k is some fixed threshold k. Such a tree T, if exists, is called a *phylogenetic kth root* of graph G. The computational complexity of PRk is open, except for $k \leq 4$. Recently, Chen *et al.* investigated PRk under a natural restriction that the maximum degree of the phylogenetic root is bounded from above by a constant. They presented a linear-time algorithm that determines if a given *connected* G has such a phylogenetic kth root, and if so, demonstrates one. In this paper, we supplement their work by presenting a linear-time algorithm for *disconnected* graphs.

1 Introduction

The reconstruction of evolutionary history for a set of species from quantitative biological data has long been a popular problem in computational biology. This evolutionary history is typically modeled by an evolutionary tree or *phylogeny*. A phylogeny is a tree where the leaves are labeled by species and each internal node represents a speciation event whereby a hypothetical ancestral species gives rise to two or more child species. Proximity within a phylogeny in general corresponds to similarity in evolutionary characteristics. Both rooted and unrooted trees have been used to describe phylogenies in the literature, although they are practically equivalent. In this paper, we will consider only unrooted phylogenies for the convenience of presentation. Note that each internal node in a phylogeny has at least 3 neighbors.

Many approaches to phylogenetic reconstruction have been proposed in the literature [8]. In particular, Lin *et al.* [4] recently suggested a graph-theoretic approach for reconstructing phylogenies from similarity data. Specifically, inter-species similarity is represented by a graph G where the vertices are the species

---

* The full version can be found at http://rnc.r.dendai.ac.jp/~chen/papers/dpr.pdf.
** Supported in part by the Grant-in-Aid for Scientific Research of the Ministry of Education of Japan, under Grant No. 14580390.

J. Hromkovič, M. Nagl, and B. Westfechtel (Eds.): WG 2004, LNCS 3353, pp. 308–319, 2004.

and the adjacency relation represents evidence of evolutionary similarity. A phylogeny is then reconstructed from G such that the leaves of the phylogeny are labeled by vertices of G (*i.e.* species) and for any two vertices of G, they are adjacent in G if and only if their corresponding leaves in the phylogeny are at most distance k apart, where k is a predetermined proximity threshold. This approach gives rise to the following algorithmic problem [4]:

PHYLOGENETIC kTH ROOT PROBLEM (PRk):
Given a graph $G = (V, E)$, find a phylogeny T with leaves labeled by the elements of V such that for each pair of vertices $u, v \in V$, $(u, v) \in E$ if and only if $d_T(u, v) \le k$, where $d_T(u, v)$ is the number of edges on the path between u and v in T.

Such a phylogeny T (if exists) is called a *phylogenetic kth root*, or a *kth root phylogeny*, of graph G. Graph G is called the kth *phylogenetic power* of T. For convenience, we denote the kth phylogenetic power of any phylogeny T as T^k. That is, $T^k = \{(u, v) \mid u$ and v are leaves of T and $d_T(u, v) \le k\}$. Thus, PRk asks for a phylogeny T such that $G = T^k$.

1.1   Previous Results on PRk

PRk was first studied in [4] where linear-time algorithms for PRk with $k \le 4$ were proposed. At present, the complexity of PRk with $k \ge 5$ is still unknown.

The hardness of PRk for large k seems to come from the unbounded degree of an internal node in the output phylogeny. On the other hand, in the practice of phylogeny reconstruction, most phylogenies considered are trees of degree 3 [8] because speciation events are usually bifurcating events in the evolutionary process. These motivated Chen *et al.* [2] to consider a restricted version of PRk where the output phylogeny is assumed to have degree at most Δ, for some fixed constant $\Delta \ge 3$. We call this restricted version the DEGREE-Δ PRk and denote it for short as ΔPRk.

Chen *et al.* [2] presented a linear-time algorithm that determines, for any input *connected* graph G and constant $\Delta \ge 3$, if G has a kth root phylogeny with degree at most Δ, and if so, demonstrates one such phylogeny. Unfortunately, their algorithm fails when the input graph G is disconnected. One of their open questions asks for a polynomial-time algorithm for disconnected graphs, because the disconnected case is real in biology.

1.2   Other Problems Related to PRk

A graph G is the kth *power* of a graph H (or equivalently, H is a kth *root* of G), if vertices u and v are adjacent in G if and only if they are at most distance k apart in H. An important special case of graph power/root problems is the TREE kTH ROOT PROBLEM (TRk): Given a graph $G = (V, E)$, we wish to find a tree $T = (V, E_T)$ such that $(u, v) \in E$ if and only if $d_T(u, v) \le k$. If T exists, then it is called a *tree kth root*, or a *kth root tree*, of graph G. There is rich literature on graph roots and powers (see [1, Section 10.6] for an overview), but few results on phylogenetic/tree roots/powers. It is NP-complete to recognize a graph power [6]; nonetheless, we can determine if a graph has a kth root tree, for any fixed k, in cubic time [3]. In particular, determining if a graph has a tree square root can be done in linear time [5]. Moreover, Nishimura et al. [7] presented a cubic time algorithm for a variant of PRk with $k \leq 4$, where internal nodes of the output phylogeny are allowed to have degree 2.

1.3   Our Contribution

Our result is a linear-time algorithm that determines, for any input *disconnected* graph G and constant $\Delta \geq 3$, if G has a kth root phylogeny with degree at most Δ, and if so, demonstrates one such phylogeny. This answers an open question in [2]. Combining this algorithm with the algorithm in [2] for connected graphs, we obtain the first linear-time algorithm for ΔPRk for any constants $\Delta \geq 3$ and $k \geq 2$. Our algorithm is complicated and it is based on hidden structures of phylogenetic kth roots of disconnected graphs. Moreover, the algorithm needs a linear-time subroutine for solving a certain optimization problem on each connected component of the input disconnected graph. The subroutine is obtained by nontrivially refining the algorithm in [2].

2   Preliminaries

We employ standard terminologies in graph theory. In particular, the subgraph of a graph G induced by a vertex set U of G is denoted by $G[U]$, the degree of a vertex v in G is denoted by $deg_G(v)$, and the distance between two vertices u and v in G is denoted by $d_G(u, v)$. Moreover, for a set W of vertices in a graph $G = (V, E)$, we write $G - W$ for $G[V - W]$. Furthermore, in a rooted tree, each vertex is both an ancestor and a descendant of itself.

For clarity, if $G = (V, E)$ is a graph and $T = (V_T, E_T)$ is a kth root phylogeny of G for some k, then we call the elements of V *vertices* and call those of V_T *nodes*.

In the remainder of this section, fix a graph $G = (V, E)$ and two integers $k \geq 4$ and $\Delta \geq 3$. A *degree-Δ kth root phylogeny* ((Δ, k)-phylogeny for short) of G is a kth root phylogeny T of G such that the maximum degree of a node in T is at most Δ.

A *degree-Δ kth root quasi-phylogeny* ((Δ, k)-QP for short) of G is a tree Q satisfying the following conditions:

- Each vertex of G is a leaf of Q and appears in Q exactly once. For convenience, we call the leaves of Q that are also vertices of G *true leaves* of Q, and call the other leaves of Q *false leaves* of Q.
- The degree of each node in Q is at most Δ.
- For every two vertices u and v in G, u and v are adjacent in G if and only if $d_Q(u, v) \leq k$.
- For each node x of Q that is a degree-2 node or a false leaf in Q, it holds that $\min_{v \in V} d_Q(x, v) \geq \lfloor \frac{k}{2} \rfloor$.

- If Q has no false leaf, then it has at least one node x such that $2 \leq deg_Q(x) \leq \Delta - 1$ and $\min_{v \in V} d_Q(x, v) \geq \lfloor \frac{k}{2} \rfloor$.

The *cost* of Q is $\max\{1, a + 2b\}$, where a is the number of degree-2 nodes in Q and b is the number of false leaves in Q. Q is an *optimal* (Δ, k)-QP of G if its cost is minimized over all (Δ, k)-QPs of G.

Lemma 1. *Suppose that $G = (V, E)$ is a connected graph. Let Q be an optimal (Δ, k)-QP of G. Then, the following hold:*

1. *Q has no node x with $\min_{v \in V} d_Q(x, v) > \lfloor \frac{k}{2} \rfloor$.*
2. *For each node x with $deg_Q(x) = 2$ or $deg_Q(x) > 3$, each connected component of $Q - \{x\}$ contains at least one true leaf of Q.*

Proof. We prove the two statements separately as follows.

Statement 1. If x were a false leaf of Q with $\min_{v \in V} d_Q(x, v) > \lfloor \frac{k}{2} \rfloor$, then the removal of x from Q would result in a new (Δ, k)-QP of G whose cost is smaller than that of Q, a contradiction. So, for every false leaf x of Q, $\min_{v \in V} d_Q(x, v) \leq \lfloor \frac{k}{2} \rfloor$. In turn, for every internal node x of Q such that one connected component of $Q - \{x\}$ contains all true leaves of Q, it holds that $\min_{v \in V} d_Q(x, v) \leq \lfloor \frac{k}{2} \rfloor$.

Now, it remains to consider those internal nodes x of Q such that no connected component of $Q - \{x\}$ contains all true leaves of Q. If among these nodes, there were one x with $\min_{v \in V} d_Q(x, v) > \lfloor \frac{k}{2} \rfloor$, then G would have no edge (u, v) such that u and v belong to different connected components of $Q - \{x\}$, contradicting the connectivity of G.

Statement 2. Let x be a node of Q with $deg_Q(x) = 2$ or $deg_Q(x) > 3$. For a contradiction, assume that some connected component C of $Q - \{x\}$ contains no true leaf of Q. If $deg_Q(x) > 3$, then the removal of C from Q results in a new (Δ, k)-QP of G whose cost is smaller than that of Q, a contradiction. If $deg_Q(x) = 2$, then by Statement 1, $\min_{v \in V} d_Q(x, v) < \lfloor \frac{k}{2} \rfloor$, a contradiction. $\square$

We classify (Δ, k)-QPs Q into four types as follows.

- Q is *helpful* if it has at most one degree-2 node and has no false leaf.
- Q is *moderate* if it has no degree-2 node but has exactly one false leaf.
- Q is *troublesome* if it has at least two degree-2 nodes but has no false leaf.
- Q is *dangerous* if it has at least one false leaf and the total number of false leaves and degree-2 nodes in Q is at least 2.

A (Δ, k)-QP Q is *unhelpful* if it is not helpful.

For a (Δ, k)-QP Q, we define its *port nodes* as follows. If Q is not helpful, then its port nodes are its false leaves and degree-2 nodes. If Q is helpful and has no degree-2 node, then its port nodes are those nodes x with $\min_{v \in V} d_Q(x, v) \geq \lfloor \frac{k}{2} \rfloor$. If Q is helpful and has a degree-2 node, then it has only one port node, namely, its unique degree-2 node.

A *nonport node* of a (Δ, k)-QP Q is a node of Q that is not a port node.

3   Algorithm for Bounded-Degree PRk

Throughout this section, fix two integers $k \geq 4$ and $\Delta \geq 3$. This section presents a linear-time algorithm for solving ΔPRk.

Let $G = (V, E)$ be the input graph. We assume that G is disconnected; otherwise, the linear-time algorithm in [2] solves the problem. Let $G_1, \ldots, G_\ell$ be the connected components of G. For each integer with $1 \leq i \leq \ell$, let V_i be the vertex set of G_i.

The following lemma can be proved by a complicated dynamic programming:

Lemma 2. *For every $i \in \{1, \ldots, \ell\}$, we can decide whether G_i has a (Δ, k)-QP, in $O(|V_i|)$ time. Moreover, if G_i has a (Δ, k)-QP, then we can compute an optimal (Δ, k)-QP of G_i in $O(|V_i|)$ time.*

Lemma 3. *If for some $i \in \{1, \ldots, \ell\}$, G_i has no (Δ, k)-QP, then G has no (Δ, k)-phylogeny.*

Proof. Suppose that G has a (Δ, k)-phylogeny T. Fix an $i \in \{1, \ldots, \ell\}$. Let Y_i be the set of all internal nodes y of T such that there is a vertex $u \in V_i$ with $d_T(u, y) \leq \lfloor \frac{k}{2} \rfloor$. Obviously, $T[V_i \cup Y_i]$ is a (Δ, k)-QP of G_i.   □

By Lemmas 2 and 3, we may assume that for each $i \in \{1, \ldots, \ell\}$, G_i has a (Δ, k)-QP. For each $i \in \{1, \ldots, \ell\}$, let Q_i be the optimal (Δ, k)-QP of G_i computed in Lemma 2.

Lemma 4. *Suppose G has a (Δ, k)-phylogeny. Then, G has a (Δ, k)-phylogeny T such that $Q_1, \ldots, Q_\ell$ all are subtrees of T.*

Proof. Let T be a (Δ, k)-phylogeny of G. For each $i \in \{1, \ldots, \ell\}$, let Y_i be as in the proof of Lemma 3. Recall that $T[V_i \cup Y_i]$ is a (Δ, k)-QP of G_i. Moreover, for every pair (i, j) with $1 \leq i \neq j \leq \ell$, $Y_i \cap Y_j = \emptyset$.

Consider the integer $i \in \{1, \ldots, \ell\}$ such that $Q_1, \ldots, Q_{i-1}$ are subtrees of T but Q_i is not. If no such i exists, then T is as required. So, assume that i exists. Let $F_1, \ldots, F_h$ be the connected components of $T - (V_i \cup Y_i)$. For each $j \in \{1, \ldots, h\}$, F_j has exactly one node z_j with $deg_{F_j}(z_j) < deg_T(z_j)$, and each leaf of F_j is a vertex in $V - V_i$. Moreover, the minimum distance from z_j to a leaf in F_j is at least $\lceil \frac{k}{2} \rceil$, because $\min_{v \in V_i} d_T(z_j, v) = \lfloor \frac{k}{2} \rfloor + 1$. Let $Z = \{z_1, \ldots, z_h\}$. Define a function $f : Z \to Y_i$ as follows. For each $j \in \{1, \ldots, h\}$, let $f(z_j)$ be the neighbor of z_j in T that is not in F_j. Let $X = \{f(z_j) \mid 1 \leq j \leq h\}$.

We claim that the cost of the (Δ, k)-QP $T[V_i \cup Y_i]$ of G_i is at most h. To see this claim, first note that X contains all false leaves and all degree-2 nodes of $T[V_i \cup Y_i]$. Moreover, for each degree-2 node x of $T[V_i \cup Y_i]$, there is at least one $z_j \in Z$ with $f(z_j) = x$. Furthermore, for each false leaf x of $T[V_i \cup Y_i]$, there are at least two $z_j \in Z$ with $f(z_j) = x$. Hence, the claim holds. By the claim, the cost of the optimal (Δ, k)-QP Q_i of G_i is at most h.

Now, we use Q_i and $F_1, \ldots, F_h$ to obtain a new (Δ, k)-phylogeny T_i of G, by performing the following steps:

1. Let $x_1, \ldots, x_a$ be the degree-2 nodes in Q_i, and let $x_{a+1}, \ldots, x_{a+b}$ be the false nodes in Q_i.
2. If $a > 0$ or $b > 0$, then set $c = a + 2b$; otherwise, set $c = 1$ and let x_1 be an (arbitrarily chosen) port node of Q_i. (Comment: c is the cost of Q_i and $h \geq c$.)
3. For all j with $1 \leq j \leq c - 2b$, add edge (x_j, z_j).
4. For all j $(1 \leq j \leq b)$, add edges $(x_{c-2b+j}, z_{c-2b+2j-1})$ and $(x_{c-2b+j}, z_{c-2b+2j})$.
5. If $h > c$, then perform the following steps:
   (a) Delete the edge between x_{c-b} and z_c.
   (b) Introduce $h - c$ new nodes $y_1, \ldots, y_{h-c}$, and connect them into a path from y_1 to y_{h-c}.
   (c) For all i with $1 \leq i \leq h - c$, add edge (y_i, z_{c+i-1}).
   (d) Add edges (y_1, x_{c-b}) and (y_{h-c}, z_h).

Obviously, T_i is a (Δ, k)-phylogeny of G and Q_i is a subtree of T_i. Moreover, for every j with $1 \leq j \leq i - 1$, Q_j remains to be a subtree of T_i, because $Y_i \cap Y_j = \emptyset$ and Q_j is a subtree of $T[V_j \cup Y_j]$ by Statement 1 in Lemma 1. Thus, T_i is a (Δ, k)-phylogeny of G such that for all $j \in \{1, \ldots, i\}$, Q_j is a subtree of T_i. Therefore, by our choice of i, we can repeat the above argument to finally obtain a (Δ, k)-phylogeny T_ℓ of G such that $Q_1, \ldots, Q_\ell$ are subtrees of T_ℓ.   □

In the remainder of this section, a (Δ, k)-phylogeny of G always means one in which $Q_1, \ldots, Q_\ell$ are subtrees. By Lemma 4, we lose no generality. For convenience, we call $Q_1, \ldots, Q_\ell$ the *unitary* (Δ, k)-QPs.

Let T be a (Δ, k)-phylogeny T of G. A *junction node* of T is a node x of T such that no unitary (Δ, k)-QP contains x. A node x of T is *over-connected*, if it satisfies one of the following conditions:

(1) $deg_T(x) > 3$ and x is a junction node of T.
(2) $deg_T(x) > 3$ and x is a port node of some unhelpful Q_i $(1 \leq i \leq \ell)$.
(3) x is a nonport node of some unhelpful Q_i $(1 \leq i \leq \ell)$ and $deg_T(x) > deg_{Q_i}(x)$.

A helpful Q_i $(1 \leq i \leq \ell)$ is *mis-connected* in T, if (i) at least one nonport node of Q_i is adjacent to a node outside Q_i in T, or (ii) there are two or more nodes x outside Q_i such that x is adjacent to a node of Q_i in T.

A (Δ, k)-phylogeny T of G is *canonical*, if it has no over-connected node and no helpful Q_i $(1 \leq i \leq \ell)$ is mis-connected in T.

Lemma 5. *If G has a (Δ, k)-phylogeny, then it has a canonical one.*

In the remainder of this section, a (Δ, k)-phylogeny of G always means a canonical one. By Lemma 5, we lose no generality.

3.1   The Case Where k Is Odd

Throughout this subsection, we assume that k is odd. A *double* (Δ, k)-QP is a tree $T_{i,j}$ obtained by combining two helpful unitary (Δ, k)-QPs Q_i and Q_j as follows:

1. Select a port node x_i of Q_i, and select a port node x_j of Q_j.
2. Introduce a junction node y, and connect it to both x_i and x_j.

Note that $T_{i,j}$ has exactly one degree-2 node (namely, the junction node y) but has no false leaf. So, $T_{i,j}$ is a helpful (Δ, k)-QP of $G[V_i \cup V_j]$. Moreover, the minimum distance from y to a true leaf in $T_{i,j}$ is exactly $\lfloor \frac{k}{2} \rfloor + 1$ (cf. Statement 1 in Lemma 1).

Lemma 6. *Suppose that each Q_i $(1 \leq i \leq \ell)$ is helpful or moderate. Then, G has a (Δ, k)-phylogeny if and only if $\ell \geq 2b + 3$, where b is the number of moderate (Δ, k)-QPs among $Q_1, \ldots, Q_\ell$.*

Proof. We prove the two directions separately as follows.

($\Longrightarrow$) Suppose that G has a (Δ, k)-phylogeny T. For each moderate Q_i, let Q_i' be the tree obtained from Q_i by deleting its unique false leaf. Let $\mathcal{T}$ be the tree obtained by modifying T as follows:

1. For each helpful Q_i $(1 \leq i \leq \ell)$, merge Q_i into a super-node s_i.
2. For each moderate Q_i $(1 \leq i \leq \ell)$, merge Q_i' into a super-node s_i'.

Obviously, the leaves of $\mathcal{T}$ are exactly the super-nodes. So, $\mathcal{T}$ has exactly ℓ leaves. Let c be the number of internal nodes of $\mathcal{T}$. Let $m_{\mathcal{T}}$ be the number of edges in $\mathcal{T}$. Note that the false leaf x_i of each moderate Q_i remains to be an internal node in $\mathcal{T}$, no neighbor of x_i in $\mathcal{T}$ is the false leaf x_j of another moderate Q_j in $\mathcal{T}$, and no neighbor of x_i in $\mathcal{T}$ is a super-node s_j corresponding to a helpful Q_j (cf. Statement 1 in Lemma 1). This implies that $m_{\mathcal{T}} \geq 3b + (\ell - b)$. Trivially, $m_{\mathcal{T}} = \ell + c - 1$, and $m_{\mathcal{T}} = \frac{3c+\ell}{2}$ because the degree of each internal node in $\mathcal{T}$ is exactly 3 (by the canonicity of T). Therefore, $\ell \geq 2b + 3$.

($\Longleftarrow$) Suppose that $\ell \geq 2b + 3$. By renumbering if necessary, we may assume that $Q_1, \ldots, Q_b$ are moderate. For each $i \in \{1, \ldots, b\}$, let y_i be the false leaf of Q_i. For each $i \in \{b+1, \ldots, \ell\}$, let z_i be an (arbitrarily chosen) port node of Q_i. We can connect $Q_1, \ldots, Q_\ell$ into a (Δ, k)-phylogeny of G as follows.

1. Introduce $\ell - b - 2$ junction nodes $x_1, \ldots, x_{\ell-b-2}$.
2. For each i with $1 \leq i \leq b$, add edges (y_i, x_i) and (y_i, x_{i+1}).
3. Add edges (x_1, z_{b+1}), (x_1, z_{b+2}), $(x_{\ell-b-2}, z_{\ell-1})$, and $(x_{\ell-b-2}, z_\ell)$. (Comment: If $\ell - b = 3$, then only three edges are added here.)
4. For each i with $b+1 \leq i \leq \ell - b - 3$, add edge (x_i, x_{i+1}).
5. For each i with $2 \leq i \leq \ell - b - 3$, add edge (x_i, z_{b+i+1}). □

In the sequel, we assume that at least one Q_i $(1 \leq i \leq \ell)$ is troublesome or dangerous (since otherwise Lemma 6 solves the problem).

Let T be a (Δ, k)-phylogeny of G. For each dangerous Q_i $(1 \leq i \leq \ell)$, we say that a false leaf x of Q_i is *active* in T, if no connected component of $T - \{x\}$ is a double (Δ, k)-QP. A dangerous Q_i $(1 \leq i \leq \ell)$ is *active* in T if at least one false leaf of Q_i is active in T.

Lemma 7. *Suppose G has a (Δ, k)-phylogeny. Then, G has a (Δ, k)-phylogeny T such that no dangerous Q_i $(1 \leq i \leq \ell)$ is active in T.*

Let I be the set of all $i \in \{1, \ldots, \ell\}$ such that Q_i is dangerous. For each $i \in I$, let t_i be the number of false leaves in Q_i. Let $t = \sum_{i \in I} t_i$. By Lemma 7, if G has a (Δ, k)-phylogeny, then there are at least $2t$ helpful unitary (Δ, k)-QPs. So, if there are less than $2t$ helpful unitary (Δ, k)-QPs, then G has no (Δ, k)-phylogeny. In the sequel, we assume that there are at least $2t$ helpful unitary (Δ, k)-QPs. Without loss of generality, we may assume that $Q_1, \ldots, Q_{2t}$ are helpful.

We connect $Q_1, \ldots, Q_{2t}$ to the dangerous unitary (Δ, k)-QPs as follows.

1. Introduce t junction nodes $x_1, \ldots, x_t$, and construct a one-to-one correspondence between them and the t false leaves of the dangerous unitary (Δ, k)-QPs.
2. For each $i \in \{1, \ldots, t\}$, add an edge from x_i to its corresponding false leaf, add an edge from x_i to an (arbitrarily chosen) port node of Q_{2i-1}, and add an edge from x_i to an (arbitrarily chosen) port node of Q_{2i}.

The above modification extends each dangerous unitary (Δ, k)-QP Q_i to a troublesome (Δ, k)-QP R_i. For convenience, let $R_i = Q_i$ for each $i \in \{2t + 1, \ldots, \ell\}$ such that Q_i is not dangerous.

Now, we are left with $R_{2t+1}, \ldots, R_\ell$; none of them is dangerous. Let τ be the number of troublesome (Δ, k)-QPs among $R_{2t+1}, \ldots, R_\ell$. Note that $\tau = |i \in \{1, \ldots, \ell\} \mid Q_i$ is troublesome or dangerous$\}$. So, $\tau \geq 1$. Without loss of generality, we may assume that $R_{2t+1}, \ldots, R_{2t+\tau}$ are troublesome.

By Lemma 7, if G has a (Δ, k)-phylogeny, then it has one in which $R_{2t+1}, \ldots$, $\ldots, R_\ell$ are subtrees. So, in the remainder of this section, a (Δ, k)-phylogeny of G always means one in which $R_{2t+1}, \ldots, R_\ell$ are subtrees.

A *bridging node* in a (Δ, k)-phylogeny T of G is a node x of T such that no R_i with $2t+1 \leq i \leq \ell$ contains x. For each (Δ, k)-phylogeny T of G and for each R_i with $2t + 1 \leq i \leq \ell$, each degree-2 node x of R_i is adjacent to exactly one bridging node y in T (by the canonicity of T); we call y the *bridging neighbor* of x in T.

For each (Δ, k)-phylogeny T of G, let $\mathcal{M}(T)$ denote the tree obtained by modifying T by merging each R_i with $2t + 1 \leq i \leq \ell$ into a super-node. For convenience, we abuse the notation to let each R_i also denote the super-node of $\mathcal{M}(T)$ corresponding to R_i. Note that each bridging node of T remains to be an internal node in $\mathcal{M}(T)$ and the leaves of $\mathcal{M}(T)$ one-to-one correspond to the helpful unitary (Δ, k)-QPs among $R_{2t+1}, \ldots, R_\ell$. Moreover, by the canonicity of T and Statement 1 in Lemma 1, no two super-nodes can be adjacent in $\mathcal{M}(T)$.

Lemma 8. *If G has a (Δ, k)-phylogeny, then it has one T such that there is a path q in $\mathcal{M}(T)$ on which $R_{2t+1}, \ldots, R_{2t+\tau}$ appear.*

Lemma 9. *If G has a (Δ, k)-phylogeny, then it has one T such that some path q in $\mathcal{M}(T)$ satisfies the following three conditions:*

1. *$R_{2t+1}, \ldots, R_{2t+\tau}$ and exactly $\tau - 1$ bridging nodes appear on q.*
2. *No two bridging nodes on q are adjacent in T.*
3. *For each bridging node x on q, there is a helpful unitary (Δ, k)-QP R_i such that x is adjacent to a port node of R_i in T.*

In the remainder of this section, a (Δ, k)-phylogeny of G always means one T such that some path q in $\mathcal{M}(T)$ satisfies the three conditions in Lemma 9. We call q the *spine* of $\mathcal{M}(T)$. The following corollary shows that it does not matter in which order $R_{2t+1}, \ldots, R_{2t+\tau}$ appear on the spine.

Corollary 1. *Let T be a (Δ, k)-phylogeny of G. Then, for every pair (R_i, R_j) of troublesome (Δ, k)-QPs, there is another (Δ, k)-phylogeny T' of G such that the spine of $\mathcal{M}(T')$ can be obtained from that of $\mathcal{M}(T)$ by exchanging the positions of R_i and R_j.*

The following corollary is obvious and shows that it does not matter via which degree-2 nodes each troublesome R_i is connected to the spine.

Corollary 2. *Let T be a (Δ, k)-phylogeny of G. Then, for every troublesome R_i and for every pair (x_1, x_2) of degree-2 nodes of R_i, we can obtain another (Δ, k)-phylogeny T' of G by deleting edges (x_1, y_1) and (x_2, y_2) and adding edges (x_1, y_2) and (x_2, y_1), where y_1 (respectively, y_2) is the bridging neighbor of x_1 (respectively, x_2) in T. Moreover, the spines of $\mathcal{M}(T)$ and $\mathcal{M}'(T)$ are the same.*

By Lemma 9, if G has a (Δ, k)-phylogeny, then there are at least $\tau - 1$ helpful unitary (Δ, k)-QPs among $R_{2t+\tau+1}, \ldots, R_\ell$. So, if there are less than $\tau - 1$ helpful unitary (Δ, k)-QPs among $R_{2t+\tau+1}, \ldots, R_\ell$, then G has no (Δ, k)-phylogeny. In the sequel, we assume that there are at least $\tau - 1$ helpful unitary (Δ, k)-QPs among $R_{2t+\tau+1}, \ldots, R_\ell$. Without loss of generality, we may assume that $R_{2t+\tau+1}, \ldots, R_{2t+2\tau-1}$ are helpful unitary (Δ, k)-QPs.

If $\tau \geq 2$, then we connect $R_{2t+1}, \ldots, R_{2t+2\tau-1}$ into a single (Δ, k)-QP $\mathcal{R}$ as follows.

1. Introduce $\tau - 1$ bridging nodes $x_1, \ldots, x_{\tau-1}$.
2. Select a degree-2 node y_{2t+1} of R_{2t+1}, and select a degree-2 node $z_{2t+\tau}$ of $R_{2t+\tau}$.
3. For each i with $2t + 2 \leq i \leq 2t + \tau - 1$, select two degree-2 nodes z_i and y_i of R_i.
4. For each i with $1 \leq i \leq \tau - 1$, add edges (x_i, y_{2t+i}) and (x_i, z_{2t+i+1}), and add an edge from x_i to an (arbitrarily chosen) port node of $R_{2t+\tau+i}$.

If $\tau = 1$, we let $\mathcal{R} = R_{2t+1}$.

Note that $\mathcal{R}$ is a troublesome (Δ, k)-QP. By Lemma 9 and Corollaries 1 and 2, if G has a (Δ, k)-phylogeny, then G has one T such that $\mathcal{R}, R_{2t+2\tau}, \ldots, R_\ell$ are subtrees of T. In the remainder of this section, a (Δ, k)-phylogeny of G always means such a tree T. Let h be the number of degree-2 nodes in $\mathcal{R}$. Let $x_1, \ldots, x_h$ be the degree-2 nodes of $\mathcal{R}$.

Lemma 10. *If G has a (Δ, k)-phylogeny, then it has one T such that for all but one $x_i \in \{x_1, \ldots, x_h\}$, the connected component of $T - \{x_i\}$ containing no node of $\mathcal{R}$ is a double (Δ, k)-QP.*

By Lemma 10, if G has a (Δ, k)-phylogeny, then there are at least $2h - 2$ helpful unitary (Δ, k)-QPs among $R_{2t+2\tau}, \ldots, R_\ell$. So, if there are less than

$2h - 2$ helpful unitary (Δ, k)-QPs among $R_{2t+2\tau}, \ldots, R_\ell$, then G has no (Δ, k)-phylogeny. In the sequel, we assume that there are at least $2h - 2$ helpful unitary (Δ, k)-QPs among $R_{2t+2\tau}, \ldots, R_\ell$. We may further assume that $R_{2t+2\tau}, \ldots,$ $R_{2t+2\tau+2h-3}$ are helpful unitary (Δ, k)-QPs. For each $i \in \{2t + 2\tau, \ldots, 2t + 2\tau + 2h - 3\}$, let z_i be an (arbitrarily chosen) port node of R_i.

We connect $\mathcal{R}, R_{2t+2\tau}, \ldots, R_{2t+2\tau+2h-3}$ into a single (helpful) (Δ, k)-QP $\mathcal{R}'$ by performing the following steps:

1. Introduce $h - 1$ bridging nodes $s_1, \ldots, s_{h-1}$.
2. For each $i \in \{1, \ldots, h-1\}$, add edges $(s_i, z_{2t+2\tau+2i-2})$, $(s_i, z_{2t+2\tau+2i-1})$, and (s_i, x_i).

Now, we are left with $\mathcal{R}', R_{2t+2\tau+2h-2}, \ldots, R_\ell$ each of which is helpful or moderate. Moreover, by Lemma 10, if G has a (Δ, k)-phylogeny, then it has one in which $\mathcal{R}', R_{2t+2\tau+2h-2}, \ldots, R_\ell$ are subtrees. So, we can modify the proof of Lemma 6 to show that G has a (Δ, k)-phylogeny if and only if $a' \geq b' + 3$, where a' (respectively, b') is the number of helpful (respectively, moderate) (Δ, k)-QPs among $\mathcal{R}', R_{2t+2\tau+2h-2}, \ldots, R_\ell$.

In summary, we have the following:

Theorem 1. *Suppose that k is odd. Then, we can decide if G has a (Δ, k)-phylogeny, and construct one if so, in linear time.*

3.2 The Case Where k Is Even

Throughout this subsection, we assume that k is even. The contents in this subsection are very similar to those in the last subsection. In particular, the lemmas in this subsection one-to-one correspond to the lemmas in the last subsection. Moreover, the proof of each lemma in this subsection is very similar to (indeed a bit simpler than) its corresponding lemma in the last subsection. So, we will omit the proofs of the lemmas.

Lemma 11. *Suppose that each Q_i $(1 \leq i \leq \ell)$ is helpful or moderate. Then, G has a (Δ, k)-phylogeny if and only if $a \geq 2$, where a is the number of helpful (Δ, k)-QPs among $Q_1, \ldots, Q_\ell$.*

In the sequel, we assume that at least one Q_i $(1 \leq i \leq \ell)$ is troublesome or dangerous (since otherwise Lemma 11 solves the problem).

Let T be a (Δ, k)-phylogeny of G. For each dangerous Q_i $(1 \leq i \leq \ell)$, we say that a false leaf x of Q_i is *active* in T, if no connected components of $T - \{x\}$ is a helpful unitary (Δ, k)-QP. A dangerous Q_i $(1 \leq i \leq \ell)$ is *active* in T if at least one false leaf of Q_i is active in T.

Lemma 12. *Suppose that G has a (Δ, k)-phylogeny. Then, G has a (Δ, k)-phylogeny T such that no dangerous unitary (Δ, k)-QP is active in T.*

Let I be the set of all $i \in \{1, \ldots, \ell\}$ such that Q_i is dangerous. For each $i \in I$, let t_i be the number of false leaves in Q_i. Let $t = \sum_{i \in I} t_i$. By Lemma 12, if G has a (Δ, k)-phylogeny, then there are at least t helpful unitary (Δ, k)-QPs. So, if there are less than t helpful unitary (Δ, k)-QPs, then G has no (Δ, k)-phylogeny. In the sequel, we assume that there are at least t helpful unitary (Δ, k)-QPs. Without loss of generality, we may assume that $Q_1, \ldots, Q_t$ are helpful.

We connect $Q_1, \ldots, Q_t$ to the dangerous unitary (Δ, k)-QPs as follows.

1. Construct a one-to-one correspondence between $Q_1, \ldots, Q_t$ and the t false leaves of the dangerous unitary (Δ, k)-QPs.
2. For each $i \in \{1, \ldots, t\}$, add an edge from an (arbitrarily chosen) port node of Q_i to the false leaf corresponding to Q_i.

The above modification extends each dangerous unitary (Δ, k)-QP Q_i to a troublesome (Δ, k)-QP R_i. For convenience, let $R_i = Q_i$ for each $i \in \{t+1, \ldots, \ell\}$ such that Q_i is not dangerous.

Now, we are left with $R_{t+1}, \ldots, R_\ell$; none of them is dangerous. Let τ be the number of troublesome (Δ, k)-QPs among $R_{t+1}, \ldots, R_\ell$. Note that $\tau = |i \in \{1, \ldots, \ell\} \mid Q_i$ is troublesome or dangerous$\}$. So, $\tau \geq 1$. Without loss of generality, we may assume that $R_{t+1}, \ldots, R_{t+\tau}$ are troublesome.

By Lemma 12, if G has a (Δ, k)-phylogeny, then it has one in which $R_{t+1}, \ldots, \ldots, R_\ell$ are subtrees. So, in the remainder of this section, a (Δ, k)-phylogeny of G always means one in which $R_{t+1}, \ldots, R_\ell$ are subtrees.

For each (Δ, k)-phylogeny T of G, let $\mathcal{M}(T)$ denote the tree obtained by modifying T by merging each R_i with $t + 1 \leq i \leq \ell$ into a super-node. For convenience, we abuse the notation to let each R_i also denote the super-node corresponding to R_i in $\mathcal{M}(T)$.

Lemma 13. *If G has a (Δ, k)-phylogeny, then it has one T such that there is a path in $\mathcal{M}(T)$ on which $R_{t+1}, \ldots, R_{t+\tau}$ appear.*

Lemma 14. *If G has a (Δ, k)-phylogeny, then it has one T such that there is a path in $\mathcal{M}(T)$ whose nodes are exactly $R_{t+1}, \ldots, R_{t+\tau}$.*

In the remainder of this section, a (Δ, k)-phylogeny of G always means one T such that there is a path q in $\mathcal{M}(T)$ whose nodes are exactly $R_{t+1}, \ldots, R_{t+\tau}$. We call q the *spine* of $\mathcal{M}(T)$. Obviously, Corollaries 1 and 2 still hold even if k is even.

If $\tau \geq 2$, then we connect $R_{t+1}, \ldots, R_{t+\tau}$ into a single (Δ, k)-QP $\mathcal{R}$ as follows.

1. Select a degree-2 node y_{t+1} of R_{t+1}, and select a degree-2 node $z_{t+\tau}$ of $R_{t+\tau}$.
2. For each i with $t + 2 \leq i \leq t + \tau - 1$, select two degree-2 nodes z_i and y_i of R_i.
3. For each i with $t + 1 \leq i \leq t + \tau - 1$, add edge (y_i, z_{i+1}).

If $\tau = 1$, we let $\mathcal{R} = R_{t+1}$.

Note that $\mathcal{R}$ is a troublesome (Δ, k)-QP. By Lemma 14 and Corollaries 1 and 2, if G has a (Δ, k)-phylogeny, then G has one T such that $\mathcal{R}, R_{t+\tau+1}, \ldots, R_\ell$ are subtrees of T. In the remainder of this section, a (Δ, k)-phylogeny of G always means such a tree T. Let h be the number of degree-2 nodes in $\mathcal{R}$. Let $x_1, \ldots, x_h$ be the degree-2 nodes of $\mathcal{R}$.

Lemma 15. *If G has a (Δ, k)-phylogeny, then it has one T such that for all but one $x_i \in \{x_1, \ldots, x_h\}$, the connected component of $T - \{x_i\}$ containing no node of $\mathcal{R}$ is a helpful unitary (Δ, k)-QP.*

By Lemma 15, if G has a (Δ, k)-phylogeny, then there are at least $h-1$ helpful unitary (Δ, k)-QPs among $R_{t+\tau+1}, \ldots, R_\ell$. So, if there are less than $h-1$ helpful unitary (Δ, k)-QPs among $R_{t+\tau+1}, \ldots, R_\ell$, then G has no (Δ, k)-phylogeny. In the sequel, we assume that there are at least $h-1$ helpful unitary (Δ, k)-QPs among $R_{t+\tau+1}, \ldots, R_\ell$. We may further assume that $R_{t+\tau+1}, \ldots, R_{t+\tau+h-1}$ are helpful unitary (Δ, k)-QPs. For each $i \in \{t+\tau+1, \ldots, t+\tau+h-1\}$, let z_i be an (arbitrarily chosen) port node of R_i.

We connect $\mathcal{R}, R_{t+\tau+1}, \ldots, R_{t+\tau+h-1}$ into a single (helpful) (Δ, k)-QP $\mathcal{R}'$ by adding edges $(x_1, z_{t+\tau+1}), \ldots, (x_{h-1}, z_{t+\tau+h-1})$.

Now, we are left with $\mathcal{R}', R_{t+\tau+h}, \ldots, R_\ell$ each of which is helpful or moderate. Moreover, by Lemma 15, if G has a (Δ, k)-phylogeny, then it has one in which $\mathcal{R}', R_{t+\tau+h}, \ldots, R_\ell$ are subtrees. So, we can modify the proof of Lemma 11 to show that G has a (Δ, k)-phylogeny if and only if $a' \geq 2$, where a' is the number of helpful (Δ, k)-QPs among $\mathcal{R}', R_{t+\tau+h}, \ldots, R_\ell$.

In summary, we have the following:

Theorem 2. *Suppose that k is even. Then, we can decide if G has a (Δ, k)-phylogeny, and construct one if so, in linear time.*

References

1. A. Brandstädt, V. B. Le, and J. P. Spinrad, *Graph Classes: a Survey*, SIAM Monographs on Discrete Mathematics and Applications, SIAM, Philadelphia, 1999.
2. Z.-Z. Chen, T. Jiang, and G.-H. Lin, *Computing phylogenetic roots with bounded degrees and errors*, SIAM Journal on Computing, 32 (2003) 864–879.
3. P. E. Kearney and D. G. Corneil, *Tree powers*, Journal of Algorithms, 29 (1998) 111–131.
4. G.-H. Lin, P. E. Kearney, and T. Jiang, *Phylogenetic k-root and Steiner k-root*, in: The 11th Annual International Symposium on Algorithms and Computation (ISAAC 2000), Lecture Notes in Computer Science, 1969 (2000) 539–551.
5. Y.-L. Lin and S. S. Skiena, *Algorithms for square roots of graphs*, SIAM Journal on Discrete Mathematics, 8 (1995) 99–118.
6. R. Motwani and M. Sudan, *Computing roots of graphs is hard*, Discrete Applied Mathematics, 54 (1994) 81–88.
7. N. Nishimura, P. Ragde, and D. M. Thilikos, *On graph powers for leaf-labeled trees*, in: Proceedings of the 7th Scandinavian Workshop on Algorithm Theory (SWAT 2000), Lecture Notes in Computer Science, 1851 (2000) 125–138.
8. D. L. Swofford, G. J. Olsen, P. J. Waddell, and D. M. Hillis, *Phylogenetic inference*, in: D. M. Hillis, C. Moritz, and B. K. Mable (Ed.), Molecular Systematics (2nd Edition), Sinauer Associates, Sunderland, Massachusetts, 1996, pp. 407–514.

Octagonal Drawings of Plane Graphs
with Prescribed Face Areas

Md. Saidur Rahman[1], Kazuyuki Miura[2], and Takao Nishizeki[2]

[1] Department of Computer Science and Engineering,
Bangladesh University of Engineering and Technology (BUET),
Dhaka 1000, Bangladesh
`saidurrahman@cse.buet.ac.bd`
[2] Graduate School of Information Sciences, Tohoku University,
Aoba-yama 05, Sendai 980-8579, Japan
`miura@nishizeki.ecei.tohoku.ac.jp`, `nishi@ecei.tohoku.ac.jp`

Abstract. An orthogonal drawing of a plane graph is called an octagonal drawing if each inner face is drawn as a rectilinear polygon of at most eight corners and the contour of the outer face is drawn as a rectangle. A slicing graph is obtained from a rectangle by repeatedly slicing it vertically and horizontally. A slicing graph is called a good slicing graph if either the upper subrectangle or the lower one obtained by any horizontal slice will never be vertically sliced. In this paper we show that any good slicing graph has an octagonal drawing with prescribed face areas, in which the area of each inner face is equal to a prescribed value. Such a drawing has practical applications in VLSI floorplanning. We also give a linear-time algorithm to find such a drawing. We furthermore present a sufficient condition for a plane graph to be a good slicing graph, and give a linear-time algorithm to find a tree-structure of slicing paths for a graph satisfying the condition.

1   Introduction

An *orthogonal drawing* of a plane graph G is a drawing of G with a given embedding such that each vertex is mapped to a point, each edge is drawn as a sequence of alternate horizontal and vertical line segments, and any two edges do not cross except at their common end as illustrated in Fig. 1(e). In an orthogonal drawing each face is drawn as a rectilinear polygon. Orthogonal drawings have attracted much attention due to their numerous practical applications in circuit schematics, data flow diagrams, entity relationship diagrams, etc. [3, 7].

In this paper we consider an orthogonal drawing of a plane graph G where the contour of the outer face of G is drawn as a rectangle, called the *outer rectangle*, and each inner face has a prescribed area. We call such an orthogonal drawing a *prescribed-area orthogonal drawing*. Figure 1(a) depicts a plane graph G where a number written in each inner face indicates a prescribed area of the face, and Fig. 1(e) depicts a prescribed-area orthogonal drawing of G. Throughout the paper the four corners a, b, c and d of an outer rectangle are drawn by white circles.

J. Hromkovič, M. Nagl, and B. Westfechtel (Eds.): WG 2004, LNCS 3353, pp. 320–331, 2004.
© Springer-Verlag Berlin Heidelberg 2004

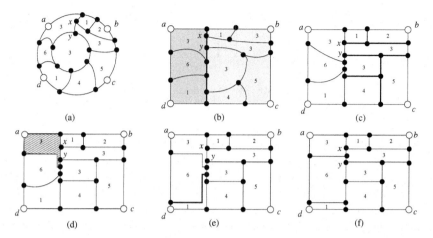

Fig. 1. Illustration of the algorithm.

A prescribed-area orthogonal drawing of a plane graph G has practical applications in VLSI floorplanning. Floorplanning is an initial step in VLSI chip design where one decides the relative locations of functional entities in a chip. A VLSI floorplan is often considered as a subdivision of a rectangle into a finite number of non-overlapping smaller rectangles, each of which corresponds to a funtional entity called a *module* [2,6]. A "slicing floorplan" is often used by VLSI design [S96, YS93, YS95]. Divide a rectangle into two smaller rectangles by slicing it vertically or horizontally, divide any subrectangle into two smaller subrectangles by slicing it vertically or horizontally, and so on, as illustrated in Figs. 2(a)–(e). The resulting floorplan like one in Fig. 2(e) is called a *slicing floorplan*. An underlying plane graph of a slicing floorplan such as one illustrated in Fig. 2(f) is called a *slicing graph G*, where the four vertices a, b, c and d of degree two on the outer face of G represent the corners of the outer rectangle. Thus a slicing graph G is a *2-3 plane graph* in which each vertex has degree two or three, and a slicing floorplan is a *rectangular drawing* of G, where each edge is drawn as a single horizontal or vertical line segment and each face is drawn as a rectangle. Since each module needs some physical area, each face of G in the drawing should satisfy some area requirements. However, when the area of each face is prescribed, there may not exist a rectangular drawing of G. We thus consider an orthogonal drawing of a slicing graph where a face is not always a rectangle as illustrated in Figs. 1(e). In VLSI floorplanning it is desirable that each inner face is drawn as a rectilinear polygon of simple shape such as a rectangle, an L-shape polygon, a T-shape polygon, etc. [1, 2, 6, 5, 9]. We thus attempt to find a prescribed-area orthogonal drawing of G keeping the shape of each inner face as simple as possible.

In this paper we consider a fairly large subclass of slicing graphs called good slicing graphs. A slicing graph is *good* if either the upper subrectangle or the lower one obtained by any horizontal slice will never be vertically sliced. The graphs in Figs. 1(a) and 2(f) are good slicing graphs. We show that any good

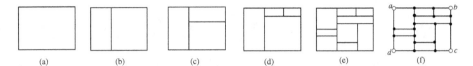

Fig. 2. Illustration of a slicing floorplan.

slicing graph has a prescribed-area orthogonal drawing in which each inner facial polygon has at most eight corners, as illustrated in Fig. 1(e). We call such a drawing an *octagonal drawing*. We also give a linear-time algorithm to find such an octagonal drawing. We furthermore present a sufficient condition for a plane graph to be a good slicing graph, and give a linear-time algorithm to find a tree-structure of slicing paths for graphs satisfying the condition. To the best of our knowledge, this is the first work on a prescribed-area octagonal drawing.

Our drawing algorithm is outlined as follows. We first draw the outer cycle of G as a rectangle with four vertices a, b, c and d as corners so that the area of the rectangle is equal to the sum of the prescribed areas of all inner faces, as illustrated in Fig. 1(b). We now embed a "slicing path" P connecting two opposite sides of the rectangle as a straight line segment so that it divides the outer rectangle into two subrectangles each of whose areas is equal to the sum of the prescribed areas of all faces inside it. In Fig. 1(b) a slicing path P is drawn by a thick line and the two subrectangles are shadded differently. We recursively find a prescribed-area orthogonal drawing of the subgraph inside each rectangle, and we obtain a drawing of G such as one illustrated in Fig. 1(f), where each inner face is drawn as a rectangle with prescribed area but the drawing is not always a drawing of G. For example, vertex x is adjacent to y in G in Fig. 1(a), but x is not adjacent to y in the drawing in Fig. 1(f). We thus need to modify the drawing in each recursive step, as illustrated in Figs. 1(c) and (d), by introducing bends on some edges, and hence some faces are drawn as rectilinear polygons instead of rectangles like the shaded face in Fig. 1(d). We finally get a prescribed-area octagonal drawing of G as illustrated in Fig. 1(e).

The rest of the paper is organized as follows. Section 2 introduces some definitions. Section 3 deals with octagonal drawings of good slicing graphs. Finally Section 5 concludes with discussions.

2   Preliminaries

In this section we give some definitions.

Let G be a plane 2-connected simple graph. We denote the set of vertices of G by $V(G)$ and the set of edges of G by $E(G)$. The *degree* of a vertex v is the number of neighbors of v in G. Since G is a plane graph, G is embedded in the plane so that no two edges intersect except at a vertex to which they are both incident. G divides the plane into connected regions called *faces*. We regard the contour of a face as a *clockwise* cycle formed by the edges on the contour. We call the contour of the outer face of G the *outer cycle* of G, and denote by $C_o(G)$

or simply C_o. A vertex on C_o is called an *outer vertex*, while a vertex not on C_o is called an *inner vertex*. An edge on C_o is called an *outer edge*, while an edge not on C_o is called an *inner edge*.

An *orthogonal drawing* of G is a drawing of G with a given embedding in which each vertex is mapped to a point, each edge is drawn as a sequence of alternate horizontal and vertical line segments, and any two edges do not cross except at their common end. A *bend* is a point where an edge changes its direction in a drawing. Each face of G is drawn as a rectilinear polygon in any orthogonal drawing of G. Every plane graph of the maximum degree at most four has an orthogonal drawing. We call an orthogonal drawing D an *octagonal drawing* if D satisfies the following two conditions (i) and (ii): (i) the outer cycle C_o is drawn in D as a rectangle; and (ii) each inner face is drawn in D as a rectilinear polygon which has at most eight corners and whose area is exactly equal to the prescribed value.

A graph G is a *2-3 plane graph* if G is a 2-connected plane graph, every vertex has degree two or three, and there are four or more outer vertices of degree two, and exactly four of them, a, b, c and d, are designated as *corners*. The four corners a, b, c and d divide C_o into four paths, the north path P_N, the east path P_E, the south path P_S, and the west path P_W, as illustrated in Fig. 3(a). A path P in G is called an *NS-path* if P starts at a vertex on P_N, ends at a vertex on P_S, and does not pass through any other outer vertex and any outer edge. An NS-path P naturally divides G into two 2-3 plane graphs G_W^P and G_E^P; G_W^P is the *west subgraph* of G including P, and G_E^P is the *east subgraph* of G including P. We call G_W^P and G_E^P the *two subgraphs corresponding to* P. Similarly, we define a *WE-path* P, the *north subgraph* G_N^P, and the *south subgraph* G_S^P.

We now present a formal recursive definition of a slicing graph. We call a 2-3 plane graph G a *slicing graph* if either it has exactly one inner face or it has an NS- or WE-path P such that each of the two subgraphs corresponding to P is a slicing graph. An NS- or WE-path P in a slicing graph G is called a *slicing path* of G if each of the two subgraphs corresponding to P is a slicing graph. The graphs in Figs. 1(a) and 2(f) are slicing graphs.

If G is a slicing graph, then all slicing paths appearing in the recursive definition can be represented by a binary tree T, called a *slicing tree*, as illustrated

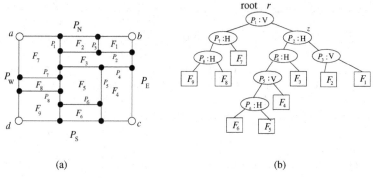

(a)   (b)

Fig. 3. (a) A slicing plane graph G, and (b) a good slicing tree T of G.

in Fig. 3 for the graph in Fig. 2(f). Each internal node u of T represents a slicing path, which is denoted by P_u. Each leaf u of T represents an inner face F_u of G. Each node u of T corresponds to a subgraph G_u of G induced by all inner faces that are leaves and are descendants of u in T. Thus $G = G_r$ for the root r of T. We classify the internal nodes of T into two types: (i) V-node and (ii) H-node. A V-node u represents an NS-slicing path P_u of G_u, while an H-node u represents a WE-slicing path P_u of G_u.

We then give a formal definition of a good slicing graph. A *face path* of a 2-3 plane graph G is a WE- or NS-path on the contour of a single inner face of G. Any face path P of a slicing graph G is a slicing path. If P is a face WE-path, then either the north subgraph G_N^P or the south subgraph G_S^P corresponding to P will never be vertically sliced. We call a slicing tree T a *good slicing tree* if P_u is a face WE-path of G_u for every H-node u in T. The tree in Figure 3(b) is a good slicing tree of the graph G in Fig. 3(a). We call a 2-3 graph a *good slicing graph* if it has a good slicing tree for an appropriate labeling of designated corners as a, b, c and d. All the graphs in Figs. 1(a), 2(f) and 3(a) are good slicing graphs. However, not every slicing graph is a good slicing graph. The definitions above imply that every vertical slice of a good slicing graph is an arbitrary "guillotine cut" but every horizontal slice must be a "guillotine cut" along a face WE-path. As we will show later in Section 3, our algorithm draws every vertical slice as a single vertical line segment, and draws every horizontal slice as either a single horizontal line segment or a sequence of three line segments, horizontal, vertical and horizontal ones, as illustrated in Fig. 1(e).

3   Octagonal Drawing

In this section we prove the following theorem as the main result of the paper. Note that a slicing graph together with its slicing tree is often given as an input in many practical applications.

Theorem 1. *A good slicing graph G has an octagonal drawing D, and the drawing D can be found in linear time if a good slicing tree T is given.*

In the rest of this section we give a constructive proof of Theorem 1. Let G be a good slicing graph. One may assume without loss of generality that all vertices of G have degree three except for the four outer vertices a, b, c and d of degree two. We will show that every inner face of G is drawn as a rectilinear polygon of at most eight corners whose shape is one of the nine's in Fig. 4. We call a rectilinear polygon of shape like in Fig. 4 an *octagon* throughout the paper. Thus a rectangle is an octagon in our terminology, because the polygon in Fig. 4(i) is a rectangle. We denote by $A(R)$ the area of an octagon R, and by $A(G)$ the sum of the prescribed areas of all inner faces of a plane graph G.

The rest of this section is organized as follows. We give our Algorithm **Octagonal-Draw** in Section 3.1. In Sections 3.2 we give the details for embedding a slicing path. In Section 3.3 we complete a proof of Theorem 1 by verifying correctness and time complexity of the algorithm.

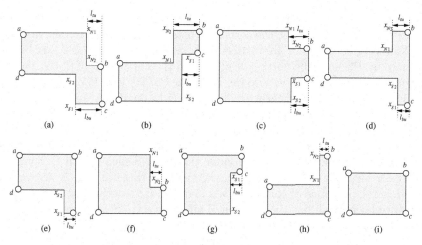

Fig. 4. Octagons.

3.1 Algorithm Octagonal-Draw

In this section we give an algorithm for finding an octagonal drawing of a good slicing graph G.

An outline of the algorithm is as follows. Let T be a good slicing tree of G. Let u be an internal node of T, let v be the right child of u, and let w be the left child of u. One may assume that if u is a V-node then its right subtree rooted at v represents the east subgraph $G_E^{P_u}$ of G and its left subtree rooted at w represents the west subgraph $G_W^{P_u}$ and hence $G_v = G_E^{P_u}$ and $G_w = G_W^{P_u}$, and that if u is an H-node then $G_v = G_N^{P_u}$ and $G_w = G_S^{P_u}$, as illustrated in Fig. 3. We now traverse T by reverse preorder traversal, that is, we first traverse the root r of T, then traverse the right subtree and finally traverse the left subtree. We thus draw the inner faces $F_1, F_2, \cdots, F_{11}$ of G in Fig. 3(a) in this order, from east to west and north to south.

Before starting the traversal from root r, we choose an arbitrary rectangle R_r of area $A(G)$. Thus $A(G) = H \times W$ if H and W are the height and width of R_r, respectively. The outer cycle $C_o(G)$ is drawn as R_r. In general, when we traverse a node u of T, we have an octagon R_u of area $A(G_u)$; $C_o(G_u)$ is drawn as R_u. If u is an internal node, then we embed the slicing path P_u inside R_u so that P_u divides R_u into two octagons R_v and R_w so that $A(R_v) = A(G_v)$ and $A(R_w) = A(G_w)$, where v is the right child and w is the left child of u. (See Fig. 5.)

We start to traverse T from root r with the following initialization. We fix the positions of four designated vertices a, b, c and d of G as the corners of the initial rectangle R_r. We then arbitrarily fix the positions of all vertices on the east path P_E^r of $G_r = G$ with preserving their relative positions. The positions of all other vertices on $C_o(G)$ are not fixed at this moment.

When we traverse an internal node u of T, we have an octagon R_u such that $A(R_u) = A(G_u)$. Four vertices of degree two on $C_o(G_u)$ have been designated as the four corner vertices a, b, c and d of G_u as illustrated in Fig. 4. Let $a, x_{N1}, x_{N2}, b, c, x_{S1}, x_{S2}, d$ be the corners of octagon R_u, some of which may not exist. Note that a, b, c and d are vertices of G_u and $x_{N1}, x_{N2}, x_{S1}, x_{S2}$ are bends. We denote by P_N^u both the north side of R_u and the north path of $C_o(G_u)$ which connects a and b. Similarly we use the notation P_E^u, P_S^u and P_W^u. The positions of vertices a, b, c and d together with all the vertices on P_E^u have been fixed, but the positions of all vertices on P_W^u and P_S^u except a, d and c have not been fixed.

We now describe the operations performed at each internal node u of T. Let v be the right child of u in T, and let w be the left child. We first consider the case where u is a V-node. One may assume that the NS-slicing path P_u connects a vertex y_N on P_N^u and a vertex y_S on P_S^u, as illustrated in Fig. 5. As we will show later, the positions of corners a, b, c and d of R_u together with all vertices on P_E^u have been fixed, but the position of all other vertices of G_u have not been fixed. The goodness and the traversal order of T are crucial in the argument. We now fix the positions of y_N and y_S and divide R_u into two octagons R_v and R_w by embedding P_u as a vertical line segment so that $A(R_v) = A(G_v)$ and $A(R_w) = A(G_w)$. Indeed R_w is always a rectangle, as illustrated in Fig. 5. We will give the detail of this step later in Section 3.2. We now designate y_N, b, c and y_S as the four corner vertices of G_v, and designate $a, y_N, y_S,$ and d as the four corner vertices of G_w.

We then consider the case where u is an H-node. Assume that the face WE-path P_u connects a vertex y_W on P_W^u and a vertex y_E on P_E^u, as illustrated in Fig. 6. The positions of all vertices on P_E^u including y_E have been fixed. We appropriately fix the position of y_W on P_W^u and divide R_u into two octagons R_v and R_w so that $A(R_v) = A(G_v)$ and $A(R_w) = A(G_w)$ by embedding P_u as either a single horizontal line segment or a sequence of three line segments, horizontal, vertical and horizontal ones, as illustrated in Fig. 6. We will give the detail of this step later in Section 3.2. We now designate a, b, y_E and y_W as the corner vertices of G_v, and designate $y_W, y_E, c,$ and d as the corner vertices of G_w.

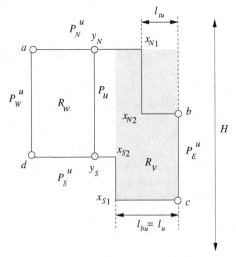

Fig. 5. Embedding of P_u in R_u for a V-node u.

We finally consider the case where we traverse a leaf node u of T. In this case u corresponds to an inner face F_u, and the embedding of F_u has been already fixed as an octagon R_u. The positions of a, b, c and d and all vertices on P_E^u and P_N^u have been fixed. We arbitrarily fix the positions of all vertices on P_W^u other than a and d, preserving their relative positions on P_W^u. If there are vertices on P_S^u other than c and d, then their positions will be fixed in some later steps.

We call the algorithm described above Algorithm **Octagonal-Draw**.

3.2   Embedding a Slicing Path

In this section we give the details of embedding a slicing path P_u inside an octagon R_u.

A polygonal vertex of R_u is called a *corner* of R_u. A corner of R_u has an interior angle 90° or 270°. A corner of an interior angle 90° is called a *convex corner* of R_u, while a corner of an interior angle 270° is called a *concave corner*. Let p and q be two consecutive polygonal vertices of R_u. We denote by pq the polygonal edge of R_u connecting p and q. We also denote by pq the straight line segment connecting two points p and q.

Let A_{min} be the area of an inner face whose prescribed area is the smallest among all inner faces of G. Let H be the height of the whole drawing, that is, the height of the initial rectangle R_r. Let f be the number of inner faces in G, and let

$$\lambda = \frac{A_{min}}{fH}. \tag{1}$$

Since $A(G) = WH$, we have $\lambda = \frac{WA_{min}}{fA(G)}$.

Let u be a node in T. Let l_{tu} be the length of line segment $x_{N2}b$ of an octagon R_u, and let l_{bu} be the length of line segment cx_{S1}, as illustrated in Fig. 4. If x_{N2} does not exist then let $l_{tu} = 0$, and if x_{S1} does not exist then let $l_{bu} = 0$. Let $l_u = \max\{l_{tu}, l_{bu}\}$. Thus $l_u = 0$ if and only if R_u is a rectangle. Let f_E^u be the number of inner faces in G_u each of which has an edge on the east path P_E^u of G_u. We call an octagon R_u a *feasible octagon* if the following eight conditions (i)–(viii) hold:

(i) $A(R_u) = A(G_u)$;
(ii) $l_u < f\lambda$;
(iii)   if x_{N2} is a concave corner as in Figs. 4(a), (c) and (f), then $l_{tu} < (f - f_E^u)\lambda$;
(iv)   if x_{S1} is a concave corner as in Figs. 4(b), (c) and (g), then $l_{bu} < (f - f_E^u)\lambda$;
(v)   if x_{N2} is a convex corner as in Figs. 4(b), (d) and (h), then $l_{tu} \geq f_E^u\lambda$;
(vi)   if x_{S1} is a convex corner as in Figs. 4(a), (d) and (e), then $l_{bu} \geq f_E^u\lambda$;
(vii)   if both x_{N2} and x_{S2} are concave corners as in Fig. 4(a), then $l_{bu} - l_{tu} \geq f_E^u\lambda$; and
(viii)   if both x_{N1} and x_{S1} are concave corners as in Fig. 4(b), then $l_{tu} - l_{bu} \geq f_E^u\lambda$.

The initial octagon R_r for the root r of T is a rectangle of area $A(G_r)$, where $G_r = G$. Since R_r is a rectangle, $x_{N1}, x_{N2}, x_{S1}, x_{S2}$ do not exist and hence $l_u = l_{tu} = l_{bu} = 0$. Therefore the rectangle R_r is a feasible octagon.

We now have the following lemma on the embedding of P_u for a V-node u.

Lemma 1. *Let u be a V-node of T, let v be the right child of u, and let w be the left child. If R_u is a feasible octagon, then the NS-slicing path P_u can be embedded inside R_u as a single vertical line segment so that R_u is divided into two feasible octagons R_v and R_w.*

Proof. We first show that P_u can be embedded as a vertical line segment inside R_u so that $A(R_v) = A(G_v)$ and $A(R_w) = A(G_w)$, and hence R_v and R_w satisfies Condition (i). One may assume that the NS-slicing path P_u connects a vertex y_N on P_N^u and a vertex y_S on P_S^u. Since T is a good slicing tree, the north path P_N^u of G_u is either on the north path P_N^r of $G_r = G$ or on a face WE-path of G_z for some H-node z which is an ancestor of u in T. Thus the part of G either above P_N^u or below P_N^u is a face of G. Therefore the positions of all vertices on P_N^u other than a and b have not been fixed although the face above P_N^u has been drawn. Since the part of G below P_S^u has not been drawn, the position of all vertices on P_S^u other than c and d have not been fixed. We can therefore embed P_u as a vertical line by sliding y_N along P_N^u together with y_S along P_S^u from west to east until the equations $A(R_v) = A(G_v)$ and $A(R_w) = A(G_w)$ hold, as illustrated in Fig. 5.

We can also show that both R_v and R_w satisfy Condition (ii)–(viii) a of feasible octagon. The detail is omitted in this extended abstract.   *Q.E.D.*

We now have the following lemma on an embedding of P_u for an H-node u.

Lemma 2. *Let u be an H-node of T, let v be the right child of u, and let w be the left child. If R_u is a feasible octagon, then the WE-slicing path P_u can be embedded inside R_u as either a single horizontal line segment or a sequence of three line segments, horizontal, vertical and horizontal ones, so that R_u is divided into two feasible octagons R_v and R_w.*

Proof. One may assume that the face WE-path P_u connects a vertex y_W on P_W^u and a vertex y_E on P_E^u, as illustrated in Fig. 6. We assume that the shape of R_u is as in Fig. 4(a). (The proof for the other shapes is similar.) In this case both x_{N2} and x_{S2} of R_u are concave corners, and hence by Condition (vii) $l_{bu} - l_{tu} \geq f_E^u \lambda > 0$. Since x_{N2} is concave, by Condition (iii) $l_{tu} \leq (f - f_E^u)\lambda$. Also $l_u < f\lambda$ by Condition (ii). The position of vertex y_E has been fixed on P_E^u when the part of G to the right of bc was drawn. The horizontal line L passing through y_E intersects either ad or $x_{S2}x_{S1}$, and hence there are the following two cases.

Case 1: L intersects ad.

Let L intersect ad at a point y', as illustrated in Figs. 6(a), (b) and (c), and let Q be the polygon $a, x_{N1}, x_{N2}, b, y_E, y'$, then we have the following three subcases.

Subcase 1(a): $A(G_v) = A(Q)$.

In this case we fix the position of vertex y_W at point y' and embed the path P_u as a single horizontal line segment $y'y_E$, as illustrated in Fig. 6(a). R_v is the octagon $a, x_{N1}, x_{N2}, b, y_E, y_W$, and R_w is the octagon $y_W, y_E, c, x_{S1}, x_{S2}, d$. R_v has the shape of a type as in Fig. 4(f), and R_w has the shape of type in Fig. 4(e).

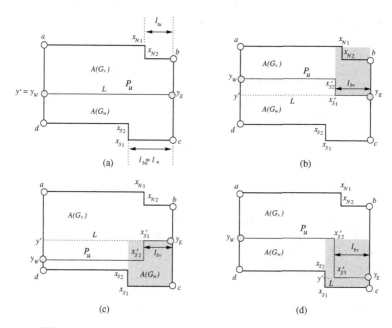

Fig. 6. Division of R_u to R_v and R_w by horizontal slice P_u.

We first show that R_v is feasible. Since $A(R_v) = A(G_v)$, Condition (i) holds for R_v. Furthermore, $l_{bv} = 0$, $l_{tv} = l_{tu}$, and hence $l_v = l_{tv} < l_u = l_{bu} < f\lambda$. Thus Condition (ii) also holds for R_v. Since $f_E^v < f_E^u$, we have $l_{tv} = l_{tu} \le (f - f_E^u)\lambda < (f - f_E^v)\lambda$ and hence Condition (iii) also holds for R_v. Since x_{S1} and x_{S2} of R_v do not exist and x_{N2} of R_v is concave, Conditions (iv)-(viii) also hold. Thus R_v is feasible.

Similarly one can show that R_w is feasible. The detail is omitted in this extended abstract.

Subcase 1(b): $A(G_v) < A(Q)$.

Clearly $f_E^v < f_E^u$. We first fix a corner y'_{S1} of R_v on L so that the horizontal line segment $y_E x'_{S1}$ has length $l_{tu} + f_E^v\lambda$ and hence $l_{bv} = l_{tu} + f_E^v\lambda$, as illustrated in Fig. 6(b). We then fix the positions of x'_{S2} and y_W so that $A(R_v) = A(G_v)$. We now claim that $y_W x'_{S2}$ is below $a x_{N1}$. By Condition (vii) $l_{bu} \ge f_E^u\lambda + l_{tu}$, and hence $l_u = l_{bu} \ge f_E^u\lambda + l_{tu} > f_E^v\lambda + l_{tu} = l_{bv}$. Since $l_u < f\lambda$ by Condition (ii), we have $l_{bv}H < l_u H < f\lambda H = A_{\min}$ by Eq. (1) and hence the shaded rectangular area of width l_{bv} and height $< H$ in Fig. 6(b) is smaller than $A_{\min}$. Since G_v has at least one inner face, we have $A_{\min} \le A(G_v)$. Therefore $y_W x'_{S2}$ is below $a x_{N1}$, and hence R_v is a (simple) octagon $a, x_{N1}, x_{N2}, b, y_E, x'_{S1}, x'_{S2}, y_W$, and R_w is an octagon $y_W, x'_{S2}, x'_{S1}, y_E, c, x_{S1}, x_{S2}, d$. Both R_v and R_w have a shape in Fig. 4(a).

We now show that R_v is feasible. Since $A(R_v) = A(G_v)$, Condition (i) holds for R_v. Clearly, $l_{tv} = l_{tu}$, $l_{bv} < l_{bu}$, and hence $l_v = l_{bv} < l_u < f\lambda$. Thus Condition (ii) also holds for R_v. Since $l_{tv} = l_{tu} \le (f - f_E^u)\lambda < (f - f_E^v)\lambda$, Condition (iii) also holds for R_v. Since $l_{bv} = l_{tu} + f_E^u \lambda \geq f_E^v \lambda$, Condition (vi) holds for R_v. Since $l_{bv} - l_{tv} = l_{bv} - l_{tu} = f_E^v \lambda$, Condition (vii) also holds. Since x_{N1} and x_{S1}' of R_v are convex corners, the other conditions also hold. Thus R_v is feasible.

Similarly one can show that R_w is a feasible octagon.

Subcase 1(c): $A(Q) < A(G_v)$.

Similar to Subcase 1(b) above. See Fig. 6(c).

Case 2: L intersects $x_{S2}x_{S1}$.

Similar to the proof of Subcase 1(b). See Fig. 6(d). $\mathcal{Q.E.D.}$

3.3   Correctness and Time Complexity

In this section we verify the correctness and time complexity of Algorithm **Octagonal-Draw**, and mention some remarks on the algorithm.

We first prove the following lemma on the correctness of Algorithm **Octagonal-Draw**.

Lemma 3. *Algorithm* **Octagonal-Draw** *finds an octagonal drawing of a good slicing graph G.*

Proof. The initial rectangle R_r at the root r of T is a feasible octagon. Assume inductively that u is an internal node of T and R_u is a feasible octagon. Let v and w be the right child and left child of u, respectively. By Lemmas 1 and 2 one can embed P_u inside R_u so that R_v and R_w are feasible octagons. Thus, after the execution of the algorithm, each inner face of G corresponding to a leaf of T is a feasible octagon. Of course, the contour of the outer face of G is the rectangle R_r. Thus Algorithm **Octagonal-Draw** finds an octagonal drawing of G. $\mathcal{Q.E.D.}$

We now have the following lemma on the time complexity of Algorithm **Octagonal-Draw** whose proof is omitted in this extended abstract.

Lemma 4. *Algorithm* **Octagonal-Draw** *runs in linear time.*

Lemmas 3 and 4 complete the proof of Theorem 1.

4   Conclusions

In this paper we showed that any good slicing graph has an octagonal drawing with prescribed face areas, and gave a linear-time algorithm to find such a drawing. We also gave a sufficient condition for a plane graph G to be a good slicing graph and gave a linear-time algorithm to construct a good slicing tree if G satisfies the condition. However, they are omitted in this extended abstract due to the page limitation. Yeap and Sarrafzadeh [10] gave a sufficient condition for a plane graph G to be a slicing graph. Although their condition is represented in terms of a dual graph of G, theirs and ours are effectively same. We, however, showed that the condition is a sufficient condition for a plane graph to be not only a slicing graph but also a *good* slicing graph. In a VLSI floorplan produced by our algorithm, the width of the narrowest part of a module is at least $\lambda = \frac{A_{\min}}{fH}$. However, one can appropriately choose a larger value as λ in many practical floorplans.

A connected graph is *cyclically k-edge connected* if the removal of any set of less than k edges leaves a graph such that exactly one of the connected components has a cycle. Let G be a 2-3 plane graph obtained from a cyclically 5-edge connected plane cubic graph by inserting four vertices a, b, c and d of degree 2 on the outer face. Thomassen [T92] showed that G has a drawing in which each edge is drawn as a single straight line segment which is not always horizontal or vertical, each inner face attains its prescribed area, and the contour of the outer face is a rectangle having the four vertices as corners. Thus, in his drawing, each inner face is drawn with a polygon which is not always rectilinear. The class of good slicing graphs is larger than the class of graphs obtained from cyclically 5-edge connected cubic plane graphs by inserting four vertices of degree 2 on the outer face.

It is remained as a future work to obtain an algorithm for finding a prescribed area orthogonal drawing for a larger class of plane graphs.

Acknowledgments

We thank Shin-ichi Nakano, Kazuyoshi Hada and Ikki Mizumura for helpful discussions at an early stage of the work.

References

1. X. He, *On floorplans of planar graphs*, SIAM Journal on Computing, 28(6), pp. 2150-2167, 1999.
2. T. Lengauer, *Combinatorial Algorithms for Integrated Circuit Layout*, John Wiley & Sons, Chichester, 1990.
3. M. S. Rahman, S. Nakano and T. Nishizeki, *A linear algorithm for bend-optimal orthogonal drawings of triconnected cubic plane graphs*, Journal of Graph. Alg. and Appl., http://jgaa.info, 3(4), pp. 31-62, 1999.
4. W. Shi, *A fast algorithm for area minimization of slicing floorplans*, IEEE Transactions on Computer-Aided Design of Integrated Circuits and Systems, 15(12), pp. 1525-1532, 1996.
5. Y. Sun and M. Sarrafzadeh, *Floorplanning by graph dualization: L-shaped modules*, Algorithmica, 10, pp. 429-456, 1998.
6. S. M. Sait and H. Youssef, *VLSI Physical Design Automation*, World Scientific, Singapore, 1999.
7. R. Tamassia, *On embedding a graph in the grid with the minimum number of bends*, SIAM J. Comput., 16, pp. 421-444, 1987.
8. C. Thomassen, *Plane cubic graphs with prescribed face areas*, Combinatorics, Probability and Computing, 1, pp. 371-381, 1992.
9. K. Yeap and M. Sarrafzadeh, *Floor-planning by graph dualization: 2-concave rectilinear modules*, SIAM J. Comput., 22(3), pp. 500-526, 1993.
10. K. H. Yeap and M. Sarrafzadeh, *Sliceable floorplanning by graph dualization*, SIAM J. Disc. Math., 8(2), pp. 258-280, 1995.

Crossing Reduction in Circular Layouts[*]

Michael Baur[1] and Ulrik Brandes[2]

[1] Department of Computer Science, University of Karlsruhe (TH), Germany
baur@ilkd.uni-karlsruhe.de
[2] Department of Computer & Information Science, University of Konstanz, Germany
Ulrik.Brandes@uni-konstanz.de

Abstract. We propose a two-phase heuristic for crossing reduction in circular layouts. While the first algorithm uses a greedy policy to build a good initial layout, an adaptation of the sifting heuristic for crossing reduction in layered layouts is used for local optimization in the second phase. Both phases are conceptually simpler than previous heuristics, and our extensive experimental results indicate that they also yield fewer crossings. An interesting feature is their straightforward generalization to the weighted case.

1 Introduction

In circular graph layout, the vertices of a graph are constrained to distinct positions along the perimeter of a circle, and an important objective is to minimize the number of edge crossings in such layouts. Since circular crossing minimization is $\mathcal{NP}$-hard [8], several heuristics have been devised [7, 3, 14]. Moreover there is a factor $O(\log^2 |V|)$ approximation algorithm [13].

We propose a two-phase approach for obtaining circular layouts with few crossings. In the first phase, vertices are iteratively added to either end of a linear layout. This leaves three degrees of freedom: the start vertex, the insertion order, and the end at which to append the next vertex. For the different strategies tried, empirical evidence suggests that a particular one outperforms both the others and previous heuristics.

For the second phase, we adapt a local optimization procedure for layered layouts, sifting [9], to the circular case. Note that, similar to 2-layer layouts, the number of crossing is completely determined by the (cyclic) ordering of vertices. The thus related one-sided crossing minimization problem in 2-layer drawings of bipartite graphs is $\mathcal{NP}$-hard as well [5], but significantly better understood. It turns out that circular sifting reduces the number of crossings both with respect to our first phase and previous heuristics.

After defining some terminology in Section 2, we describe our greedy append and circular sifting algorithms for the phases in Sections 3 and 4. Both are evaluated experimentally in Section 5.

---

[*] Research partially supported by DFG under grants Wa 654/13-2 and Br 2158/1-2.

J. Hromkovič, M. Nagl, and B. Westfechtel (Eds.): WG 2004, LNCS 3353, pp. 332–343, 2004.
© Springer-Verlag Berlin Heidelberg 2004

2   Preliminaries

Throughout this paper, let $G = (V, E)$ be a simple undirected graph with $n = |V|$ vertices and $m = |E|$ edges. Furthermore, let $N(v) = \{u \in V : \{u, v\} \in E\}$ denote the neighborhood of a vertex $v \in V$. A *circular layout* of G is a bijection $\pi : V \to \{0, \dots, n-1\}$, interpreted as a clockwise sequence of distinct positions on the circumference of a circle. By selecting a reference vertex $s \in V$ we obtain linear orders $\prec_s^\pi$ from π by defining

$$u \prec_s^\pi v \iff (\pi(u) - \pi(s) \mod n) < (\pi(v) - \pi(s) \mod n)$$

for all $u, v \in V$, i.e. u is encountered before v in a cyclic traversal starting from s. We say that $u, v \in V$ are *consecutive*, denoted by $u \curvearrowright_\pi v$, if $\pi(v) - \pi(u) \equiv 1$ mod n. A subset $W \subset V$ is *consecutive*, if there is an ordering of the vertices of W so that $w_0 \curvearrowright_\pi w_1 \curvearrowright_\pi \dots \curvearrowright_\pi w_{|W|-1}$, $w_i \in W$.

Let

$$\chi_\pi(\{u_1, v_1\}, \{u_2, v_2\}) = \begin{cases} 1 & \text{if } u_1 \prec_{u_1}^\pi u_2 \prec_{u_1}^\pi v_1 \prec_{u_1}^\pi v_2 \\ 0 & \text{otherwise .} \end{cases} \tag{1}$$

for all $\{u_1, v_1\}, \{u_2, v_2\} \in E$ and w.l.o.g. $\pi(u_i) < \pi(v_i)$. We say that $e_1, e_2 \in E$ *cross* in π, iff $\chi_\pi(e_1, e_2) = 1$, i.e. the endvertices of e_1, e_2 are encountered alternately in a cyclic traversal. The *crossing number* of a circular layout π is $\chi(\pi) = \sum_{e_1, e_2 \in E} \chi_\pi(e_1, e_2)$ and $\chi(G) = \min_\pi \chi(\pi)$ is called the *circular crossing number* of G. We will omit π from our notation whenever the layout is clear from context.

Theorem 1 ([8]). *Circular crossing minimization is $\mathcal{NP}$-hard.*

On the other hand, a graph has a circular layout with no crossings, if and only if it is outerplanar. A linear time recognition algorithm for outerplanar graphs [11] is easily extended to yield a crossing-free circular layout [14].

Since, in particular, trees have circular layouts with no crossings, it is possible to consider the biconnected components of a graph separately, and insert their circular layouts into a crossing-free layout of the block-cutpoint-tree without producing additional crossings (see Fig. 1). Hence, only biconnected graphs are used in the experimental evaluation summarized in Section 5.

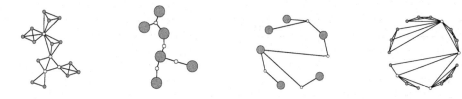

Fig. 1. The circular crossing number of a graph is the sum of those of its biconnected components (cutpoints shown in lighter color).

3   Initial Layout

Our approach for an initial layout is inspired by a heuristic algorithm for the minimum total edge length problem in circular layouts [7]. This problem is somewhat related to crossing minimization, since shorter edges tend to cross few other edges.

The basic idea is simple: start with a layout consisting of a single vertex and place the other vertices, one at a time, at either end of the current (linear) layout (see Algorithm 1). After all vertices are inserted, the final layout is considered to be circular. This method leaves us with three parameters to choose:

- the start vertex s,
- the processing sequence, and
- the end to append the next vertex at.

Note that the processing sequence need not to be fixed in the beginning, but may be determined while the algorithm proceeds. Since, in our experiments, the rules for choosing a start vertex had little influence on the final result, it is chosen at random. In the following we describe instantiations for the other two parameters.

During the algorithm some vertices are already placed while others are not. An edge is called *open*, if it connects a placed vertex with an unplaced one, and *closed*, if both its vertices have been inserted.

Four rules for determining an insertion order are investigated. The rationale behind these heuristics is to keep the number of open edges low, because they tend to result in crossings later on.

1. *Degree.* Vertices are inserted in non-increasing order of their degree.
2. *Inward Connectivity.* At each step, a vertex with the largest number of already placed neighbors is selected, i.e. a vertex which closes the most open edges.
3. *Outward Connectivity.* At each step, a vertex with the least number of unplaced neighbors is selected, i.e. a vertex which opens the fewest new edges.
4. *Connectivity.* At each step, a vertex with the least number of unplaced neighbors is selected, where ties are broken in favor of vertices with fewer unplaced neighbors.

The other degree of freedom left is the selection of an end of the current layout at which to append the next vertex. Again, four rules of choice are investigated.

Algorithm 1: Greedy-Append Heuristic.

---

place start vertex $s \in V$ arbitrarily;
$V \leftarrow V \setminus \{s\}$;
while $V \neq \emptyset$ **do**
  greedily choose $v \in V$;
  append v at either end of the current layout;
  $V \leftarrow V \setminus \{v\}$;

---

1. *Random.* Select the end at which to append randomly each time.
2. *Fixed.* Always append to the same end.
3. *Length.* Append each vertex to the end that yields the smaller increase in total edge length.
4. *Crossings.* Append each vertex to the end that yields fewer crossing of edges being closed with open edges. In Fig. 2, there are eight such crossings for the left end and only six for the right end. Note that crossings with closed edges not incident to the currently inserted vertex need not be considered because they are the same for both sides. It should also be noted that crossings with open edges are independent of the positions at which the unplaced vertex will eventually be placed.

Fig. 2. Incident edges of v cross open edges.

The experiments outlined in Section 5 show that the combination of the *Connectivity* insertion order with *Crossings* outperforms all other combinations, and it can be implemented efficiently.

Theorem 2. *The Greedy-Append heuristic with* Connectivity *insertion order and end-to-append selection based on* Crossings *can be implemented to run in* $\mathcal{O}((n + m) \log n)$ *time.*

Proof. The insertion sequence can be realized by storing all unplaced vertices in a two-dimensional priority queue, in which the first key gives the number of already placed neighbors and the second the number of unplaced neighbors. With an efficient implementation, update and extract operations require $\mathcal{O}(\log n)$ time. Since each vertex is extracted once, and each edge triggers exactly one update, the total running time for determining the insertion order is $\mathcal{O}((n + m) \log n)$.

The number of crossings with open edges can be determined from prefix and suffix sums over vertices already in the layout. These can be maintained efficiently using a balanced binary tree storing in its leaves the number of open edges incident to a placed vertex, and in its inner nodes the sum of the values of its two children. The prefix sum at a vertex is the sum of all values in left children of nodes on the path from the corresponding leaf to the root. The suffix sum is determined symmetrically. Insertion of a vertex thus requires $\mathcal{O}(\log n)$ time to determine the crossing numbers from prefix and suffix sums and $\mathcal{O}(d(v) \log n)$ for updating the tree. The total is again $\mathcal{O}((n + m) \log n)$.   □

Note that the heuristic is easily generalized to weighted graphs. In the next section we show how to further reduce the number of crossings, given an initial layout.

4   Improvement by Circular Sifting

Sifting was originally introduced as a heuristic for vertex minimization in ordered binary decision diagrams [12] and later adapted for the one-sided crossing minimization problem [9]. The idea is to keep track of the objective function while moving a vertex along a fixed ordering of all other vertices. The vertex is then placed in its (locally) optimal position. The method is thus an extension of the greedy-switch heuristic [4].

For crossing reduction the objective function is the number of crossings between the edges incident to the vertex under consideration and all other edges. The efficient computation of crossing numbers in sifting for layered layouts is based on the crossing matrix. Its entries correspond to the number of crossings caused by pairs of vertices in a particular linear ordering and are computed easily in advance. Whenever a vertex is placed in a new position only a smallish number of updates is necessary.

It is not possible to adapt the crossing matrix to the circular case, since two vertices cannot be said to be in a (linear) order generally. Thus we define the crossing number

$$c_{uv}(\pi) = \sum_{x \in N(u)} \sum_{y \in N(v)} \chi_\pi(\{u, x\}, \{v, y\}) \tag{2}$$

only for pairs of consecutive vertices $u \curvearrowright v \in V$ and use the following exchange property, which is the basis for sifting and holds nevertheless.

Lemma 1. *Let $u \curvearrowright v \in V$ be consecutive vertices in a circular layout π, and let π' be the layout with their positions swapped, then*

$$\chi(\pi') = \chi(\pi) - c_{uv}(\pi) + c_{vu}(\pi')$$
$$= \chi(\pi) - \sum_{x \in N(u)} |\{y \in N(v) : y \prec_x^\pi u\}| + \sum_{y \in N(v)} \left|\{x \in N(u) : x \prec_y^{\pi'} v\}\right|$$

Proof. Since u and v are consecutive, edges incident to neither u nor v do not change their crossing status. The first equality follows immediately. For the second equality, observe that the sums are obtained from (2) by inserting (1). See Fig. 3 for an illustration.   □

Based on the above lemma, the locally optimal position of a single vertex can be found by iteratively swapping the vertex with its neighbor and recording the change in crossing count, which is computed by considering only edges incident to one of these two vertices. After the vertex has been moved past every other vertex, it is placed where the intermediary crossing counts reached their minimum. Repositioning each vertex once in this way is called a *round of circular sifting*.

If adjacency lists are ordered according to the current layout, the sums in Lemma 1 are over suffix lengths in these lists. Updating the crossing count therefore corresponds to merging the adjacency lists, where the length of the remaining suffix is added or subtracted.

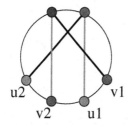 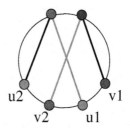

Fig. 3. After swapping consecutive vertices $u \curvearrowright v$, exactly those pairs of edges cross that did not before.

Theorem 3. *One round of circular sifting takes $\mathcal{O}(nm)$ time.*

Proof. Sorting the adjacency lists according to the vertex order is easily done in $\mathcal{O}(m)$ time (traverse the vertices in order, and add each to the adjacency lists of its neighbors). If adjacency lists are stored cyclically, a head pointer yields $\prec_v$ for arbitrary v, i.e. the adjacency lists need not be reordered before a swap. The final relocation of u takes time $\mathcal{O}(1)$.

When swapping u with neighbor v_k the adjacency lists are traversed in time $\mathcal{O}(d_G(u) + d_G(v_k))$. Since

$$\sum_{u \in V} \sum_{v \in V} \left(d_G(u) + d_G(v) \right) = \sum_{u \in V} \sum_{v \in V} d_G(u) + \sum_{u \in V} \sum_{v \in V} d_G(v) = 2 \cdot n \cdot 2m$$

the total running time is in $\mathcal{O}(nm)$. □

At the end of the outer loop each vertex is placed at its locally optimal position, so that circular sifting can only decrease the number of crossings. Our experiments outlined in the next section suggest that a few rounds of sifting suffice to reach a local minimum.

Note that in edge-weighted graphs we can define the *weighted crossing number* by counting each crossing with the product of the two edge weights involved. If suffix cardinalities are replaced by suffix sums of weights, Lemma 1 generalizes to the weighted case. Modifying the algorithm accordingly is straightforward.

5   Experimental Evaluation

We performed extensive experiments to determine the relative behavior of the different variants of our heuristics. As a base reference we use CIRCULAR [14], the currently most effective heuristic for circular crossing minimization. CIRCULAR consists of two phases as well: an initial placement (CIRCULAR 1) derived from a recognition algorithm for outerplanar graphs [11], and a subsequent improvement phase (CIRCULAR 2) that probes alternative positions for each vertex and relocates if the number of crossings is reduced. While the second phase appears to be similar to circular sifting, it differs in that a vertex

---

Algorithm 2: Circular sifting.

---

for $(u \in V)$ **do**

    let $v_0 = u \prec_u v_1 \prec_u \ldots \prec_u v_{n-1}$ denote the current layout;

    for $(v \in V)$ **do**

        sort adjacency list of v according to the current layout;

    $\chi \leftarrow 0$; $\chi^* \leftarrow 0$; $v^* \leftarrow v_{n-1}$;

    for $(k \leftarrow 1, \ldots, n-1)$ **do**

        let $x_0 \prec_{v_k} \ldots \prec_{v_k} x_{r-1}$ denote the adjacency list of u without v_k;

        let $y_0 \prec_{v_k} \ldots \prec_{v_k} y_{s-1}$ denote the adjacency list of v_k without u;

        $c \leftarrow 0$; $i \leftarrow 0$; $j \leftarrow 0$;

        while $(i < r$ **and** $j < s)$ **do**

            if $(x_i \prec_{v_k} y_j)$ **then**

                $c \leftarrow c - (s-j)$; $i \leftarrow i+1$;

            else if $(y_j \prec_{v_k} x_i)$ **then**

                $c \leftarrow c + (r-i)$; $j \leftarrow j+1$;

            else

                $c \leftarrow c - (s-j) + (r-i)$; $i \leftarrow i+1$; $j \leftarrow j+1$;

        $\chi \leftarrow \chi + c$;

        if $(\chi < \chi^*)$ **then** $\chi^* \leftarrow \chi$; $v^* \leftarrow v_k$;

    move u so that $v^* \curvearrowright u$;

---

is moved to fewer candidate positions and may thus miss good positions. Note also that CIRCULAR 2 actually counts crossings (rather than just changes) so that its running time depends on the number of crossings. When restricting replacements to a subset of positions, circular sifting simulates CIRCULAR 2 with an improved worst-case performance, but in our experiments we rather implemented an improved method for counting crossings, since realistic graphs have relatively few crossings anyway.

All algorithms have been implemented by the same person in C++ using LEDA [10]. Our experiments were carried out on a standard desktop computer with 1.5 GHz and 512 MB running Linux. Each data point is the average of 10 runs with different internal initializations (in particular, permuted adjacency lists).

The experiments were run on three families of undirected, biconnected graphs (recall from Section 2 that crossings between edges in different biconnected components can be avoided altogether):

- *Rome graphs.* A set of 10 541 biconnected components with 10 to 80 vertices used in [2]. These are sparse real-world graphs with $m \approx 1.3n$.
- *Fixed average degree.* Three sets of random graphs with 10 to 200 vertices and variable edge probability of $\frac{3}{n-1}$, $\frac{5}{n-1}$, and $\frac{10}{n-1}$, resulting in graphs with expected average degree of 3, 5, and 10.
- *Fixed density.* Three sets of random graphs with 10 to 200 vertices and fixed edge probability of 0.02, 0.05, and 0.1, resulting in graphs with expected density of 2, 5, and 10 percent.

A selection of results is given in the appendix. For a comprehensive list of figures see [1]. We here summarize our conclusions.

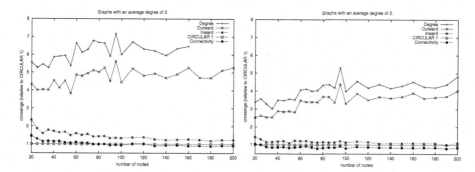

Fig. 4. Greedy append: insertion orders combined with *Fixed* (left) and *Crossings* (right) placement rules.

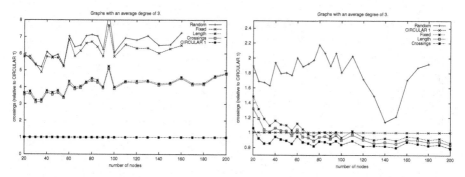

Fig. 5. Greedy append: placement rules combined with *Degree* (left) and *Connectivity* (right) insertion orders.

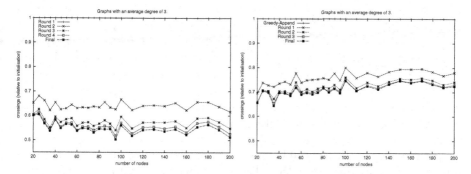

Fig. 6. Circular sifting: improvement after various rounds without initialization (left) and with greedy append (right).

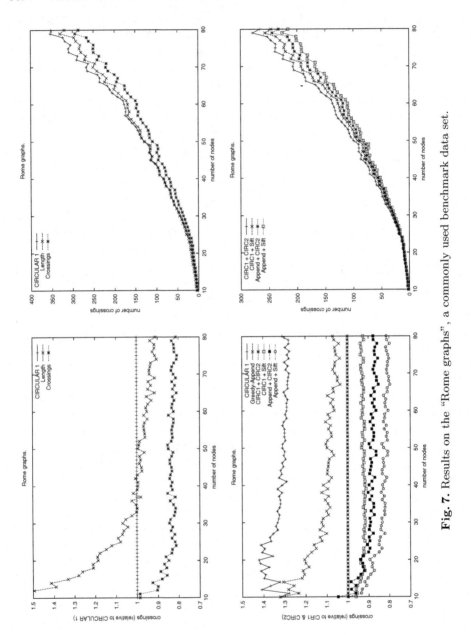

Fig. 7. Results on the "Rome graphs", a commonly used benchmark data set.

5.1 Initialization Using Greedy Append

The performance of various combinations of insertion orders for greedy append is shown in Fig. 4 relative to CIRCULAR 1. While for some rules of choice the results depend on number of edges in the graph, the *Connectivity* variant consistently outperforms all others, including CIRCULAR 1. The results in Fig. 5

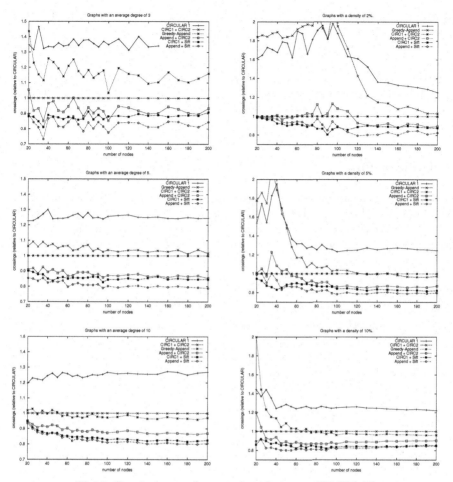

Fig. 8. Results on random graphs relative to CIRCULAR.

indicate that appropriate placement is indeed important, but has a much smaller effect than the insertion order. On random graphs, the combination of *Connectivity* insertion with *Length* or *Crossings* perform almost equally well, with a slight advantage for *Crossings*.

The two best combinations, *Connectivity* with *Length* or *Crossings*, compare favorably with CIRCULAR 1 in terms of the resulting number of crossings (see Figs. 7). Note that the running time of the initialization methods is negligible, especially when compared to the improvement strategies.

5.2   Subsequent Improvement Using Circular Sifting

Circular sifting reaches a local minimum in few rounds. As can be expected, the improvement is larger in early rounds, and the number of rounds required depends on the initial configuration (see Fig. 6). It can be concluded that the improvement algorithms (circular sifting and CIRCULAR 2) should not be used by themselves, but only in combination with a good initialization method.

With any of the good initialization strategies identified in the previous subsection, circular sifting is able to further reduce the number of crossings produced by CIRCULAR 2 as can be seen in Figs. 7 and 8 and is also confirmed by an independent study of He and Sýkora [6]. This suggests that the additional positions considered for relocation indeed pay off. However, there is a slight runtime penalty if sifting is run until there is no further improvement.

6   Conclusion

We have presented an approach for circular graph layout with few crossings. It consists of two phases: in the first phase, we greedily append vertices to either end of a partial (linear) layout according to some criteria, and in the second we further reduce the number of crossings by repeatedly sifting each vertex to a locally optimal position.

Our experimental evaluation clearly shows that the method of choice is to initialize circular sifting with a greedy-append approach using the *Connectivity* insertion order with the *Crossings* placement rule and that this combination consistently outperforms previous heuristics. They also show that both phases are necessary. While circular sifting yields a substantial improvement over the initial layouts, a good initialization significantly reduces the number of rounds required and thus the overall running time at essentially no extra cost.

References

1. M. Baur and U. Brandes. Crossing reduction in circular layouts. *Technical Report 2004-14*, Universität Karlsruhe (TH), Fakultät für Informatik, August 2004
2. G. Di Battista, A. Garg, G. Liotta, R. Tamassia, E. Tassinari, and F. Vargiu. An experimental comparison of four graph drawing algorithms. *Computational Geometry: Theory and Applications*, 7:303–326, 1997.
3. U. Doğrusöz, B. Madden, and P. Madden. Circular layout in the Graph Layout Toolkit. *Proc. 4th Intl. Symp. Graph Drawing (GD '96)*, LNCS 1190:92–100. Springer, 1996
4. P. Eades and D. Kelly. Heuristics for reducing crossings in 2-layered networks. *Ars Combinatoria*, 21(A):89–98, 1986.
5. P. Eades and N. C. Wormald. Edge crossings in drawings of bipartite graphs. *Algorithmica*, 11:379–403, 1994.
6. H. He and O. Sýkora. New circular drawing algorithms. Unpublished manuscript.
7. E. Mäkinen. On circular layouts. *International Journal of Computer Mathematics*, 24:29–37, 1988.
8. S. Masuda, T. Kashiwabara, K. Nakajima, and T. Fujisawa. On the $\mathcal{NP}$-completeness of a computer network layout problem. *Proc. IEEE Intl. Symp. Circuits and Systems*, pages 292–295, 1987.
9. C. Matuszewski, R. Schönfeld, and P. Molitor. Using sifting for k-layer straight-line crossing minimization. *Proc. 7th Intl. Symp. Graph Drawing (GD '99)*, LNCS 1731:217–224. Springer, 1999.

10. K. Mehlhorn and S. Näher. *The LEDA Platform of Combinatorial and Geometric Computing.* Cambridge University Press, 1999.

11. S. L. Mitchell. Linear algorithms to recognize outerplanar and maximal outerplanar graphs. *Information Processing Letters*, 9(5):229–232, 1979.

12. R. Rudell. Dynamic variable ordering for ordered binary decision diagrams. *Proc. IEEE Intl. Conf. Computer Aided Design (ICCAD '93)*, pages 42–47, 1993.

13. F. Shahrokhi, O. Sýkora, László,L. A. Székely, I. Vrto. Book embeddings and crossing numbers. *Proc. 20th Workshop on Graph-Theoretic Concepts in Computer Science*, WG '94, LNCS 903:256–268, Springer, 1995.

14. J. M. Six and I. G. Tollis. Circular drawings of biconnected graphs. *Proc. 1st Workshop Algorithm Engineering and Experimentation (ALENEX '99)*, LNCS 1619:57–73. Springer, 1999.

Characterization and Recognition
of Generalized Clique-Helly Graphs

Mitre Costa Dourado[1,*], Fábio Protti[2,*], and Jayme Luiz Szwarcfiter[3,*]

[1] COPPE-Sistemas
Universidade Federal do Rio de Janeiro
Caixa Postal 68511, 21945-970, Rio de Janeiro, RJ, Brasil
mitre@cos.ufrj.br
[2] Instituto de Matemática and Núcleo de Computação Eletrônica
Universidade Federal do Rio de Janeiro
Caixa Postal 2324, 20001-970, Rio de Janeiro, RJ, Brasil
fabiop@nce.ufrj.br
[3] Instituto de Matemática, Núcleo de Computação Eletrônica and COPPE-Sistemas
Universidade Federal do Rio de Janeiro
Caixa Postal 68511, 21945-970, Rio de Janeiro, RJ, Brasil
jayme@nce.ufrj.br

Abstract. Let $p \geq 1$ and $q \geq 0$ be integers. A family of sets $\mathscr{F}$ is (p,q)-*intersecting* when every subfamily $\mathscr{F}' \subseteq \mathscr{F}$ formed by p or less members has total intersection of cardinality at least q. A family of sets $\mathscr{F}$ is (p,q)-*Helly* when every (p,q)-intersecting subfamily $\mathscr{F}' \subseteq \mathscr{F}$ has total intersection of cardinality at least q. A graph G is a (p,q)-*clique-Helly graph* when its family of (maximal) cliques is (p,q)-Helly. According to this terminology, the usual Helly property and the clique-Helly graphs correspond to the case $p = 2, q = 1$. In this work we present a characterization for (p,q)-clique-Helly graphs. For fixed p, q, this characterization leads to a polynomial-time recognition algorithm. When p or q is not fixed, it is shown that the recognition of (p,q)-clique-Helly graphs is NP-hard.

1 Introduction

A well known result by Helly published in 1923 [6, 13] states that if there are given n convex subsets of a d-dimensional euclidean space with $n > d$ and if each family formed by $d + 1$ of these subsets has a point in common, then the n subsets contain a common point.

This result inspired the definition of the *Helly property* for families of sets in general, a concept that has been extensively studied in many contexts (see e.g. [9]). We say that a family $\mathscr{F}$ of sets has the Helly property (or *is Helly*) when every subfamily $\mathscr{F}' \subseteq \mathscr{F}$ of pairwise intersecting sets has non-empty total intersection.

* Partially supported by Conselho Nacional de Desenvolvimento Científico e Tecnológico – CNPq and Fundação de Amparo à Pesquisa do Estado do Rio de Janeiro – FAPERJ, Brazilian research agencies.

J. Hromkovič, M. Nagl, and B. Westfechtel (Eds.): WG 2004, LNCS 3353, pp. 344–354, 2004.
© Springer-Verlag Berlin Heidelberg 2004

When the family of (maximal) cliques of a graph G satisfies the Helly property, we say that G is a *clique-Helly* graph. Clique-Helly graphs were characterized via the notion of *extended triangle* [11, 17], defined as an induced subgraph consisting of a triangle T together with the vertices which form a triangle with at least one edge of T. This characterization leads to a straightforward recognition algorithm for clique-Helly graphs with time complexity $O((n + t)m)$, where t is the number of triangles of the input graph.

The more general *p-Helly property* holds when every $\mathscr{F}' \subseteq \mathscr{F}$ of p-wise intersecting sets has non-empty total intersection. Thus, the original result of Helly may be restated by simply saying that any family of convex subsets of a d-dimensional euclidean space is $(d + 1)$-Helly.

In this work we focus on a generalization of the p-Helly property, by considering the intersection sizes. Following [18, 19], we require that the subfamilies $\mathscr{F}' \subseteq \mathscr{F}$ ought to satisfy the following property: "if every collection of p or less members of $\mathscr{F}'$ have q elements in common, then $\mathscr{F}'$ has total intersection of cardinality at least q." This leads to the concept of the *(p, q)-Helly property* for general families of sets. Such families have been studied in [10]. In this work, we investigate the (p, q)-Helly property applied to the family of cliques of a graph, naturally conducting to the *(p, q)-clique-Helly graphs*. We describe a characterization for this class and a recognition algorithm based on it. The algorithm terminates within polyniomial time, for fixed p, q. Note that such an algorithm does not follow directly from the definition of the class. Further, we prove that recognizing (p, q)-clique-Helly graphs is NP-hard, whenever p or q is not fixed.

2 Preliminaries

Let G be a graph. A vertex $w \in V(G)$ is *universal* when w is adjacent to every other vertex of G. If $S \subseteq V(G)$, then we denote by $G[S]$ the subgraph of G induced by S. A subgraph H of G is a *spanning subgraph* of G when $V(H) = V(G)$. A *complete set* is a subset of pairwise adjacent vertices. A *clique* is a maximal complete set.

Let $\mathscr{F}$ be a subfamily of cliques of G. The *clique subgraph induced by $\mathscr{F}$* in G, denoted by $G_c[\mathscr{F}]$, is the subgraph of G formed exactly by the vertices and edges belonging to the cliques of $\mathscr{F}$.

The *core* of a family of sets $\mathscr{F}$ is defined as $core(\mathscr{F}) = \cap_{S \in \mathscr{F}} S$. We also define $V(\mathscr{F}) = \cup_{S \in \mathscr{F}} S$.

We say that a set S is a *q-set* when $|S| = q$, a *q^--set* when $|S| \leq q$, and a *q^+-set* when $|S| \geq q$. This notation will also be applied to families, cores, complete sets, cliques, etc.

The next definition is a first step towards the definition of the (p, q)-Helly property:

Definition 1. *Let $p \geq 1$ and $q \geq 0$. We say that a family of sets $\mathscr{F}$ is (p, q)-intersecting when every p^--subfamily $\mathscr{F}' \subseteq \mathscr{F}$ has a q^+-core.*

By the above definition, if $\mathscr{F}$ is (p, q)-intersecting (for $p > 1$ and $q > 0$) then it is also $(p - 1, q)$-intersecting and $(p, q - 1)$-intersecting.

Now, let us state the (p, q)-Helly property for families of sets in general, as in [18, 19]. This definiton is a generalization of the usual Helly property, which corresponds to the case $p = 2, q = 1$.

Definition 2. *Let $p \geq 1$ and $q \geq 0$. We say that a family of sets $\mathscr{F}$ satisfies the (p, q)-Helly property when every (p, q)-intersecting subfamily $\mathscr{F}' \subseteq \mathscr{F}$ has a q^+-core. In this case, we also say that $\mathscr{F}$ is (p, q)-Helly.*

It is easy to see that if a family of sets $\mathscr{F}$ is (p, q)-Helly then $\mathscr{F}$ is also $(p + 1, q)$-Helly.

The following theorem gives a characterization of (p, q)-Helly families of sets:

Theorem 1. *[10] Let $p > 1$ and $q > 0$. A family of sets $\mathscr{F}$ is (p, q)-Helly if and only if for every $(p + 1)$-family $\mathcal{Q}$ of q-subsets of $V(\mathscr{F})$, the subfamily $\{S \in \mathscr{F} \mid S$ contains at least p members of $\mathcal{Q}\}$ has a q^+-core.*

Now let us apply Definition 2 to the family of cliques of a graph:

Definition 3. *Let $p \geq 1$ and $q \geq 0$. We say that a graph G is (p, q)-clique-Helly when its family of cliques is (p, q)-Helly.*

In the remainder of this work, we will assume that $p > 1$ and $q > 0$, unless otherwise stated.

According to the definition above, $(2, 1)$-clique-Helly graphs are exactly the clique-Helly graphs. A characterization of $(2, 2)$-clique-Helly graphs by means of clique-Helly graphs was described in [7].

A first characterization of (p, q)-clique-Helly graphs is a direct consequence of Theorem 1:

Observation 1. *Let $p > 1$ and $q > 0$. A graph G is (p, q)-clique-Helly if and only if for every $(p + 1)$-family $\mathcal{Q}$ of q-complete sets contained in a common clique C of G, the subfamily of cliques of G that contain at least p members of $\mathcal{Q}$ has a q^+-core.* $\square$

However, we will present in Section 3 a more useful characterization than the above one, in the sense that it will lead to a polynomial-time recognition algorithm for fixed p, q.

Let us analyze some containment relations involving classes of (p, q)-clique-Helly graphs by means of the following example:

Example 1. Define the graph $G_{p,q}$ as follows: $V(G_{p,q})$ is formed by a $(q - 1)$-complete set Q, a p-complete set $Z = \{z_1, \ldots, z_p\}$, and a p-independent set $W = \{w_1, \ldots, w_p\}$. Furthermore, there exist the edges (z_i, w_j), for $i \neq j$, and the edges (q, x), for $q \in Q$ and $x \in Z \cup W$.

The graph $G_{p,q}$ contains exactly $p + 1$ cliques of size $p + q - 1$ each: $Q \cup \{z_1, \ldots, z_p\}$ and $Q \cup (Z \backslash \{z_i\}) \cup \{w_i\}$, for $1 \leq i \leq p$.

Observe that $G_{p,q}$ is (p, q)-clique-Helly, but it is not $(p - 1, q)$-clique-Helly. Therefore, $G_{p,q}$ is (t, q)-clique-Helly for $t \geq p$, and not (t, q)-clique-Helly for $t < p$.

Moreover, $G_{p+1,q}$ is not (p, q)-clique-Helly, but it is (p, t)-clique-Helly for any $t \neq q$. Consequently, for distinct q and t, (p, q)-clique-Helly graphs and (p, t)-clique-Helly graphs are incomparable classes. $\square$

To conclude this section, we relate K_r-free graphs and (p,q)-clique-Helly graphs. (A graph G is K_r-free when the size of a maximum clique of G is at most $r-1$.) An interesting fact derived from Definition 3 is that every K_r-free graph is (p,q)-clique-Helly for $p+q \geq r$.

First, we need the following lemma:

Lemma 1. *Let $\mathcal{Q}$ be a $(p+1)$-family of q-complete sets of a graph G. If all sets of $\mathcal{Q}$ are contained in a common $(p+q-1)^-$-complete set of G, then the cliques of G that contain at least p members of $\mathcal{Q}$ have a q^+-core.*

Proof. Let $\mathcal{Q}$ be a $(p+1)$-family of q-complete sets contained in a $(p+q-1)^-$-complete set C, and let $\mathcal{F}$ be the subfamily of cliques of G that contain at least p members of $\mathcal{Q}$. Observe that if a vertex x of C belongs to two members of $\mathcal{Q}$, then x belongs to all the cliques of $\mathcal{F}$. We will show that there exist at least q vertices in C belonging simultaneously to at least two members of $\mathcal{Q}$, which proves the lemma.

Suppose the contrary. Thus at most $q-1$ vertices of C belong simultaneously to more than one member of $\mathcal{Q}$. Assume initially that $|C| = p+q-1$. Then at least $p+q-1-(q-1) = p$ vertices of C have the property of belonging to exactly one member of $\mathcal{Q}$. Let X be the set formed by such vertices, where $|X| = p+r, 0 \leq r \leq q-1$. Observe that every member of $\mathcal{Q}$ must contain at least $r+1$ vertices belonging to X. This implies $|X| \geq (p+1)(r+1) = p+r+pr+1 > p+r$, a contradiction.

If C contains less than $p+q-1$ vertices, the same argument above could be used. $\square$

Theorem 2. *Let p,q,r such that $p > 1$, $q > 0$, $r > 1$ and $p+q \geq r$. If G is a K_r-free graph then G is (p,q)-clique-Helly.*

Proof. Let $\mathcal{Q}$ be a $(p+1)$-family of q-complete sets contained in a clique C of G. By Observation 1, we have to show that the subfamily $\mathcal{F}$ of cliques of G that contain at least p members of $\mathcal{Q}$ has a q^+-core. Since G is $K_{(p+q)}$-free, all members of $\mathcal{Q}$ are contained in a common $(p+q-1)^-$-complete set of G. By Lemma 1, $\mathcal{F}$ has a q^+-core. $\square$

3   The Characterization

The following definitions and lemmas will be useful.

Definition 4. *Let G be a graph and C a p-complete set of G. The p-expansion relative to C is the subgraph of G induced by the vertices w such that w is adjacent to at least $p-1$ vertices of C.*

We remark that the p-expansion for $p = 3$ has been used for characterizing clique-Helly graphs [11, 17]. It is clear that constructing a p-expansion relative to a given p-complete set can be done in polynomial time.

Lemma 2. *Let G be a graph, C a p-complete set of it, H the p-expansion of G relative to C, and $\mathcal{C}$ the subfamily of cliques of G that contain at least $p-1$ vertices of C. Then $G_c[\mathcal{C}]$ is a spanning subgraph of H.*

Proof. We have to show that $V(G_c[\mathscr{C}]) = V(H)$. Let $v \in V(H)$. Then v is adjacent to at least $p - 1$ vertices of C. Hence, v together with those $p - 1$ vertices form a p-complete set, which is contained in a clique that contains at least $p - 1$ vertices of C. Therefore, $v \in V(G_c[\mathscr{C}])$. Now, consider $v \in V(G_c[\mathscr{C}])$. Then v belongs to some clique containing $p - 1$ vertices of C. That is, v is adjacent to at least $p - 1$ vertices of C, and hence $v \in V(H)$. Consequently, $V(G_c[\mathscr{C}]) = V(H)$. Furthermore, both H and $G_c[\mathscr{C}]$ are subgraphs of G, but H is induced. Thus $E(G_c[\mathscr{C}]) \subseteq E(H)$. $\square$

Definition 5. *Let G be a graph. The graph $\Phi_q(G)$ is defined in the following way: the vertices of $\Phi_q(G)$ correspond to the q-complete sets of G, two vertices being adjacent in $\Phi_q(G)$ if the corresponding q-complete sets in G are contained in a common clique.*

Notice that $\Phi_q(G)$ can be constructed in polynomial time, for fixed q. We remark that Φ_q is precisely the operator $\Phi_{q,2q}$ described in [15], p.136, and the graph $\Phi_2(G)$ is the *edge clique graph* of G, introduced in [1].

An interesting property of Φ_q is that it preserves the subfamily of q^+-cliques of G:

Lemma 3. (Clique Preservation Property) *Let G be a graph. Then there exists a bijection φ_q between the subfamily of q^+-cliques of G and the family of cliques of $\Phi_q(G)$.*

Proof. Let C be a q^+-clique of G, and let $c = |C|$. Consider all the q-complete sets of G contained in C. These sets clearly correspond to a $\binom{c}{q}$-complete set C' of $\Phi_q(G)$. Assume that C' is not maximal. Then there exists $x \in V(\Phi_q(G))$, $x \notin C'$, such that x is adjacent to all the vertices of C'. But x corresponds to a q-complete set Q of G such that for every q-complete set $Q_1 \subseteq C$, both Q and Q_1 are contained in a common q^+-clique of G. This implies that every vertex v of Q is adjacent to every vertex $w \neq v$ of C. Since $x \notin C'$, Q must necessarily contain at least one vertex not belonging to C. In other words, C is not maximal, a contradiction. Hence, C' is a clique of $\Phi_q(G)$ and $C' = \varphi_q(C)$.

Conversely, let C' be a clique of $\Phi_q(G)$ and $\mathscr{F}$ be the family of q-complete sets of G corresponding to the vertices of C'. Since any two vertices of C' are adjacent, any two complete sets of $\mathscr{F}$ are contained in a common q^+-clique of G. Hence, the union of the q-complete sets of $\mathscr{F}$ is a q^+-complete set C of G.

Suppose by contradiction that C is not maximal. Thus, there exists a vertex $u \notin C$ which is adjacent to all the vertices of C. Consider $v_1, v_2, ..., v_{q-1} \in C$. It is clear that $Q = \{u, v_1, v_2, ..., v_{q-1}\}$ is a q-complete set of G, and for every Q_1 in $\mathscr{F}$, both Q and Q_1 are contained in a common q^+-clique of G. Since $u \notin C$, $Q \notin \mathscr{F}$, and this means that Q corresponds to a vertex $x \in V(\Phi_q(G))$ such that $x \notin C'$ and x is adjacent to all the vertices of C'. This implies that C' is not maximal, a contradiction. $\square$

It is worth remarking that the above lemma was already shown for the case $q = 2$ in [1, 7].

The following definition is possible due to the Clique Preservation Property:

Definition 6. *Let G be a graph. If $\mathscr{F}$ is a subfamily of q^+-cliques of G, define $\varphi_q(\mathscr{F}) = \{\varphi_q(C) \mid C \in \mathscr{F}\}$. If $\mathscr{C}$ is a subfamily of cliques of $\Phi_q(G)$, define $\varphi_q^{-1}(\mathscr{C}) = \{\varphi_q^{-1}(C) \mid C \in \mathscr{C}\}$.*

Lemma 4. *Let G be a graph, $\mathscr{F}$ a subfamily of q^+-cliques of it, $\mathscr{C} = \varphi_q(\mathscr{F})$, and $H = \Phi_q(G)$. Then $H_c[\mathscr{C}]$ contains a universal vertex if and only if $G_c[\mathscr{F}]$ contains q universal vertices.*

Proof. If $H_c[\mathscr{C}]$ contains a universal vertex x, then every clique of $\mathscr{F}$ contains the q-complete set of G that corresponds to x, that is, $G_c[\mathscr{F}]$ contains q universal vertices. Conversely, if $G_c[\mathscr{F}]$ contains q universal vertices forming a q-complete set Q of G, then every clique of $\mathscr{C}$ contains the vertex of H that corresponds to Q, that is, $H_c[\mathscr{C}]$ contains a universal vertex. $\square$

Lemma 5. *Let C be a $(p + 1)$-complete set of a graph G, and let $\mathscr{C}$ be a p^--subfamily of cliques of G such that every clique of $\mathscr{C}$ contains at least p vertices of C. Then $core(\mathscr{C}) \neq \emptyset$.*

Proof. Trivial. $\square$

We now are able to present a characterization for (p, q)-clique-Helly graphs. The cases $p = 1$ and $p > 1$ will be dealt with separately.

Theorem 3. *Let G be a graph, and let W be the union of the q^+-cliques of G. Then G is a $(1, q)$-clique-Helly graph if and only if $G[W]$ contains q universal vertices.*

Proof. Assume that G is a $(1, q)$-clique-Helly graph. Consider the subfamily $\mathscr{F}$ formed by the q^+-cliques of G.

If $w \in W$, then w clearly belongs to a q^+-clique of G. This implies that $w \in V(G_c[\mathscr{F}])$. On the other hand, if $w' \in V(G_c[\mathscr{F}])$, then w' belongs to a q^+-clique of G, and therefore $w' \in W$. This shows that $G_c[\mathscr{F}]$ is a spanning subgraph of $G[W]$.

Since $\mathscr{F}$ is $(1, q)$-intersecting by hypothesis, it has a q^+-core. This means that $G_c[\mathscr{F}]$ contains (at least) q universal vertices. Hence, $G[W]$ contains q universal vertices.

Conversely, assume that $G[W]$ contains q universal vertices forming a q-complete set Q. Let $\mathscr{F} = \{C_1, \ldots, C_k\}$ be a $(1, q)$-intersecting subfamily of cliques of G. Then $|C_i| \geq q$, that is, every $w \in C_i$ is contained in a q-complete set of G, for $i = 1, \ldots, k$. This implies that every C_i is an induced subgraph of $G[W]$. Therefore, every $u \in Q$ is adjacent to all the vertices of $C_i \setminus \{u\}$. By the maximality of C_i, it contains all the vertices $u \in Q$, for $i = 1, \ldots, k$. Hence, $\mathscr{F}$ has a q^+-core, as required. $\square$

Theorem 4. *Let $p > 1$ be an integer. Then a graph G is (p, q)-clique-Helly if and only if every $(p + 1)$-expansion of $\Phi_q(G)$ contains a universal vertex.*

Proof. Suppose that G is a (p, q)-clique-Helly graph and there exists a $(p + 1)$-expansion T, relative to a $(p+1)$-complete set C of $\Phi_q(G)$, such that T contains no universal vertex.

Let $\mathscr{C}$ be the subfamily of cliques of $H = \Phi_q(G)$ that contain at least p vertices of C. Let $\mathscr{F} = \varphi_q^{-1}(\mathscr{C})$. Consider a p^--subfamily $\mathscr{F}' \subseteq \mathscr{F}$. Let $\mathscr{C}' = \varphi_q(\mathscr{F}')$. By Lemma 5, $core(\mathscr{C}') \neq \emptyset$. That is, $H_c[\mathscr{C}']$ contains a universal vertex. This implies, by Lemma 4, that $G_c[\mathscr{F}']$ contains q universal vertices. Thus, $\mathscr{F}'$ has a q^+-core, that is, $\mathscr{F}$ is (p,q)-intersecting. Since G is (p,q)-clique-Helly, we conclude that $\mathscr{F}$ has a q^+-core and $G_c[\mathscr{F}]$ contains q universal vertices. By using Lemma 4 again, $H_c[\mathscr{C}]$ contains a universal vertex. Moreover, by Lemma 2, $H_c[\mathscr{C}]$ is a spanning subgraph of T. However, T contains no universal vertex. This is a contradiction. Therefore, every $(p+1)$-expansion of $H = \Phi_q(G)$ contains a universal vertex.

Conversely, assume by contradiction that G is not (p,q)-clique-Helly. Let $\mathscr{F} = \{C_1, \ldots, C_k\}$ be a minimal (p,q)-intersecting subfamily of cliques of G which does not have a q^+-core. Clearly, $k > p$.

The minimality of $\mathscr{F}$ implies that there exists a q-subset $Q_i \subseteq core(\mathscr{F}\backslash\{C_i\})$, for $i = 1, \ldots, k$. It is clear that $Q_i \nsubseteq C_i$. Moreover, every two distinct Q_i, Q_j are contained in a common clique, since $k \geq 3$. Hence the sets $Q_1, Q_2, \ldots, Q_{p+1}$ correspond to a $(p+1)$-complete set C in $\Phi_q(G)$.

Let $\mathscr{C}$ be the subfamily of cliques of $H = \Phi_q(G)$ that contain at least p vertices of C. Let $\mathscr{C}' = \varphi_q(\mathscr{F})$. Since every $C_i \in \mathscr{F}$ contains at least p sets from $Q_1, Q_2, \ldots, Q_{p+1}$, it is clear that the clique $\varphi_q(C_i) \in \mathscr{C}'$ contains at least p vertices of C. Therefore, $\varphi_q(C_i) \in \mathscr{C}$, for $i = 1, \ldots, k$.

Let T be the $(p+1)$-expansion of H relative to C. By Lemma 2, $H_c[\mathscr{C}]$ is a spanning subgraph of T. Therefore, $Q \subseteq V(T)$, for every $Q \in \mathscr{C}$. In particular, $V(\varphi_q(C_i)) \subseteq V(T)$, for $i = 1, \ldots, k$. By hypothesis, T contains a universal vertex x. Then x is adjacent to all the vertices of $\varphi_q(C_i)\backslash\{x\}$, for $i = 1, \ldots, k$. This implies that $\varphi_q(C_i)$ contains x, otherwise $\varphi_q(C_i)$ would not be maximal. Thus, $core(\mathscr{C}') \neq \emptyset$ and $H_c[\mathscr{C}']$ contains a universal vertex. By Lemma 4, $G_c[\mathscr{F}]$ contains q universal vertices, that is, $\mathscr{F}$ has a q^+-core. This contradicts the assumption for $\mathscr{F}$. Hence, G is a (p,q)-clique-Helly graph. $\square$

4   Complexity Aspects

Let p and q be fixed positive integers. If $p = 1$, testing whether the subgraph induced by the union of the q^+-cliques of G contains q universal vertices (Theorem 3) can be easily done in polynomial time, by considering the subgraph of G induced by the union of all the q-complete subsets of G.

If $p > 1$, testing the existence of a universal vertex in every $(p+1)$-expansion of $\Phi_q(G)$ (Theorem 4) can also be done in polynomial time, since the number of such $(p+1)$-expansions is $O(|V(G)|^{q(p+1)})$. Thus:

Corollary 1. *For fixed positive integers p, q, there exists a polynomial time algorithm for recognizing (p,q)-clique-Helly graphs.* $\square$

Now we will show that when p or q is not fixed, the problem of deciding whether a given graph is (p,q)-clique-Helly is NP-hard. We first recall the following NP-complete problems [8]:

SATISFIABILITY: Given a boolean expression $\mathscr{E}$ in the conjunctive normal form, is there a truth assignment for $\mathscr{E}$?

CLIQUE: Given a graph G and a positive integer k, is there a k^{+}-clique in G?

The NP-hardness of CLIQUE can be proved by a transformation from SATISFIABILITY [8]: given a boolean expression $\mathscr{E}$ with m clauses in the conjunctive normal form, first construct the graph $\mathscr{G}(\mathscr{E})$ by defining a vertex for each occurrence of a literal in $\mathscr{E}$, and by creating an edge between two vertices if and only if the corresponding literals lie in distinct clauses and one is not the negation of the other. In addition, set $k = m$. Then the following applies:

Fact 1. [8] *The boolean expression $\mathscr{E}$ with m clauses in the conjunctive normal form is satisfiable if and only if the graph $\mathscr{G}(\mathscr{E})$ contains an m-clique.*

Consider now the recognition of (p, q)-clique-Helly graphs, for p or q variable. Let us first show the NP-hardness proof when p is fixed and q is variable:

Theorem 5. *Let p be a fixed positive integer. Given a graph G and a positive integer q, the problem of deciding whether G is (p, q)-clique-Helly is NP-hard.*

Proof. Transformation from CLIQUE. Given a graph G and a positive integer k, construct the graph G' by adding $2p + 2$ new vertices forming a $(p+1)$-complete set $Z = \{z_1, z_2, \ldots, z_{p+1}\}$ and a $(p+1)$-independent set $W = \{w_1, w_2, \ldots, w_{p+1}\}$. Add the edges (z_i, w_j), for $i \neq j$, and the edges (v, u), for $v \in V(G)$ and $u \in Z \cup W$. The construction of G' is completed. Figure 1 shows the construction, where non-edges between Z and W are represented by dashed lines linking z_i to w_i.

Define $q = k + 1$. We will show that G contains a $(q - 1)$-clique if and only if G' is not (p, q)-clique-Helly. Assume first that G contains a $(q - 1)$-clique C. Consider the following $p + 1$ cliques of G':

$$C \cup \{w_j\} \cup (Z \setminus \{z_j\}), \text{ for } 1 \leq j \leq p + 1.$$

These cliques are (p, q)-intersecting, but do not have a q^{+}-core. Therefore, G' is not (p, q)-clique-Helly.

Conversely, assume that the cliques of G have size at most $q - 2$. Since $G'[Z \cup W]$ is $K_{(p+2)}$-free, cliques of G' have size at most $(q-2)+(p+1) = q+p-1$, that is, G' is $K_{(p+q)}$-free. By Theorem 2, G' is (p, q)-clique-Helly, as desired. $\square$

Now we prove the NP-hardness in the case where q is fixed and p is variable:

Theorem 6. *Let q be a fixed positive integer. Given a graph G and a positive integer p, the problem of deciding whether G is (p, q)-clique-Helly is NP-hard.*

Proof. Transformation from SATISFIABILITY. Given a boolean expression $\mathscr{E} = (\mathscr{E}_1, \ldots, \mathscr{E}_m)$ in the conjunctive normal form, let us construct a graph G', as follows.

First, construct the graph $\mathscr{G}(\mathscr{E})$ above described. Define $\mathscr{V}_i$ as the subset of vertices of $V(\mathscr{G}(\mathscr{E}))$ corresponding to the ocurrences of literals in clause $\mathscr{E}_i$, $1 \leq i \leq m$.

Next, add m new vertices, one for each $\mathscr{E}_i$, forming an m-independent set $W = \{w_1, w_2, ..., w_m\}$. For $i = 1, \ldots, m$, add the edges (w_i, v) where $v \in V(\mathscr{G}(\mathscr{E}))$ and $v \notin \mathscr{V}_i$.

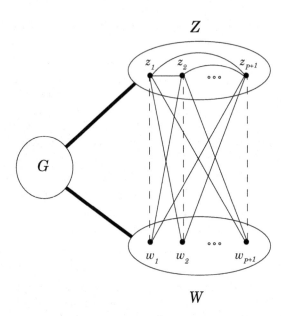

Fig. 1. The graph G' for Theorem 5

Finally, add $q - 1$ new vertices forming a $(q - 1)$-complete subset $Z = \{z_1, ..., z_{q-1}\}$, and add the edges (z, v), for $z \in Z$ and $v \in W \cup \mathcal{G}(\mathcal{E})$. The construction of G' is finished. Clearly, every vertex of Z is universal in G', and every clique of G' contains these $q - 1$ vertices. Figure 2 shows a scheme of the construction, where the dashed lines mean that w_i is not adjacent to the vertices of $\mathcal{V}_i$, for $1 \le i \le m$.

Set $p = m - 1$. We will show that $\mathcal{E}$ is satisfiable if and only if G' is not (p, q)-clique-Helly. Assume first that $\mathcal{E}$ is satisfiable. By Fact 1, $\mathcal{G}(\mathcal{E})$ contains a $(p+1)$-clique $K = \{v_1, v_2, ..., v_{p+1}\}$, where $v_j \in \mathcal{V}_j$. By the construction of G', it contains the $(p+q)$-cliques

$$K_j = (K \backslash \{v_j\}) \cup \{w_j\} \cup Z, \text{ for } 1 \le j \le p+1.$$

These $p + 1$ cliques are (p, q)-intersecting, but do not have a q^+-core. Thus, G' is not (p, q)-clique-Helly.

Conversely, assume that $\mathcal{E}$ is not satisfiable. In this case, by Fact 1, $\mathcal{G}(\mathcal{E})$ is $K_{(p+1)}$-free. Thus, every clique of G' contains exactly a vertex of W, since for any p^--subset $S \subseteq V(\mathcal{G}(\mathcal{E}))$, there exists at least one vertex of W adjacent to all the vertices of S.

Let $\mathcal{Q}$ be a $(p+1)$-family of q-complete sets contained in a common clique of G', and let $\mathcal{F}$ be the subfamily of cliques of G' that contain at least p members of $\mathcal{Q}$. By Observation 1, we need to prove that $\mathcal{F}$ has a q^+-core. (Recall that $\mathcal{F}$ contains Z, that is, $|core(\mathcal{F})| \ge q - 1$.)

If $V(\mathcal{Q})$ is contained in a $(p+q-1)^-$-complete set of G', Lemma 1 guarantees that $\mathcal{F}$ has a q^+-core, and nothing remains to prove. Hence, let us assume that $V(\mathcal{Q})$ is a $(p+q)^+$-complete set of G'.

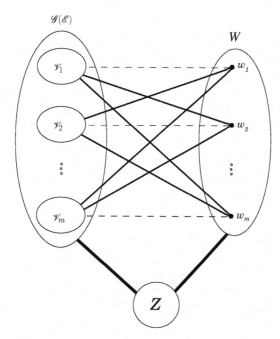

Fig. 2. The graph G' for Theorem 6

Since $\mathcal{G}(\mathcal{E})$ is $K_{(p+1)}$-free, a maximum clique C' of G' is of size at most $(q-1)+1+p = p+q$. Therefore, $V(\mathcal{Q})$ is in fact a $(p+q)$-clique of G'.

Write $C' = V(\mathcal{Q})$. Then C' is of the form $C' = Z \cup \{w_k\} \cup P$, where $k \in \{1,\ldots,p+1\}$ and P is a p-complete set contained in $V(\mathcal{G}(\mathcal{E}))$. It is clear that the ocurrences of literals corresponding to the vertices of P lie in distinct clauses of $\mathcal{E}$. This means that there is exactly one vertex $v \in P \cap \mathcal{V}_j$, for every $j \in \{1,\ldots,p+1\}\setminus\{k\}$. Thus, write $P = \{v_1,\ldots,v_{k-1},v_{k+1},\ldots,v_{p+1}\}$, where $v_j \in \mathcal{V}_j$ for $j \in \{1,\ldots,p+1\}\setminus\{k\}$.

Let $v \in \{w_k\} \cup P$. If v belongs simultaneously to two members of $\mathcal{Q}$, then v belongs to all the members of $\mathcal{F}$. In other words, $Z \cup \{v\}$ is a q-core of $\mathcal{F}$, as desired. Therefore, it only remains to analyze the case in which

$$\mathcal{Q} = \{ Z \cup \{v_j\} \mid 1 \le j \le p+1, j \ne k \} \cup \{ Z \cup \{w_k\} \}.$$

In this case, let us show that w_k belongs to every member of $\mathcal{F}$. Suppose that some $C'' \in \mathcal{F}$ does not contain w_k. Recall that C'' contains a vertex $w_j, j \ne k$. Moreover, $v_j \in P$ is not adjacent to w_j. This implies that C'' cannot contain the member of $\mathcal{Q}$ which v_j belongs to. Since C'' does not contain w_k, C'' can neither contain the member of $\mathcal{Q}$ which w_k belongs to. A contradiction arises, since C'' should contain p members of $\mathcal{Q}$. Thus, w_k indeed belongs to every member of $\mathcal{F}$, and $Z \cup \{w_k\}$ is a q-core of $\mathcal{F}$, as desired. $\square$

From Theorems 5 and 6, we conclude:

Corollary 2. *The recognition of (p,q)-clique-Helly graphs, for p or q variable, is NP-hard.* $\square$

5 Conclusions

The study of graph classes and their characterizations is an important subject in graph theory in general, and in particular in graph algorithms. See [5]. In this paper, we have described and characterized (p, q)-clique-Helly graphs, a class arising from a generalization of the Helly property and which generalizes clique-Helly graphs. The characterization leads to a recognition algorithm for (p, q)-clique-Helly graphs whose complexity is polynomial, whenever p and q are fixed. In contrast, we have shown that the recognition problem is NP-hard for arbitrary p or q.

References

1. M. O. Albertson and K. L. Collins. Duality and perfection for edges in cliques. *Journal of Combinatorial Theory Series B* 36 (1984) 298–309.
2. C. Berge. *Graphs and Hypergraphs*. North-Holland, Amsterdam, 1973.
3. C. Berge. *Hypergraphs*. Elsevier Science Publishers B. V., Amsterdam, 1989.
4. C. Berge and P. Duchet. A generalization of Gilmore's theorem. *Recent Advances in Graph Theory* (M. Fiedler, ed.), Acad. Praha, Prague 1975, 49–55.
5. A. Brandstädt , V. B. Lee and J. Spinrad. *Graph Classes: A Survey*. SIAM Monographs on Discrete Mathematics and Applications, vol. 3, SIAM, Philadelphia, 1999.
6. P. L. Butzer, R. J. Nessel, and E. L. Stark. Eduard Helly (1884-1943): In memoriam. *Resultate der Mathematik* 7 (1984).
7. M. R. Cerioli. Edge-clique Graphs (in Portuguese). PhD Thesis, Federal University of Rio de Janeiro, 1999.
8. S. A. Cook. The complexity of theorem-proving procedures. Proc. 3rd Ann. ACM Symp. on Theory of Computing (1971), New York, 151-158.
9. L. Danzer, B. Grünbaum, and V. L. Klee. Helly's theorem and its relatives. *Proc. Symp. on Pure Math AMS* Vol.7 (1963) 101–180.
10. M. C. Dourado, F. Protti, and J. L. Szwarcfiter. Complexity Aspects of Generalized Helly Hypergraphs. Submitted.
11. F. F. Dragan. *Centers of Graphs and the Helly Property* (in Russian). Doctoral Thesis, Moldava State University, Chisinău, 1989.
12. M. C. Golumbic and R. E. Jamison. The edge intersection graphs of paths in a tree. *J. Comb. Theory Series B* 38 (1985) 8–22.
13. E. Helly. Ueber Mengen konvexer Koerper mit gemeinschaftlichen Punkter, Jahresber. *Math.-Verein.* 32 (1923) 175–176.
14. E. Prisner. Hereditary clique-Helly graphs. *Journal of Combinatorial Mathematics and Combinatorial Computing* 14 (1993) 216–220.
15. E. Prisner. *Graph Dynamics*. Pitman Research Notes in Mathematics 338, Longman, London (1995).
16. F. Protti and J. L. Szwarcfiter. Clique-inverse graphs of K_3-free and K_4-free graphs. *Journal of Graph Theory* 35 (2000) 257–272.
17. J. L. Szwarcfiter. Recognizing clique-Helly graphs. *Ars Combinatoria* 45 (1997) 29–32.
18. Zs. Tuza. Extremal bi-Helly families. *Discrete Mathematics* 213 (2000) 321–331.
19. V. I. Voloshin. On the upper chromatic number of a hypergraph. *Australas. J. Combin.* 11 (1995) 25–45.

Edge-Connectivity Augmentation
and Network Matrices*

Michele Conforti[1], Anna Galluccio[2], and Guido Proietti[2,3]

[1] Dipartimento di Matematica, Università di Padova,
via Belzoni 7, 35131 Padova, Italy
conforti@math.unipd.it
[2] Istituto di Analisi dei Sistemi ed Informatica, CNR,
Viale Manzoni 30, 00185 Roma, Italy
galluccio@iasi.rm.cnr.it
[3] Dipartimento di Informatica, Università di L'Aquila,
Via Vetoio, 67010 L'Aquila, Italy
proietti@di.univaq.it

Abstract. We study the following NP-hard graph augmentation problem: Given a weighted graph G and a connected spanning subgraph H of G, find a minimum weight set of edges of G to be added to H so that H becomes 2-edge-connected. We provide a formulation of the problem as a set covering problem, and we analyze the conditions for which the linear programming relaxation of our formulation always gives an integer solution. This yields instances of the problem that can be solved in polynomial time. As we will show in the paper, these particular instances have not only theoretical but also practical interest, since they model a wide range of survivability problems in communication networks.

1 Introduction

Several network design problems have been formulated as integer programming problems, and in many cases mathematical programming techniques revealed very effective to understand and to solve these problems. These techniques had a leading role in designing efficient approximation algorithms for network problems, but they were also crucial in enlightening the structural properties of these problems. In particular, they help to understand where the hardness of a problem lies by characterizing those instances which are polynomial-time solvable.

Here, we consider the problem of strengthening the edge-connectivity of an existing network to fulfill increasing communication reliability requirements. Such a strengthening is realized through the addition of a set of new links, in such a way that the original degree of (either edge or node)-connectivity of the network is increased. Generally, these additional links must be selected from a set of *potential* links, where each potential link has a certain cost, based on some standard measure (e.g., the set-up cost). Therefore, besides from increasing the

---

* This work has been partially supported by the Research Project GRID.IT, funded by the Italian Ministry of Education, University and Research, and by the CNR-Agenzia 2000 Program, under Grant No. CNRC00CAB8.

J. Hromkovič, M. Nagl, and B. Westfechtel (Eds.): WG 2004, LNCS 3353, pp. 355–364, 2004.

connectivity degree, one wants also to minimize some objective function of these link costs.

From a theoretical point of view (and for the edge-connectivity case, which is of interest for this paper), this gives rise to a family of *augmentation problems* on graphs [1]. Among them, one of the most basic problems can be formulated as follows: Given an undirected, 2-edge-connected and real weighted graph $G = (V, F)$, and given a connected spanning subgraph $H = (V, E)$ of G, find a minimum-weight subset of $F \setminus E$, say $\text{AUG}_2(G, H)$, such that $H' = (V, E \cup \text{AUG}_2(G, H))$ is 2-edge-connected.

This problem is NP-hard [4], and thus most of the research in the past focused on the design of approximation algorithms for solving it. To this respect, efficient approximation algorithms are known, with performance ratio 2 [4, 10]. For the unweighted case, improving the approximation ratio below 2 has been a long standing open problem. Just recently, Nagamochi [12] developed a $(1.875 + \epsilon)$-approximation algorithm, for any constant $\epsilon > 0$, afterwards improved to $3/2$ by Even *et al.* [2]. Analogous versions of augmentation problems for node-connectivity and for directed graphs have been widely studied, and we refer the interested reader to the following comprehensive papers [3, 11].

The problem of characterizing polynomially solvable cases of the 2-edge-connectivity augmentation problem has been deeply investigated as well. The polynomiality of the problem can be achieved by giving additional constraints on the structure of the pair (G, H). First, Eswaran and Tarjan [1] proved that $\text{AUG}_2(G, H)$ can be found in polynomial time if G is complete and all edges in G have weight 1, i.e., all potential links between sites may be activated at the same cost. Afterwards, Watanabe and Nakamura extended this result to any desired edge-connectivity value [15], and faster algorithms in this case have been proposed in [5]. From another perspective, the structure of H can be equally relevant in order to have polynomial time algorithms. For example, $\text{AUG}_2(G, H)$ can be found in polynomial time if G is any weighted graph and H is a spanning tree of G which can be rooted in a node $r \in V$ in such a way that for every edge $uv \in F \setminus E$, either u is an ancestor of v, or v is an ancestor of u in the rooted tree [6].

In this paper we move one significant step towards the characterization of the intrinsic complexity of the 2-edge-connectivity augmentation problem. In particular, we formulate the problem as a *set covering* problem on special matrices defined by the pair (G, H), and we provide sufficient conditions on (G, H) so that the linear programming relaxation of our formulation always gives an integer solution. This yields instances of the problem that can be solved efficiently. As we will show in the paper, these particular instances have not only theoretical but also practical interest, since they model several survivability problems in communication networks. Specifically, we shall provide an application in many-to-one networks (i.e., networks where data flow from multiple sources towards a single sink).

The paper is organized as follows: in Section 2, we recall some basic results and definitions from integer programming and graph theory. In Section 3, we formulate the 2-edge-connectivity augmentation problem as a set covering problem. In Section 4, we provide sufficient conditions that guarantee the polynomial solvability of the problem. Then, in Section 5, we explain in detail how to transform the 2-edge-connectivity augmentation problem into a min-cost circulation problem in a directed graph and, as a consequence, we provide an effective polynomial time algorithm to solve the 2-edge-connectivity augmentation problem in a large class of graphs. Finally, in Section 6 we show the practical impact of our results, by providing a realistic scenario in which they find an application.

2   Preliminaries

Let A be a $(0,1)$-matrix with n rows and m columns and let $w \in \mathbb{R}^m$ and $b \in \mathbb{R}^n$. An *integer programming (IP)* problem is formulated as follows:

$$
\begin{aligned}
& \min w^T x \\
& \text{s.t. } Ax \geq b \\
& \quad x \text{ integer.}
\end{aligned}
\tag{1}
$$

Being $\mathcal{P} = \{x \in \mathbb{R}^m : Ax \geq b, x \geq 0\}$ a polyhedron, an IP problem consists in finding an integral vector $x^* \in \mathcal{P}$ that minimizes the linear function $w^T x$ over the integral vectors of $\mathcal{P}$. This problem is NP-hard, in general, but some instances can be solved as *linear programming (LP)* problems, for all objective functions. For example, when the polyhedron $\mathcal{P}$ is integral, i.e., all of its extreme points have integer coordinates, the integrality constraint in (1) can be replaced by $x \geq 0$, and the optimum of (1) can be found by any polynomial time algorithm developed for linear programming.

A $(0, \pm 1)$-matrix A is *totally unimodular* (TU, for short) if $\det(A') = 0, \pm 1$ for every square submatrix A' of A. A classical result of Hoffman a Kruskal [9] shows that $\{x \in \mathbb{R}^m : Ax \geq b, x \geq 0\}$ is an integral polyhedron for every integral vector b if and only if A is a TU matrix.

The case when the elements of A, b and x in (1) are restricted to be 0 or 1, is usually known as the *set covering problem* and it is one of the most studied IP problems since many combinatorial optimization problems can be formulated in this form. The integrality of the set covering polyhedron $\{x \in \mathbb{R}^m : Ax \geq 1, x \geq 0\}$ is guaranteed by the weaker property of A of being *balanced*. A $(0,1)$-matrix M is balanced if and only if it contains no square submatrix of odd order whose row and column sums are all 2. Since the determinant of a square matrix of odd order whose row and column sums are all 2 is ± 2, it follows that every TU matrix is balanced. This containment in general is known to be strict.

We complete the section with some graph theory definitions. Let $G = (V, F)$ be an undirected graph with node set V and edge set F. Graph G is said to be *weighted* if it is given a real function $w : F \mapsto \mathbb{R}$. Given two vertices $v_0, v_k \in V$, a (simple) $v_0 v_k$-*path* in G is a sequence $\langle v_0, e_1, v_1, \ldots, e_k, v_k \rangle$, where v_i's are distinct vertices in V, and each e_i is an edge of F with endvertices v_{i-1} and v_i. A graph is *connected* if every pair of vertices is joined by a path. A graph G is *2-edge-connected* if the removal of any edge from G leaves G connected.

Given a node $v \in V$, the *star* of v in G, denoted by $\delta_G(v)$, is the set of edges $\{e_1, e_2, \ldots, e_k\}$ of F having v as endnode.

If $G = (V, A)$ is a *directed* graph (*digraph*, for short), then we denote by $\delta_G^+(v)$ ($\delta_G^-(v)$, respectively) the set of arcs in A leaving (entering, respectively) a node $v \in V$. Given two vertices $v_0, v_k \in V$, a $v_0 v_k$-path in G is a sequence $\langle v_0, a_1, v_1, \ldots, a_k, v_k \rangle$, where v_i's are distinct vertices in V, and each a_i is either an arc $v_{i-1} v_i$ or an arc $v_i v_{i-1}$ of A. A $v_0 v_k$-path in G is *directed* if every arc a_i along the path is oriented from v_{i-1} to v_i.

3   Formulating the Augmentation Problem

Given an undirected, 2-edge-connected, real weighted graph $G = (V, F)$, and given a connected spanning subgraph $H = (V, E)$ of G, the problem of finding a 2-*edge-connectivity augmentation* of H with respect to G asks for selecting a minimum weight set of edges in $F \setminus E$, denoted as $\text{AUG}_2(G, H)$, such that the graph $H' = (V, E \cup \text{AUG}_2(G, H))$ is 2-edge-connected.

Notice that if H is connected, then, without loss of generality, we can assume that H is a tree. Indeed, each 2-edge-connected component of H can be contracted into a single *vertex*. This transforms the graph G into a multigraph $\mathcal{G}$, and the graph H into a tree T, whose edges are the bridges of H. It is then easy to see that finding $\text{AUG}_2(G, H)$ is equivalent to finding $\text{AUG}_2(\mathcal{G}, T)$. Based on that, we will restrict ourselves to the problem of finding 2-edge-connectivity augmentation of trees.

Let then be given a real weighted graph $G = (V, F)$ and a tree $T = (V, E)$ defined over the same vertex set. Let $F = \{f_1, \ldots, f_m\}$ and $E = \{e_1, \ldots, e_n\}$ with $E \cap F = \emptyset$. The *graphic matrix* associated with the pair (G, T) is the $(0, 1)$-matrix M whose rows and columns are indexed by the edges in E and F, respectively, and whose elements are defined as follows: for any $f_j = uv \in F$, we have

$$m_{ij} = \begin{cases} 1 & \text{if the unique } uv\text{-path in } T \text{ contains } e_i; \\ 0 & \text{otherwise.} \end{cases} \tag{2}$$

The 2-edge-connectivity augmentation problem can be formulated as the following integer program:

$$\min \left\{ w^T x : Mx \geq 1, x \in \{0, 1\}^m \right\}. \tag{3}$$

Indeed, if we let w to be the vector of weights of the edges of G, it is easy to see that an optimal vector for (3) is the incidence vector of a minimum weight subset $F' \subseteq F$ such that $T' = (V, E \cup F')$ is 2-edge-connected.

The problem (3) is a *set covering* problem. This problem is NP-hard even if M is a graphic matrix, but if M is totally unimodular, and hence balanced, then all the basic feasible solutions (in the linear programming sense) of the constrains system $\{Mx \geq 1, x \geq 0\}$ are integer-valued and the problem becomes polynomial time solvable.

In the following, we explore which graphic matrices are TU, and we show that balancedness and totally unimodularity coincide in the class of graphic matrices.

4 Network Matrices and Graphic Matrices

Let $D = (V, F)$ be a digraph and let $T = (V, E)$ be a directed tree on the same node set, with $F \cap E = \emptyset$. The *network matrix* (see [14]) associated with the pair (D, T) is the $(0, \pm 1)$-matrix N whose rows and columns are indexed by the edges in E and F, respectively, and whose elements are defined as follows: for any $f_j = uv \in F$, we have

$$
n_{ij} = \begin{cases}
+1 & \text{if the unique } uv\text{-path in } T \text{ contains } e_i \text{ as a forward arc;} \\
-1 & \text{if the unique } uv\text{-path in } T \text{ contains } e_i \text{ as a backward arc;} \quad (4) \\
0 & \text{otherwise.}
\end{cases}
$$

The next two lemmas are known facts on TU and network matrices (see Schrijver [14] and Nemhauser, Wolsey [13] for more details).

Lemma 1. *Let A be a $(0, \pm 1)$-matrix having at most one $+1$ and one -1 in every column. Then A is TU.* □

Let A_T be the node-arc incidence matrix of a directed tree $T = (V, E)$ and let A_D be the node-arc incidence matrix of a directed graph $D = (V, F)$ on the same node set. Finally, let N be the network matrix associated with the pair (D, T). Let $f_i = uv$ be an arc of D, and let N^{f_i} be the column of N indexed by f_i. Since N^{f_i} is the incidence vector of the uv-path in T with forward edges having $+$ sign and backward edges having $-$ sign, then $A_T N^{f_i}$ is a vector indexed on V whose j-th component is defined as follows:

$$
(A_T N^{f_i})_j = \begin{cases}
+1 & \text{if } j = u; \\
-1 & \text{if } j = v; \quad (5) \\
0 & \text{otherwise.}
\end{cases}
$$

This shows that $A_T N = A_D$.

The matrices A_T and A_D are not full rank matrices. Let $\tilde{A}_T$ and $\tilde{A}_D$ arise from A_T and A_D by deleting one row (corresponding to the same node). By using the fact that a spanning tree always has a node of degree 1, it is easy to prove that the columns of $\tilde{A}_T$ are independent, and therefore $\tilde{A}_T$ is a square nonsingular matrix where $\text{rank}(A_T) = \text{rank}(\tilde{A}_T)$. So the system $A_T N = A_D$ implies that $N = \tilde{A}_T^{-1} \tilde{A}_D$.

Lemma 2. *Let N be a network matrix. Then N is TU.* □

It follows from the definitions of graphic and network matrices that a graphic matrix M associated with an undirected graph G and a tree T is also a network matrix if and only if there is an orientation G of G and T of T such that for each arc uv of G the uv-path in T is a directed path from u to v (i.e. all the edges are forward edges). We call these orientations *good*.

Note that if (G, T) is a good orientation of G, T, then also the orientations obtained from G, T by reversing the directions of all the arcs of G and T are good. This shows that, in order to find a good orientation if one exists, one can orient an edge of G or T arbitrarily. Note now that the orientation of an edge e of T forces the orientation of all the edges of G that contain e in the subpath of T between their endnodes. Furthermore the orientation of an edge f of G forces the orientations of every edge of T that is contained in the subpath of T between the endnodes of f. It is immediate now to see that this rule provides a fast algorithm to find a good orientation of G, T if one exists.

We now provide a characterization of the graphic matrices that are also network matrices, i.e., we identify which pairs (G, T) admit a good orientation. A node v of T is a *junction* if $|\delta_T(v)| \geq 3$. For any junction v with $\delta_T(v) = \{e_1, e_2, \ldots, e_k\}$, let G_v be the graph whose node set $v_{e_1}, v_{e_2}, \ldots, v_{e_k}$ represents edges $e_1, e_2, \ldots, e_k$, and $v_{e_i} v_{e_j} \in E(G_v)$ if and only if e_i, e_j belong to an xy-subpath identified by an edge $f = xy$ of G.

Theorem 1. *A graphic matrix M associated with G and T is a network matrix if and only if for every junction v of T, the graph G_v is bipartite.*

Proof. (Necessity) Assume G_v is not bipartite for some $v \in V$: Then G_v contains an odd chordless cycle $v_{e_1}, v_{e_2}, \ldots, v_{e_j}$. This shows that T contains distinct edges $e_1, e_2, \ldots, e_j$, j odd, having v as endnodes and G contains distinct edges $f_1, f_2, \ldots, f_j$ such that for $1 \leq i \leq j - 1$, e_i, e_{i+1} are consecutive edges of the subpath of T between the endnodes of f_i and e_j, e_1 are consecutive edges of the subpath of T between the endnodes of f_j. Let B be the square submatrix of M whose rows and columns are indexed by $e_1, e_2, \ldots, e_j$ and $f_1, f_2, \ldots, f_j$. Now B is a square matrix of odd order j and by the above argument, B has two 1's per row and column. By the cofactor expansion formula, it is easy to see that $\det(B) = \pm 2$. So M is not TU and by Lemma 2, M is not a network matrix.

(Sufficiency) Let (G, T) be a pair satisfying the hypothesis of the theorem. The proof is by induction on the number of junctions of T.

Assume first that T has only one junction v, and let E_1, E_2 be a partition of the edges in $\delta_T(v)$ corresponding to the bipartition of the nodes of G_v. Orient the edges in $\delta_T(v)$ so that $\delta_T^-(v) = E_1$, $\delta_T^+(v) = E_2$. This orientation forces the orientation of all the edges of G and all the other edges of T in some subpath between endnodes of edges of G. Since every such subpath contains an edge of E_1 and an edge of E_2, the orientation thus obtained is good.

So T contains at least two junctions. Let V_1, V_2 be a partition of V such that only one edge e of T is in $\delta_T(V_1)$ and the subtrees T_1, T_2, obtained from T by contracting V_2, V_1 into single nodes v_2, v_1 and removing loops, both contain less junctions than T. Let G_1, G_2, obtained from G by contracting V_2, V_1 into single nodes v_2, v_1. Note that both pairs (G_1, T_1) and (G_2, T_2) satisfy the hypothesis of the theorem. So, by the inductive hypothesis, both pairs (G_1, T_1) and (G_2, T_2) admit good orientations, and we can assume without loss of generality that e is oriented the same way in T_1 and T_2. This forces those edges of G that are common to G_1 and G_2 to get the same orientation. It is now immediate to combine the two orientations to obtain a good orientation of (G, T). $\square$

From the proof of the necessity part of Theorem 1 and from Lemma 2 we can deduce the following:

Corollary 1. *Let M be a graphic matrix. Then M is balanced if and only if M is a network matrix.* $\square$

5   Augmentation as Circulation

The immediate consequence of the results presented so far is that:

Theorem 2. *For any pair (G,T) such that G_v is bipartite for any junction $v \in T$, $\text{AUG}_2(G,T)$ can be found in polynomial time.* □

Notice also that it is checkable in polynomial time whether a pair (G,T) satisfies the hypothesis of Theorem 2.

Examples of trees satisfying the conditions of Theorem 2 are the depth-first-search (DFS) trees for which efficient polynomial time algorithms were already found [10,6]. Indeed, for DFS-trees each graph G_v consists of a star plus a (possibly empty) set of singletons.

In the remaining of the section, we describe in detail an algorithm that finds $\text{AUG}_2(G,T)$ when (G,T) satisfies the hypothesis of Theorem 2. In particular, we show that the set covering problem with constraint matrices that are network matrices can be solved by flow techniques. Consider the linear programming relaxation

$$\min\{w^T x : Mx \geq 1,\ x \geq 0\} \tag{6}$$

of the set covering problem, and rewrite it in standard form by adding a vector s of slack variables:

$$\min\{w^T x : Mx - Is = 1,\ x, s \geq 0\}. \tag{7}$$

Now, suppose that M is a network matrix, associated with the pair (G,T). Then by Lemma 2 M is a TU matrix and by the Hoffman-Kruskal's theorem every basic feasible solution (x,s) of the above linear program is an integral vector. Thus an optimal solution of the above linear program solves the set covering problem.

Furthermore, if M is a network matrix, the pair (G,T) admits a good orientation $(\boldsymbol{G},\boldsymbol{T})$. Hence, $M = \tilde{A}_T^{-1}\tilde{A}_G$ and we obtain the following equivalent linear program:

$$\min\{w^T x : \tilde{A}_G x - \tilde{A}_T s = \tilde{A}_T 1,\ x, s \geq 0\}. \tag{8}$$

Note that the matrix $-A_T$ is the incidence matrix of the directed tree $\boldsymbol{T}'$ obtained from $\boldsymbol{T}$ by reversing the orientations of all its arcs. Hence, by substituting the vector s with the vector $s' = s + 1$, the linear program (8) can be written as the *min-cost circulation problem*:

$$\min\{w^T x : \tilde{A}_G x + \tilde{A}_{T'} s' = 0,\ x \geq 0, s' \geq 1\}, \tag{9}$$

and an integral optimal solution can be found by flow techniques, see e.g. [13].

The equivalence between the above set covering problem and the circulation problem can be seen directly as follows. In the sequel, for any set S and vector $p \in \mathbb{R}^{|S|}$ and for any $U \subseteq S$, we use $p(U)$ to abbreviate $\sum_{i \in U} p_i$. The equality constraints of the above linear system are:

$$x\big(\delta_G^+(v)\big) - x\big(\delta_G^-(v)\big) + s'\big(\delta_{T'}^+(v)\big) - s'\big(\delta_{T'}^-(v)\big) = 0 \quad \forall v \in V. \tag{10}$$

For any edge e_j of $\boldsymbol{T}$, let $S_j \subset V$ be such that $\delta^+_{\boldsymbol{T'}}(S_j) = e_j$, where $\delta^+_{\boldsymbol{T'}}(S_j) = \bigcup_{v \in S_j} \delta^+_{\boldsymbol{T'}}(v)$. If $x \geq 0, s' \geq 1$ is a vector satisfying (10), then

$$x\big(\delta^+_{\boldsymbol{G}}(S_j)\big) - x\big(\delta^-_{\boldsymbol{G}}(S_j)\big) + s'\big(\delta^+_{\boldsymbol{T'}}(S_j)\big) - s'\big(\delta^-_{\boldsymbol{T'}}(S_j)\big) = 0.$$

Since $s'\big(\delta^+_{\boldsymbol{T'}}(S_j)\big) \geq 1$ and $\delta^-_{\boldsymbol{T'}}(S_j) = \emptyset$, we have that $x\big(\delta^-_{\boldsymbol{G}}(S_j)\big) \geq 1$ and the constraint of (6) associated with e_j is satisfied.

On the other hand, assume that x is a solution of (6). Let $F(e_i) \subseteq F$ denote the set of edges containing e_i on the path in T between their endnodes, and let

$$s'_{e_i} = \sum_{f_j \in F(e_i)} x_{f_j}.$$

Then, it is straightforward to check that if x is an augmenting set, then $s'_{e_i} \geq 1$, for each $e_i \in E$, and (x, s') is a flow vector in a good orientation of (G, T).

It is worth noticing that the results of this section can also be generalized in another direction. Indeed, the totally unimodularity of a graphic matrix M implies a more general result (by the Hoffmann-Kruskal theorem [9]), namely that the optimal solution of the following linear program

$$\min\{w^T x : Mx \geq b, c \geq x \geq 0\} \tag{11}$$

is integral for any b and c integral vectors.

Thus, depending on the values of b and c, it is possible to identify interesting network problems that can be solved in polynomial time when M satisfies conditions of Theorem 1. One of this problems is presented in the following section.

6   An Application to Many-to-One Networks

We consider a survivability problem arising in networks in which the communication occurs in a *many-to-one* fashion. In these networks, we have a set of *source* nodes which transmit messages to a single destination node, named the *sink*. Messages are routed from the sources to the sink through a set of intermediate nodes, which serve as *routers*. As a classic example of networks using a many-to-one communication protocol, we mention *sensor networks*, where data of interest (e.g., temperature, pollution index, etc.) are gathered at different locations (sources), and are transmitted to a single destination point (sink), where they can be stored and analyzed.

Given the inherent hierarchical structure, the typical topology of a many-to-one communication network is a directed rooted tree, with the sink as the root, the sources as the leaves, the routers as the internal nodes, and all the arcs oriented towards the root. In this tree, an arc uv represents the communication channel between nodes u and v. Hence, depending on the boundary conditions, such a link will afford a given amount of traffic, named its *load*. In the simplest situation, each link handles an amount of traffic which is proportional to the number of source nodes using it as a bridge towards the root. Therefore, the communication tree can be modelled as a tree $T(r) = (V, E)$ rooted at the source node r, where with each $e \in E$ is associated an integer $b(e)$ representing the link load.

Under normal working conditions, the tree links guarantee the data flow from the sources to the sink. However, networks usually undergo link failures, and therefore one might be interested in increasing the network reliability by creating alternative paths from the leaves towards the root. This can be accomplished by adding new links to the tree. Given the underlying hierarchical structure, we can assume that these links join *related* nodes in $T(r)$, i.e., nodes which belong to a given root-leaf path (this can be motivated, for instance, by constraints over the terrain where the network is deployed). Each of these additional links f is characterized by two main features: (1) a weight $w(f)$, which summarizes the cost to route a message through that link, and (2) a link capacity, say $c(f)$, which denotes the maximum amount of traffic load that can be routed through f. Let $G = (V, F)$ be the corresponding capacitated weighted graph.

Then, in the depicted scenario, it makes sense to define the following problem: given the pair $(G, T(r))$, find a minimum-cost set of edges in G whose addition to T guarantees that the traffic load $b(e)$, previously carried on e, can be re-routed on the additional edges, for every $e \in E$. Indeed, after the failure of any $e = uv$ of $T(r)$, an optimal solution for this problem allows to maintain (with a minimum set-up cost) the traffic between the leaves and the root, by means of the following simple changes to the communication protocol: let $f_1, \ldots, f_k$ be the set of additional edges associated with e; then, collect in v all the messages sent from the leaves below v in $T(r)$, and redistribute them among the replacement edges (according to their capacity) by letting the messages descend $T(r)$ to reach the starting nodes of $f_1, \ldots, f_k$, from where they will be routed above u (see Figure 1).

From a linear programming point of view, the above augmentation problem can be formulated as follows:

$$\min\{w^T x : Mx \geq b, c \geq x \geq 0, \} \tag{12}$$

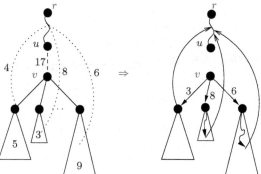

Fig. 1. An edge uv in $T(r)$ is removed, and messages are redirected through replacement edges (dotted). On the left, values in triangles denote the number of sources in the subtree, the value on edge uv denotes the load, while values on replacement edges denote capacities; on the right, values on tree edges denote the number of redirected messages.

where w, b and c are defined as above, and M is the graphic matrix associated with G and T, considering this latter as undirected. Since $T(r)$ is a tree and edges in G joins related nodes of $T(r)$, it is easy to see that for each non-leaf node $v \in V, v \neq r$, G_v is a forest (more precisely, it is a star plus possibly some isolated nodes). Thus, Theorem 1 holds and the matrix M is TU. Therefore, problem (12) is solvable in polynomial time.

References

1. K.P. Eswaran and R.E. Tarjan, Augmentation problems, *SIAM Journal on Computing*, **5**(4) (1976) 653–665.
2. G. Even, J. Feldman, G. Kortsarz, and Z. Nutov, A 3/2-approximation algorithm for augmenting the edge-connectivity of a graph from 1 to 2 using a subset of a given edge set, *4th Int. Workshop on Approximation Algorithms for Combinatorial Optimization (APPROX'01)*, Vol. 2129 of Lecture Notes in Computer Science, Springer-Verlag, 90–101.
3. A. Frank, Augmenting graphs to meet edge-connectivity requirements, *SIAM Journal on Discrete Mathematics*, **5**(1) (1992) 25–53.
4. G.N. Frederickson and J. Jájá, Approximation algorithm for several graph augmentation problems, *SIAM Journal on Computing*, **10**(2) (1981) 270–283.
5. H.N. Gabow, Application of a poset representation to edge-connectivity and graph rigidity, *Proc. 32nd Ann. IEEE Symp. on Foundations of Computer Science (FOCS'91)*, IEEE Computer Society, 812–821.
6. A. Galluccio and G. Proietti, Polynomial time algorithms for edge-connectivity augmentation problems, *Algorithmica*, **36**(4) (2003) 361–374.
7. D.S. Hochbaum, Approximating covering and packing problems: set cover, vertex cover, independent set and related problems, in *Approximation Algorithms for NP-Hard Problems*, Dorit S. Hochbaum Eds., PWS Publishing Company, Boston, MA, 1996.
8. D.S. Hochbaum, N. Megiddo, J.S. Naor, and A. Tamir, Tight bounds and 2-approximation algorithms for integer programs with two variables per inequality, *Mathematical Programming*, **62** (1993) 69–83.
9. A.J. Hoffman and J.B. Kruskal, Integral boundary points of convex polyhedra, in *Linear Inequalities and Related Systems*, H.W. Kuhn and A.W. Tucker Eds., Princeton University Press, New Jersey (1956), 223-246.
10. S. Khuller and R. Thurimella, Approximation algorithms for graph augmentation, *Journal of Algorithms*, **14**(2) (1993) 214–225.
11. S. Khuller, Approximation algorithms for finding highly connected subgraphs, in *Approximation Algorithms for NP-Hard Problems*, Dorit S. Hochbaum Eds., PWS Publishing Company, Boston, MA, 1996.
12. H. Nagamochi, An approximation for finding a smallest 2-edge-connected subgraph containing a specified spanning tree, *Discrete Applied Mathematics*, **126**(1) (2003) 83–113.
13. G.L. Nemhauser and L.A. Wolsey, *Integer and Combinatorial Optimization*, J. Wiley & Sons, (1986).
14. A. Schrijver, *Theory of Linear and Integer Programming*, J. Wiley & Sons, (1986).
15. T. Watanabe and A. Nakamura, Edge-connectivity augmentation problems, *Journal of Computer and System Sciences*, **35**(1) (1987) 96–144.

Partitioning a Weighted Graph
to Connected Subgraphs of Almost Uniform Size

Takehiro Ito, Xiao Zhou, and Takao Nishizeki

Graduate School of Information Sciences, Tohoku University
Aoba-yama 05, Sendai, 980-8579, Japan
take@nishizeki.ecei.tohoku.ac.jp, {zhou,nishi}@ecei.tohoku.ac.jp

Abstract. Assume that each vertex of a graph G is assigned a nonnegative integer weight and that l and u are nonnegative integers. One wish to partition G into connected components by deleting edges from G so that the total weight of each component is at least l and at most u. Such an "almost uniform" partition is called an (l, u)-partition. We deal with three problems to find an (l, u)-partition of a given graph. The minimum partition problem is to find an (l, u)-partition with the minimum number of components. The maximum partition problem is defined similarly. The p-partition problem is to find an (l, u)-partition with a fixed number p of components. All these problems are NP-complete or NP-hard even for series-parallel graphs. In this paper we show that both the minimum partition problem and the maximum partition problem can be solved in time $O(u^4 n)$ and the p-partition problem can be solved in time $O(p^2 u^4 n)$ for any series-parallel graph of n vertices. The algorithms can be easily extended for partial k-trees, that is, graphs with bounded tree-width.

1 Introduction

Let $G = (V, E)$ be an undirected graph with vertex set V and edge set E, and let $|V| = n$. Assume that each vertex $v \in V$ is assigned a nonnegative integer $\omega(v)$, called the *weight* of v. Let l and u be nonnegative integers, called the *lower bound* and *upper bound* on component size, respectively. We wish to partition G into connected components by deleting edges from G so that the total weights of all components are almost uniform, that is, the sum of weights of all vertices in each component is at least l and at most u for some bounds l and u with small $u - l$. We call such an almost uniform partition an (l, u)-*partition* of G. Figures 1(a) and (b) illustrate two $(10, 20)$-partitions of the same graph, where each vertex is drawn by a circle, the weight of each vertex is written inside the circle, and the deleted edges are drawn by dotted lines. In this paper we deal with three partition problems to find an (l, u)-partition of a given graph G. The *minimum partition problem* is to find an (l, u)-partition of G with the minimum number of components. The minimum number is denoted by $p_{\min}(G)$. The *maximum partition problem* is defined similarly. The *p-partition problem* is to find an (l, u)-partition of G with a fixed number p of components. The

J. Hromkovič, M. Nagl, and B. Westfechtel (Eds.): WG 2004, LNCS 3353, pp. 365–376, 2004.
© Springer-Verlag Berlin Heidelberg 2004

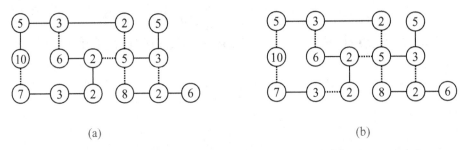

(a)   (b)

Fig. 1. (a) Solution for the minimum partition problem, and (b) solution for the maximum partition problem, where $l = 10$ and $u = 20$.

$(10, 20)$-partition with four components in Fig. 1(a) is a solution for the minimum partition problem, and hence $p_{\min}(G) = 4$ for the graph G in Fig. 1(a). The $(10, 20)$-partition with six components in Fig. 1(b) is a solution for the maximum partition problem.

The three partition problems often appear in many practical situations such as the image processing [5, 7], the paging system of operation system [10], and the political districting [3, 11]. Consider a map of a country, which is divided into several regions. Let G be a dual graph of the map. Each vertex v of G represents a region, and the weight $\omega(v)$ represents the number of voters in region v. Each edge (u, v) of G represents the adjacency of the two regions u and v. For the political districting, one wishes to divide the country into electoral zones. Each zone must consist of connected regions, that is, the regions in each zone must induce a connected subgraph of G. There must be an almost equal number of voters in each zone, that is, the sum of $\omega(v)$ for all regions v in each zone is at least l and at most u for some bounds l and u with small $u - l$. Such electoral zoning corresponds to an (l, u)-partition of the plane graph G.

Two related problems have been studied for trees. One is to partition a tree into the maximum number of subtrees so that the total weight of each subtree is at least l [8]. The other is to partition a tree into the minimum number of subtrees so that the total weight of each subtree is at most u [6]. Both can be solved for trees in linear time. Our three partition problems are generalizations of these problems. One may expect that there would exist efficient algorithms for the three partition problems on trees, but our problems are more difficult than the two problems in [6, 8], except for paths; all the three partition problems can be solved for paths in linear time [7].

An NP-complete problem, called the set partition problem [4], can be easily reduced in linear time to our problems for a complete bipartite graph $K_{2,n-2}$, and $K_{2,n-2}$ is a series-parallel graph. (A definition of a series-parallel graph will be given in Section 2.) Therefore, the p-partition problem for general p is NP-complete and both the minimum partition problem and the maximum partition problem for general l and u are NP-hard even for series-parallel graphs. Hence, it is very unlikely that the three partition problems can be solved for series-parallel graphs in polynomial time, although a number of combinatorial problems including many NP-complete problems on general graphs can be solved for series-parallel graphs and partial k-trees in polynomial time or even in linear time [1, 2, 9]. One can also observe from the reduction above that, for any $\varepsilon > 0$, there is no polynomial-time ε-approximation algorithm for the minimum partition problem or the maximum partition problem on series-parallel graphs unless P = NP.

In this paper we first obtain pseudo-polynomial-time algorithms to solve the three partition problems for series-parallel graphs. More precisely, we show that both the minimum partition problem and the maximum partition problem can be solved in time $O(u^4 n)$ and hence in time $O(n)$ for any bounded constant u, and that the p-partition problem can be solved in time $O(p^2 u^4 n)$. We then show that our algorithms can be easily extended for partial k-trees, that is, graphs with bounded tree-width [1, 2]. (A definition of a partial k-tree will be given in Section 5.)

2 Terminology and Definitions

In this section we give some definitions.

A *(two-terminal) series-parallel graph* is defined recursively as follows [9]:

(1) A graph G of a single edge is a series-parallel graph. The ends of the edge are called the *terminals* of G and denoted by $s(G)$ and $t(G)$. (See Fig. 2(a).)

(2) Let G' be a series-parallel graph with terminals $s(G')$ and $t(G')$, and let G'' be a series-parallel graph with terminals $s(G'')$ and $t(G'')$.

    (a) A graph G obtained from G' and G'' by identifying vertex $t(G')$ with vertex $s(G'')$ is a series-parallel graph, whose terminals are $s(G) = s(G')$ and $t(G) = t(G'')$. Such a connection is called a *series connection*, and G is denoted by $G = G' \bullet G''$. (See Fig. 2(b).)

    (b) A graph G obtained from G' and G'' by identifying $s(G')$ with $s(G'')$ and identifying $t(G')$ with $t(G'')$ is a series-parallel graph, whose terminals are $s(G) = s(G') = s(G'')$ and $t(G) = t(G') = t(G'')$. Such a connection is called a *parallel connection*, and G is denoted by $G = G' \parallel G''$. (See Fig. 2(c).)

The terminals $s(G)$ and $t(G)$ of G are often denoted simply by s and t, respectively. Since we deal with the partition problems, we may assume without loss of generality that G is a simple graph and hence G has no multiple edges.

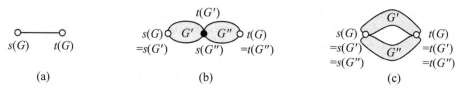

$$(a) \qquad\qquad\qquad (b) \qquad\qquad\qquad\qquad (c)$$

Fig. 2. (a) A series-parallel graph of a single edge, (b) series connection, and (c) parallel connection.

A series-parallel graph G can be represented by a "binary decomposition tree" [9]. Figure 3 illustrates a series-parallel graph G and its binary decomposition tree T. Labels s and p attached to internal nodes in T indicate series and parallel connections, respectively. Nodes labeled s and p are called s- and p-nodes, respectively. Every leaf of T represents a subgraph of G induced by a single edge. Each node v of T corresponds to a subgraph G_v of G induced by all edges represented by the leaves that are descendants of v in T. Thus G_v is a series-parallel graph for each node v of T, and $G = G_r$ for the root r of T. Since a binary decomposition tree of a given series-parallel graph G can be found in linear time [9], we may assume that a series-parallel graph G and its binary decomposition tree T are given. We solve the three partition problems by a dynamic programming approach based on a decomposition tree T.

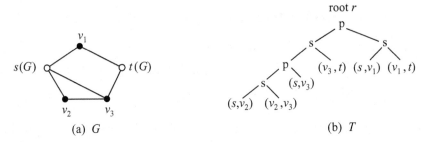

Fig. 3. (a) A series-parallel graph G, and (b) its binary decomposition tree T.

3   Minimum and Maximum Partition Problems

In this section we have the following theorem.

Theorem 1. *Both the minimum partition problem and the maximum partition problem can be solved for any series-parallel graph G in time $O(u^4 n)$, where n is the number of vertices in G and u is the upper bound on component size.*

In the remainder of this section we give an algorithm to solve the minimum partition problem as a proof of Theorem 1, because the maximum partition problem can be similarly solved. We indeed show only how to compute the minimum number $p_{\min}(G)$. It is easy to modify our algorithm so that it actually finds an (l, u)-partition having the minimum number $p_{\min}(G)$ of components.

Every (l, u)-partition of a series-parallel graph G naturally induces a partition of its subgraph G_v for a node v of a decomposition tree T of G. The induced partition is not always an (l, u)-partition of G_v but is either a "connected partition" or a "separated partition" of G_v, which are illustrated in Fig. 4 and will be formally defined later. Roughly speaking, two functions $f(G_v, x)$ and $h(G_v, x, y)$, $0 \le x, y \le u$, represent the minimum number of components without terminals in connected partitions and separated partitions of G_v, respectively, and x and y represent the total weight of non-terminal vertices in a component with a terminal. Our idea is to compute $f(G_v, x)$ and $h(G_v, x, y)$ from leaves v to the root r of T by means of dynamic programming.

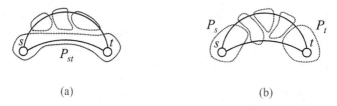

(a)   (b)

Fig. 4. (a) A connected partition, and (b) a separated partition.

We now formally define the connected partition and the separated partition of a series-parallel graph $G = (V, E)$. Let $\mathcal{P} = \{P_1, P_2, \cdots, P_r\}$ be a partition of vertex set V of G into r nonempty subsets $P_1, P_2, \cdots, P_r$ for some integer $r \geq 1$. Thus $|\mathcal{P}| = r$. The partition $\mathcal{P}$ of V is called a *partition of G* if P_i induces a connected subgraph of G for each index $i, 1 \leq i \leq r$. For a set $P \subseteq V$, we denote by $w(P)$ the total weight of vertices in P, that is, $w(P) = \sum_{v \in P} w(v)$. Let $w_{st}(G) = w(s) + w(t)$. We call a partition $\mathcal{P}$ of G a *connected partition* if $\mathcal{P}$ satisfies the following two conditions (see Fig. 4(a)):

(a) there exists a set $P_{st} \in \mathcal{P}$ such that $s, t \in P_{st}$ and $w(P_{st}) \leq u$; and
(b) $l \leq w(P) \leq u$ for each set $P \in \mathcal{P} - \{P_{st}\}$.

Note that the equation $l \leq w(P_{st})$ does not necessarily hold for P_{st}. For a connected partition $\mathcal{P}$, we always denote by P_{st} the set in $\mathcal{P}$ containing both s and t. A partition $\mathcal{P}$ of G is called a *separated partition* if $\mathcal{P}$ satisfies the following two conditions (see Fig. 4(b)):

(a) there exist two distinct sets $P_s, P_t \in \mathcal{P}$ such that $s \in P_s$, $t \in P_t$, $w(P_s) \leq u$, and $w(P_t) \leq u$; and
(b) $l \leq w(P) \leq u$ for each set $P \in \mathcal{P} - \{P_s, P_t\}$.

Note that the equations $l \leq w(P_s)$ and $l \leq w(P_t)$ do not always hold for P_s and P_t. For a separated partition $\mathcal{P}$, we always denote by P_s the set in $\mathcal{P}$ containing s and by P_t the set in $\mathcal{P}$ containing t.

We then formally define a function $f(G, x)$ for a series-parallel graph G and an integer $x, 0 \leq x \leq u$, as follows:

$$f(G, x) = \min\{q \geq 0 \mid G \text{ has a connected partition } \mathcal{P} \text{ such that}$$
$$x = w(P_{st}) - w_{st}(G) \text{ and } q = |\mathcal{P}| - 1\}. \quad (1)$$

If G has no connected partition $\mathcal{P}$ such that $w(P_{st}) - w_{st}(G) = x$, then let $f(G, x) = +\infty$. We now formally define a function $h(G, x, y)$ for a series-parallel graph G and a pair $(x, y), 0 \leq x, y \leq u$, as follows:

$$h(G, x, y) = \min\{q \geq 0 \mid G \text{ has a separated partition } \mathcal{P} \text{ such that}$$
$$x = w(P_s) - w(s), y = w(P_t) - w(t) \text{ and } q = |\mathcal{P}| - 2\}. \quad (2)$$

If G has no separated partition $\mathcal{P}$ such that $w(P_s) - w(s) = x$ and $w(P_t) - w(t) = y$, then let $h(G, x, y) = +\infty$.

Our algorithm computes $f(G_v, x)$ and $h(G_v, x, y)$ for each node v of a binary decomposition tree T of a given series-parallel graph G from leaves to the root r of T by means of dynamic programming. Since $G = G_r$, one can compute the minimum number $p_{\min}(G)$ of components from $f(G, x)$ and $h(G, x, y)$ as follows:

$$p_{\min}(G) = \min\Big\{\min\{f(G, x) + 1 \mid l \le x + w_{st}(G) \le u\},$$
$$\min\{h(G, x, y) + 2 \mid l \le x + w(s) \le u, l \le y + w(t) \le u\}\Big\}. \quad (3)$$

Note that $p_{\min}(G) = +\infty$ if G has no (l, u)-partition.

We first compute $f(G_v, x)$ and $h(G_v, x, y)$ for each leaf v of T, for which the subgraph G_v contains exactly one edge. For $x = 0$

$$f(G_v, 0) = 0, \quad (4)$$

and for $(x, y) = (0, 0)$

$$h(G_v, 0, 0) = 0. \quad (5)$$

For each integer $x, 1 \le x \le u$,

$$f(G_v, x) = +\infty, \quad (6)$$

and for each pair $(x, y), 1 \le x, y \le u$,

$$h(G_v, x, y) = +\infty. \quad (7)$$

By Eqs. (4)–(7) one can compute $f(G_v, x)$ in time $O(u)$ for each leaf v of T and all integers $x \le u$, and compute $h(G_v, x, y)$ in time $O(u^2)$ for each leaf v and all pairs (x, y) with $x, y \le u$. Since G is a simple series-parallel graph, the number of edges in G is at most $2n - 3$ and hence the number of leaves in T is at most $2n - 3$. Thus one can compute $f(G_v, x)$ and $h(G_v, x, y)$ for all leaves v of T in time $O(u^2 n)$.

We next compute $f(G_v, x)$ and $h(G_v, x, y)$ for each internal node v of T from the counterparts of the two children of v in T. We first consider a parallel connection. Let $G_v = G' \parallel G''$, and let $s = s(G_v)$ and $t = t(G_v)$. (See Figs. 2(c) and 5.)

We first explain how to compute $h(G_v, x, y)$. The definitions of a separated partition and $h(G, x, y)$ imply that if $w(P_s) = x + w(s) > u$ or $w(P_t) = y + w(t) > u$ then $h(G_v, x, y) = +\infty$. One may thus assume that $x + w(s) \le u$ and $y + w(t) \le u$. Then every separated partition $\mathcal{P}$ of G_v can be obtained by combining a separated partition $\mathcal{P}'$ of G' with a separated partition $\mathcal{P}''$ of G'' as illustrated in Fig. 5(a). We thus have

$$h(G_v, x, y) = \min\{h(G', x', y') + h(G'', x - x', y - y') \mid 0 \le x', y' \le u\}. \quad (8)$$

We next explain how to compute $f(G_v, x)$. If $w(P_{st}) = x + w_{st}(G_v) > u$, then $f(G_v, x) = +\infty$. One may thus assume that $x + w_{st}(G_v) \le u$. Then every

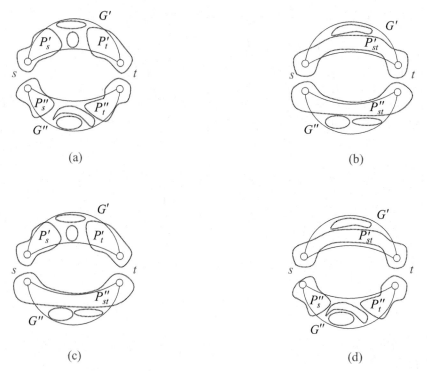

(a)   (b)

(c)   (d)

Fig. 5. The combinations of a partition $\mathcal{P}'$ of G' and a partition $\mathcal{P}''$ of G'' for a partition $\mathcal{P}$ of $G_v = G' \parallel G''$.

connected partition $\mathcal{P}$ of G_v can be obtained by combining a partition $\mathcal{P}'$ of G' with a partition $\mathcal{P}''$ of G'', as illustrated in Figs. 5(b), (c) and (d). There are the following three Cases (a)–(c), and we define three functions $f^a(G_v, x)$, $f^b(G_v, x)$ and $f^c(G_v, x)$ for the three cases, respectively.

Case (a): *both $\mathcal{P}'$ and $\mathcal{P}''$ are connected partitions.* (See Fig. 5(b).)
Let

$$f^a(G_v, x) = \min\{f(G', x') + f(G'', x - x') \mid 0 \le x' \le u\}. \tag{9}$$

Case (b): *$\mathcal{P}'$ is a separated partition, and $\mathcal{P}''$ is a connected partition.* (See Fig. 5(c).)
Let

$$f^b(G_v, x) = \min\{h(G', x', y') + f(G'', x - x' - y') \mid 0 \le x', y' \le u\}. \tag{10}$$

Case (c): *$\mathcal{P}'$ is a connected partition, and $\mathcal{P}''$ is a separated partition.* (See Fig. 5(d).)
Let

$$f^c(G_v, x) = \min\{f(G', x - x'' - y'') + h(G'', x'', y'') \mid 0 \le x'', y'' \le u\}. \tag{11}$$

From f^a, f^b and f^c above, one can compute $f(G_v, x)$ as follows:

$$f(G_v, x) = \min\{f^a(G_v, x), f^b(G_v, x), f^c(G_v, x)\}. \tag{12}$$

By Eq. (8) one can compute the function $h(G_v, x, y)$ for all pairs $(x, y), 0 \le x, y \le u$, in time $O(u^4)$, and by Eqs. (9)–(12) one can compute the function $f(G_v, x)$ for all integers $x, 0 \le x \le u$, in time $O(u^3)$. Thus one can compute the functions $f(G_v, x)$ and $h(G_v, x, y)$ for each p-node v of T in time $O(u^4)$.

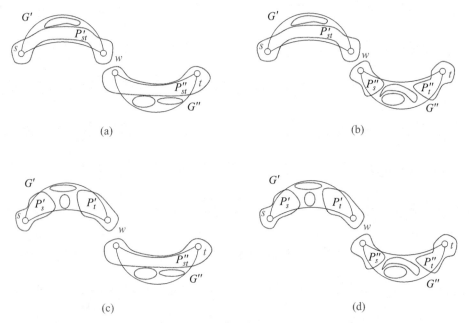

(a) $\qquad\qquad\qquad\qquad\qquad$ (b)

(c) $\qquad\qquad\qquad\qquad\qquad$ (d)

Fig. 6. The combinations of a partition $\mathcal{P}'$ of G' and a partition $\mathcal{P}''$ of G'' for a partition $\mathcal{P}$ of $G_v = G' \bullet G''$.

We next consider a series connection. Let $G_v = G' \bullet G''$, and let w be the vertex of G identified by the series connection, that is, $w = t(G') = s(G'')$. (See Figs. 2(b) and 6.)

We first explain how to compute $f(G_v, x)$. If $x + \omega_{st}(G_v) > u$, then $f(G_v, x) = +\infty$. One may thus assume that $x + \omega_{st}(G_v) \le u$. Then every connected partition $\mathcal{P}$ of G_v can be obtained by combining a connected partition $\mathcal{P}'$ of G' with a connected partition $\mathcal{P}''$ of G'' as illustrated in Fig. 6(a). We thus have

$$f(G_v, x) = \min\{f(G', x') + f(G'', x'') \mid 0 \le x', x'' \le u,$$
$$x' + x'' + \omega(w) = x\}. \tag{13}$$

We next explain how to compute $h(G_v, x, y)$. If $x + \omega(s) > u$ or $y + \omega(t) > u$, then $h(G_v, x, y) = +\infty$. One may thus assume that $x + \omega(s) \le u$ and $y + \omega(t) \le u$. Then every separated partition $\mathcal{P}$ of G_v can be obtained by combining a partition $\mathcal{P}'$ of G' with a partition $\mathcal{P}''$ of G'', as illustrated in Figs. 6(b), (c) and (d). There are the following three Cases (a)–(c), and we define three functions $h^a(G_v, x, y), h^b(G_v, x, y)$ and $h^c(G_v, x, y)$ for the three cases, respectively.

Case (a): $\mathcal{P}'$ is a connected partition, and $\mathcal{P}''$ is a separated partition. (See Fig. 6(b).)
   Let

$$h^a(G_v, x, y) = \min\{f(G', x') + h(G'', x'', y) \mid 0 \leq x', x'' \leq u,$$
$$x' + x'' + \omega(w) = x\}. \quad (14)$$

Case (b): $\mathcal{P}'$ is a separated partition, and $\mathcal{P}''$ is a connected partition. (See Fig. 6(c).)
   Let

$$h^b(G_v, x, y) = \min\{h(G', x, y') + f(G'', x'') \mid 0 \leq y', x'' \leq u,$$
$$y' + x'' + \omega(w) = y\}. \quad (15)$$

Case (c): both $\mathcal{P}'$ and $\mathcal{P}''$ are separated partitions. (See Fig. 6(c).)
   Let

$$h^c(G_v, x, y) = \min\{h(G', x, y') + h(G'', x'', y) + 1 \mid 0 \leq y', x'' \leq u,$$
$$l \leq y' + x'' + \omega(w) \leq u\}. \quad (16)$$

From h^a, h^b and h^c above one can compute $h(G_v, x, y)$ as follows:

$$h(G_v, x, y) = \min\{h^a(G_v, x, y), h^b(G_v, x, y), h^c(G_v, x, y)\}. \quad (17)$$

By Eq. (13) one can compute the function $f(G_v, x)$ for all integers $x, 0 \leq x \leq u$, in time $O(u^2)$, and by Eqs. (14)–(17) one can compute the function $h(G_v, x, y)$ for all pairs $(x, y), 0 \leq x, y \leq u$, in time $O(u^4)$. Thus one can compute the functions $f(G_v, x)$ and $h(G_v, x, y)$ for each s-node v of T in time $O(u^4)$.

In this way one can compute the functions $f(G_v, x)$ and $h(G_v, x, y)$ for each internal node v of T in time $O(u^4)$. Since T is a binary tree and has at most $2n - 3$ leaves, T has at most $2n - 4$ internal nodes. Since $G = G_r$ for the root r of T, one can compute the functions $f(G, x)$ and $h(G, x, y)$ in time $O(u^4 n)$. By Eq. (3) one can compute the minimum number $p_{\min}(G)$ of components in an (l, u)-partition of G from the functions $f(G, x)$ and $h(G, x, y)$ in time $O(u^2)$. Thus the minimum partition problem can be solved in time $O(u^4 n)$. This completes a proof of Theorem 1.

4   *p*-Partition Problem

In this section we have the following theorem.

Theorem 2. *The p-partition problem can be solved for any series-parallel graph G in time $O(p^2 u^4 n)$, where n is the number of vertices in G, u is the upper bound on component size, and p is the fixed number of components.*

The algorithm for the p-partition problem is similar to the algorithm for the minimum partition problem in the previous section. So we present only an outline.

For a series-parallel graph G and an integer $q, 0 \leq q \leq p - 1$, we define a set $F(G, q)$ of nonnegative integers x as follows:

$$F(G, q) = \{x \geq 0 \mid G \text{ has a connected partition } \mathcal{P}$$
$$\text{such that } x = w(P_{st}) - w_{st}(G) \text{ and } q = |\mathcal{P}| - 1\}.$$

For a series-parallel graph G and an integer $q, 0 \leq q \leq p - 2$, we define a set $H(G, q)$ of pairs of nonnegative integers x and y as follows:

$$H(G, q) = \{(x, y) \mid G \text{ has a separated partition } \mathcal{P} \text{ such that}$$
$$x = w(P_s) - w(s), y = w(P_t) - w(t) \text{ and } q = |\mathcal{P}| - 2\}.$$

Clearly $|F(G, q)| \leq u + 1$ and $|H(G, q)| \leq (u + 1)^2$.

We compute $F(G_v, q)$ and $H(G_v, q)$ for each node v of a binary decomposition tree T of a given series-parallel graph G from leaves to the root r of T by means of dynamic programming. Since $G = G_r$, the following lemma clearly holds.

Lemma 1. *A series-parallel graph G has an (l, u)-partition with p components if and only if the following condition (a) or (b) holds:*

(a) *$F(G, p-1)$ contains at least one integer x such that $l \leq x + w_{st}(G) \leq u$; and*

(b) *$H(G, p-2)$ contains at least one pair of integers (x, y) such that $l \leq x + w(s) \leq u$ and $l \leq y + w(t) \leq u$.*

One can compute in time $O(p)$ the sets $F(G_v, q)$ and $H(G_v, q)$ for each leaf v of T and all integers $q(\leq p-1)$, and compute in time $O(p^2 u^4)$ the sets $F(G_v, q)$ and $H(G_v, q)$ for each internal node v of T and all integers $q(\leq p-1)$ from the counterparts of the two children of v in T. Since $G = G_r$ for the root r of T, one can compute the sets $F(G, p-1)$ and $H(G, p-2)$ in time $O(p^2 u^4 n)$. By Lemma 1 one can know from the sets in time $O(u^2)$ whether G has an (l, u)-partition with p components. Thus the p-partition problem can be solved in time $O(p^2 u^4 n)$.

5 Partial k-Trees

In this section we have the following theorem.

Theorem 3. *The minimum and maximum partition problems can be solved in time $O(u^{2(k+1)} n)$ and the p-partition problem can be solved in time $O(p^2 u^{2(k+1)} n)$ for any partial k-trees, where $k = O(1)$.*

The algorithm for partial k-trees is similar to those for series-parallel graphs in the previous sections. So we present only an outline of the algorithm for the minimum partition problem.

A graph G is a k-*tree* if either it is a complete graph on k vertices or it has a vertex v whose neighbors induce a clique of size k and $G - \{v\}$ is again a k-tree. A graph is a *partial k-tree* if it is a subgraph of a k-tree. A series-parallel graph is a partial 2-tree. A partial k-tree G can be decomposed into pieces forming a tree structure with at most $k + 1$ vertices per piece. The tree structure is called a binary decomposition tree T of G [1,2]. Each node v of T corresponds to a set $V(v)$ of $k + 1$ or fewer vertices of G, and corresponds to a subgraph G_v of G. For a series-parallel graph, it suffices to consider only two kinds of partitions, a connected partition and a separated partition, while for a partial k-tree we have to consider many kinds of partitions of G_v. Let π be the number of all partitions of set $V(v)$ into pairwise disjoint nonempty subsets. Then $\pi \leq (2^{k+1})^{k+1} = O(1)$ since we assume $k = O(1)$ in the paper. For a partial k-tree G, we consider π kinds of partitions of G_v. Let $\mathcal{V}_i, 1 \leq i \leq \pi$, be the ith partition of set $V(v)$, let $\rho(i)$ be the number of subsets in the partition $\mathcal{V}_i$, and let $\mathcal{V}_i = \{V_1, V_2, \cdots, V_{\rho(i)}\}$. Clearly $1 \leq \rho(i) \leq k+1$. In every partition of G_v of the ith kind, its jth connected component, $1 \leq j \leq \rho(i)$, contains all the vertices in the jth subset $V_j (\subseteq V(v))$ in $\mathcal{V}_i$. We consider a set of functions $h_i(G_v, x_1, x_2, \cdots, x_{\rho(i)}), 1 \leq i \leq \pi$, defined similarly to Eqs. (1) and (2). Variable $x_j, 1 \leq j \leq \rho(i)$, represents the sum of weights of all vertices in the jth component except for the vertices in V_j. Thus $0 \leq x_j \leq u$. One can observe that the set of functions for G_v for an internal node v can be computed from the counterparts of the two children of v in T in time $O((u+1)^{2(k+1)})$. Thus the set of functions for G can be computed in time $O((u+1)^{2(k+1)}n)$. The hidden coefficient in the complexity is $\pi^2 (\leq 2^{2(k+1)^2})$.

6 Conclusions

In this paper we first obtained pseudo-polynomial-time algorithms for three partition problems on series-parallel graphs. Both the minimum partition problem and the maximum partition problem can be solved in time $O(u^4 n)$, and hence they can be solved in time $O(n)$ if $u = O(1)$. On the other hand, the p-partition problem can be solved in time $O(p^2 u^4 n)$. Thus these algorithms take polynomial time if u is bounded by a polynomial in n.

We then showed that our algorithms for series-parallel graphs can be easily extended for partial k-trees, that is, graphs of bounded tree-width. The extended algorithm takes time $O(u^{2(k+1)}n)$ for the minimum and maximum partition problems, and takes time $O(p^2 u^{2(k+1)}n)$ for the p-partition problem.

We finally remark that, for ordinary trees, one can solve the minimum and maximum partition problems in time $O(u^2 n)$ and the p-partition problem in time $O(p^2 u^2 n)$ or $O(_n C_{p-1} + n)$.

Acknowledgments

We thank Takeshi Tokuyama for suggesting us the partition problems.

References

1. S. Arnborg and J. Lagergren. Easy problem for tree-decomposable graphs. *J. Algorithms*, Vol. 12, No. 2, pp. 308–340, 1991.
2. H. L. Bodlaender. Polynomial algorithms for graph isomorphism and chromatic index on partial k-trees. *J. Algorithms*, Vol. 11, No. 4, pp. 631–643, 1990.
3. B. Bozkaya, E. Erkut and G. Laporte. A tabu search heuristic and adaptive memory procedure for political districting. *European J. Operational Research*, Vol. 144, pp. 12–26, 2003.
4. M. R. Garey and D. S. Johnson. Computers and Intractability: A Guide to the Theory of NP-Completeness. *Freeman, San Francisco, CA*, 1979.
5. R. C. Gonzales and P. Wintz. Digital Image Processing. *Addison-Wesley, Reading, MA*, 1977.
6. S. Kundu and J. Misra. A linear tree-partitioning algorithm. *SIAM J. Comput.*, Vol. 6, pp. 131–134, 1977.
7. M. Lucertini, Y. Perl and B. Simeone. Most uniform path partitioning and its use in image processing. *Discrete Applied Mathematics*, Vol. 42, pp. 227–256, 1993.
8. Y. Perl and S. R. Schach. Max-min tree-partitioning. *J. ACM*, Vol. 28, pp. 5–15, 1981.
9. K. Takamizawa, T. Nishizeki and N. Saito. Linear-time computability of combinatorial problems on series-parallel graphs. *J. ACM*, Vol. 29, No. 3, pp. 623–641, 1982.
10. D. C. Tsichritzis and P. A. Bernstein. Operating Systems. *Academic Press*, New York, 1981.
11. J. C. Williams Jr. Political redistricting: a review. *Papers in Regional Science*, Vol. 74, pp. 12–40, 1995.

The Hypocoloring Problem: Complexity and Approximability Results when the Chromatic Number Is Small

Dominique de Werra[1], Marc Demange[2],
Jerome Monnot[3], and Vangelis Th. Paschos[3]

[1] Ecole Polytechnique Fédérale de Lausanne, Switzerland
dewerra.ima@epfl.ch
[2] ESSEC, Dept. SID, France
demange@essec.fr
[3] Université Paris Dauphine, LAMSADE, CNRS UMR 7024, 75016 Paris, France
{monnot,paschos}@lamsade.dauphine.fr

Abstract. We consider a weighted version of the subcoloring problem that we call the hypocoloring problem: given a weighted graph $G = (V, E; w)$ where $w(v) \geq 0$, the goal consists in finding a partition $\mathcal{S} = (S_1, \ldots, S_k)$ of the node set of G into hypostable sets and minimizing $\sum_{i=1}^{k} w(S_i)$ where an hypostable S is a subset of nodes which generates a collection of node disjoint cliques K. The weight of S is defined as $\max\{\sum_{v \in K} w(v) \mid K \in S\}$. Properties of hypocolorings are stated; complexity and approximability results are presented in some graph classes. The associated decision problem is shown to be **NP-complete** for bipartite graphs and triangle-free planar graphs with maximum degree 3. Polynomial algorithms are given for graphs with maximum degree 2 and for trees with maximum degree Δ.

1 Introduction

Chromatic scheduling is the domain of scheduling problems which can be formulated in terms of graph coloring or more precisely of generalized graph coloring (i.e., coloring with a few additional requirements). These generalizations appear in [16,11,9,5] and are called *conditional coloring* of G with respect to a graph theoretical property $\mathcal{P}$; the conditional (or $\mathcal{P}$) chromatic number $\chi_{\mathcal{P}}(G)$ is the minimum integer k such that there is a partition of the nodes into k sets such that the subgraph induced by each set has the property $\mathcal{P}$. Note that $\chi(G)$ corresponds to the case $\mathcal{P}(V') = true$ iff the subgraph induced by V' does not contain an induced P_2 (i.e., chain of length 1). An important application of conditional coloring is the circuit manufacturing problem and is defined by $\mathcal{P}(V') = true$ iff the subgraph induced by V' is planar (see [18] for a survey). To our knowledge the weighted case has not been studied specifically until now.

In particular the concept of weighted coloring has been introduced in [10] to generalize classical coloring models and to handle situations where operations occur with possibly different processing times in some types of batch scheduling problems.

J. Hromkovič, M. Nagl, and B. Westfechtel (Eds.): WG 2004, LNCS 3353, pp. 377–388, 2004.

Our generalized weighted coloring model can be described in terms of conditional coloring where property $\mathcal{P}$ is defined by $\mathcal{P}(V') = true$ iff the subgraph induced by V' does not contain an induced P_3. This induces the so-called *subcoloring* problem that has been studied in [1, 13, 7]. An alternate definition consists of finding a partition $\mathcal{S} = (S_1, \ldots, S_k)$ of the node set into *hypostable* sets minimizing k. We shall say that a subset $S = \{K_j : j \in J\}$ of nodes is a hypostable set in G if it induces a collection of node-disjoint cliques (with no edges between them). However, since we study a weighted model, the weight of a hypostable set $S = \{K_j : j \in J\}$ will be $w(S) = \max\{w(K_j) : j \in J\}$ and our problem, called MIN HYPOCOLORING, consists of finding a hypocoloring $(S_1, \ldots, S_k)$ of the nodes of G, i.e., a partition of the node set into hypostable sets such that:

$$opt = \sum_{i=1}^{k} w(S_i) \text{ is minimum} \tag{1}$$

In terms of batch scheduling, there exist many situations where operations have to be assigned to batches (of compatible operations) that are processed sequentially ([6]). Examples in satellite communication and in production have also been modelled as special cases of the above batch scheduling problem (see [19, 6]). In current model, all operations in a batch are assigned to different processors and processed simultaneously. The processing time of a batch S is limited by the largest processing time of the operations in S. If the processing times may take different values, it may be worthwhile to assign two (or more) incompatible operations v with small processing times $w(v)$ to the same batch; they will be processed consecutively on the same processor. This will not increase the processing time $w(S)$ of the batch S as long as the sum of processing times of these operations do not exceed the longest processing time $w(v)$ in S. In order to allow this possibility in our model, a natural way to define weight $w(K_j)$ is $w(K_j) = \sum_{v \in K_j} w(v)$. Since K_j corresponds to incompatible operations (assigned to the same processor), the processing time of all operations in K_j will be the sum of all processing times.

The MIN HYPOCOLORING problem may also be used for representing some machine scheduling problems: for instance, we are given a collection of jobs v with processing times $w(v)$ in a flexible manufacturing system; we link the nodes representing two jobs, if they share a certain number of tools; thus, it will be interesting to assign these jobs to the same machine on which the appropriate tools will be installed. A batch will consist of an assignment of jobs to some machines; in such an assignment, we try to assign to a same machine jobs sharing the same tools. Since there exists only a limited number of tools of each type, we will try to assign to different machines jobs that do not need the same tools. Hence a batch will be represented by a hypostable set in the graph of compatibilities (common tools) and the processing time of a batch will be the maximum load of a machine (maximum of the sums of processing times of jobs assigned to the same machine). We will focus on this model of weighted hypocoloring which is motivated in a natural way by the batch scheduling context.

In this paper, the *neighborhood* of node v will be denoted by $N(v)$, the *degree* of v by $d(v)$ or $d_{G_i}(v)$ when the particular graph G_i in which it is considered has to be emphasized, the *maximum degree* by $\Delta(G)$ or Δ and the *subgraph of G induced* by S by $G[S]$. The size of hypocoloring $S = (S_1, \ldots, S_k)$ will be denoted by $|S| = k$ and, finally, the number of different values of weights w by $|w|$. For graph-theoretical terms not defined here, the reader is referred to [3]. Moreover, we always assume that S is sorted by non-increasing weights (i.e., $w(S_1) \geq \ldots \geq w(S_k)$) and, without additional specification, we assume that $w(v) > 0, \forall v \in V$.

2   Elementary Properties

We will derive here some properties which are based on the fact that hypocolorings are in some sense extensions of node colorings;

Lemma 1. *Any optimal hypocoloring S satisfies $|S| \leq \Delta(G) + 1$.*

Proof. Let $S = (S_1, \ldots, S_k)$ be an optimal hypocoloring and let $v \in S_k$. If $k > \Delta(G) + 1$ then there exists color $c \leq \Delta(G) + 1$ such that $N(v) \cap S_c = \emptyset$. So, we can recolor v with color c without increasing the value of S.

This bound is not the best possible; by analogy with the theorem of Brooks [8], we could try to get a bound of $\Delta(G)$ instead of $\Delta(G) + 1$.

Proposition 1. *There exists an optimal hypocoloring S satisfying the following:*

(i) $\forall i \leq k$, $\forall v \in S_i$, $d_{G_{i,v}}(v) \geq i - 1$ where $G_{i,v} = G[S_1 \cup \ldots \cup S_{i-1} \cup \{v\}]$.
(ii) $\forall i \leq k$, S_i contains no $K_{\Delta(G)+3-i}$.
(iii) $|S| \leq \Delta(G)$.

Proof. Let $S = (S_1, \ldots, S_k)$ be an optimal hypocoloring. For (i): If $d_{G_{i,v}}(v) < i - 1$, then we can recolor node v with some color missing in $\{1, \ldots, i-1\}$. For (ii): Assume that S_i contains a $K_{\Delta(G)+3-i}$ and let $v \in K_{\Delta(G)+3-i}$. We deduce that $d_{G_{i,v}}(v) \leq i - 2$ which gives a contradiction with (i). For (iii): Let $v \in S_k$. From Lemma 1, we can assume $k \leq \Delta(G) + 1$ and $\exists u \in N(v) \cap S_{\Delta(G)}$. Moreover, using (i) with node u, we have $N(u) \cap S_{\Delta(G)} = \emptyset$. So we can recolor v with color $\Delta(G)$ (at this stage, the solution may have a greater value). By repeating this as long as $S_k \neq \emptyset$, we obtain another optimal hypocoloring.

Note that it is always possible to find in polynomial time a hypocoloring which verifies Proposition 1. We can also obtain a bound of the number of different colors used in any optimal coloring S^* using a relation between $|w|$ and the chromatic number $\chi(G)$.

Proposition 2. *Any optimal hypocoloring S satisfies: $|S| \leq 1 + |w|(\chi(G) - 1)$.*

Proof. The proof is by induction on $|w|$. Let $S = (S_1, \ldots, S_k)$ be an optimal hypocoloring and let $t = \max\{i : w(S_i) \geq \max_{v \in V} w(v)\}$. Remark that $t \leq \chi(G)$

(otherwise, an optimal coloring gives a better solution); moreover, if $t = \chi(G)$, then $t = |S|$ (for the same reason). So, if $t = |S|$ then we have $|S| = t \leq \chi(G) \leq 1 + |w|(\chi(G) - 1)$. Now, assume $|S| > t$ (we deduce $t \leq \chi(G) - 1$); then $(S_{t+1}, \ldots, S_k)$ is an optimal hypocoloring on $G' = G[S_{t+1} \cup \ldots \cup S_k]$ and using an inductive hypothesis, we deduce $|S| - t \leq 1 + (|w| - 1)(\chi(G') - 1) \leq 1 + (|w| - 1)(\chi(G) - 1)$ and the result follows.

3   Complexity Results

We will now show that MIN HYPOCOLORING is close to MIN COLORING in some cases; more precisely, we prove that MIN HYPOCOLORING is at least as hard to approximate as MIN COLORING. Let Ψ be a class of graphs.

Theorem 1. *There exists a approximation preserving reduction from MIN COL-ORING restricted to Ψ-graphs to MIN HYPOCOLORING restricted to Ψ-graphs.*

Thus, using results of [12], we deduce that MIN HYPOCOLORING is not approximable within n^ε for any $\varepsilon > 0$, unless **ZPP=NP**. It also follows that MIN HYPOCOLORING is **NP-hard** for graphs with $\Delta(G) \geq 4$, even if these graphs do not contain triangles, or for planar graphs, etc. Moreover, when $\Delta(G) = 3$ and $|w| = 1$, the previous proof and Brooks theorem [8] show that this case is polynomial. On the other hand, we now prove that when G is a triangle-free graph with $\Delta(G) = 3$ and $|w| = 2$, then MIN HYPOCOLORING becomes difficult.

Theorem 2. MIN HYPOCOLORING *is* **strongly NP-hard** *even for triangle-free graphs with $\Delta(G) = 3$.*

Proof. We shall reduce 1-IN 3SAT, proved to be **NP-complete** in [20] to our problem. This problem is defined as follows: Given a collection $\mathcal{C} = (C_1, \ldots, C_m)$ of m clauses over the set $X = \{x_1, \ldots, x_n\}$ of n Boolean variables such that each clause C_j has exactly three literals (i.e., $C_j = x \vee y \vee z$), is there a truth assignment f satisfying $\mathcal{C}$ such that each clause in $\mathcal{C}$ has exactly one true literal?

From instance $I = (\mathcal{C}, X)$ of 1-IN 3SAT, we construct an instance $I' = (G, w)$ of MIN HYPOCOLORING such that the answer of I is yes iff $opt(I) \leq 3$. We use gadget clauses and gadget variables. From the clause $C_i = x \vee y \vee z$, we build the graph F_i in Figure 1 and the weights are given by $w(v_1(F_i)) = 2$ and $\forall v \in V(F_i) \setminus \{v_1(F_i)\}$, $w(v) = 1$.

From the variable x_j, we build the graph H_j in Figure 2 where the weights of nodes are one (i.e., $w(z) = 1$, $\forall z \in V(H_j)$). In addition, we link theses different graphs in the following way: if variable x_j appears positively in clause C_i, then we add edge $[x_i(H_j), x(F_i)]$ and otherwise, we add edge $[\overline{x_i(H_j)}, x(F_i)]$.

This graph G satisfies $\Delta(G) = 3$. Let f be a truth assignment of $I = (\mathcal{C}, X)$. The hypostable sets S_1 and S_2 are given by:

$$S_1 = \cup_{i=1}^m [\ \{v_1(F_i), v_4(F_i), v_5(F_i)\} \cup \{x(F_i) : f(x) = 1\} \cup \{\overline{x(F_i)} : f(x) = 0\}\ ]$$
$$\cup_{j=1}^n \cup_{k=1}^m \{x_k(H_j), v_{2k}(H_j) : f(x) = 0\} \cup \{\overline{x_k(H_j)}, v_{2k-1}(H_j) : f(x) = 1\}$$
$$S_2 = V(G) \setminus S_1$$

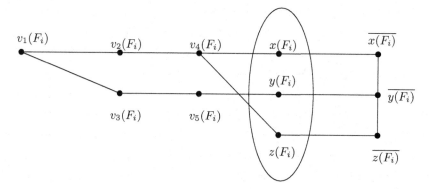

Fig. 1. Gadget clause F_i.

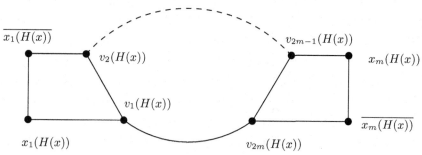

Fig. 2. Gadget variable H_j.

It is easy to verify that $\mathcal{S} = (S_1, S_2)$ satisfies $w(S_1) = 2$ and $w(S_2) = 1$; thus $opt(I) \leq 3$. Conversely, let $\mathcal{S}$ be a hypocoloring of I' such that $val(\mathcal{S}) \leq 3$. We can observe that: (i) $\mathcal{S} = (S_1, S_2)$ with $w(S_1) = 2$ and $w(S_2) = 1$, (ii) S_2 is a stable set and $\forall i \leq m$, $|S_1 \cap \{x(F_i), y(F_i), z(F_i)\}| = 1$, (iii) $\forall j \leq n$, $H_j \cap S_1$ and $H_j \cap S_2$ are stable sets and (iv) $x_i(H_j)$ (resp. $\overline{x_i(H_j)}$) and $x(F_i)$ have two distinct colors if these nodes are linked.

So, we can exhibit a truth assignment f of I by taking $f(x) = 1$ iff $x \in S_1$.

Theorem 3. MIN HYPOCOLORING *is **strongly NP-hard** for triangle-free planar graphs with* $\Delta(G) = 3$.

Proof. In the previous theorem, all gadgets F_i and H_j are planar and then only edges $[x_l(F_i), x_p(H_j)]$ may create some problems since they may cross each other. In this case, we apply the *crossover* technique, [14] which consists of replacing each edge crossing by a planar gadget. First, we embed the graph G' of Theorem 2 in the plane in such a way that every edge is a straight line and the crossing edge occurs only between two edges $[x_l(F_i), x_p(H_j)]$. Second, we replace each crossing edge by the gadget (L, w) indicated in Figure 3. This graph contains 8 particular nodes $x_1, x'_1, y_1, y'_1, x_2, x'_2, y_2, y'_2$. The weight of any node is 1 except for x'_1, y'_1, x'_2, y'_2 which are weighted by 2. It is easy to see that we have the following properties for any hypocoloring $\mathcal{S} = (S_1, S_2)$ with $val(\mathcal{S}) \leq 3$:

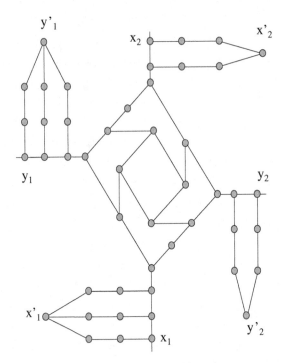

Fig. 3. Planar gadget (L, w).

(*i*) $\{x_1', x_2', y_1', y_2'\} \subseteq S_1$ and S_2 is a stable set, (*ii*) $\forall x = x_1, x_2, y_1, y_2$, the neighbors outside of the gadget (L, w) have not the same color as x, and (*iii*) x_1 and x_2 (resp., y_1 and y_2) have not the same color.

Using these properties, we deduce that there exists a hypocoloring $\mathcal{S}$ of G' with $val(\mathcal{S}) \leq 3$ if and only if there exists a hypocoloring $\mathcal{S}'$ of G'' with $val(\mathcal{S}') \leq 3$.

Now, we deal with bipartite graphs. Surprisingly, in some cases an optimal hypocoloring is just a coloring and such a coloring is difficult to compute.

Theorem 4. MIN HYPOCOLORING *is **strongly NP-hard** for bipartite graphs with $\Delta(G) = 39$.*

Proof. We polynomially transform the pre-extension coloring problem 1-PREXT (proved to be **NP-complete** in [4]) into the hypocoloring problem in bipartite graphs; 1-PREXT can be described as follows: given a bipartite graph $G = (L, R; E)$ with $|L| \geq 3$ and $\Delta(G) = 12$ and three nodes v_1, v_2, v_3 in L, does there exist a 3-coloring (S_1, S_2, S_3) such that $v_i \in S_i$ for $i = 1, 2, 3$?

Let $G = (L, R; E)$ be a bipartite graph and let $\{v_1, v_2, v_3\} \subseteq L$ be a set of three nodes. We polynomially construct a new bipartite graph G' such that there exists a hypocoloring $\mathcal{S}$ of G' with $val(\mathcal{S}) \leq 7$ iff there exists a coloring (S_1, S_2, S_3) of G with $v_i \in S_i$, $i = 1, 2, 3$. In order to do that, we use the two following gadgets:

- The weighted bipartite graph $H_0 = (L_0, R_0; E_0, w)$ on 12 nodes with $l_i, l_i' \in L_0$ and $r_i, r_i' \in R_0$ for $i = 1, 2, 3$. Moreover, only edges $[l_i, r_i], [l_i, r_i'], [l_i', r_i]$ and $[l_i', r_i']$ for $i = 1, 2, 3$ are missing in H_0. The weights are $w(l_i) = w(l_i') = w(r_i) = w(r_i') = 2^{3-i}$ for $i = 1, 2, 3$.
- The complete weighted bipartite graph $K_{3,2}$ with two specified nodes x and y (x in the left set and y in the right set). The weights are $w(x) = w(y) = 1$ and $w(v) = 2$ otherwise.

Now, $I = (G', w)$ is built in the following way: starting from G, we add a copy of H_0 and we identify nodes v_1, v_2, v_3 of G with nodes l_1, l_2, l_3 of H_0. Moreover, for each edge $e = [l, r]$ of G, we introduce a copy of $K_{3,2}$ and we identify nodes l, r with nodes x_e, y_e respectively.

Let $S = (S_1, S_2, S_3)$ be a 3-coloring of G with $v_i \in S_i$, $i = 1, 2, 3$; then we extend S into a coloring S' of G' by the following process: we start with $S_i' = (S_i \setminus \{v_i\}) \cup \{l_i, l_i', r_i, r_i'\}$. For each edge $e = [l, r]$ of G with $l \in L$ and $r \in R$, if $l \in S_j$ (resp. $r \in S_j$) with $j = 1, 2$ then we add $L_e \setminus \{x_e\}$ (resp., $R_e \setminus \{y_e\}$) to S_i' else ($j = 3$) and we add $L_e \setminus \{x_e\}$ (resp., $R_e \setminus \{y_e\}$) to S_i' where $r \in S_i$ (resp., $l \in S_i$). S' is a coloring of G' (thus a hypocoloring) and satisfies $val(S') = w(S_1') + w(S_2') + w(S_3') = 7$.

Conversely let S' be a hypocoloring satisfying $val(S') \leq 7$. It is easy to prove that: (i) $\forall i = 1, 2, 3$, $\{l_i, l_i', r_i, r_i'\} \subseteq S_i'$, (ii) $S' = (S_1', S_2', S_3')$ with $w(S_1') = 4$, $w(S_2') = 2$, $w(S_3') = 1$, and (iii) the restriction of S' to graph G is a coloring. Thus, using these properties the result follows.

4 Approximability of Some Cases of Hypocoloring

We shall present here approximation algorithms for hypocolorings when coloring is easy and the chromatic number is small; formally, we denote by Ψ_k a class of graphs verifying: (i) for any G' subgraph of G, if $G \in \Psi_k$, then $G' \in \Psi_k$, (ii) $\forall G \in \Psi_k$, $\chi(G) \leq k$ and (iii) coloring on Ψ_k-graphs is polynomial. For instance, the set of forests is a Ψ_2-class. Assume that the nodes are ordered according to their non-increasing weights ($w(v_1) \geq \ldots \geq w(v_n)$) and let $G_i = G[\{v_1, \ldots, v_i\}]$. Moreover, j_0 denotes the smallest index i such that G_i contains an induced P_3 (if $G_n = G$ does not contain it, we set $j_0 = n + 1$). Finally, $Colo(V')$ denotes an optimal coloring on $G[V']$ (i.e., $|colo(V')| = \chi(G[V'])$). A trivial bound of the approximability on Ψ_k-graphs is k and consists of computing an optimal coloring in the entire graph. We now propose an algorithm achieving a better constant approximation ratio.

---

1. Sort the nodes of G in non-increasing weight order;
2. Compute $j_0 = min\{i : G_i$ contains an induced $P_3\}$ and $V_{j_0} = \{v_1, \ldots, v_{j_0}\}$;
3. For $i = 1$ to j_0 do
   3.1 $S_1^i = V_{j_0} \setminus \{v_i, \ldots, v_{j_0}\}$;
   3.2 Compute $Colo(V \setminus S_1^i)$ and define hypocoloring S^i by $(S_1^i, Colo(V \setminus S_1^i))$;
4. Compute $S = argmin\{val(S^i) : i = 1, \ldots, j_0\}$;

Theorem 5. MIN HYPOCOLORING *is* $\frac{k^2}{2k-1}$*-approximable in* Ψ_k*-graphs.*

Proof. let $I = (G, w)$ with $G \in \Psi_k$ be an instance of MIN HYPOCOLORING and let $\mathcal{S}^* = (S_1^*, \ldots, S_l^*)$ be an optimal hypocoloring. We can assume $j_0 \leq n$ and then $l > 1$. If $v_1 \notin S_1^*$, then $val(\mathcal{S}) \leq val(\mathcal{S}^1) \leq k\, w(v_1)$ and $opt \geq 2w(v_1)$. Now, set $i_0 = \max\{i : \{v_1, \ldots, v_i\} \subseteq S_1^*\}$ and examine solution $\mathcal{S}^{i_0}$.

If $w(S_2^{i_0}) \leq w(S_1^{i_0})\frac{k-1}{k}$, then for any $r \geq 3$, $w(S_r^{i_0}) \leq \frac{k-1}{2k-1}w(S_2^{i_0}) + \frac{k}{2k-1}w(S_2^{i_0})$
$\leq \frac{k-1}{2k-1}(w(S_1^{i_0}) + w(S_2^{i_0}))$. Summing these inequalities, we deduce:
$val(\mathcal{S}) \leq \frac{k^2}{2k-1}(w(S_1^{i_0}) + w(S_2^{i_0})) \leq \frac{k^2}{2k-1}opt$.

If $w(S_2^{i_0}) \geq w(S_1^{i_0})\frac{k-1}{k}$, then $opt \geq w(S_1^{i_0})\frac{2k-1}{k}$ and the result follows.

Using Grotzsch theorem [15], Brooks theorem [8] (with its constructive proof given by Lovasz), since we can assume without loss of generality that G does not contain any copy of $K_{\Delta(G)+1}$, and the previous theorem, we obtain:

Corollary 1. MIN HYPOCOLORING *is* $\frac{9}{5}$*-approximable if* G *satisfies* $\Delta(G) \leq 3$ *or if* G *is triangle-free and planar; it is* $\frac{4}{3}$*-approximable if* G *is bipartite.*

We can also establish lower bounds on the approximability of these types of graphs by using the proofs of Theorem 3 and Theorem 4.

Proposition 3. *Unless* **P=NP**, MIN HYPOCOLORING *is not* $(\frac{4}{3} - \varepsilon)$*-approximable, if* G *is triangle free and planar, and not* $(\frac{8}{7} - \varepsilon)$*-approximable, if* G *is bipartite, for any* $\varepsilon > 0$.

There is another simple approximation algorithm which works for any value of $\Delta(G)$. This algorithm uses a decomposition of G into at most $s = \lceil \frac{\Delta(G)+1}{3} \rceil$ subgraphs G_i satisfying $\Delta(G_i) \leq 2$ by applying a result of [17]. Then, for each $i = 1, \ldots, s$, we compute an optimum hypocoloring $\mathcal{S}_i^*$ on G_i by using the algorithm presented in subsection 5.2 (Proposition 6) and we color the corresponding solution with new colors. Finally, the solution $\mathcal{S}$ is the juxtaposition of these hypocolorings $\mathcal{S}_i^*$.

Theorem 6. MIN HYPOCOLORING *is* $\lceil \frac{\Delta(G)+1}{3} \rceil$*-approximable.*

Proof. We have $val(\mathcal{S}) = \sum_{i=1}^{s} opt(G_i)$ and $opt \geq opt(G_i)$ for any $i = 1, \ldots, s$. Then, $val(\mathcal{S}) \leq s \times opt$.

5   Polynomial Cases

In this section, we consider two polynomial cases of MIN HYPOCOLORING: when the input is a tree with maximum degree at most Δ and when the input is a 2-regular graph. For sake of convenience, we assume that $w(v) \geq 0, \forall v$ (so, it may exist some nodes v with $w(v) = 0$); in this case, as we will show, the first case is equivalent to MIN HYPOCOLORING in forests with degree at most Δ whereas the second case is equivalent to MIN HYPOCOLORING in graphs with $\Delta(G) = 2$. Thus, since a tree is a particular bipartite graph, we have a boundary for the hardness of MIN HYPOCOLORING between trees with maximum degree at most

39 and bipartite graphs with maximum degree at most 39. Finally, there is also another hardness gap for general graphs between graphs with maximum degree at least 3 and graphs with maximum degree at most 2.

Before establishing these results, we shall give some results on MIN HYPO-COLORING in $(t+1)$-clique free graphs. For a hypostable set S, the *characteristic value* will be the integer number q such that $q = w(S)$. More generally, for a hypocoloring $\mathcal{S} = (S_1, \ldots, S_k)$ with $w(S_1) \geq \ldots \geq w(S_k)$ we call *vector of characteristic values*, the vector $(q_1, \ldots, q_k)$ such that for any $i \leq k, q_i = w(S_i)$. The MIN HYPOCOLORING problem is close to the LIST-HYPOCOLORING$_t$ problem.

LIST-HYPOCOLORING$_t$:
Instance: a graph $G = (V, E)$, a set $\mathcal{C}$ of colors and, for every clique K with size at most t, $C_K \subseteq \mathcal{C}$ is a set of colors such that each one of them may occur on some nodes of the clique K but not on all nodes at a time.
Question: does G admit a hypocoloring such that for any clique K, not all the nodes of K have the same color i with $i \in C_K$?

Clearly, we must have $C_K \subseteq C_{K'}$ when $K \subseteq K'$ and LIST-HYPOCOLORING$_t$ polynomially reduces to LIST-HYPOCOLORING$_{t'}$ when $t \leq t'$. Moreover, we have:

Proposition 4. *In the graphs with maximum degree Δ, MIN HYPOCOLORING polynomially reduces to* LIST-HYPOCOLORING$_{\Delta+1}$.

Proof. A minimum hypocoloring can be computed by the following algorithm:

---

1. For every vector $(q_1, \ldots, q_\Delta)$ with $q_1 \geq \ldots \geq q_\Delta$ and such that $q_i = \sum_{v \in V(K_i)} w(v)$ for some clique K_i of G do
   1.1 Solve the related LIST-HYPOCOLORING$_{\Delta+1}$ instance;
   1.2 If the answer is yes, construct such a hypocoloring;
2. Select a minimum weight hypocoloring among feasible hypocolorings computed during an execution of step 1.2;

The complexity-time of this algorithm is $O(n^{\Delta^2})$ times the complexity-time to solve LIST-HYPOCOLORING$_{\Delta+1}$.

Corollary 2. *If* LIST-HYPOCOLORING$_{\Delta+1}$ *is polynomial on Ψ-graphs, then* MIN HYPOCOLORING *is also polynomial on Ψ-graphs.*

Using Proposition 2 and a slight modification of Proposition 4, we deduce:

Corollary 3. *Let us consider a class Ψ of $(t + 1)$-clique free graphs satisfying $\chi(G) \leq k$ and such that* LIST-HYPOCOLORING$_t$ *is polynomial on Ψ. Then,* MIN HYPOCOLORING *is also polynomial on Ψ when $|w|$ is bounded by a constant.*

5.1   Trees with Maximum Degree Δ

In trees, there are at most $2n - 1$ characteristic values for the different hypostable sets. Thus, the complexity of the algorithm of Proposition 4 is in this case in $O(n^\Delta)$ times the complexity-time of LIST-HYPOCOLORING$_{\Delta+1}$. We now show how we can solve LIST-HYPOCOLORING$_{\Delta+1}$ in trees by using dynamic programming. Let $\mathcal{C} = \{1, \ldots, \Delta\}$ be the set of colors. Let us then consider $(T = (V, E); (C_K)_{K \in V \cup E})$ an instance of LIST-HYPOCOLORING$_{\Delta+1}$ where T is a tree. Given a node v, we respectively denote by $H_v(T)$ and $H'_v(T)$ the sets of colors defined by:

> $h \in H_v(T)$ (resp.$H'_v(T)$) if and only if there is a feasible hypocoloring for which v is colored by h and no (resp. exactly one) neighbor of v is colored by h.

We denote by $v_1, \ldots, v_d$ the neighbors of v. The deletion of v induces a forest with d connected components $T_1, \ldots, T_d$ where T_i is the subtree containing v_i.

Lemma 2. *For $h \in \mathcal{C}$, we have:*

- $h \in H_v(T) \Leftrightarrow h \notin C_v$ and $\forall j, (H_{v_j}(T_j) \cup H'_{v_j}(T_j)) \setminus [(H_{v_j}(T_j) \cup H'_{v_j}(T_j)) \cap \{h\}] \neq \emptyset$.
- $h \in H'_v(T) \Leftrightarrow h \notin C_v$ and $\exists j \leq d, h \in H_{v_j}(T_j) \setminus C_{[v,v_j]}$ and $\forall j' \neq j, (H_{v_{j'}}(T_{j'}) \cup H_{v_{j'}}(T_{j'})) \setminus [(H_{v_{j'}}(T_{j'}) \cup H'_{v_{j'}}(T_{j'})) \cap \{h\}] \neq \emptyset$.

Proposition 5. *For any $t \geq 2$, LIST-HYPOCOLORING$_t$ in trees is polynomial.*

Proof. Let us consider the following polynomial-time algorithm:

---

1. Choose a root $r \in V$ and orient the tree from r to leaves (T_v denotes the subtree induced by v and its successors);
2. Compute, for every node v and from leaves to the root, sets $H_v(T_v)$ and $H'_v(T_v)$ (by using Lemma 2);
3. For every color in $H_r(T) \cup H'_r(T)$ compute a feasible hypocoloring by using Lemma 2 (from the root to leaves);

5.2   Graphs with Maximum Degree Two

We shall examine here the special situation where the graph G has maximum degree $\Delta(G) = 2$ (the case $\Delta(G) = 1$ being trivial). From Proposition 1, there exists an optimal hypocoloring $\mathcal{S} = (S_1, S_2)$ of G with $w(S_1) \geq w(S_2)$. The case $S_2 = \emptyset$ is trivial and can be solved in linear-time. Thus, we will suppose $S_i \neq \emptyset$ for $i = 1, 2$. We prove by a technique similar of the one described earlier that the case of maximum degree two is also polynomial. However, the method presented here is slightly more involved than the previous one. First, observe that solving MIN HYPOCOLORING in graphs with $\Delta(G) = 2$ or in 2-regular graphs are equivalent. As a consequence, we may restrict our attention to graphs $G = (V, E)$ whose connected components are cycles; let $n = |V| = |E|$. We will define the weight $w(e)$ of an edge $e = [x, y]$ as the sum $w(x) + w(y)$. From (ii) of Proposition 1, we know that S_2 does not contain any K_3; then, we notice that there are at most $n + t + 1$ possible values for $w(S_1)$ where t is the number of triangles of G and $2n$ possible values for $w(S_2)$. It is important to notice that we cannot solve separately the problem in each connected component.

The algorithm is the following: starting with the smallest possible value of p and the smallest possible value of $q \leq p$, we apply Properties 1 to 4 (given below) to get the smallest q for which a solution (S_1, S_2) exists such that $w(S_1) = p$ and $w(S_2) = q$. If such a hypocoloring can be found, we store the current solution $S = (S_1, S_2)$ with $val(S) = p + q$ if it is better than the best solution found so far. Whenever such a solution has been found, we increase p to the next possible value and we start again with the minimum q. An optimal hypocoloring (S_1, S_2) will be given by the solution stored.

Property 1. If $w(v) > q$, then $v \in S_1$; if x, y, z are three consecutive nodes on an induced P_3 with $x, y \in S_i$ then $z \in S_{3-i}$ for $i = 1, 2$.

Property 2. If for some edge $e = [x, y]$, we have $w(e) > p$, then x, y are neither both in S_1 nor both in S_2; if $w(e) > q$, then x, y are not both in S_2. In such situations, we shall simply say that the color i is not feasible for edge $e = [x, y]$.

Starting from G with given values p, q we will apply the above properties as long as possible to derive consequences on the colors to be assigned to the nodes and to the edges of G. If we arrive to a situation where no solution exists then, we increase the value of p. Now, each cycle C_i has at least one node with a fixed color. We can describe C_i by the sequence $(F_1, D_1, F_2, \ldots, F_k, D_k)$ where F_i and D_i are chains. Moreover, for any i, all nodes of F_i have a fixed color and each D_i has two endpoints with a fixed color and all intermediate nodes are uncolored. Let $a_1, \ldots, a_s$ be the nodes of the chain D_i.

Property 3. If $a_1, a_s \in S_j$ with s odd or $a_1 \in S_j$, $a_s \in S_{3-j}$ with s even for some $j = 1, 2$, then we can alternate the colors 1 and 2 in D_i.

Property 4. If $a_1, a_s \in S_j$ with s even or $a_1 \in S_i$, $a_s \in S_{3-j}$ with s odd for some $j = 1, 2$, then $[a_1, a_2]$ gets one of its feasible colors.

By applying Properties 3 and 4 for each chain D_i, we color properly the remaining cycles. Now, when a value of p is fixed, we observe that the consequence of Properties 1 and 2 can be obtained in $O(n^2)$ steps and this gives a feasible value of q (if there exists). Then again in $O(n)$ steps, we can apply Properties 3 and 4 to determine a 2-hypocoloring. It should be observed that cases where no solution can be found occur only when consequence of Properties 1 and 2 are drawn.

Proposition 6. *The previous algorithm solves* MIN HYPOCOLORING *in graphs with $\Delta(G) \leq 2$ in $O(n^3)$ time.*

References

1. M. O. Albertson, R. E. Jamison, S. T. Hedetniemi, and S. C. Locke, *The subchromatic number of a graph*, Discrete Math. **74** (1989), 33-49.
2. L. W. Beineke and A. T. White, *Selected topics in graph theory*, Academic press, London, 1978.
3. C. Berge, *Graphs and Hypergraphs*, North Holland, Amsterdam, 1973.

4. H. L. Bodlaender, K. Jansen, and G. J. Woeginger, *Scheduling with incompatible jobs*, Discrete Appl. Math. **55** (1994), 219-232.

5. M. Borowiecki, I. Broere, M. Frick, P. Mihok and G. Semanisin, *Survey of hereditary properties of graphs*, Discussiones Mathematicae-Graph Theory **17** (1997), 5-50.

6. M. Boudhar and G. Finke, *Scheduling on a batch machine with job compatibilities*, Jorbel **40** (2000), 69-80.

7. H. Broersma, F. V. Fomin, J. Nesetril, and G. J. Woeginger, *More about subcolorings (extended abstract)*, Proc. WG 02, LNCS **2573** (2002), 68-79.

8. R. L. Brooks, *On colouring the nodes of a network*, Proc. Cambridge Phil. Soc. **37** (1941), 194-197.

9. J. L. Brown and D. G. Corneil, *On generalized graph colorings*, J. Graph Theory **11** (1987), 87-99.

10. M. Demange, D. de Werra, J. Monnot, and V. Th. Paschos, *Weighted node coloring: when stable sets are expensive*, Proc. WG 02, LNCS **2573** (2002), 114-125.

11. M. R. Dillon, *Conditionnal coloring*, Ph.D. thesis, University of Colorado at Denver, 1998.

12. U. Feige and J. Kilian, *Zero knowledge and the chromatic number*, J. Comput. System Sci. **57** (1998), 187-199.

13. J. Fiala, K. Jansen, V. B. Le, and E. Seidel, *Graph subcolorings: Complexity and algorithms*, Proc. WG 01 LNCS **2204** (2001), 154-165.

14. M. R. Garey and D. S. Johnson, *Computers and intractability. a guide to the theory of NP-completeness*, CA, Freeman, 1979.

15. H. Grotzsch, *Ein dreifarbensatz fur dreikreisfreie netze auf der kugel*, Wiss. Z. Martin Luther Univ. Halle-Wittenberg, Math. Naturwiss Reihe **8** (1959), 109-120.

16. F. Harary, *Conditional colorability in graphs*, in Graphs and Applications, Proc. First Colo. Symp. graph theory (F. Harary and J. Maybee eds), Wiley intersci., Publ. N.Y., 1985.

17. L. Lovász, *On decomposition of graphs*, Stud. Sci. Math. Hung. **1** (1966), 237-238.

18. P. Mutzel, T. ODENTHAL, and M. SCHARBRODT, *The thickness of graphs: A survey*, (1998).

19. F. Rendl, *On the complexity of decomposing matrices arising in satellite communication*, Operations Research Letters **4** (1985), 5-8.

20. T. J. Schaefer, *The complexity of satisfiability problems*, Proc. STOC (1978), 216-226.

Core Stability of Minimum Coloring Games

Thomas Bietenhader and Yoshio Okamoto*

Department of Computer Science, ETH Zurich, CH-8092 Zurich, Switzerland
thomasbi@student.ethz.ch, okamotoy@inf.ethz.ch

Abstract. In cooperative game theory, a characterization of games with stable cores is known as one of the most notorious open problems. We study this problem for a special case of the minimum coloring games, introduced by Deng, Ibaraki & Nagamochi, which arises from a cost allocation problem when the players are involved in conflict. In this paper, we show that the minimum coloring game on a perfect graph has a stable core if and only if every vertex of the graph belongs to a maximum clique. We also consider the problem on the core largeness, the extendability, and the exactness of minimum coloring games.

1 Introduction

One of the scopes of cooperative game theory is to establish the criterion of how to distribute a given revenue or cost among the agents in a fair manner when they work in cooperation. Since the effect of cooperation is usually non-linear and non-additive, the proportional division might not be considered fair. Several criteria, called solutions, are proposed by many researchers. When game theory was founded, von Neumann & Morgenstern [24] proposed a solution called a stable set, which turned out to be very useful for the analysis of a lot of bargaining situations but also turned out to be too difficult to reveal some fundamental properties. Much easier to investigate is the core, due to Gillies [11]. So, people are interested in when the core and the stable set coincide, namely when the core is stable. This question is known as one of the most notorious problems. So far, there are some necessary or sufficient conditions known (see, e.g., [23]), but they are far from a characterization of cooperative games with stable cores. From the computational point of view, the problem around stable sets is also eccentric. Deng & Papadimitriou [8] pointed out that determining the existence of a stable set for a given cooperative game is not known to be computable, and it is still unsolved.

Since combinatorial optimization problems can be found in several real-world situations, naturally they also raise some revenue/cost allocation problems. A *combinatorial optimization game* is a cooperative game which arises from a combinatorial optimization problem. There are many kinds of combinatorial optimization games proposed and studied, according to the underlying combinatorial

---

* Supported by the Berlin-Zurich Joint Graduate Program "Combinatorics, Geometry, and Computation" (CGC), financed by ETH Zurich and the German Science Foundation (DFG).

J. Hromkovič, M. Nagl, and B. Westfechtel (Eds.): WG 2004, LNCS 3353, pp. 389–401, 2004.

optimization problems. However, as far as the core stability is concerned, almost nothing is studied. The only exception is a work by Solymosi & Raghavan [22] on assignment games.

In this paper, we study core stability of minimum coloring games introduced by Deng, Ibaraki & Nagamochi [6], which arise from cost allocation problems when the agents are involved in conflict [18]. The reason that we restrict to perfect graphs is that it is NP-complete to decide whether a given graph yields a minimum coloring game with a nonempty core [6] (meaning that there seems no good characterization of minimum coloring games with nonempty cores) and that a graph G is perfect if and only if the minimum coloring game on G is totally balanced [7], where the total balancedness is a quite nice property. We prove that the minimum coloring game on a perfect graph has a stable core if and only if every vertex belongs to a maximum clique. We also consider the problem on the extendability, the largeness, and the exactness of cores, which are concepts related to core stability. We prove that they are equivalent for the minimum coloring game on a perfect graph, and also equivalent to that every clique is contained in a maximum clique.

Armed with our characterizations, we also study algorithmic aspects of these properties. First we give a polynomial-time algorithm to determine whether a given perfect graph yields a minimum coloring game with stable core or not. On the other hand, we prove that it is hard (or coNP-complete, technically speacking) to determine whether a given perfect graph yields a minimum coloring game which is extendable, exact or with large core. To the best of our knowledge, this is the first computational intractability result for extendability, exactness and core largeness of cooperative games.

2   Preliminaries

2.1   Notation

Throughout the paper, for a vector $\boldsymbol{x} \in \mathbb{R}^N$ and $S \subseteq N$, we write $\boldsymbol{x}(S) := \sum \{x_i \mid i \in S\}$. When $S = \emptyset$, set $\boldsymbol{x}(S) := 0$. For a subset $S \subseteq N$ of a finite set N, the *characteristic vector* of S is a vector $\mathbb{1}_S \in \{0,1\}^N$ defined as $(\mathbb{1}_S)_i = 1$ if $i \in S$ and $(\mathbb{1}_S)_i = 0$ otherwise. Note that for $S, T \subseteq N$ it holds that $\mathbb{1}_S(T) = \sum \{(\mathbb{1}_S)_i \mid i \in T\} = |S \cap T|$. We use the notation $A \subset B$ to mean that "A is a proper subset of B."

2.2   Graphs

A *graph* G is a pair $G = (V, E)$ of a finite set V, called the set of *vertices*, and a set $E \subseteq \binom{V}{2}$ of 2-element subsets of V, called the set of *edges*. For $U \subseteq V$, the subgraph of G induced by U is denoted by $G[U]$, where the vertices of $G[U]$ are the elements of U and the edges of $G[U]$ are the edges of G which are also 2-element subsets of U. The *complement* of $G = (V, E)$ is a graph with vertex set V and edge set the complement of E. A *clique* is a vertex subset inducing a graph with every pair being an edge (such a graph is called *complete*). A clique is *maximal* if none of its proper supersets is a clique. A clique is *maximum* if it has a maximum size among all cliques. The size of a maximum clique of G is denoted by $\omega(G)$. An *independent set* is a vertex subset inducing a graph with no edge. A *coloring* of $G = (V, E)$ is a map $c : V \rightarrow \mathbb{N}$ such that $c(u) \neq c(v)$ for every $\{u, v\} \in E$. A *minimum coloring* of G is a coloring with minimum possible $|c(V)|$. The *chromatic number* of G is $|c(V)|$ of a minimum coloring c of G and denoted by $\chi(G)$. Conventionally, the chromatic number of a graph with no vertex is defined to be zero. A graph $G = (V, E)$ is *perfect* if $\omega(G[U]) = \chi(G[U])$ for every $U \subseteq V$. A prominent example of non-perfect graphs is a cycle of length five.

2.3   Cooperative Games

A *cooperative game* (or simply a *game*) is a pair (N, γ) of a nonempty finite set N and a function $\gamma : 2^N \rightarrow \mathbb{R}$ satisfying $\gamma(\emptyset) = 0$. An element of N is called a *player* of the game, and γ is called the *characteristic function* of the game. Furthermore, each subset $S \subseteq N$ is called a *coalition*. Literally, for $S \subseteq N$ the value $\gamma(S)$ is interpreted as the total profit (or the total cost) for the players in S when they work in cooperation. In particular, $\gamma(N)$ represents the total profit (or cost) for the whole players when they all agree on working together. When γ represents a profit, we call the game a *profit game*. On the other hand, when γ represents a cost, we call the game a *cost game*. (Thus, the terms "profit game" and "cost game" are not mathematically determined. They are just determined by the interpretation of a game.) In this paper, we will mainly consider a certain class of cost games.

One of the aims of cooperative game theory is to provide a concept of "fairness," namely, how to allocate the total cost (or profit) $\gamma(N)$ to each player in a "fair" manner when we take all the $\gamma(S)$'s into account. Now, we concentrate on cost games, and define some cost allocations which are considered fair in cooperative game theory. Formally, a cost allocation is defined as a preimputation in the terminology of cooperative game theory. A *preimputation* of a cost game (N, γ) is a vector $\boldsymbol{x} \in \mathbb{R}^N$ satisfying $\boldsymbol{x}(N) = \gamma(N)$. Each component x_i expresses how much the player $i \in N$ should owe according to the cost allocation $\boldsymbol{x}$.

Let (N, γ) be a cost game. A vector $\boldsymbol{x} \in \mathbb{R}^N$ is called an *imputation* if $\boldsymbol{x}$ satisfies the following conditions: $\boldsymbol{x}$ is a preimputation of (N, γ) and $x_i \leq \gamma(\{i\})$ for every $i \in N$. The set of all imputations of (N, γ) is denoted by $\mathsf{Imp}(N, \gamma)$. A vector $\boldsymbol{x} \in \mathbb{R}^N$ is called a *core allocation* if $\boldsymbol{x}$ satisfies the following conditions: $\boldsymbol{x}$ is an imputation of (N, γ) and $\boldsymbol{x}(S) \leq \gamma(S)$ for all $S \subseteq N$. The set of all core allocations of (N, γ) is called the *core* of (N, γ) and denote by $\mathsf{Core}(N, \gamma)$. The core was introduced by Gillies [11].

Note that $\mathsf{Core}(N, \gamma) \subseteq \mathsf{Imp}(N, \gamma)$ and both can be empty. Therefore, a cost game with a nonempty core is especially interesting, and such a cost game is called *balanced*. Moreover, we call a cost game *totally balanced* if each of the subgames is balanced. (Here, a *subgame* of a cost game (N, γ) is a cost game $(T, \gamma^{(T)})$ for some nonempty $T \subseteq N$ defined as $\gamma^{(T)}(S) = \gamma(S)$ for each $S \subseteq T$.)

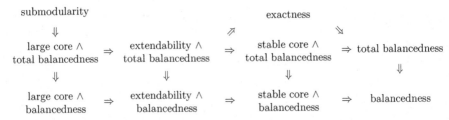

Fig. 1. Implication relationship. The symbol "∧" represents "and".

Naturally, a totally balanced game is also balanced. A special subclass of the totally balanced games consists of submodular games (Shapley [20]), where a cost game (N, γ) is called *submodular* (or *concave*) if it satisfies $\gamma(S) + \gamma(T) \geq \gamma(S \cup T) + \gamma(S \cap T)$ for all $S, T \subseteq N$. Therefore, we have a chain of implications "submodularity ⇒ total balancedness ⇒ balancedness," which are fundamental in cooperative game theory.

Let (N, γ) be a balanced cost game. The core $\mathsf{Core}(N, \gamma)$ is called *stable* if for every $y \in \mathsf{Imp}(N, \gamma) \setminus \mathsf{Core}(N, \gamma)$ there exist a core allocation $x \in \mathsf{Core}(N, \gamma)$ and a nonempty coalition $S \subset N$ such that $x(S) = \gamma(S)$ and $x_i < y_i$ for each $i \in S$. (The concept of stability is due to von Neumann & Morgenstern [24].) The core $\mathsf{Core}(N, \gamma)$ is called *large* if for every $y \in \mathbb{R}^N$ satisfying that $y(S) \leq \gamma(S)$ for all $S \subseteq N$ there exists $x \in \mathsf{Core}(N, \gamma)$ such that $y \leq x$. (The largeness was introduced by Sharkey [21].) The game (N, γ) is *extendable* if for every nonempty $S \subseteq N$ and every $y \in \mathsf{Core}(S, \gamma^{(S)})$ there exists $x \in \mathsf{Core}(N, \gamma)$ such that $x_i = y_i$ for all $i \in S$. (The extendability was introduced by Kikuta & Shapley [13], and named by van Gellekom, Potters & Reijnierse [23].) The game (N, γ) is called *exact* if for every $S \subset N$ there exists $x \in \mathsf{Core}(N, \gamma)$ such that $x(S) = \gamma(S)$. (The exactness was first defined by Schmeidler [19].) Note that an exact game is always totally balanced.

Here, we summarize the known relationships among these classes of games. See also Fig. 1. Sharkey [21] showed that if a game is submodular then it has a large core. Kikuta & Shapley [13] showed that if a balanced game has a large core then it is extendable, and if a balanced game is extendable then it has a stable core. Sharkey [21] showed that if a totally balanced game has a large core then it is exact. Biswas, Parthasarathy, Potters & Voorneveld [1] pointed out that he actually proved that extendability implies exactness. The reverse directions in Fig. 1 do not hold in general. (Some of them are explained by van Gellekom, Potters and Reijnierse [23].)

3   Minimum Coloring Games

Let $G = (V, E)$ be a graph. The *minimum coloring game* on G is a cost game (V, χ_G) where $\chi_G : 2^V \to \mathbb{R}$ is defined as $\chi_G(S) := \chi(G[S])$ for all $S \subseteq V$. Furthermore, we always assume that $V \neq \emptyset$ when we consider the minimum coloring game, so that the minimum coloring game meets the definition of a cooperative game.

Let us first make some easy observations.

Observation 1. *Let $G = (V, E)$ be a graph and (V, χ_G) be the minimum coloring game on G.*

(a) For every $S \subseteq T \subseteq V$, it holds that $\chi_G(S) \leq \chi_G(T)$.
(b) For every nonempty independent set $I \subseteq V$ of G it holds that $\chi_G(I) = 1$. In particular, $\chi_G(\{v\}) = 1$ for each $v \in V$.
(c) If $x \in \mathsf{Core}(V, \chi_G)$, then it holds that $0 \leq x_v \leq 1$ for every $v \in V$.

Proof. (a) Since $S \subseteq T$, we have $\chi(G[S]) \leq \chi(G[T])$. The claim follows from the definition of χ_G.

(b) For a nonempty independent set I, we have $\chi(G[I]) = 1$.

(c) Let $x \in \mathsf{Core}(V, \chi_G)$. By the definition of the core and the part (b), we have that $x_v \leq \chi_G(\{v\}) = 1$. Suppose that $x_v < 0$ for contradiction. Then, it holds that $\chi_G(V) < \chi_G(V) - x_v$. Furthermore, by part (a) we have $\chi_G(V \setminus \{v\}) \leq \chi_G(V)$, and also we have $x(V) = \chi_G(V)$ since $x \in \mathsf{Core}(V, \chi_G)$. Therefore, we obtain $\chi_G(V \setminus \{v\}) \leq \chi_G(V) < \chi_G(V) - x_v = x(V) - x_v = x(V \setminus \{v\})$. This is a contradiction to $x \in \mathsf{Core}(V, \chi_G)$.   □

Deng, Nagamochi & Ibaraki [6] proved that it is NP-complete to decide whether the minimum coloring game on a given graph is balanced. Subsequently, Deng, Ibaraki, Nagamochi & Zang [7] showed that the minimum coloring game on a graph G is totally balanced if and only if G is perfect. So the decision problem on the total balancedness of a minimum coloring game is as hard as recognizing perfect graphs, which was found to be solved in polynomial time [2, 4]. Furthermore, Okamoto [17] showed that the minimum coloring game on a graph G is submodular if and only if G is complete multipartite. So we can decide whether a given graph yields a submodular minimum coloring game in polynomial time. The following proposition due to Okamoto [18] characterizes the core of the minimum coloring game on a perfect graph. This will be used nicely in a later investigation.

Proposition 1 (Okamoto [18]). *Let $G = (V, E)$ be a perfect graph. Then, the core of the minimum coloring game (V, χ_G) is the convex hull of the characteristic vectors of maximum cliques of G.*

4   Results

4.1   Core Stability

The following theorem characterizes totally balanced minimum coloring games with stable cores.

Theorem 2. *Let $G = (V, E)$ be a perfect graph. Then, the minimum coloring game (V, χ_G) has a stable core if and only if every vertex $v \in V$ belongs to a maximum clique of G.*

First we prove the only-if part of the theorem. The proof uses the following lemma.

Lemma 1. *Let $G = (V, E)$ be a graph such that the minimum coloring game (V, χ_G) is balanced. If (V, χ_G) has a stable core, then for every $v \in V$ there exists a core allocation $\boldsymbol{x} \in \mathsf{Core}(V, \chi_G)$ such that $x_v \neq 0$.*

Proof. Assume that $\mathsf{Core}(V, \chi_G)$ is stable, and suppose, for the contradiction, there exists a vertex $v \in V$ such that

$$x_v = 0 \quad \text{for all } \boldsymbol{x} \in \mathsf{Core}(V, \chi_G). \tag{1}$$

(Particularly $V \neq \emptyset$.) Take such a vertex v. Let $\hat{\boldsymbol{x}} \in \mathsf{Core}(V, \chi_G)$ be an arbitrary core allocation. Since $V \neq \emptyset$, it holds that $\chi_G(V) > 0$. So, there exists $w \in V$ such that $\hat{x}_w > 0$. Now, define $\boldsymbol{y} \in \mathbb{R}^V$ as

$$y_u := \begin{cases} \hat{x}_u & \text{if } u \notin \{v, w\}, \\ \hat{x}_w & \text{if } u = v, \\ 0 & \text{if } u = w. \end{cases}$$

Namely, $\boldsymbol{y}$ is obtained from $\hat{\boldsymbol{x}}$ by interchanging the v-th component and the w-th component. Then, $\boldsymbol{y}$ is an imputation of (V, χ_G). Since $y_v = \hat{x}_w > 0$, due to (1), we can see that $\boldsymbol{y}$ is not a core allocation. Hence, $\boldsymbol{y} \in \mathsf{Imp}(V, \chi_G) \setminus \mathsf{Core}(V, \chi_G)$.

Since $\mathsf{Core}(V, \chi_G)$ is stable, there exist a nonempty set $S \subset V$ and a core allocation $\overline{\boldsymbol{x}} \in \mathsf{Core}(V, \chi_G)$ such that $\overline{\boldsymbol{x}}(S) = \chi_G(S)$ and $\overline{x}_u < y_u$ for every $u \in S$. Now we cliam that $S \setminus \{v\} \neq \emptyset$. To show this, suppose not, i.e., $S \setminus \{v\} = \emptyset$. Since $S \neq \emptyset$, we have that $S = \{v\}$. Then, it follows that

$$
\begin{aligned}
\chi_G(\{v\}) &= \overline{x}_v && (\text{since } \chi_G(S) = \overline{\boldsymbol{x}}(S)) \\
&< y_v && (\text{since } \overline{x}_u < y_u \text{ for every } u \in S) \\
&\leq \chi_G(\{v\}) && (\text{since } \boldsymbol{y} \in \mathsf{Imp}(V, \chi_G)).
\end{aligned}
$$

This is a contradiction, hence the claim follows.

Going back to the proof of Lemma 1, we obtain

$$
\begin{aligned}
\chi_G(S) &= \overline{\boldsymbol{x}}(S) \\
&= \overline{\boldsymbol{x}}(S \setminus \{v\}) && (\text{by (1)}) \\
&< \boldsymbol{y}(S \setminus \{v\}) && (\text{by the choice of } \overline{\boldsymbol{x}} \text{ and Claim above}) \\
&\leq \hat{\boldsymbol{x}}(S \setminus \{v\}) && (\text{by the construction of } \boldsymbol{y}) \\
&\leq \chi_G(S \setminus \{v\}) && (\text{since } \hat{\boldsymbol{x}} \in \mathsf{Core}(V, \chi_G)) \\
&\leq \chi_G(S) && (\text{by Observation 1(a)}).
\end{aligned}
$$

This is a contradiction. □

Then, let us prove the only-if part of the theorem.

Proof (of the only-if part of Theorem 2). Assume that (V, χ_G) has a stable core. By Lemma 1, for every $v \in V$ there exists a core allocation $\boldsymbol{x} \in \mathsf{Core}(V, \chi_G)$ such that $x_v > 0$. On the other hand, by Proposition 1, $\boldsymbol{x}$ is a convex combination of the characteristic vectors of maximum cliques of G. Therefore, at least one maximum clique of G must contain v. □

In order to prove the if part, we need some more lemmas.

Lemma 2. *Let $G = (V, E)$ be a graph with $\chi(G) = \omega(G)$. Then, there exists a nonempty independent set $I \subseteq V$ such that $K \cap I \neq \emptyset$ for every maximum clique K of G.*

Proof. Consider a minimum coloring of G and take the vertices colored by an identical color. Denote by I the set of these vertices. By the construction, I is an independent set. On the other hand, in each maximum clique K of G all colors used to color G can be found since $\chi(G) = \omega(G) = |K|$. Namely, every maximum clique intersects I. Thus, I is a desired independent set. □

Here is another lemma.

Lemma 3. *Let $G = (V, E)$ be a perfect graph, and consider the minimum coloring game (V, χ_G). Then, for every $\mathbf{y} \in \mathsf{Imp}(V, \chi_G) \setminus \mathsf{Core}(V, \chi_G)$ there exists a nonempty independent set $I \subseteq V$ such that $\mathbf{y}(I) > \chi_G(I)$ and $y_v > 0$ for every $v \in I$.*

Proof. Fix $\mathbf{y} \in \mathsf{Imp}(V, \chi_G) \setminus \mathsf{Core}(V, \chi_G)$ arbitrary, and define $\mathcal{S} := \{S \subseteq V \mid \mathbf{y}(S) > \chi_G(S) \text{ and } y_v > 0 \text{ for every } v \in S\}$.

First, note that $\mathcal{S} \neq \emptyset$. To see this, since $\mathbf{y} \in \mathsf{Imp}(V, \chi_G) \setminus \mathsf{Core}(V, \chi_G)$, there exists $T' \subseteq V$ such that $\mathbf{y}(T') > \chi_G(T')$. Let $T := T' \setminus \{v \in T' \mid y_v \leq 0\}$. Then, it holds that $\mathbf{y}(T) \geq \mathbf{y}(T') > \chi_G(T') \geq \chi_G(T)$. (The last inequality is due to $T \subseteq T'$ and Observation 1(a).) Since $y_v > 0$ for each $v \in T$, it follows that $T \in \mathcal{S}$. This implies that $\mathcal{S}$ is nonempty.

Choose $S \in \mathcal{S}$ of minimum size. Since G is perfect, we have that $\chi(G[S]) = \omega(G[S])$. By Lemma 2, there exists a nonempty independent set $I \subseteq S$ such that for every maximum clique K of $G[S]$ we have $K \cap I \neq \emptyset$. Now, we claim that $I \in \mathcal{S}$. (This proves the lemma.) First of all, since $I \subseteq S$ it holds that $y_v > 0$ for every $v \in I$. So it suffices to show that $\mathbf{y}(I) > \chi_G(I)$.

Since I intersects with every maximum clique of $G[S]$, we can see that $\omega(G[S \setminus I]) < \omega(G[S])$. Since G is perfect, this means that

$$\chi_G(S \setminus I) < \chi_G(S). \tag{2}$$

Since I is nonempty, we have $|S \setminus I| < |S|$. By the minimality of S, it holds that

$$\mathbf{y}(S \setminus I) \leq \chi_G(S \setminus I). \tag{3}$$

Now, we obtain the following.

$$
\begin{aligned}
\mathbf{y}(I) &= \mathbf{y}(S) - \mathbf{y}(S \setminus I) & & (I \subseteq S) \\
&> \chi_G(S) - \chi_G(S \setminus I) & & (S \in \mathcal{S} \text{ and } (3)) \\
&\geq 1 & & ((2) \text{ and the integrality of } \chi_G) \\
&= \chi_G(I) & & (\text{Observation 1(b)}).
\end{aligned}
$$

This concludes the proof. □

Now, we are ready to prove the if part of Theorem 2.

Proof (of the if part of Theorem 2). Let $\boldsymbol{y} \in \mathsf{Imp}(V, \chi_G) \setminus \mathsf{Core}(V, \chi_G)$. Then, by Lemma 3, there exists a nonempty independent set $I \subseteq V$ such that $\boldsymbol{y}(I) > \chi_G(I) = 1$ and $y_v > 0$ for every $v \in I$. Denote by $\mathcal{K}$ the set of maximum cliques of G. To every vertex $v \in I$, we assign a maximum clique $K(v) \in \mathcal{K}$ such that $v \in K(v)$, and fix this assignment. By our assumption, this assignment is well-defined. Since I is an independent set, this assignment is injective.

For every $K \in \mathcal{K}$, let

$$
\lambda_K := \begin{cases} \dfrac{y_v}{\boldsymbol{y}(I)} & \text{if } K = K(v) \text{ for some } v \in I \\ 0 & \text{otherwise.} \end{cases}
$$

Since the assignment $v \mapsto K(v)$ is injective, the value λ_K is well-deined. Then, for each $K \in \mathcal{K}$, we have that $0 \leq \lambda_K \leq 1$ (since $y_v > 0$ for every $v \in I$ and $\boldsymbol{y}(I) > 1$ by the choice of I with Lemma 3, and $y_v \leq 1$ for every $v \in I$ by Observation 1(b) and the definition of an imputation). Furthermore, we can check that $\sum_{K \in \mathcal{K}} \lambda_K = 1$. Therefore, if we let $\boldsymbol{x} := \sum_{K \in \mathcal{K}} \lambda_K \mathbb{1}_K$, by Proposition 1, it holds that $\boldsymbol{x} \in \mathsf{Core}(V, \chi_G)$.

If $v \in I$ then $x_v = \lambda_{K(v)}$. This is because $v \notin K(u)$ for $u \in I \setminus \{v\}$. Therefore, if $v \in I$, then

$$
x_v = \lambda_{K(v)} = \frac{y_v}{\boldsymbol{y}(I)} < y_v,
$$

since $\boldsymbol{y}(I) > 1$. Furthermore, it holds that

$$
\boldsymbol{x}(I) = \sum_{u \in I} x_u = \sum_{u \in I} \lambda_{K(u)} = 1 = \chi_G(I).
$$

Thus, $\boldsymbol{x}$ is an appropriate core allocation and hence the core is stable. □

4.2   Exactness, Extendability, and Core Largeness

We prove that exactness, extendability and core largeness are equivalent for minimum coloring games on perfect graphs. This is also characterized in terms of graphs, and summarized as the following theorem.

Theorem 3. *Let $G = (V, E)$ be a perfect graph. Then, the following conditions are equivalent.*

(1) The minimum coloring game (V, χ_G) is exact.
(2) The minimum coloring game (V, χ_G) is extendable.
(3) The core $\mathsf{Core}(V, \chi_G)$ is large.
(4) Every clique of G is contained in a maximum clique of G.

First remark that the implication "(3) $\Rightarrow$ (2) $\Rightarrow$ (1)" is true for any kinds of games [13]. It remains to prove "(1) $\Rightarrow$ (4)" and "(4) $\Rightarrow$ (3)."

Let us first prove "(1) $\Rightarrow$ (4)."

Proof (of (1) $\Rightarrow$ (4)). Let $G = (V, E)$ be a perfect graph such that (V, χ_G) is exact. Let S be a clique of G. Then, by exactness, there exists $\boldsymbol{x} \in \mathsf{Core}(V, \chi_G)$ such that $\boldsymbol{x}(S) = \chi_G(S) = |S|$. Denoting by $\mathcal{K}$ the set of maximum cliques of G, by Proposition 1, we can express $\boldsymbol{x}$ as

$$\boldsymbol{x} = \sum_{K \in \mathcal{K}} \lambda_K \mathbb{1}_K, \qquad (4)$$

where $\lambda_K \geq 0$ for every $K \in \mathcal{K}$ and $\sum_{K \in \mathcal{K}} \lambda_K = 1$. Then, it holds that

$$
\begin{aligned}
|S| = \boldsymbol{x}(S) = \sum_{K \in \mathcal{K}} \lambda_K \mathbb{1}_K(S) \qquad &\text{(by (4))}\\
= \sum_{K \in \mathcal{K}} \lambda_K |S \cap K| \\
\leq \sum_{K \in \mathcal{K}} \lambda_K |S| \qquad &\text{(since } S \cap K \subseteq S\text{)}\\
= |S| \sum_{K \in \mathcal{K}} \lambda_K = |S| \qquad &\text{(since } \sum_{K \in \mathcal{K}} \lambda_K = 1\text{).}
\end{aligned}
$$

So, the equality holds throughout the expressions, meaning that $S \cap K = S$ for each $K \in \mathcal{K}$ with $\lambda_K > 0$. Thus, S is contained in a maximum clique of G.   □

To show "(4) $\Rightarrow$ (3)," we use some more facts. The first one is due to van Gellekom, Potters & Reijnierse [23]. For a cost game (N, γ), let

$$L(N, \gamma) := \{ \boldsymbol{y} \in \mathbb{R}^N \mid \boldsymbol{y}(S) \leq \gamma(S) \text{ for every } S \subseteq N \},$$

and call it the set of *lower vectors*.

Lemma 4 (van Gellekom, Potters & Reijnierse [23]). *Let (N, γ) be a balanced cost game. Then (N, γ) has a large core if and only if $\boldsymbol{y}(N) \geq \gamma(N)$ for all extreme points $\boldsymbol{y}$ of $L(N, \gamma)$.*

In order to apply Lemma 4 to our setting, we have to know the extreme points of $L(V, \chi_G)$ for a perfect graph G. The following lemma can be shown with a similar method to the proof of the weak perfect graph conjecture due to Lovász [16].

Lemma 5. *Let $G = (V, E)$ be a perfect graph. Then, each extreme point of $L(V, \chi_G)$ is the characteristic vector of a maximal clique of G.*

Armed with Lemmas 4 and 5, we are able to show "(4) $\Rightarrow$ (3)."

Proof (of (4) $\Rightarrow$ (3)). Let G be a perfect graph such that every clique is contained in a maximum clique of G. Choose an extreme point of $L(V, \chi_G)$. By Lemma 5, this is the characteristic vector of some maximal clique K of G. Namely, this extreme point is $\mathbb{1}_K$. By our assumption, K is a maximum clique of G. Therefore, it holds that $\mathbb{1}_K(V) = |K| = \omega(G) = \chi_G(V)$. Hence, by Lemma 4, the core is large.   □

This completes the whole proof of Theorem 3.

5 Algorithmic Aspects

In this section, using the theorems we have obtained already, we discuss the algorithmic issues for minimum coloring games. The first problem we consider is the following.

Problem: CORE STABILITY FOR PERFECT GRAPHS

Instance: A perfect graph $G = (V, E)$

Question: Does the minimum coloring game (V, χ_G) have a stable core?

Now, we describe an algorithm which shows the following theorem.

Theorem 4. *The problem* CORE STABILITY FOR PERFECT GRAPHS *can be solved in polynomial time.*

Proof. Consider the algorithm in Algorithm 1.

Algorithm 1: A polynomial-time algorithm for CORE STABILITY FOR PERFECT GRAPHS.

Input: a perfect graph $G = (V, E)$.

Output: "Yes" if $(V, \chi(G))$ has a stable core; "No" otherwise.

1 $\omega(G) \leftarrow$ the weight of a maximum clique in G;
2 $M \leftarrow |V|$; ·
3 **foreach** *vertex* $v \in V$ **do**
4   Set a weight vector $\boldsymbol{w} \in \mathbb{R}^V$ as $w_v = M$ and $w_u = 1$ $(u \in V \setminus \{v\})$;
5      $\omega(G, \boldsymbol{w}) \leftarrow$ the maximum weight of a clique in G with respect to $\boldsymbol{w}$;
6   **if** $\omega(G, \boldsymbol{w}) - \omega(G) < M - 1$ **then**
7   **return** *"No"*;
     end
   end
8 **return** *"Yes"*.

Let us prove that Algorithm 1 is correct. The first observation is that in each "foreach" loop we compute a clique K_v of maximum size which contains v. That is just because M is huge. Now, if $|K_v| < \omega(G)$, then we can see that a maximum clique containing v is not a maximum clique of G. Namely, v is not contained in any maximum clique of G. Then, by Theorem 2, the game does not have a stable core. Therefore, we have to check that $|K_v| < \omega(G)$ if and only if $\omega(G, \boldsymbol{w}) - \omega(G) < M - 1$ (i.e., the condition in Line 6 is true). First of all, we can see that $|K_v| = \omega(G, \boldsymbol{w}) - M + 1$. So, we have that $|K_v| - \omega(G) = \omega(G, \boldsymbol{w}) - \omega(G) + M - 1$. Hence, $|K_v| < \omega(G)$ holds if and only if $\omega(G, \boldsymbol{w}) - \omega(G) < M - 1$. This completes the proof of the correctness.

Now, we discuss the running time of Algorithm 1. Computing a maximum weight clique in a perfect graph can be done in polynomial time [12]. So, Lines 1 and 5 can be executed in polynomial time. Line 2 is also fine. In the "foreach" loop, Line 4 can be done swiftly. The condition check in Line 6 is easy. The number of iterations of the foreach loop is at most $|V|$. Hence, the overall running time is polynomial in the size of input.   □

Next, we discuss the following three problems.

Problem: EXTENDABILITY FOR PERFECT GRAPHS
Instance: A perfect graph $G = (V, E)$
Question: Is the minimum coloring game (V, χ_G) extendable?

Problem: EXACTNESS FOR PERFECT GRAPHS
Instance: A perfect graph $G = (V, E)$
Question: Is the minimum coloring game (V, χ_G) exact?

Problem: CORE LARGENESS FOR PERFECT GRAPHS
Instance: A perfect graph $G = (V, E)$
Question: Does the minimum coloring game (V, χ_G) have a large core?

Thanks to Theorem 3, these problems are equivalent to the following problem.

Problem: SIZE EQUALITY OF A MAXIMUM CLIQUE AND A MINIMUM MAXIMAL CLIQUE IN PERFECT GRAPHS
Instance: A perfect graph $G = (V, E)$
Question: Do a maximum clique and a minimum maximal clique in G have the same size?

This problem turns out to be coNP-complete.

Theorem 5. *The problem* SIZE EQUALITY OF A MAXIMUM CLIQUE AND A MINIMUM MAXIMAL CLIQUE IN PERFECT GRAPHS *is coNP-complete. Consequently,* EXTENDABILITY FOR PERFECT GRAPHS, EXACTNESS FOR PERFECT GRAPHS *and* CORE LARGENESS FOR PERFECT GRAPHS *are coNP-complete.*

Proof. The membership in coNP is immediate. The coNP-hardness follows from a result due to Zverovich [25]. □

Theorem 5 deals with perfect graphs in general. Now, let us discuss some special cases for which the problem can be solved in polynomial time. Observe that, due to Theorem 3, it suffices to compute a minimum maximal clique in a given perfect graph. If it is also a maximum clique in the graph, then all maximal cliques are maximum cliques. Then, the condition (4) in Theorem 3 holds. If not, then this maximal clique is not contained in a maximum clique, meaning that the condition (4) is violated. Namely, we consider the following optimization problem.

Problem: MINIMUM MAXIMAL CLIQUE
Instance: A graph G
Feasible solution: A maximal clique K of G
Objective: Minimize $|K|$.

There are some classes of perfect graphs for which we can solve MINIMUM MAXIMAL CLIQUE in polynomial time. They include the bipartite graphs (easy), the comparability graphs [15], the chordal graphs [10], and the complements of chordal graphs [9]. (See also an article by Kratsch [14].) For these classes of graphs, as we already observed, we can conclude the following.

Theorem 6. *Consider a class of perfect graphs for which* MINIMUM MAXIMAL CLIQUE *can be solved in polynomial time. For this class of graphs,* EXTENDABIL-ITY FOR PERFECT GRAPHS, EXACTNESS FOR PERFECT GRAPHS *and* CORE LARGENESS FOR PERFECT GRAPHS *can be solved in polynomial time.*

6 Summary

We discussed the core stability problem for minimum coloring games, introduced by Deng, Ibaraki & Nagamochi [6], of perfect graphs. We obtained a good characterization for a minimum coloring game with stable core (Theorem 2), and this led us to a polynomial-time algorithm for the corresponding decision problem (Theorem 4). We also discussed the extendability, the exactness and the core largeness for minimum coloring games of perfect graphs, and characterized them in terms of a property of graphs (Theorem 3). With this characterization, we showed that it is coNP-complete to determine whether a given perfect graph yields the minimum coloring game which is extendable, exact, or with large core (Theorem 5). For some subclasses of perfect graphs, we know that there exists a polynomial-time algorithm for this problem (Theorem 6).

Little is known about core stability of cooperative games. This paper expanded the knowledge of this problem, and also gave rise to some algorithmic perspectives.

Acknowledgements

The authors thank anonymous referees for helpful comments.

References

1. A.K. Biswas, T. Parthasarathy, J.A.M. Potters and M. Voorneveld. Large cores and exactness. Games and Economic Behavior **28** (1999) 1–12.
2. M. Chudnovsky and P. Seymour. Recognizing Berge graphs. Submitted.
3. D.G. Corneil and Y. Perl. Clustering and domination in perfect graphs. Discrete Applied Mathematics **9** (1984) 27–39.
4. G. Cornuéjols, X. Liu and K. Vušković. A polynomial algorithm for recognizing perfect graphs. Proc. 44th FOCS (2003) 20–27.
5. I.J. Curiel. Cooperative Game Theory and Applications: Cooperative Games Arising from Combinatorial Optimization Problems. Kluwer Academic Publishers, Dordrecht, 1997.
6. X. Deng, T. Ibaraki and H. Nagamochi. Algorithmic aspects of the core of combinatorial optimization games. Math. Oper. Res. **24** (1999) 751–766.
7. X. Deng, T. Ibaraki, H. Nagamochi and W. Zang. Totally balanced combinatorial optimization games. Math. Program. **87** (2000) 441–452.
8. X. Deng and Ch.H. Papadimitriou. On the complexity of cooperative solution concepts. Math. Oper. Res. **19** (1994) 257–266.
9. M. Farber. Independent domination in chordal graphs. Oper. Res. Lett. **1** (1982) 134–138.
10. D.R. Fulkerson and O.A. Gross. Incidence matrices and interval graphs. Pacific J. Math. **15** (1965) 835–855.

11. D.B. Gillies. Some theorems on n-person games. Ph.D. Thesis, Princeton University, 1953.

12. M. Grötschel, L. Lovász and A. Schrijver. Geometric algorithms and combinatorial optimization. Second edition. Springer-Verlag, Berlin, 1993.

13. K. Kikuta and L.S. Shapley. Core stability in n-person games. Manuscript, 1986.

14. D. Kratsch. Algorithms. In: T.W. Haynes, S.T. Hedetniemi, P.J. Slater eds., Domination in Graphs (Advanced Topics), Marcel Dekker Inc, New York, 1998, pp. 191–231.

15. D. Kratsch and L. Stewart. Domination on cocomparability graphs. SIAM J. Discrete Math. **6** (1993) 400–417.

16. L. Lovász. Normal hypergraphs and the perfect graph conjecture. Discrete Math. **2** (1972) 253–267.

17. Y. Okamoto. Submodularity of some classes of the combinatorial optimization games. Math. Methods Oper. Res. **58** (2003) 131–139.

18. Y. Okamoto. Fair cost allocations under conflicts — a game-theoretic point of view —. Proc. 14th ISAAC, Lect. Notes Comp. Sci. **2906** (2003) 686–695.

19. D. Schmeidler. Cores of exact games I. J. Math. Anal. Appl. **40** (1972) 214–225.

20. L.S. Shapley. Cores of convex games. Internat. J. Game Theory **1** (1971) 11–26. Errata is in the same volume, 1972, pp. 199.

21. W.W. Sharkey. Cooperative games with large cores. Internat. J. Game Theory **11** (1982) 175–182.

22. T. Solymosi and T.E.S. Raghavan. Assignment games with stable cores. Internat. J. Game Theory **30** (2001) 177–185.

23. J.R.G. van Gellekom, J.A.M. Potters and J.H. Reijnierse. Prosperity properties of TU-games. Internat. J. Game Theory **28** (1999) 211–227.

24. J. von Neumann and O. Morgenstern. Theory of Games and Economic Behaviour. Princeton University Press, Princeton, 1944.

25. I.E. Zverovich. Independent domination on $2P_3$-free perfect graphs. DIMACS Technical Report 2003-22, 2003.

Author Index

Lecture Notes in Computer Science

For information about Vols. 1–3253

please contact your bookseller or Springer

Vol. 3298: S.A. McIlraith, D. Plexousakis, F. van Harmelen (Eds.), The Semantic Web – ISWC 2004. XXI, 841 pages. 2004.

Vol. 3296: L. Bougé, V.K. Prasanna (Eds.), High Performance Computing - HiPC 2004. XXV, 530 pages. 2004.

Vol. 3295: P. Markopoulos, B. Eggen, E. Aarts, J.L. Crowley (Eds.), Ambient Intelligence. XIII, 388 pages. 2004.

Vol. 3294: C.N. Dean, R.T. Boute (Eds.), Teaching Formal Methods. X, 249 pages. 2004.

Vol. 3293: C.-H. Chi, M. van Steen, C. Wills (Eds.), Web Content Caching and Distribution. IX, 283 pages. 2004.

Vol. 3292: R. Meersman, Z. Tari, A. Corsaro (Eds.), On the Move to Meaningful Internet Systems 2004: OTM 2004 Workshops. XXIII, 885 pages. 2004.

Vol. 3291: R. Meersman, Z. Tari (Eds.), On the Move to Meaningful Internet Systems 2004: CoopIS, DOA, and ODBASE, Part II. XXV, 824 pages. 2004.

Vol. 3290: R. Meersman, Z. Tari (Eds.), On the Move to Meaningful Internet Systems 2004: CoopIS, DOA, and ODBASE, Part I. XXV, 823 pages. 2004.

Vol. 3289: S. Wang, K. Tanaka, S. Zhou, T.W. Ling, J. Guan, D. Yang, F. Grandi, E. Mangina, I.-Y. Song, H.C. Mayr (Eds.), Conceptual Modeling for Advanced Application Domains. XXII, 692 pages. 2004.

Vol. 3288: P. Atzeni, W. Chu, H. Lu, S. Zhou, T.W. Ling (Eds.), Conceptual Modeling – ER 2004. XXI, 869 pages. 2004.

Vol. 3287: A. Sanfeliu, J.F. Martínez Trinidad, J.A. Carrasco Ochoa (Eds.), Progress in Pattern Recognition, Image Analysis and Applications. XVII, 703 pages. 2004.

Vol. 3286: G. Karsai, E. Visser (Eds.), Generative Programming and Component Engineering. XIII, 491 pages. 2004.

Vol. 3285: S. Manandhar, J. Austin, U.B. Desai, Y. Oyanagi, A. Talukder (Eds.), Applied Computing. XII, 334 pages. 2004.

Vol. 3284: A. Karmouch, L. Korba, E.R.M. Madeira (Eds.), Mobility Aware Technologies and Applications. XII, 382 pages. 2004.

Vol. 3283: F.A. Aagesen, C. Anutariya, V. Wuwongse (Eds.), Intelligence in Communication Systems. XIII, 327 pages. 2004.

Vol. 3282: V. Guruswami, List Decoding of Error-Correcting Codes. XIX, 350 pages. 2004.

Vol. 3281: T. Dingsøyr (Ed.), Software Process Improvement. X, 207 pages. 2004.

Vol. 3280: C. Aykanat, T. Dayar, İ. Körpeoğlu (Eds.), Computer and Information Sciences - ISCIS 2004. XVIII, 1009 pages. 2004.

Vol. 3279: G.M. Voelker, S. Shenker (Eds.), Peer-to-Peer Systems III. XI, 300 pages. 2004.

Vol. 3278: A. Sahai, F. Wu (Eds.), Utility Computing. XI, 272 pages. 2004.

Vol. 3275: P. Perner (Ed.), Advances in Data Mining. VIII, 173 pages. 2004. (Subseries LNAI).

Vol. 3274: R. Guerraoui (Ed.), Distributed Computing. XIII, 465 pages. 2004.

Vol. 3273: T. Baar, A. Strohmeier, A. Moreira, S.J. Mellor (Eds.), <<UML>> 2004 - The Unified Modelling Language. XIII, 454 pages. 2004.

Vol. 3272: L. Baresi, S. Dustdar, H. Gall, M. Matera (Eds.), Ubiquitous Mobile Information and Collaboration Systems. VIII, 197 pages. 2004.

Vol. 3271: J. Vicente, D. Hutchison (Eds.), Management of Multimedia Networks and Services. XIII, 335 pages. 2004.

Vol. 3270: M. Jeckle, R. Kowalczyk, P. Braun (Eds.), Grid Services Engineering and Management. X, 165 pages. 2004.

Vol. 3269: J. Lopez, S. Qing, E. Okamoto (Eds.), Information and Communications Security. XI, 564 pages. 2004.

Vol. 3268: W. Lindner, M. Mesiti, C. Türker, Y. Tzitzikas, A. Vakali (Eds.), Current Trends in Database Technology - EDBT 2004 Workshops. XVIII, 608 pages. 2004.

Vol. 3267: C. Priami, P. Quaglia (Eds.), Global Computing. VIII, 377 pages. 2004.

Vol. 3266: J. Solé-Pareta, M. Smirnov, P.V. Mieghem, J. Domingo-Pascual, E. Monteiro, P. Reichl, B. Stiller, R.J. Gibbens (Eds.), Quality of Service in the Emerging Networking Panorama. XVI, 390 pages. 2004.

Vol. 3265: R.E. Frederking, K.B. Taylor (Eds.), Machine Translation: From Real Users to Research. XI, 392 pages. 2004. (Subseries LNAI).

Vol. 3264: G. Paliouras, Y. Sakakibara (Eds.), Grammatical Inference: Algorithms and Applications. XI, 291 pages. 2004. (Subseries LNAI).

Vol. 3263: M. Weske, P. Liggesmeyer (Eds.), Object-Oriented and Internet-Based Technologies. XII, 239 pages. 2004.

Vol. 3262: M.M. Freire, P. Chemouil, P. Lorenz, A. Gravey (Eds.), Universal Multiservice Networks. XIII, 556 pages. 2004.

Vol. 3261: T. Yakhno (Ed.), Advances in Information Systems. XIV, 617 pages. 2004.

Vol. 3260: I.G.M.M. Niemegeers, S.H. de Groot (Eds.), Personal Wireless Communications. XIV, 478 pages. 2004.

Vol. 3259: J. Dix, J. Leite (Eds.), Computational Logic in Multi-Agent Systems. XII, 251 pages. 2004. (Subseries LNAI).

Vol. 3258: M. Wallace (Ed.), Principles and Practice of Constraint Programming – CP 2004. XVII, 822 pages. 2004.

Vol. 3257: E. Motta, N.R. Shadbolt, A. Stutt, N. Gibbins (Eds.), Engineering Knowledge in the Age of the Semantic Web. XVII, 517 pages. 2004. (Subseries LNAI).

Vol. 3256: H. Ehrig, G. Engels, F. Parisi-Presicce, G. Rozenberg (Eds.), Graph Transformations. XII, 451 pages. 2004.

Vol. 3255: A. Benczúr, J. Demetrovics, G. Gottlob (Eds.), Advances in Databases and Information Systems. XI, 423 pages. 2004.

Vol. 3254: E. Macii, V. Paliouras, O. Koufopavlou (Eds.), Integrated Circuit and System Design. XVI, 910 pages. 2004.